THE DON SMITH MAGNETIC RESONANCE ENERGY CRAFTING SYSTEMATIC INDEX

First book in this series of Rensonance by the Editor Richard Friedrich in 2018:
Exploring the Resonance Inductive Coupler Kit: *in the unlimited open context of crafting true negative resistance by triggering sharp disruptive discharges through low resistance impedances, ampliflying and reciprocating with sympathetic resonance, in freely reactive multi-body dynamic networks.*

A book that comes with the very kit Don Smith first learned from and said to start with.
http://www.r-charge.net/product.asp?itemid=305

A special thanks to Alex Kolb for making this book possible in providing original manuscripts, VHS recordings, and notes from over 20 years ago.

and to

Bruno VK who helped with much of the preliminary transcribing.

as well as

Jason Poteet who provided constructive editorial and technical assistance.

The Don Smith Magnetic Resonance Energy Crafting Systematic Index.

Compiled and Edited

by

Richard Friedrich

The Don Smith Magnetic Resonance Energy Crafting Systematic Index.

Published in the United States of American in 2018 by
Alethea In Heart Publishing, for
Renaissance Charge, LLC
R-Charge.com

ISBN: 1-932370-84-6

Author: Donald Lee Smith
Compiler, Editor, Transcriber, Illustrator, Cover Design: Richard Friedrich

Printed in the United States of America

Table of Contents Outline

With contextual comments.

I. Table of Contents Outline v

II. Illustrations and Pictures xv

III. Don Smith Energy Generation Models xvi

IV. Preface xvii

Reasons for Publication xvii

Don Limited his Revelations by Commercial Commitments xxii

Don was the Real Deal xxiii

Human Nature Generally Contradictory and Prejudiced xxiv

Grounded Reality Model vs Floating Around Falling Delusion Model xxvi

Criteria of True and False Systems of Science xxviii

People Looking for a Trick not Foundation xxviii

Grounded Reality Model Applied to Electron Path xxxii

Processing Don's Pages xxxiii

Transcribing Don's Audio and Videos xxxiv

Purpose and Result xxxv

Where and How to Begin, with What Level of Expectation xxxv

V. Editor's Introduction xxxvii

The Reasons People Fail in Don Smith Replications xxxviii

Two Obstacles in the way of Over Unity Energy Harvesting xxxviii

Useful Energy is Defined as that which is Other than Ambient xxxviii

Electron Spin Instrumental Cause of Useable Electrical Energy xxxix

Perpetual Electron Cycle xl

What is Fundamentally Wrong with Conventional Generators? xlv

Differences between Under and Over Unity Generators xlvi

Bloch Wall Interrupting Magnet Generator xlviii

Conventional Power Grid System Simplified xlix

Don Smith Power System Outline li

Free Energy Harvesting Requirements l

Summary of Magnetic Resonant Stages l

Kirchhhoff's Law is for the Birds! li

Corrected Schematic and Avramenko Plug lii

The Simple Klystron Amplifier liv

The Klystron Compared with Don Smith Resonant Energy System lv

Frank Wyatt Prentice Electrical Power Accumulator lvii

Similarity between Don Smith Ground System and Frank Prentice Electrical Power Accumulator lviii

VI. Important Statements by Don Smith lxi

These are key stand-alone statements by Don Smith from the printed book and videos that should be read by themselves.

1. Miscellaneous Printed Publications by Donald Lee Smith from 1994 to 2006 1

Useful Circuits from Nikola Tesla 1

Essential circuits to begin understanding Don Smith systems, and are often referred to.

Contents

Resonate Electrical Power System 2

Pages 2-6 begin to unfold theory and details of the Resonate System.

Non-Ionic Electrons, Electron Related Energy 5

Useful Circuits, Suggestions for Building an Operational Unit 6

Showing the Plasma Ball System with a larger coil than the one shown on page 155.

Suggestions 7

Replicators have mostly failed to pay attention to these details, and that inductors, capacitors, and resistors all have inductance, capacitance and resistance, which combined, all play a part in relation to tuning multiple parts together. You learn this by experience.

Page 8 begins to reveal the multiple ground system that is enlarged upon later in the diagrams. The next point, a main theme of the book, is that the Don Smith systems are divided into modules where only the magnetic flux is transferred, not electrons.

Electrical Energy Related - Plasma thru Solids 9

The last statement begins to show how the energy is increased with electrons being drawn into the system. The point missed by most everyone is to not let them immediately return.

Resonate Electromagnetic Power System 10

This is more of a general diagram which lacks some of the isolation and grounding details.

Comments 11

These beginning pages were together at the start of several of the early editions where you can see the original page numbers above the page numbering of this book.

Ultimate Energy Sources 12

The next few pages show environmental concerns with conventional energy production.

Decline of Oxygen in the Earth's Atmosphere Ratio of Oxygen to Nitrogen 13

These numbers are concerning but may be a little off. Sampling USA today shows 20.9%.

Page 15 should be read with pages 50 and 42 diagrams.

Pages 16 and 17 categorize and outline some of Don's main patented processes .

Energy as Relates, Chemistry Vs Physics 18

This compares exothermic free energy in chemistry to above ambient energy in physics while providing a basic formula.

Free Energy Harvesting Requirements 19

In relation to electron dynamics.

Observations of an Inventor 20

About various over energy systems with some history.

Ultimate Energy Systems 21

Physics of Avalanche 22

Fundamental processes of open systems contrasted with closed systems.

An Answer to America's Energy Deficit, 1995 25

These several pages enlarge upon the electron theory, measuring it in the earth, and some history. The specific explanation here about the electron spin, with the magnetic and electric phasing, is one of the most important things to understand in this book and which is why it is on the front cover.

Electrical Energy Reference Points, Electrical Energy Generating System 31

This is the lecture from video 1. The first two pages are almost the same as earlier pages.

Functions of Electrons 34

More specifics on electron theory and gravity. Read with pages 41-43.

Contents

Ohm's Law with Corrections 35
Examples of Overunity 36
Expected Results 37
Resistance becomes zero in resonance above 20kHz. VAR (Volts Amperes Reactive) System.
Examples of Overunity Devices 38
Here we see the electron analogy of the room filled with random or aligned ping pong balls, followed by an important Summary.
Electrical Energy with Associated Phenomena 40
Directional energy lines of force.
Electrical Energy Diagram 41
Important mass directional diagram in relation to electric, magnetic and gravity
Energy Acquired by Magnetic and Electrical Impulsing diagram 42
Essential diagram showing dual output and phasing from electron spin
Path of Electrons diagram from Alex K during Don Smith Workshop 43
Showing 3D infinity loop spin and phasing of electrons.
Earth Electrical System II, Information & Concept (booklet) 44
Abstract 45
The Evidence against Under Unity 46
E. E. S. II, Background Information & Concept (read with pages 50-52 diagrams) 47
EES II, Modular Units 48
Here we start to see specific modules in a system (High Voltage Induction, Air Core Induction Coil, and Inverter). In the Systematic Index at the end of the book we will see quotations arraged generally as six different possible sections of Don Smith systems that are organized in the way of functions or processes more than parts on a module board.
Alternating Impact Energy Transfer Analogy (diagram of swinging steel balls) 50
Earth Electrical System II, Domestic Use Range Module A (notice polarity references) 51
Earth Electrical System II, Domestic Use Range Module B (notice grounding polarity directions) 52
Unique Aspects of E.E.S. II 53
Electron Related Energy 54
TransWorld Energy Letter 55
Electrical Energy Generating System 56
Electrical Energy Systems Preface 57
Electrical Energy Systems Method 58
Ambient Energy Sources 59
Electrical Energy System 60
Derivation of Magnetic and Electrical power diagram 63
Shows bloch wall of magnet in relation to coils. Page 64 relates this to diodes.
Tesla Coil Geometry 65
Explains how position of L1 over L2 relates to voltage and amperage output.
Power Triangle 66
Electrical Energy Generating System schematics 67
Second version early schematic with less parts. Explains input charging with 1/4 wavelength
Electrical Energy Generating System schematics 68
Enlarged diagram of full early schematic showing frequency adjustments and wave shaping. Second half only on page 69 was part of the suitcase model. Shows basic inverter.

Contents

Induced Electrical Energy System 70

Showing flux lines squared symbolized. Page 71 shows Capacitor and Inductor formulas.

Electrical Power Generation / Points of Reference (with Useful Energy Formula) 73

Benefits of the Inventor's System are summarized 75

Abundant Energy Every Where Present. User's Guide. Ambient Energy Generator 76

Page 76 shows basic diagram. On page 329 I have added the details described on page 77 along with the other control parts implied in Don's general teaching.

Air Core Induction Coil Builder's Guide 78

Magnetic Resonance Power System. Basic Sonar Power System 80

Unity (Ambient) Vs Greater Than Ambient (Over unity) 82

Dipole Transformer Generator Description (patent papers show plasma column also) 83

Ultimate Energy Sources 92

Reactance Charts 96

Negative Resistance Trumpet Wave, Transient Response of an RLC Circuit 98

Shown in the first three videos. 1. Positive damped wave, 2. Standing wave, 3. Negative resistance increasing. http://www.tubebooks.org/Books/Spangenberg_vacuum_tubes.pdf

Schematic of Flat Coil Induction System (different size than model shown in pictures) 100

Schematic of Cobalt Nickel Core Toroid Induction Device 6-1 102

Shaped Current Force Field Generators diagrams with Bloch Wall switching 103

This is from the Don Smith CD and relates to Device 9.

Alex K 1998 Don Smith Workshop Notes (two videos are not available) 104

How to build modified wave power inverters 109

These three diagrams were in Don's books and from *Encyclopedia of Electronic Circuits Vol 3* 1990. Page 109 shows the inverter with the input having the correct polarity. Page 110 has the inverter Don copied from, which has the input polarity backwards, and which made it's way into Don's schematic. I have a corrected version of that schematic at the end of the Introduction. Page 111 shows the mosfet version of this inverter.

Matrix Transformer 112

How to Specify Power Transformer & Filter Ratings 113

Comments on Capacitor & Diode Selection 115

High Voltage Dividers 116

High Voltage Devices 117

Marx Generator and Notes about High Amperage 119

Here we see some of Don's notes in the pages.

Impulse Gernerators 120

Simple Tesla Coil 126

Marx Generators 129

Electrical Principles, Terminology & Safety 134

2. Don Smith 2001 TransWorld Website Information and Pictures 145

Bruce P. Introduction to Don.

Website Introduction, Electromagnetic Spectrum Energy 146

Resonance Energy Concept 147

Magnetic Resonance Energy Spectrum 150

Sample Website Prototypes: Device 1, Tesla Flat Coil Transformer 151

Device 2, Magnetic Flux Resonance Tesla Receiver coils multiplied 153

Device 3, Magnetic Flux Amplified 154
Device 4, Magnetic Flux from Plasma Globe 155
Device 5, Self Powered System for Vending Machines in Japan 156
Device 6, Cobalt Nickel Core Toroid Power System for Brazil 157
Device 8, 14" Diameter Tesla Coil System 159
Device 9, Intermittent Bloch Wall Switching Motor for Brazil and Saudi Arabia 160
Additional Don Smith Prototype Pictures, Barium Titinate Transformer 163
Cockroft-Walton Device 164
Plasma Column Dipole 165
45KW Commercial Prototype 166
Japanese Prototype, Resonate Transformer predating the MEG 167
4KW Generator 168
Large Transformer for Commercial System 170
Don Smith Device-2006 with Ignition Coil and fluxed Metglas transformer 172
3. Transcripts of Audio and Video Recordings from 1994 to 2005. 173
Video 1 1994 173

This video starts off as a second part of a presentation that began in the morning. No copy of that beginning part was located for this book. Don shares miscellaneous information from the book as part of a continuation from the morning. At 38 minutes he shares how the Department of Energy assigned the name Earth's Electrical System II to his Energy System idea but did nothing else for him regarding it. At 42 minutes he wraps up the printed lecture and goes into the model which is shown on page 196. While drawing the diagram he shows a large microwave oven transformer. Then he answers a question about how many amps can these systems run. He said that the load will determine the amount of amperage so that you can run just about anything. At the one hour and 24 minute mark he begins to focus on the flat pancake coil system (Device 1). He was not able to get it running other than for it to flicker a few times and attempted to change out the protection diode a few times, and looked at other parts of the circuit, in order to make it run. He didn't remember that with that module it ran on 24V rather than 12V as shown in the schematic of this system on page 100. These modules will still run on lower voltages, but they would not be able to run that system when it was tuned for 24V. This is why you see some flickering when it was switched on or off at times. As someone who has done many conventions I have often experienced this kind of lapse in memory about something so simple when you are in front of many people with limited time and with many objectives filling up your mind. Don didn't have time to set these things up beforehand, with no one around, which would have solved this problem. After spending a lot of time on that, and answering a few questions, he moved on to the 4 coil Tesla coil setup, which was his first (Device 2) model. After struggling to get that running he then shows it powering each secondary coil with their lightning arrestors. He explains that from that point you would put your loading module. You could have as many of these secondaries duplicating the input coil without depleting the input energy. He then tries to explain that he does not have the back end of the circuit to run any direct loads on. This causes one person to mock him. Then he explains how you can make your own inverter or wave shaping device, while he uses one to run a fan from a 12V battery.

Video 1 schematic of drawn instructions 196

Video 2 1995 197

Don starts the lecture at the foundational point of electron theory and develops the content from his booklet *An Answer to America's Energy Deficit*. He shows the Resonance kit as well has typical L1 and L2 coils while explaining their relationships to each other and resistors. After going over more of the booklet he explains the 14" Tesla coil model 6. From there he goes over the displayed model, which does not show the capacitor 12, big resistors, and transformer 16 that is in the schematic explained (see pages 67-69 for both). Then he goes back to pages in the book. The first part of the video is a short lecture that actually was later in the day.

The second part of the video, starting at the 55 minute mark, is from the morning. This is the most important lecture in all these videos. Video 1 is about equally important. Both of these provide the proper context of understanding Don's book and later videos. Don starts off talking about his book and how it's going to be around 400 pages and look real nice when it gets finished. Well, here it is at over 500 pages and it looks real nice. Then he talks about the flood of people contacting him after the lecture last year and how he had to drive on group away. Then he begins the lecture examining the claim from people that you have have high volts with no amps and therefore no useful energy. After this he gives some basic history about the laws of energy. Then he talks about using magnetic flux to measure the voltage of a system. Then he goes on about electron theory and the term unity. Then he talks about why people fail in replications. The reason being because of failure to understand that parts have capacitance, resistance and inductance and people usually only consider one of these characteristics. Then he continues with the book material while interjecting practical tips and additional facts. After the first break he begins to show actual devices on slides, starting with the 14" Tesla coil. He then moves to models 29, 2, and 30 while giving details. Then he goes over in detail page 1 showing Tesla's coils. He moves on to spark gaps just before the next break. Then he continues on with Marx generators and high voltage devices. For the last thirty minutes he opens up the floor for questions while putting up the schematic of what I have named model 29 setup. In the last eleven minutes he answers about the source of energy and various subjects.

Video 3 1996 237

Don starts off the lecture/demonstration with an important statement that signals a new direction for him with these meetings: "Since we met last year I realized that there are a lot of people that had trouble understanding things unless you could go turn the wall switch on and that was the extent of it. So this particular speech is built around that particular aspect of it." He then shares how the lecture will go over what was published in the recent Tesla magazine which is part of the book. As with the previous lectures, he goes through similar slides while interjecting facts and anecdotes. He starts with the commercial 45kW model 13. Then he goes over model 3 which is the display model. Both of these are in the schematics page following this video on page 268. He mentions that the neon sign transformer used is an over energy device by itself. At minute 42 he goes into the electromedical background that lead to this technology, starting with D'Arsonval. At the first hour mark Don enlarges upon his point that batteries do not last very long and have relatively few cycles compared with capacitors. This is generally true with conventional usage but with proper charging, as we do with the Renaissance Charge chargers, they can last indefinitely. Don experienced this with his batteries that stayed charged for years due to

receiving charging by radio frequency transmission from the system they were powering. Then he moves on to the very important pages from the *Vacuum Tubes* book showing the trumpet wave from the negative resistance process relating to the klystron.
After the first break Don continued to show and explain slides. At 1:12 he shows a patent application diagram which is included on page 268 and is also shown as an important schematic in the Introduction. After this he continues with slides from the book. At 1:39 he finishes the slide show and moves towards demonstrating his suitcase model. He required an earth grounding to run it and used the center plug of the wall socket. He starts out running a regular drill and then asks if anyone wants to test it with any meters. A guy comes up but doesn't understand how to use his meter. Later on another guy looks at the wave of the load while running 400W of bulbs. Later he replaces the bulbs with 1000W of bulbs and measures again. Don spends the rest of the time answering questions about the availability of the technology, and other topics. He shares how his extension of JJ Thomson's electron theory is Nobel Prize worthy. He ends with a convincing demonstration of powering 1000W of bulbs (from the skeptic) for over 3 minutes, when it had already run for over ten minutes at 400W previously. The skeptic then said he was conveniced after he reasoned that a 6AH battery could not sustain an 83A load for any length of time.

Video 3 schematic of drawn instructions 268

Video 4 1998 269

This short video is one of three (the other two we were not able to get copies of). Don is in his home office where many people visited him. This is a very important video packed full of important summary and specific information. The video was a back and forth question and answer dialogue that was done well. Don mostly answers all of the questions asked. He explains how the energy works and also some of the parts he used (including two modules). The Device 3 is again shown and explained, along with the 45kW commercial model. At the end of the video he gives the details about how to build your own device.

Video 4 schematic and Instructions of How to Build your own Device 279

Video 5 2001 280

This is the first *Inventor's Meeting* recording we have of Don. He showed more signs of slight impairment due to strokes, which he seemed to be aware of. He would sometimes change his subject in mid-sentence, and also expressed being disoriented at times. He would slow down at times and then come back and be very sharp. He forgot his slides and ended up just talking about the three resonant energy devices he brought in. The first one was his first four coil Tesla coil setup which was meant to demonstrate free energy through radio transmission. The second was the suitcase model, where he actually got shocked from it because of radio transmission from the plasma column running. Then he talked in detail about the Device 3 on the big board. Don also talks about various meters to use as well as Spice computer simulation programs.. There are many important bits of information in this video. It is the last one showing Don having firm clarity on the subjects he covers. At the end Don tells an interesting story as and example of the kinds of deals people wanted to make with him of which he turned down millions of dollars of cash from China.

Audio 2003 302

This short audio recording was done over the phone because Don had just had a major stroke which left his one arm paralyzed. Don had a very hard time hearing the questions being asked so that almost no questions were specifically answered. There is very little new information in this interview, but enough useful content to justify including it. Some statements on gravity and magnetorestrictive materials were significant. Most of Don's remarks were in general. This exchange reveals some biographical information about Don's early years in the petroleum industry along with his present physical struggles. It was very hard to transcribe this recording so there are a few places left out as indicated. Most of the last few minutes were removed as they pertained to people contacting each other.

Video 6 2005 310

This video starts out in mid-sentence and focuses in on the magnetic flux switching bloch wall magnet 9.1 device. Don mentions that this was replicated four times the size and patented by the Russian Academy of Science for him. He works through 70 slides making important comments about most pictures shown. One that he doesn't comment on is rather important, and found at the bottom of page 169 which I have added to the video 1 schematic page. At the 22 minute mark Don brings up Von Däniken as if he somehow proved that the Egyptians had lightbulbs. This is one minor example of where Don was mistaken. Von Däniken's mistake was based upon ignorance of the ancient religious symbols as documented here: www.ancientaliensdebunked.com A few minutes after this Don shows the worst example of memory failure in back to back statements about the 1996 Tesla Symposium where he thought it was maybe in 1979 and that the bulbs were on for hours. It may be that the bulbs were left on after the recording took place as those meetings went on for days. Later on in this video Don's son shares how he really didn't want his father to be coming out to these meetings. While there is a lot of important information in these last two videos, and the previous audio, it is clear that some of the details given are probably not very accurate. Don continues going through the slides showing what has been shown before and a few new pictures. He spent some time talking about being commissioned by Japan to make devices to power their vending machines. Later he mentions attempting to help the Brazilian government but that since the exchange rate kept fluxuating the deal couldn't be finalized. After the break at minute 56 Professor Denis Lee (of MIT I think) does an important short demonstration of electron pairing with two different size slinky coils. At the half-way point Don talks about the Casimir effect with a capacitor. He then demonstrates with a dielectric tester and a large plate capacitor how the ambient provides an energy duplicate on the reverse side of the capacitor. See page 328 for illustrations. After the demonstration someone askes where can they buy a Don Smith device. Don explains why they are not available to the general public. Then his son Mark gives a long explanation about Don's commitments with leading industry and how they didn't want him to give away the technology in these meetings. But Don was committed to helping people learn how to use this technology even though he may not have given all the specific details about each part number (although this book puts it all together for you in that way). The full technology in products would only be gradually used based upon market saturation. After more miscellaneous questions and answers that are unrelated to the question (because Don didn't hear the question right) Bruce P. says the following important testimony: "You know, Don's practically handed it to you on a plate. The schematics have been up there for a few years and there have been some people that have duplicated it and contacted me. It works." Then Don talks about how people in eastern Europe have been the most open and successful at replicating his devices. Later he explains more details about his magnet motor and says Denis Lee copied his devices except that he does it in a way that creates too much heat and which ruins the magnets fast. The last point is one of the most significant where he talks about activating magnets at their natural frequencies in the gigahertz range for millions of watts output.

Video 6 Diagram: Demonstration of Ambient Energy Generator Don Smith Effect. 329

Video 7 2005 330

The video starts off with a Don Smith company video on Resonant Energy Systems which demonstrates the copper sheet transformer device powering about 4000W off a car battery while maintaining it's charge. Don then goes through some slides exactly like video 6 until he gets to his dipole patent which he presents for the first time. He let's his assistant read and explain it. After this Don goes over some points about pages in the book very much like video 6. Then he ends the presentation with the same demonstration from the previous video, while answering a few questions after.

4. Relevant Emails from the NuEnergy Yahoo forum. 342

We begin with the NuEnergy list followed by the Free-Energy list emails. Don covers many topics and answers important questions very specifically. There is a lot of new material in these emails that is not in the book editions or in the videos. While Don's spelling and grammar are fairly bad at times the material given is valuable. Some of Don's emails make more sense as we look at the context of questions and statements made by others before his response. So we have included these in part and in whole. Since the emails still can be viewed online, anyone can read these in context. But they will have to go through tens of thousands of emails as I have done for this book. Some examples of topics covered in no particular order are the MEG device, magnetorestrictive materials and devices, circuit simulation software, and just about every subject in the book. This list was a lot more open and constructive the than Free-Energy list which was highly skeptical. This list never organized a proper attempt to replicating any of Don's devices while they had access to him for years. Bruce said in a later video that Don's systems work but the email group never appeared to have any goals or means of coming to practical conclusions.

Relevant Emails from the Free-Energy Yahoo forum. 373

This group was highly critical of Don to the point of people calling him a fraud. The attacks revealed a lack of attention to detail and the standard assumptions of conventional theory mindsets. Unfortunately, due to his age and the nearness of his major strokes, Don shows less ability to fully respond to various questions, and his spelling and grammar really suffer. Don apologizes for this and writes: Please excuse the garble , I do not have the use of my hands ." Because of his strokes and sometimes lack of caring about precision in relaying numbers so as to avoid all ambiguity, his few mistakes in figures were seized upon by his critics as a means to completely dismiss him. These harsh critics assumed, or attempted to portray him as someone who was ignorant of basic science and math. This was obviously not the case. But Don didn't really care about what people thought of him, or about being palatable to those who demanded appearances of formalities and credentials. He was going against the flow and as a humble man he was only there to pass along information that would help those who were open enough to consider it.

6. Systematic Index. 387

This Index quotes Don Smith according to the subject under each heading in the order they appear in this book as indicated by the page numbers that follow each paragraph. This gives the reader the ability to study in one place what Don wrote or said about a certain subject.

There was a lot of repetition so it is not exhaustive.
General Historical Development 388
Don Smith Historical Development 389
Don Smith Electron Theory 391
Types of Over Unity 397
Example Proofs of Over Unity 398
Illucidations of Over Unity 400
Sectional Processes in Over Unity Systems 403
These are quotations arranged generally as five different possible sections of Don Smith systems that are organized in the way of functions or processes more than parts on a module board.
1. Agitation, Imbalance, Disruption of Ambient Equilibrium 404
2. Asymmetrical Engineering the Imbalance 405
3. Gathering and Collection of the Potential in Capacitor, Inductor, Ground 407
4. Waveshaping Conversion to Make Useful 410
5. Inverting Energy as Needed. 412
Devices and Materials Mentioned 414
Using Tools and Making Calculations 415
Benefits of Magnetic Resonance 417
Consequences of Under Unity Systems 417
Where Conventional Theory is Mistaken 418
Hindrances to Over Unity and Why Replicators Fail 419
7. Bibliography 420
8. Index Glossary of Key Words, Ideas, and People by Don Smith 428

Illustrations and Diagrams

Page
xxxix. Electron Spin Instrumental Cause of Useable Energy
xl. Perpetual Electron Cycle
xlv. What is Wrong with Conventional Generators?
xlvi. Differences: Under Over and Unity Generators
xlxviii. Conventional Power Grid System Simplified
xlix. Don Smith Power System Outline
li. Kirchhoff's Law is for the Birds!
lii. Corrected Schematic and Avramenko Plug
liv. The Simple Klystron Amplifier
lv. The Klystron Compared with the Don Smith System
lvii. Frank Wyatt Prentice Electrical Power Accumulator
lviii. Similarity between Smith and Prentice Ground System
1. Useful Circuits from Nikola Tesla
10. Resonate Electromagnetic Power System
13. Decline of Oxygen in the Earth's Atmosphere
40. Electrical Energy with Associated Phenomena
41. Electrical Energy
42. Energy Acquired by Magnetic and Electrical Impulsing
50. Alternating Impact Energy Transfer Analogy
51-51. Earth Electrical System II. Domestic Use Module
63-64. Derivation of Magnetic and Electrical Power
65. Tesla Coil Geometry
66. Power Triangle
67-69. Electrical Energy Generating System
70. Induced Electrical Energy System
76. Ambient Energy Generator.
81. Magnetic Resonance Power System for Water Systems
83-91. Dipole Transformer Generator
96-97. Reactance Charts
98-99. Transient response of and RLC circuit
100-101. Flat Coil circuit (model 1)
102. Don Smith System (model 10)
103. Neutral Center Generator Lab Notes
104-108. Alex Kolb's Don Smith Workshop Notes
109-111. Power Inverter circuits
112-133. Various high frequency circuit diagrams
134-144. Electrical Principles, Terminology & Safety
145. Radio Sky 408 MHz
149. Electromagnetic Spectrum Energy
151-152. Tesla Flat Coil Transformer System (model 1)
153. Energy Duplication by Magnetic Flux Resonance
154. Resonant Energy Device (model 3)
155. Magnetic Flux from The Plasma
156. Self Powered System for Vending Machines in Japan
157-158. Tesla Coil wound on Cobalt Nickel Core
159. 14" Diameter Tesla Coil System
160-162. Bloch Wall Interrupting Magnet Generator
162. 50F Capacitor Bank Module
163-164. Barium Titinate Transformer
164. Cockroft-Walton Device
165. Plasma Column Dipole
165. Electrical Energy Generating System (video 2)
166. 45kW Commercial Prototype
167. Japan Vending Machine Prototype
167. Resonate Transformer from Suitcase
167. Capacity Coupling in Radio Amplifying Transformers
168-169. 4kW Generator (with two toroids)
169, 196. Transformer module (video 1)
170-171. Custom Copper sheet 110-440V Transformer
172. Ambient Energy System Metglas transformer
172. Terfanol-D Rod Dipole with sheet capacitors
196. Video 1 Resonant Induction Generation System
268. 45kW Commercial Prototype Schematic
268. Resonant Energy Device Schematic (model 3)
268. Resonance Energy Generation System (patent)
279. Build Your Own Device (video 4)
329. Demonstration of Ambient Energy Generator. DSE
329. Tesla Impedance Circuit (called 'Hairpin Circuit')
329. Ambient Energy Generator, Don Smith Effect

Don Smith Energy Generation Models

Device

1. Tesla Flat Coil Transformer System, 152, 331, 100-101, 189
2. Energy Duplication by Magnetic Flux Resonance (4 Tesla coils, first device), 153, 204, 219, 352
3. Resonant Energy Device (3" adjustable L1 coil), 154, 239
4. Magnetic Flux from The Plasma, 6, 155
5. Self Powered System for Vending Machines in Japan, 156
6. Tesla Coil L-1 and L-2 wound on large diameter Cobalt Nickel Core, 102, 157-158
7. [shows the plasma ball again]
8. 14" Diameter Tesla Coil System, 159, 203, 215, 249
9. Bloch Wall Interrupting Magnet Generator, 160-162, 327, 374
10. Barium Titinate Transformer, 163
11. Cockroft-Walton Device, 164
12. Plasma Column Dipole, 165
13. 45kW Commercial Prototype, 166, 237
14. Japan Vending Machine Prototype, 167
15. Resonate Transformer from Suitcase, 167, 255, 345
16. 4kW Generator (with two toroids), 168-169, 323
17. Video 1 Schematic and Transformer module, 169
18. Custom Copper sheet 110-440V Transformer, 170-171, 330
19. Ambient Energy System with Ignition coil and MetGlas Transformer, 172
20. Terfanol-D Rod Dipole with sheet capacitors, 172, 333
21. Direct Energy from tapping into a Magnet's natural frequency, 327
22. Resonate Electrodynamic Power System (general concept),10
23. Earth Electrical System II, Domestic Use Range Module, Plan A, 51, 173
24. Earth Electrical System II, Domestic Use Range Module, Plan B, 51,
25. Magnetic Resonance Power System for Water Systems, 80-81
26. Video 1 Resonant Induction Generation System, 181-187, 196
27. Beam Energy System, 364
28. Florescent Light Gaseous Tube Electrical Generator, 366
29. Electrical Energy Generating System, 67-69, 165, 205, 218
30. Resonance Induction Coupler Kit, 198, 219, 420
31. Build Your Own Device (video 4), 279

PREFACE

The making of this book has been both arduous and rewarding. It was quite a different thing to go from the casual perusal of Don's writings and videos to then commence the task of carefully analyzing every word and image with the end to systematize it *for the benefit* of both the beginner and the learned. This demanded maximum attention to details while keeping maybe fifty trains of thought before my mind continually for many months. It was necessary to do this in order to put myself in his shoes to fully understand his teaching, and also to be able to properly evaluate the priority and importance of various facts and processes, the organization of which can lead people to successful replications. It is satisfying knowing that I have done my best to do this with the time I could afford. As a result of this it is also fulfilling to have all the facts, from beginning to end, organized in my mind. This book is a partial production of that realization. More help will follow to encourage practical application of these teachings in the real world, with hands on experience with building of additional Kits we hope to shortly provide, along with various workbooks and exercises. That addition has already started with the preceding book and *Resonance Kit*. That was the first part of this process of helping people to follow Don's instructions. It gave people a tool to start at the place he said to start, because it was where he first started.

Reasons for Publication

There really was not enough time given to this work to give it the justice it should be given. If sufficient effort by those better qualified had been the case this could very well have been one of the most important publications on energy. There are so many things that could be added to this book, but one must decide when to end it. At 500 pages it is already overwhelming. I have always been one to 'do something useful rather than nothing at all.' I am satisified that this book is useful and is better than doing nothing. There are many pages from Donald Lee Smith's various early book editions that are not available online and would have been lost to the general public forever. This was one reason for publishing the work before you.

Over the last 26 years I have traveled the vast United States in search of very important documents and publications of great historical, scientific, and spiritual importance to this country and it's culture. This started before the Internet as we experience it today, and before interlibrary loans were common. I made hundreds of thousands of photocopies of hard-to-find documents, and some of them literally fell apart in my hands as I preserved the text from possibly the last remaining copy. These were the days before Google digital books, where a few souls would attempt to make these old works available for the modern world on CDs and in reprints. On my website TruthinHeart.com I have hosted hundreds of these for many years now. While that particular ministry has been mostly idle for ten or so years, the work has continued in regard to energy harvesting related subjects in the same kind of way. I have sometimes reproduced Tesla's lectures and patents that needed to be transcribed and noticed again. A few years ago I collected Carlos F. Benitez various patents from 1914 to 1918. Recently these have been made available in print through our Renaissance Charge website at R-Charge.com Before that I helped John Bedini and friends in processing similar valuable documents while pushing forward a technology that had remained undeveloped and in the fringes. In time I learned that the technology came from others these people had copied. Even Don will bring that out a little here and there in this book. It's not that important who first came up with a good idea as much as who can effectively make such *practically* available for the public good.

Earlier this year a customer of mine by the name of Alex Kolb, who was a friend of Don Smith, lent me what original materials he had from Don. Don's autograph note to Alex is found on the second page of this book. After considering the amount and value of this new material it became obvious that there needed to be a special publication of this material. Also consider what Don said in one of these vastly important early videos: "That's the front page on the book that we were talking about. It's growing. There's going to be about 400 pages in the book and it's going to look real nice when it gets finished." See page 208 in this book for the context. Don never got it up to 200 pages, or made it a really nice publication. So I felt compelled to at least improve upon what was done. We will look later at what specifically is added to this publication from that which is found in a PDF from the CD that he sold in the early 2000's.

After reviewing the new materials it not only became evident that a book needed to be done with all the material in one place, but that there needed to be an entirely new and more organized structure. There are various Don Smith pages floating around here and there, and all this important material needed to be organized and at least partly explained. But something more than a basic compiling of these pages was needed so that it could do greater good. Once I decided to transcribe all the videos included in the book it was obvious that a regular index would be very important for people to easily find all the places that Don referred to anything. Then the idea of making the book include a systematic development of various essential topics came about as the most important addition to the book. The regular index soon became a sort of mixture between an index and a glossary of entries consisting of a few words, to several sentences long, of quotations from Don himself. Aside from my lengthy Introduction and various schematics and illustration pages, as well as some bibliography entries, the only places where the content is not from Don is where I have indicated editorial notices between square brackets [ ]. These brackets are also used to differentiate other people from Don in the videos asking questions, or from others talking.

Another reason for publishing this is to counteract the misleading publication of Patrick Kelly's large work on free energy systems of numerous inventors (of which he distorts Don Smith). The idea of his work is good, but the execution of it is more than seriously flawed. In my observation over the years of that growing publication I have seen many obvious changes to the original text of inventors which subtly and/or seriously distorted the meaning of such inventors. As an editor and publisher I have witnessed this sort of thing with major publishers and have published on this subject an exposure in my republications of Charles Finney's *Lectures on Revivals of Religion*. While there may not be revisionist history going on to the extent of what some people like to imagine, I have carefully noticed and documented this actually happening to some concerning degree. Sometimes editors and publishers just want to modernize spelling, but many times they wish to tone down an author here and there. Unless someone has an original, and goes through word by word through the entire book–which is a very hard process indeed–all subsequent generations may never have the original words again. As a foundationalist it is very important to me that I not only have the real words of an author, but also for me to try and put myself in their context as best I can in order to rightly understand their meaning. Of all the books written, very few are really important. So it has been important for me to attempt to preserve the actual words of the books that really matter.

The distortion of Don Smith's words, claims, and intentions is an ongoing investigation for me, which overlaps my work with related people in the fringe 'free energy' or 'over unity' branch of alternative energy. In a way, this subject is of more importance to me than the research into free energy itself. Because if there is a manifest attempt to supress this information, keeping those who think they know from actually knowing what they need to know, then that is a bigger problem than possessing the information itself. While that may sound like something out of a movie, and be too much for many people to believe, it is not my purpose here to try and convince anyone of such things, or to distract people from the purpose of this book by enlarging upon it too much. But as people want to know why many have not succeeded in replicating Don Smith's work in the United States, it is important to consider several points as to why this has happened.

In the case of Don Smith there is not a lot of original material that is purposely distorted, as is the case with other historically important people and their publications. What we find with Patrick Kelly is him changing a word or sentence here or there while failing to consistently make the same sort of changes elsewhere. What he did with the Carlos Benitez patents was far worse than with Don's. While there are few changes to Don's work, Kelly's work cannot be trusted at all, and the reader cannot sufficiently tell when he is interjecting his own opinions into the text, or whether it is part of the original wording. The opinions of Mr. Kelly are in most cases wrong, and reveal that he is not understanding by experience the things he is talking about. You have to somewhat judge the production by the effect it has upon the masses who read it. I have had the fortunate pleasure of talking to thousands of people in the free energy community over the last 14 years. After careful observation–and in many cases, extensive questioning–the testimony I have from such people is that they are *overwhelmed with information that they don't know how to process*. A good number of these people have been so overwhelmed for 15 or 30 years. It is granted that much fault is with the people themselves for scattering themselves out so thin, and not working from a foundation up. But it is my understanding that those who claim to be leaders ought to have taken more pains to help people go from a proper foundation to a conclusion. Instead we find this world full of false promises, fanciful stories, undisclosed devices that are claimed to do extraordinary things. The people in general haven't known *where to begin*, *what is and what is not important*, and *what they should expect*. So the best way to keep such people from succeeding was to feed them with more and more supposedly useful information, while always leaving them to hope that one day something will be the piece of the puzzle they are looking for to make sense of it all. Meanwhile, many of these good folks will tell you that they know such and such system works over here, and such and system works over yonder. All words shared seem to imply sufficient verification of such claims until specific questions are asked which reveal that the person does not have anything actually producing what is claimed in general. They have things that do something, but nothing that practically supplies power at all. They have been conned by the claims of inventors who have not given sufficient help or who have purposely deceived them. Or they have been unorganized in their attempt to replicate such devices. At any rate, an effective way to ensure that most people will not succeed is to overwhelm them with sensational information while not stressing a foundational approach.

In the third video from 1996, Don says: "I realized that there are a lot of people that had trouble understanding things unless you could go turn the wall switch on, and that was the extent of it." This was not an insult, but a reality in our day when people either did not care to learn this stuff (which is perfectly fine) or were not capable of thinking for themselves. Nevertheless, many such people dabbled in this or that system because they were under the impression from various inventors that it was almost turnkey or plug and play. Not having any proper foundation people do not know how to evaluate such systems. Nevertheless, they are repeatedly convinced that the next charismatic guy that comes along surely is the real deal, so they buy into it for a season. I hate to see all of this, and it moved me over the last few years to try and make a change to this general experience. For the last four years I have had small meetings with customers and have tried to build a foundation. We start with real products that are demonstrated, and easy to replicate, and we progressively walk through the whole process of understanding what they are and how to use them. However, I saw over the last year how people were still easily overwhelmed by too much information. So in the *Resonance* book I wrote earlier this year, I opted for a *themes approach* rather than a specific fact approach. This book will continue to do that with the various *systematic categories* and various illustrations illuminating the core and foundational ideas. Some of these Don provided already in pages that will be new to you. Others, I have provided in this Introduction as I have had time for. Unless this was done I fear many would have just got lost again with too much information to process. On the contrary, this book is made to be a foundational approach, which would not be the case unless I added this content and made a rigorous effort to systematize it. I still see that so much more could be done so I will leave the rest to be done in future workbooks and illustrative kits.

While most of Don's words published online were not changed, the attack upon Don's message and publications or claims came more recently in another more effective way. *In addition to the information overload, by setting up and/or hijacking forums that start out favorable towards Don, and rapidly filling them with good and bad information, the result of such forums ends up giving the impression that no one has succeeded in replicating anything at all.* One such person, appearing to be open, came in early on a popular forum, and then provided source material apparently no one else had of Don's systems. As a result the people on the forum were excited. But gradually this person changed tune and eventually revealed their source: a friend of Don's, and who promoted his technology while he was alive (and who said then that "it works"). In the end these two gradually cast doubt about Don and ended up saying there is nothing to Don's claims. So that produces a far worse effect upon the credibility of the technology than the changing of Don's words would produce upon people. Hook people into thinking there is something with loads of information and hopes, and then pull the rug out from under their feet and say *we know* there is nothing here as we are the *inside club*. Now there is a lot more to the story than what can be shared, but if you simply compare what was said before and after, you will realize the manifest contradictions.

The truth is that most people only really want something they can buy. So even though there are concerted efforts to fool people in such ways, the truth is that people in general today, even technical people, are not able to follow basic instructions and think for themselves. Don was of the HAM radio days where people had a hobby of playing with these sort of things. He was a teacher and succeeded in teaching these things to the cubscouts. So he left us homework assignments. The fact is that he did tell everyone where to start and only a few people bothered to follow those instructions. He first learned these things from a basic radio transmission/receiver kit as we will hear him tell the story several times in this book. We decided last year to make this kit available with more parts, and with a large book. It is called the *Resonance Induction Coupler Kit*, or RICK as the homeschooler who makes it calls it (I did not come up with the name as it was already a resonance induction coupler kit before I ever knew of it). As we shall see progressively in this book, failure to experimentally understand resonance will ensure failure in attempting to replicate Don Smith systems.

We recognized this to be a key problem so we attempted to solve that problem with a more affordable kit, which has really helped many people over the last 6 months. But just like this problem, there are several other obstacles in the way of people's success. These are not Don's fault at all, as we shall see. They spring purely from inattention to detail. If we don't follow the instructions then there's no telling what part we ignored is critical. So this book is me jumping up and down at times stressing what is important (with illustration pages) while organizing the various sections of the process together so that you can exhaustively hear Don out on the matter in word and pictures. This really is spoon-feeding and I hope by doing this I do not counter my main purpose for doing this book. *My goal is to try and get people to think for themselves, and think honestly.* By stressing certain things so that you don't miss them, I hope that you can look more carefully at everything else Don says in these matters when you realize you may have overlooked what he said in these matters before.

Don was Limited in his Revelations by Commercial Commitments

Don showed up to the Tesla Symposium in 1994 and created quite a stir as can be seen in the video I have posted on my rickfriedrich YouTube channel, and the transcript in this book. Like most new inventors he showed more at the beginning than later on after commercial commitments were started. While Don was later limited in showing details he seems to have designed many of these devices around HAM radio parts so that such people would be able to replicate them. He claimed that they were the most successful in Eastern Europe. His teaching and presentation seemed to deliberately poke at the establishment, and for that reason such people did not take him seriously in the US. He was in three wars, and started this research at the end of a very successful life as a geologist. He didn't care to play the *prestige jargon* game. It seems he purposely wrote at times so as to stumble the high and lofty types. In the end Don was going to make you work for it. Don didn't demonstrate the lofty attitude of some inventors that look down upon others as not worthy of free energy. But he did give people homework assignments if they wanted to take his course and open up a new world to themselves. They may not have had to *sell everything to buy that field with the pearl of great price*, but they did need to pay attention to details, and put some time and effort into it. If people had actually done that, and listened to what he was saying, then they could have seen how resonance multiplies the energy, and very easily results in energy gains as we show with the starter kit. I will not repeat in this book most of what is shown in the Resonance Kit book. That is like a volume 1 in this teaching. Most importantly in that book, besides letting the themes settle in, is the learning of resonance *by experience*, as it relates to one wire and wireless transfer, and grounding providing the amperage into the system.

In the 6th video in 2006 (on page 322), Don's son Mark Smith gives some detail about the conflict of interest Don had in giving out this information to the public: "My father is trying to partner with industry partners to construct devices like this. They don't like it when he comes and does these conventions. They talk about, 'well what are you going to say?' 'How much you going to give away?' Because the reality is if he stands up here and gives a workshop, and says now take coil A, and insert it into slot B, those industry partners will disappear like mist. People then say, 'We're not going to buy these things.' Well the reality is, most of the industry partners, in fact all of them to this point, that have partnered with my father, and he developed something for them, and they take it in-house. They're not turning around and selling Donald Smith technologies overunity devices. ... So the reality is, they take his device. They improve an existing process, and you'll see, you might see a new whatever widget X come out that's ten percent more efficient than it was a year ago, that may be based on things my father has done. But you will not see which widget X that will power itself forever. So my father is trapped between two worlds. He needs industry partners to take his devices worldwide, and to try to make an impact. By the same token he'd like to just unleash this to everyone, to the world. So he comes and says as much as he can at these conventions so that those of you who want to try to follow along with this, and do this on your own, can pick up from what he's talking about. But don't ever expect him to give you the whole nine yards, because even if he wanted to, I wouldn't let him. Because the reality is, we have had enough people come to the table at our house and try to steal his intellectual property, and I'm sick of that. To be frank, after he had his heart attack and his stroke that left him half paralyzed, I offered to move him with my money, if he wouldn't use his money, to New Zealand to the country. Stop doing this, it's too much of a headache. But he's still here, and he's still presenting to you guys, because he would like to see the world benefit from what he's doing and it not just become a closet technology that someone sits on." Don didn't just want to give the technology to the world, he actually did do that, so long as people pay attention to details. This book makes that process a whole lot easier as I have organized and illustrated all the important details. It appears to me that Don *progressively* revealed pictures, schematics, and truths so that he left nothing out. He was like that crafty grandfather who lured his grandchild to figure out important mysteries by hands-on experience. He wasn't about to take *that* important process away from anyone.

Don was The Real Deal

There was not enough time to give a comprehensive revelation of Donald Lee Smith *the man*. It didn't seem critical to me after I carefully considered every word available that he spoke and wrote on these subjects, to realize that he knew exactly what he was talking about, and gave no indication of misleading people. I know several people who have been in his home for more than a day, even repeatedly. But I am approaching this testimony at face value because it checks out. I have demonstrated many things which he revealed, and his teaching perfectly consists with what I have been showing for years with the motors, and especially with what has been called my 'third stage process.' Having said that, we are "to call no one master," and only use his revelations as a tool to perfect our way in these ends. I'm not saying Don had everything perfect. For this reason I have added to my illustration pages the following disclaimers: "The author does not hold to the meaning of words like "electron" in the same way as conventional physics, and does not necessarily agree with Don Smith's views on that or some other parts of his theory. These diagrams, and this book, present Don's words *and* basic truths while revealing inconsistencies in conventional theory." It is not my object in this book to try and correct any mistakes I think that Don made, other than the obvious mistakes in his memory at times. It is my intention in this book to merely give you a tool to better understand Don in his own words, as he deserves our attention in that respect. His object was for people to learn the various concepts in a way that resulted in them finding energy independence. He really wasn't interested in dotting his i's and crossing his t's for eloquent mathematicians or engineers.

While Don didn't claim to be a saint, and sometimes was too graphic in what he expressed against others, he gave all appearance of being a good-ol-boy American (which type has mostly been lost). In his day a man was generally educated, and learned to be self-sufficient and independent. It appears he was really only reaching out to those kinds of folks with his presentation. As for his commercial dealings and family details, it is manifest that he really didn't want people to know that much about such, so it has not been my objective to invade that privacy. Don's son made it clear that if it was up to him that he would not have had his dad teach at those meetings. People ask me what I know and this is all I am going to share about it. Don allowed many people to visit him while he was alive, and spent countless hours on the phone and in email with people all over the world. He did not owe that to anyone. While I am nowhere near the level of knowing the things he knew, I do understand what it is like to be talking that much to all these people. There is no other inventor in the free energy community that has done anything like that. It appears to me that Don has done far more to progress free energy research than any one else, including Tesla. Perhaps as much as everyone else combined, if you look at how he has indirectly affected so many others. Don's system is not one but many different kinds of systems, with variety in each kind. His theory is also foundational and simple. It has vast implications. It also practically accords with existing technology over the last 100 years. While Don joked here and there, he was no joke. *He was the real deal.*

Human Nature Generally Contradictory and Prejudiced

In video 6 a question is asked: "We, by changing the way we're thinking, and helping each other to get to abundance for everyone, can change. How do we do that?" Don replies: "Unfortunately human nature is constructed in such a way that it's gonna be very difficult. And although I think what I'm doing is for the benefit of humanity, in the long run it's going to be very difficult to implement." People are notoriously *inconsistent* and *contradictory*. It is important to learn how to have *sound judgment*, and to learn when contradictions are incompatible with reality as opposed to when they are not because they merely apply to different environments or circumstances. For example, Kirchhoff's laws are a special case of Faraday's law. A *seeming contradiction* is solved when one considers *the bigger picture*. Kirchhoff's law is like *a general truth* that applies *most of the time*. But it is not universal. It is only a universal truth when it relates to a constant current circuit. If we draw a circle (as illustrated in the diagram in the Introduction) and write Kirchhoff's law in it, and say this law is truth, we then have to draw another circle around it and name that Faraday's law. The one circle is within the larger circle because it is a special case of the larger context. Failure to properly categorize facts of life leads to fundamentally mistaken conclusions.

Big or little assumptions, category mistakes, omissions, nonsensical prestige jargon terminology, and mathematical gynastics mystifies the world and keeps the world in slavish confusion, ever learning but never able to come to the knowledge of the truth.

Don and others claimed that Morgan and Edison took control of the institutions early on, so that, on the one hand, *one would be considered esteemed to belong to the credited club* of engineers, while on the other hand, *they would be slightly mislead about particulars relating to energy conservation*. There was to be much prestige for those who paid to join the club. As an all-pervading experience, no matter what the subject, the universal principle underlying all reasoning for those who are part of the club is, that when it comes to energy harvesting: 1. *The input can never be greater than the output.* 2. *There must be losses in any transfer of energy.* 3. *There can be no true gain, so we must work hard*. The slightly misleading particulars in regard to this overly simplified points, will be a focus of this book. The fact is that these statements are only true in a certain context and depend on the defining of terms.

1. It is true that effects are not greater than the Ultimate Cause of all being, which is God. However, when we wrongly attribute an effect to a mistaken cause, we may be mistaken about how great the effect can be. If the real cause, or in our case, source of energy, is a much greater donor than what people think then our estimate of output possibilities may be wrong. This is the case with the spiritual gifts of God in the same way. People reject God's earthly and heavenly gifts all the time. In the same way they fail to consider the facts of life, because of prejudiced ideas or dogma, and limit themselves in energy harvesting gains.

2. While there may be losses in having to create the mechanisms to harvest energy, and with the very loads we want to run, the losses in potential energy produced from source charge of energy can easily be recovered and can actually be ignored when we more fully tap into all the real potential. The subtle deception introduced long ago is that only certain methods of energy harvesting are possible, while these short out the vast potential. With a slight misunderstanding of the cause and causal relationships

people limit their expectations and ambiguously define loss to be something much worse than it really needs to be. They have made themselves colorblind to the full potential of the colorful source charge, and only admit of black and white pixelation.

3. The idea of gain is also slightly off when the real context is understood. There can be no gain over the input, it is said, with the idea that there can be nothing self-sustaining as that would violate conservation laws. But again, what if the source is much more giving than you think? What if it liberally supplies like the sun? What if we figure out the natural processes God created, and learn to fly, so to speak, by learning from the wings of birds? If the electrons are in perpetual motion and never wear out, are you open to the idea that maybe we can get them to work for us in more efficient ways? The truth is that we all have already done this but have not realized it. *In the modern world many products have been used by the masses that actually do take advantage of electrons in ways that would be considered over unity if the purpose was for electrical energy harvesting rather than for producing the other forms or applications of energy.* You just didn't realize what you had all along. Don talks about this in this book as you will see. To take it all the way to the foundation, he said that anything powered above or below ambient, so *anything powered is over unity by definition*. You just have to see the real context. But if you limit yourself to the small circle or box, that the selfish establishment has created for you, then they will freely feed off of your *wasted potential*.

Grounded Reality Model vs Floating Around Falling Delusion Model

Are you grounded in reality or floating around and falling in delusion?

You may have a coherent system that is internally self-consistent, but if it does not correspond to reality then it is merely a pathetic delusion, sustained by prejudice and lack of integrity, in willfully neglecting to do thorough and honest research.

Reality is the true essence and nature of things, and in relation to everything that exists. Selective delusions are fostered by popular culture when people, for example, immerse themselves in the particular *universes* of the movies and shows they are molded by. There are many other means to make people comfortable with the reality-is-what-you-make-it mindset so they become like the frog being slowly boiled to death in water. It is also fostered through narcotics and everyday drugs in our regular environment. It is fostered through educating children to create their own reality, as well as with false religion mythology and party politics. Shakespeare put the nail in the coffin when he revealed the inconsistency of the masses in modern culture that think there is nothing beyond the grave, on the one hand, but then act like their life means something, on the other hand:

Tomorrow, and tomorrow, and tomorrow,
Creeps in this petty pace from day to day,
To the last syllable of recorded time;
And all our yesterdays have lighted fools
The way to dusty death. Out, out, brief candle!
Life's but a walking shadow, a poor player,

That struts and frets his hour upon the stage,
And then is heard no more. It is a tale
Told by an idiot, full of sound and fury,
Signifying nothing.

This nihilistic philosophy from Macbeth does three things in relation to my point. It shows *the absurdity of not being consistent* in such important matters of life. It reveals that nihilism itself is but a delusion that *ignores the bigger picture of reality* that the masses have no choice but to believe. Finally, it illustrates the truth that universes in the theaters, movies, and shows signify nothing actually real! The same skepticism in philosophy is used against over unity realities. But the point being made here in the bold print is, that just because people have a self-consistent system, does not mean it actually embraces *all* of reality and is true. Conventional methods sometimes forget the foundation they're building *from* and act like they are attempting to pull the ladder up to themselves while suspended in mid air at the top (as in the Cover illustration). In the same way, *attempting to make an over unity device without knowing the core method is like considering only the ladder (technique) and after climbing it (building it) you attempt to get off into the air and to pull it up to yourself (use it) while forgetting (being ignorant) of the ground (reality) holding up the ladder (method and reason that makes the technique possible or work).*

When I was 10 years old I used to jog 10km every morning before school so I could have the highest score in long distance running. When I did that I used to avoid stepping on the sidewalk divisions as a sort of game, or distraction, and perhaps it was a superstition. This took some persistent and strenuous effort that I suppose gave me some entertaining diversity, and helped pass the time. At the same time I eventually had dreams where I imagined that if I tried hard enough I could hover over the sidewalk and not have to even touch it. The dreams seemed so real considering how much I was jogging that it almost seemed that I could have done it, or that I already had done it. In the same way this was like climbing up a very long ladder and forgetting about the ground it's based on, and then at the top trying to pull it up to yourself *as if you are a ground in yourself.* This is what everyone does when they *make themselves the measure of all things* and look to a technique (technology, trick) to support (justify) their delusion that has no REAL foundation. Before we look at some examples of how people misunderstand techniques by considering them out of context with reality, please consider some important points made from Asa Mahan in 1883 from a book I have republished some time ago. In *A Critical history of Philosophy* (truthinheart.com/EarlyOberlinCD/CD/Mahan/CH/intro.html):

"We shall attempt, not only to expose the errors of 'science, falsely so-called,' but at the same time to render plain the track on which real science conducts to the domain of truth itself. Truth, when fully apprehended, not only demonstrates to the mind its own validity as truth, but at the same time makes equally manifest error as it is, error as constituted wholly of 'vain imaginings,' wild assumptions, and false deductions. Naked error is powerless to deceive, and borrows all its effectiveness from the fragments of truth with which it is associated. Error always starts upon the track of truth, and at particular points takes its departure from that path. Philosophy will never have completed its heaven-appointed mission until it shall have fully disclosed, not only the line on which truth leads, but shall have shown, with equal distinctness, where and why error, in all its forms, takes its departure from that line...."

"**Criteria of True and False Systems of Science.**"

"When systems are constructed exclusively from principles and facts which are the objects of intuitional knowledge, and from deductions necessarily implied by such principles and facts, we then, and only then, have true science. When, on the other hand, mere assumptions, opinions, beliefs, conjectures, or guesses, with deductions from the same, pass over and lie at the foundation, or enter as constituent elements into the structure of systems of affirmed knowledge, we then have 'science, falsely so-called.' We have here undeniably universal and infallible tests or criteria, by which, of affirmed systems of knowledge or science, we are to distinguish the valid from the invalid, the true from the false. Systems of the former classes are to be esteemed as rightfully having place in the sphere of true science. Those of the latter class are to be regarded as having place nowhere but within the circle of logical fiction. Deductions of the former class are to be regarded as truths of science; those of the latter, as the lawless sophisms or wild guesses of false science."

People looking for a Trick not Foundation

People live haphazardly. In life they are looking for a *trick* to get by on. Just like people are looking for a *technique* to give them the answer, *rather than them knowing the foundation, and building upon it*. If they generally live in contradiction then the tricks may not work. This applies in the following sample areas:

Free energy: People want to know a trick to get free energy. Since they don't understand what the process is really doing they are vulnerable to misusing it so that it may not work for them. An obvious example would be to use a solar panel at night. Another would be to make a magnet motor with a steel frame where the magnets drag against the frame. In our case we will look at reasons why people have misused Don's instructions and have built something else. For example, I never noticed anyone get Don's grounding the way he instructed, or at least to understand how he said it worked. For this reason I made special illustrations. If you want to benefit from this book then do not look for a mere technique to solve your problems. *This book is about the big picture.*

Engineering in general: The same applies to engineering and science in general. An example I show at my conventions is where some bridges were built by following a technique, but without considering the bigger picture of the local environment. They failed to guard against other forces that can create a sympathetic resonance with the bridges. In one case the bridge failed. In another case 60 years later, the engineers made excuses as to why they could not have foreseen such problems. In other words, we need to properly do our homework and not limit ourselves to a specific function. We need to consider the entire environment.

Martial Arts: I was trained in Aikido and Hapkido, the most advanced martial arts. Not that I recommend martial arts anymore. The same points relating to techniques apply there as well. If you try and merely do the technique it will not work at all. The *how* of the technique is very subtle and on an entirely different level. In fact people used to the hard styles, which depend on strength, will assume those techniques are fake and practiced dances. It parallels with the free energy systems that do not depend upon pushing current, but rather magnetic resonance. The object in such non-conflicting, non-agressive arts is to rather gently disable the agressor by using their own force against them. But no one will be able to merely do the techniques unless they first understand how it works. To be more specific, for example, if you grab someone's wrist, they will naturally pull back and resist. This becomes a tension and usually the stronger person will prevail. But in Aikido, and other advanced levels of martial arts, the result comes not from what is outwardly observed, but from the full method adopted. The first point is to be *grounded* so that you are fully balanced and locked into the ground consistent with actual gravity. The second point is that in this situation, when the agressor grabs you, you do not pull back or cause any noticeable resistance. The agressor will always be moving, and one can subtly move with them, and slightly redirect them at the same time, without them even realizing it. So in this case they will move into the agressor many times while acting as a dead weight, and in a split second the agressor will be deflected or subdued because they have lost their balance in attacking. The observer cannot just copy this technique without understanding how it works and spending a lot of time *unlearning* the resistive responses, etc. In the same way, those experienced with conventional electrical systems find it very difficult to follow instructions and overlook the reason for various processes, and thus fail to do them as directed. The above example parallels perfectly with these energy processes.

Health: Diets are a common example of people looking to, as they say, die (first three letters). Various techniques are relied upon to get a slim body. Most of them are not healthy and fail to look at *the big picture*. There are a great many factors relating to our health that are not so cut and dry. One can be doing many very good things for one's body and just make one mistake and experience bad health. The key is in doing what it takes to keep your immune system strong, avoid toxic environments, and live as natural a life as possible. It is too difficult to cover this in a paragraph, but the point is that fixating on one or a few techniques is not going to solve your real problems. Many people always assume that some one thing caused someone to be sick, when it usually is more like one thing was the final straw that broke the camel's back. In the same way there are many reasons why people fail at making these energy systems work when you are merely trying to copy something without considering the local environment and how to tune things.

Love: We have all heard the claims that if you follow so many steps (techniques) that you will find love or will have a solid relationship. While it is true that if you actually have *real virtue*, and you are in a relationship with a virtuous person (which is extremely rare these days), then listing a bunch of virtues to engage in would be *valid techniques* (however, virtue is of the heart and not merely outward). But such articles written in today's 'modern progressive world' merely hint at being entirely virtuous and so completely fail in the same way we have been considering. They miss the heart of the matter about first being completely right with God and then that love will naturally overflow into proper love towards your mate. Most people are fake at their core, and all their attempts to do good works are actually selfish when their ultimate intentions in life are wrong (*ungrounded* in reality). This is easily seen in close relationships. The foundation in this case is having the *same* character as God and actually being holy as He is holy. The same opposition to over unity possibilities from conventionally minded experts exist with the vast majority of religious and non-religous people in considering it possible to actually be holy as God is holy. The amazing thing is that the religious people completely disregard their own bibles where it manifestly calls them to be holy and gives them "exceeding and precious promises where they can partake in the divine nature." "So then, dear friends, since we have these promises, let us cleanse ourselves from every impurity of the flesh and spirit, perfecting holiness in the fear of God." In the same way that Christians and others disregard the foundational truths of love and morality (or the living waters of the divine nature and fellowship of the Spirit motivating and moving them to give to all without insincerity), and substitute dead outward works (techniques), so engineers design synthetic processes, contrary to what is naturally revealed, that produce the heat death and ruin mankind in so many ways. What is proposed as a good and loving thing in our world today, is often very bad when you look at the big picture of reality.

Law: The form of law can be used as a tool (technique) to manipulate people for one's own advantage contrary to the spirit of a good law. A circumstantial law can be enforced as a universal and people can be wrongly cited for violation. In the same way in electrical matters people will say that over unity claims violate Kirchhoff's laws, when such laws are circumstantial and not universal. Kirchhoff's laws becomes sophistry to silence claims. Or people can attempt to adhere to the special circumstances that Kirchhoff's 'rule' pertains to in attempting to replicate a free energy device. Naturally they kill the beneficial processes because their attempt at replication (technique) was wholly mistaken and *out of context*.

Finances: People can steal from others (as a technique that works) even though it is not beneficial for society or for them in the end. Pragmatism is false as a philosophy and practice. What works is not what is true when we consider *the big picture*. Making energy for others can also harm others and the environment as is the general practice. People can also make financial decisions (techniques) that result in them being deceived because they did not know the full picture. Therefore they can lose their money. In the same way there are many people taking people's money, time and faith with false free energy claims. It is the main object of this book to counter that tendency by not promoting one or a few techniques but rather *the big picture*.

Education: People go through the first quarter of their lives in school to get a degree (technique) that will give them recognition, experience and knowledge. While the idea of being educated is good, if the big picture is lost sight of, then the education is worse than useless and is positively destructive to such people and the world. If people become skilled in prejudice they will not process truth fairly and will use this technique against others. Education today in general is justified as diverse, internally consistent, and relevant. And yet it has lost it's entire purpose of educating students to be Universal (consistent with the big picture) and Moral (the purpose of life). In the same way that teachers are fired for their proper teaching of students about morals and reality, they are also fired with regard to over-energy systems in the institutions.

Philosophy/Theology: People often study (technique) the most important subjects that pertain to life with a heart that is inconsistent with the truths being studied. When this happens it results in *sophistry*, a perversion of logic. This is like the lawyer in studying logic who uses it against others. Or like the theology student who develops a hardened heart towards God in reducing Him to propositions, or relates to Him only historically as a mere force in the past. In a similar way one jumping into free energy research easily goes in without considering their physical health and easily harms themselves in the rush for such treasures. All these result from misunderstanding the foundational purpose for such things.

Mission: If you merely carry out the mission outwardly to the letter (technique), without understanding the real purpose for that mission, you may really fail in it. You give free energy to the world for good reasons. You give someone a free energy device and they use it for evil. It is a common assumption that giving free energy to the world will change everything for the better. This is the deception that Tesla was under when he said that the machine was our savior. Since those days 100 years ago the world has not become better. Technology (technique) has not made the world a better place to live in. The environment is rapidly being ruined, the health of the world is getting worse in general, the virtue and intelligence of people is lost, etc., etc. People generally pervert good things. So my objective in this book is to get people to see *the big picture* and live consistently with God and His creation, and not to learn some techniques to manipulate the natural world, themselves, or others.

Life Purpose: Is life merely about surviving (technique)? What is the big picture? Is it just about existing? We go to school to get an education, to get a job, to pay our bills, that allow us to exist. We work towards our retirement and then die. From dust to dust. What is our purpose? Today people are self-absorbed. But we were made in God's image and to be just like Him in character and mission. We therefore only find fulfillment when we understand the big picture of the techniques of life. We find that it is better to give than to receive. Just as happiness is found in seeking the good of others as much as we seek our own good, rather than seeking our own good more than others. But the bigger picture is seeking God's good higher than our own, and that is where we can only find the end of that internal tension, because that is how He made our nature. The big picture of our nature and relations is that this is the fundamental law of all creation. Many things are *opposite to what we expect*. We expect to find true fulfillment in getting, and the more we get the worse we feel inside. But if that gets reversed then we are surprised to find that we are happy for the first time. Any other happiness that is commonly experienced is just a transitory or fake happiness and is not the experience of "rivers of living water" flooding our souls. Many scientists practically worship the laws of nature. Can you imagine living consistently with the most important law of nature? Can you imagine living inconsistently with the most important law in the universe and heaven? But that is the case all the time. In the same way people assume the opposite things with free energy systems. They try to push things rather than let the energy do it's thing. They place one particular rule over a universal. So we have to humbly accept the truth whatever the outcome. I get many phone calls and have met many people in this research from all over the world. While most have been good experiences, occasionally I meet someone who has it all figured out, but really doesn't. Poor souls let their pride stand in the way.

Asa Mahan said some 164 years ago: "As is the method of a philosopher, so will be his system; and the adoption of a method, decides the destiny of a philosophy." Is your method consistent with the big picture of reality? I really don't care if you find free energy. What I care about is that you learn to seek out the big picture which helps you find free energy and many other more important things. *What you go into anything with will be what you get out of the experience.* If your method is partial your result will be partial. If your method is not absolutely honest then your results will do you no real good. This you know to be true deep down. It is true that people accidentally stumble upon truth, but a truth does us no good ultimately if we don't live consistently with *all* truth perceived.

Grounded Reality Method in Electron Path

This very lengthy focus on techniques is to encourage you to seek out and adopt the right *Grounded Reality Method* for all areas of your life. In this book we will specifically apply it to over-energy harvesting. *These words were specifically chosen because they directly relate to a key process that is the most important energy theme of this book:* ***Getting energy from actual grounding.*** Just as Mahan said about the method of a philosopher determines the destiny of his philosophy, so the method of grounding (technique) must be rightly done. Rightly understood to do it right. Because in these systems proper grounding is an essential process. No grounding, or improper grounding, results in failure for most people. As I was pondering this walking by corn in the field I thought of the crops coming from the ground, producing fruit, and then returning to the ground just as we come from dust and return to dust in this life. This is the same for the Don Smith *Resonant Energy Systems*. They pull in electrons from the ground that spin for a season and produce much fruit for free and then return back to ground at relative rest. In the same way we come from the ground and can produce real fruit that lasts forever, or we can produce harmful fruit that ruins forever before we return back to the ground. Just as we cannot return to the ground in the way which we came from it, so *with these energy systems the electron path needs to go back to ground through the path which bears fruit*. If you don't understand this secret you will bear no real fruit that you are after. Your techniques will not work because they miss the big picture. *If you miss this point in this book it is your own fault!* This travel is a loving path circuit similar to my so-called second stage process and Carlos Benitez Switches. This realization takes you beyond looking at parts as a technique and makes you see instead how you can use parts to apply *the* proper method. This will make you interested in seeing how other technologies use this same process for harvesting other types of energy in their workings. It will also make conventional theory make sense as a limited understanding of this process. My various illustration pages bring out these points. So there is "one thing needful." To keep "your eye single," and to stay focused and consistent with the foundational truths of reality. Build upon a foundation and never go backwards to foolishly try and make the foundation what you *want* it to be. How silly indeed! Count the cost of the life on the narrow road and you will not be disappointed in the end. Stop trying to pull the ladder up to yourself in mid air because you will fall to the ground soon enough. "There is no foundation other than that which is already laid." You know the rest...

The modern man's thinking is as if they have climbed up a ladder and after getting to the top they attempt to get off into the empty space of their own fancy and pull the ladder up to them so that they can climb again higher. But the ladder was on a real foundation. And as soon as the fool steps off of that which is real into the fanciful, he learns eventually that there are real consequences.

Processing Don's Pages

The following changes were the only changes made to Don's printouts. It is never my practice to change an author's words but to keep them exactly as they were given. So in this case I have copied the book pages and have only changed them in the way of centering them (as many were crooked) and removing photocopy lines and blotches. Any other changes to the pages are mentioned between the added square [ ] brackets. In a few cases I made two pages into one. There are plenty of spelling, grammar and punctuation mistakes which I have not changed at all. Some of the pages show images from the other side of the page. In other cases I did not copy Don's pages but found the original book from which he copied from and so I did a better copy while providing the link to the whole book. I have not held back any of the material given to me, or any details about Don Smith, that was given to me. This was a choice I made even though I could have profited a lot more by keeping this all to myself. This would have happened if I merely reproduced all the papers without creating the indexes and helps.

The order of the pages from 1 to 100 is not necessarily a perfect or systematic order. I attempted to follow the order of Don's largest book. But the order was somewhat different from year to year. You can see the original page numbers as he used. There were several booklets along with many single pages. The order is satisfactory. You will be able to see a progression.

Trascribing Don's Audio and Videos

It was a very tedious task to transcribe Don's words in 8 meetings over 14 hours in length. We started with the YouTube converter that gave no punctuation and had serious mistakes in every sentence. It was of interest to see some examples of words or phrases that were added to the text that could not be accounted for by normal programming. So if such was read by people they would have been intentionally mislead. With the raw YouTube transcription two others helped me to correct obvious mistakes. I ended up doing the Audio and video 6 recordings completely, and video 1 almost entirely. Bruno graciously helped with videos 2, 3, 4, 5, and 7 which saved me weeks of work (as each hour of video took many hours to process). I wrote Bruno the following about why some people are not understanding Don properly in the videos: The problem is that unless you are familiar with English in America you will naturally mistake many words or not understand various ways we say things. Don talked with sort of a slang accent where his contractions would easily be mistaken by people not familiar with that kind of English. It is Southern English as well, which is very far from British English that you would be familiar with. Also, as you can see, towards the end of his life his speech was more and more impaired due to the strokes, so he cut off his thoughts in mid sentences (as I do myself a lot) and went on to another point. Finally, only someone very familiar with this technology will understand the hard-to-hear words by understanding the context. What I have done is near perfect transcribing these words of Don. I didn't have the time to make it 100% perfect where it didn't matter, but I did where it does matter. There are probably only a handful of words in 200 pages that I may have got wrong. The hardest one was the 2003 Audio text. That was done over the phone. I had to use ear plug speakers to get the words, and this is where I may have got a few mistaken, or lost a little word here or there. But that was the least important transcript as he had just recovered from a major stroke and was not really following what people were asking him. I also removed most of the filler words as mentioned before. This is not according to my regular policy. To leave in otherwise would make for a very painful reading as he often said ah, um, etc. several times in a sentence. The readers would be very irritated and not be able to follow his meaning. So this will help you also to better focus on what he said.

YouTube had already removed all the "ah" and a few other filler words anyway.

Sometimes it was difficult to hear what Don was saying, especially as he got older and when he didn't have a microphone. It was also difficult to transcribe while I was also trying to process all the words for indexing. So there will be some mistakes here and there. But since the videos are preserved you can follow along with the book and have these corrected. Having said all that, the transcribing of the videos was more important than I expected. It resulted in correcting my understanding on several different things, and showed many subtle points that were no doubt lost by everyone. Those who are more used to the technology and accent will have understood more. But I can see that Europeans that only know British English may mistake almost half of what he was saying. I suspect that I misunderstood/lost at least 5% no matter how many times I watched these videos. The point is, that unless you actually stop, and write out, and process each point, you will miss a lot of things. So the transcripts really help with this by clearing up all the words you missed, and removing all the unnecessary filler words like "ah, um, so, okay, now, like, and", etc. These filler words were in almost every sentence.

Purpose and Result

My purpose with this book was simply to have the same one as Don, and improve upon what he did. Organize Don's presentation so that it is clearly understood, all in one place. There really are very few additional ideas from me in this other than my methods of pointing out the particular details and the major themes people just didn't get due to lack of focus. The book in general, and especially this Preface and Introduction, provides an important context, several different kinds of context, that brings it all together. Don's work was vastly important. But because he did not put enough effort into making it what it could have been the book ended up falling by the wayside with the effect of being generally ambiguous to the reader who gets lost in so much information. What I found over the years, and as I tried to point out in the last meeting, is that people need to be settled in the important points *one at a time*. By providing too much information, even good information, without settling each point, while people may be in awe of it, it also brings out a frustration deep within as people today don't really know how to settle anything or any point. Each point needs to be settled, and each in the context of a big picture, so that hope in the end result is justified as each part leads to the a completion. I hope I have addressed these foundational problems so that people can learn how to come to grips with the truth in clarity and confidence.

Where to Begin?

Where you begin will depend on what you are after. This book has something for everyone. Some people may only want the theory. Others want just the practical. You will not get the theory unless you actually process these ideas in the real world. You also will not be able to understand how to make these processes work without paying attention to the theory as mentioned above.

In the Introduction I will go over the details of the various illustrations I have created. Here I will mention that you need to go into this with the right attitudes and expectations and tools. The attitudes have been covered already. The expectations will depend on your attitudes and focus. The problem is not so much with learning a new thing as much as unlearning bad habits and prejudices. The first tool you need, the most important tool, is a pen with notepad. Any reading of this book without resorting to those tools is pointless. So the first assignment is to go back and reread everything so far and make points that you later ponder and process with a clear mind. The next tool is a mirror so that you can look at yourself and point the finger. Learn to make sure you are sincere in your research before you judge others as not being such. Physician heal thyself. Unless you can overcome the pre-judice principle that you have been trained by, you will over-and-under-believe yourself and others and get nowhere. Other tools you will need after the first Resonance Kit level of experience are an LCR meter, a Milligauss meter, an oscilloscope and anything else that a qualified RF technician would recommend for safety protection. I cannot recommend unqualified people work with these bigger systems as they can hurt themselves. The little first kit allows for safe experimentation, but the systems Don shows in this book could easily develop very dangerous levels of power. The teachings in this book can be applied to the little kit without having to make bigger systems. We use several of them already in the kit and will be adding more feature options to the kit shortly. After you process this book you may be in a place where you can collect and use wasted energy from existing devices in your own home. This is one of my hopes for each reader.

The compiling and editing of this book has been an exhausting process that I am happy to be done with. Nevertheless, as I came to the end of processing all the material I felt some sadness about never hearing anything more from Don again. It was my pleasure to learn many things from him over the years, and as I came to the end of the emails it hit me that he was gone and there would be no more helps from him anymore. This was compounded by the fact that the Free-Energy forum he was on (as the last correspondance we have of him) was mostly disrespectful to him while he still persisted in trying to help people to the end. I cannot attest to Don's personal character, but he was a gentleman to people I know that spent time with him, he appeared that way in the videos, and his teachings work for me. Naturally I was compelled to make this production. I hope it blesses everyone who reads it as it has me.

September 25, 2018

Editor's Introduction

"The Disk [CD] is a goodwill effort at sharing information which I consider relevant, as is the Book. The Book is largely my theories going through maturation and the Disk relates to the same developing process. Together they represent a lifetime of thinking by someone who is in the know."

Don Smith left us with a wealth of relevant information for reasonably experiencing energy independence. This fact becomes self-evident after attentively reviewing the details. This publication lays out the details exhaustively, systematically, and with illustrations to help the reader to realize this, and experience what Don wished for. In this Introduction we will look at the core details that most people are not aware of, but first consider generally and specifically why some replicators have failed. The main reasons for this publication are to prevent such failures, and help others overcome their mistakes. But as mentioned in the Preface, there are many people that either cannot or will not use this information to do this. Don addresses these problems in all the different sections of the book. For example:

"Paul, thanks for the kind words. The sole purpose of the website is awaken people to the fact that there are ways [to] survive the enslavement of special interest. After many years of trying to cooperate with the US agencies it became apparent that [they] were the problem which resulted in my activities being relocated offshore. I have a huge following world wide with immediate success. Many people realize the necessity for what I am doing and most want results only if someone else hands it to them on a platter with no effort on their part. This causes a vitriolic reaction which is difficult to manage. Again Thanks, Don Smith"

The organized information in this book is not going to call forth and possess parts to come together to produce a certain outcome on their own. Time must be given for this information to be noticed, processed, and consistently applied. One must consistently apply what is to be done and avoid doing what prevents success. Rushing this will only lead to misunderstanding and frustration. Time is needed to process the information and directions, and to allow people to consider different ways of looking at things. For unless people correct their mistakes that presently limit them to an under unity experience how can they expect to rise above that? One has to change the way they think about parts, processes and the environment. This book should enlarge the persective of looking at *individual parts* as isolated two-dimensional entities to seeing them as three-dimensional organs with multiple characteristics dynamically interacting with the device and the local environment. This relates to why people fail in replications:

"Now since last year [1994] there have been hundreds of people that have tried to duplicate what I did. There's about 12, I guess, out of several hundred that were successful.

The main reason most of them failed was the impedance problem. Because when they started putting capacitors and resistors and you know various other things in there, which controlled the timing circuit, so that the different coils could talk to each other, but basically they blew themselves out of the water. And none of them, or very few of them ever bothered to check to see if there was something there other than resistance or capacitance or whatever. Or even in your L1 and your L2 coils."

Don is talking about how parts relate to each other and the fact that each of these main parts has capacitance, resistance, and inductance while people generally only consider each part as having just one of these attributes. He is another example:

"Your understanding of electricity is limited to 60 cycle understanding. High frequencies are a different world. See Tesla's Book 'My Inventions, Researches and Writings'. Any and all of My Devices Work exactly as I state. I can not be responsible [for] the short comings of others. Regards, Don Smith"

The reasons people fail in Don Smith Replications

1. All processes are only considered, made and tested in pre-judgment with a close-looped, constant current, non-resonant linear perspective, where the person prevents themselves from experiencing gains.
2. Trying to copy a physical model while not understanding how it works, so as to not understand how to tune it in resonance to a particular load in a particular environment.
3. Overlooking the fact that inductors, capacitors and resistors all have inductance, capacitance and resistance which all affect the tuning.
4. They fail to consider details like how and where to ground the system, because they fail to understand what those details mean even though Don did explain them.

Two Obstacles in the way of Over Unity Energy Harvesting

1. People don't recognize the two forms of energy and how to measure them.
2. Therefore they don't know how to make use of them.

This is where we go beyond merely slapping parts together in the hope that we get some desired result. We must get into the theory, which is a large part of this book. One doesn't have to agree with all the words to get the desired results. They just have to get a feel for the big picture and operate consistently within that context. This again means not over-simplifying things or ignoring details. People have generally ignored the magnetic side of energy, which is a special focus in this book. They have also ignored the source, path, and spin of electrons. These fundamentals in Don Smith systems are not understood by those who fail in their replications.

"Useful Energy is Defined as that which is Other than Ambient."

This is one of the most profound and simple ideas in this book. This is the foundation point from which Don builds. Without this perspective you cannot progress.

Electron Spin Instrumental Cause of Useable Electrical Energy*

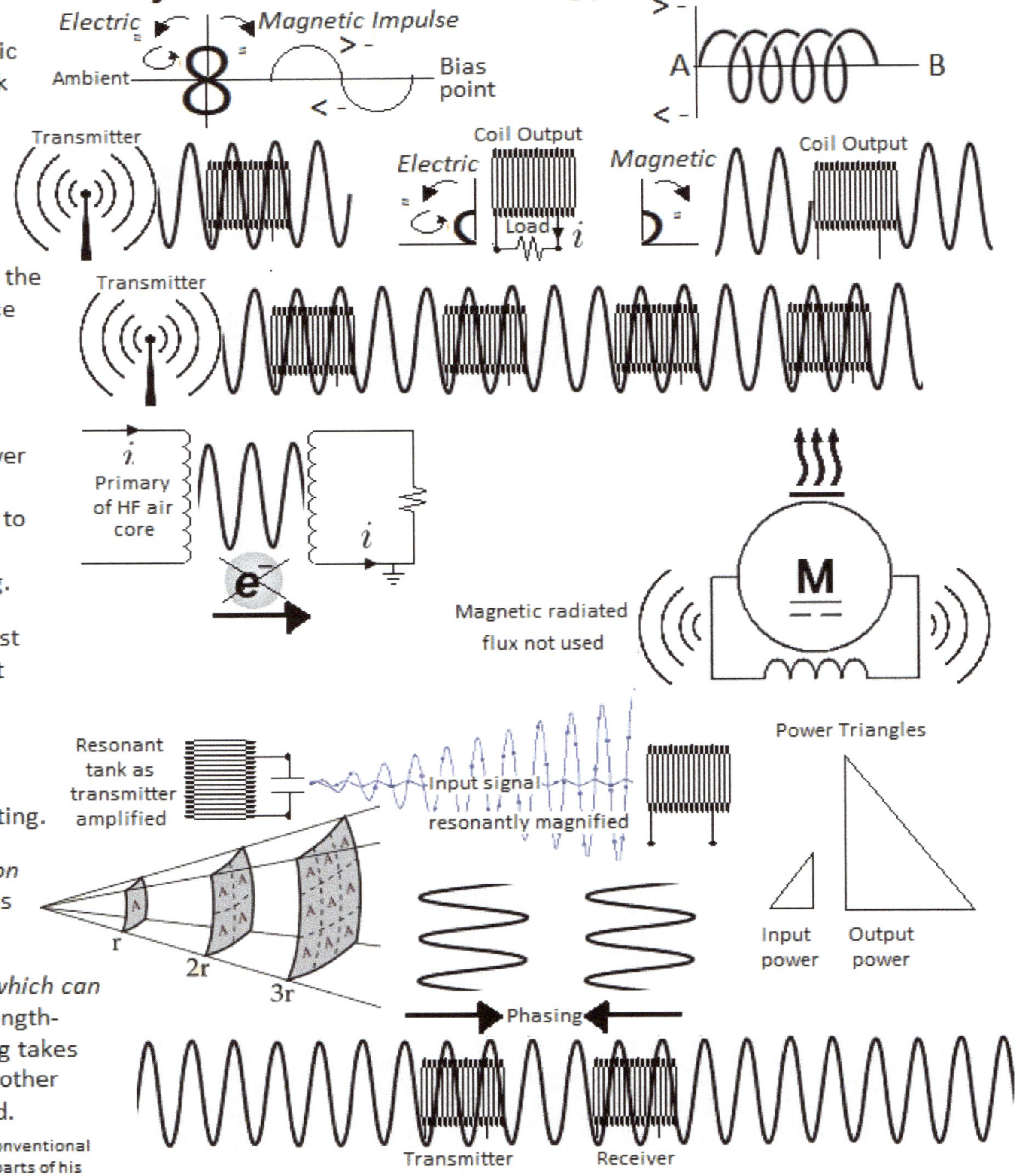

- *When electrons* are electrically impulsed they spin* from ambient towards a less negative potential giving off magnetic energy, then they give off electrical energy on their spin back to ambient, the result being two outputs for one input.
- *Electrons* can be electrically or magnetically impulsed,* resulting in this two-fold output spin. If magnetically pulsed they produce electrical energy and then magnetic on return.
- *Magnetic resonant impulsing or fluxing is preferred* because the magnetic lines or waves: a) do not degrade, b) their influence is in an open-system context where the input body is not diminished by the independent recipient bodies, c) and can have multiplied recipients without limiting each other.
- *Magnetic fluxing is not a transfer of electrons* so that in power transfer through isolation transformers no electricity is transferred but only the magnetic is passing from one place to the other. This means each side is an independent body creating its own energy by electron spin from remote fluxing.
- *All electrical power systems depend on electron spin* but most do not take advantage of these essential facts and ignore at least half of the potential energy available.
- *In resonant tank circuits the circulating current or voltage is multiplied* from the input current with the result that output fluxing is equally multiplied and available for energy harvesting.
- *As the frequency and/or potential increases the multiplication of output flux comes under the square laws* so that enormous gains can be experienced very fast.
- *Recipient bodies on their return spin magnetically impulse which can result in a sympathetic resonance condition* with the wavelength-matching donor transmitter where phase conjugate mirroring takes place as the two magnetically dance together feeding each other with no other input necessary, while multiplying flux beyond.

* The author does not hold to the meaning of words like "electron" in the same way as conventional physics, and does not necessarily agree with Don Smith's views on that or some other parts of his theory. These diagrams, and this book, present Don's words *and* basic truths while revealing inconsistencies in conventional theory.

PERPETUAL ELECTRON CYCLE*

Can be compared to various natural and artificial water systems

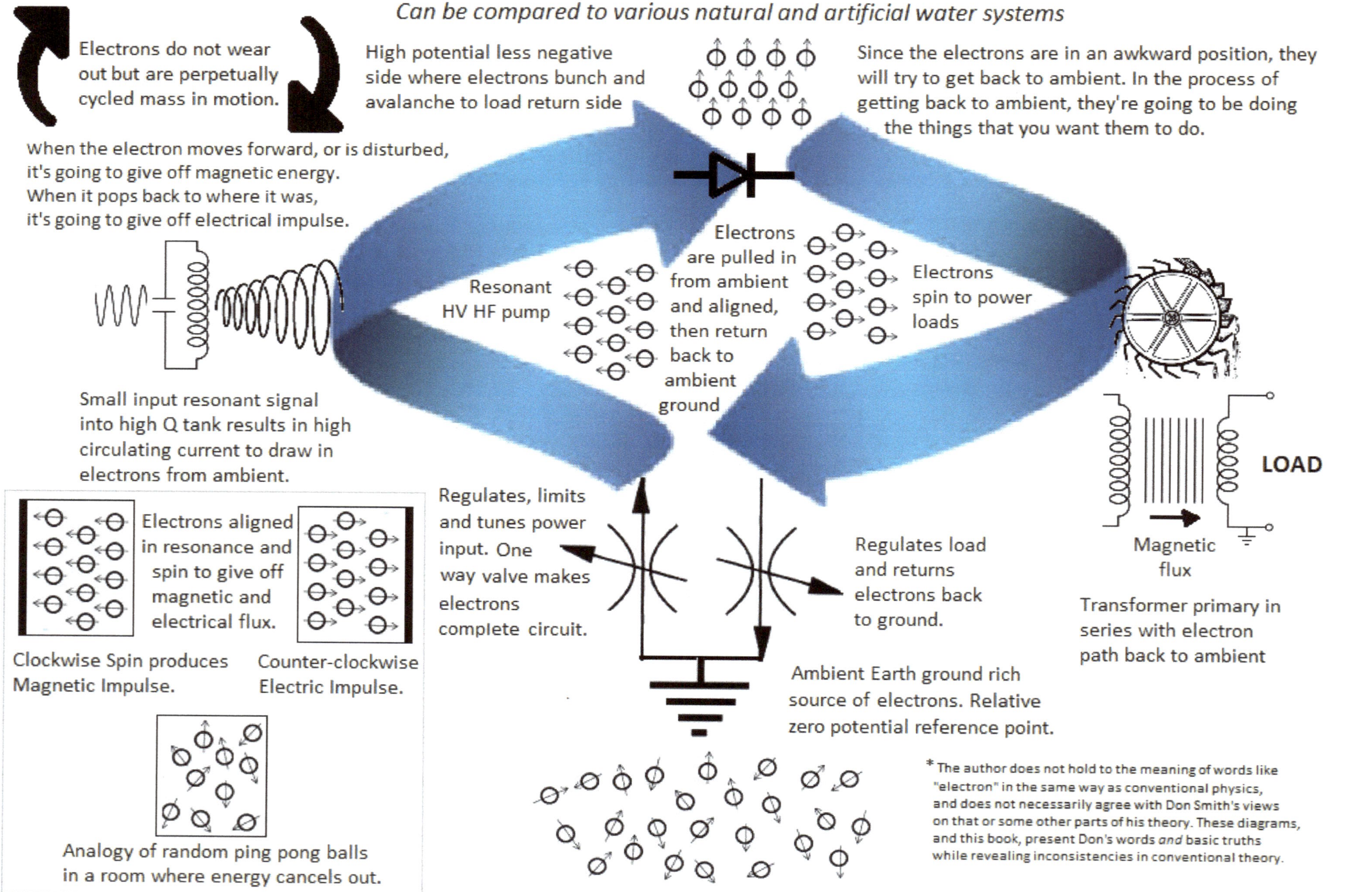

* The author does not hold to the meaning of words like "electron" in the same way as conventional physics, and does not necessarily agree with Don Smith's views on that or some other parts of his theory. These diagrams, and this book, present Don's words *and* basic truths while revealing inconsistencies in conventional theory.

The ***Electron Spin*** and ***Electron Cycle*** pages illustrate and summarize core ideas in Don's book flowing from the foundational idea of *useful energy is different from ambient*. Don gave many similar summaries, of which the following is essential:

"The basic unit of electricity (the electron) upon encountering a moving magnetic field (or wave) spins, giving off an electric impulse. When this impulse collapses, it spins back to it's natural position, giving off a magnetic impulse. Therefore, magnetic and electric are two sides of the same coin. When the magnetic side is pulsed, it yields electricity and conversely, pulsing of the electrical side yields a magnetic field. Moving one in relation to the other produces useful energy. When done consecutively, each cycle pushes (current) forward, while pulling electrons into the system...in much the same way as a water pump moves water. These electrons are obtained from Earth and air grounding."

The *Pertetual Electron Cycle* page is an attempt to diagram these ideas and should be carefully studied and referred back to while examining this book. A key point in this is that the path of the electron is in a controlled loop from ambient ground back to ground *another way*. People have failed to produce any significant gains in trying to do replications because they have had a ground connection that allows the electrons to return back to ground the *same way* they have come in and which does not *maneuver them to freely power the loads*. The next thing to notice is that there are *two output impulses* when you spin electrons. The order is opposite depending on how they are spun (electrically or magnetically). People do not usually think of electron spin producing two useful outputs and they tend to ignore the usefulness of either the electric or magnetic output.

In the ***Renaissance Charge Motor Energizer Systems*** we use both of these in either forms. When we pulse the motor coils (electrical impulse) there is first a magnetic impulse followed by an electrical impulse that results from electron spin according to this theory. In this situation, normally the magnetic output is all that is desired in conventional motor or solenoid systems, while the electrical output (from the electron spin) is suppressed, shunted out, and considered merely a problem to be removed. We however, use that to charge batteries, power transformers or other significant loads.

On the other hand, we do the reverse order with the generator and EMP receiver coils. When the motor coils produce a resulting magnetic impulse, the electrons associated with these receiving coils will first produce an output electrical impulse followed by a magnetic output. Normally people would not look for, or expect, a magnetic output, so they do not engineer around it. I discovered this advantage many years ago as I designed motors, and have been showing these unique results, where the motor speeds up while the generator coils are shorted or loaded. More recently I have demonstrated what I have called EMP receiver coils receiving the magnetic impulsing from the generator coils producing charge in them as well. These all have to be carefully placed so that the relationships between these three coils are not a regular transformer coupling but are distant enough. The generator coils also must be positioned so that they advance the rotation of the rotor when they magnetically impulse.

The air core, or high frequency core systems, are similar (except they have much higher gains due to resonance and frequency). They also have to *position parts in a way that is consistent with this electron spin* with resulting electric and magnetic output impulsing. With the motors it is easier to see how geometry contrary to these facts results in poor or no useful performance. But with the magnetic resonant induction systems it is harder to figure all this out without the right theory and ability to measure everything. Without understanding these core facts it is like a blind man trying to paint a color picture without knowing that there is such a thing as color.

The same is true for *cycling the electrons in loops*. We started with useful energy resulting from a potential difference between ambient ground and anything else. The key is to pull them into the high potential on the intake, and then let them work for you on the return back to ambient.

"First stage of the electron cycle is the contraction stage [page 33] where electrons come together, the second stage is the expanding stage where they move apart. The second stage is negative resistance, and is where you apply your load."

It is rather *critical to not load on the intake* expanding stage. The electrons want to go back to ambient so they will work for you on their return. It is relatively easy to excite them and divide them (as positive and negative pairs) with magnetic fluxing. The loading is on the negative resistance side. These, and many other such details, Don shared from time to time, but few seemed to pay attention.

Resonance amplifies these actions and *draws in more electrons into the system from the available source* (like earth ground). If you make any of these systems, like Tesla coils without proper engineering and grounds, then they will draw electrons from the local environment air grounding and will draw them from *you*. You will feel worse and ill because of that. If you ground properly, as Tesla and Don did, then you may be able to produce an electron pump that can give you more energy (see *Earthing* book). These facts have prevented me from promoting these kinds of high voltage and high frequency processes to the general public. Being *in the presence of electron robbers* is one thing, but being *a path to ground* on the other hand, can be deadly. At this point many people need to realize this sort of research and experimentation is not for them as it is not worth the risk. There does not seem to be many people warning about these things because of the bad influence of people like Tesla who encouraged people to jump into the unknown without proper consideration.

The *Electron Spin* page gives many summary illustrations and points relating to resonant magnetic fluxing. The **first point** refers to pages 42 and 43 of this book. This is the infinity spin of electrons in relation to their phasing in electric and magnetic impulsing, and also in 3D.

The **second point** illustrates impulsing more specifically with a trasmitter and receiver. The magnetic output from the receiver coil is overlooked by most people. This is especially important in relation to repeater coils, where we extend signals, or in phase conjugate mirroring, where we can lock the transmitter and receiver into a sympathetic relationship where they reciprocate the energy in perfect phasing so that they feed and power each other. The latter is the **last point**.

The **third point** illustrates how the magnetic waves can influence multiple receivers without the input drawing more energy, and without diminishing the wave strength passing through each receiver. The receivers are not actually absorbing the waves but rather the electrons are being spun by the magnetic fields, waves, or rays. This was the major point Don realized with his first model [2, see page 153] after he started with the *Resonance Induction Coupler Kit*. He said repeately that you could multiply identical coils. This was one of the mystery points to be figured out in our book in the Resonance Kit. Keep in mind, however, the further out from the transmitter you go the less waves will be intersecting with your receivers. Some mistake this for waves deminishing, but it is simply a matter of waves spreading out due to the inverse square laws. If repeater coils are used then you can conceive of adding as many coils as desired. This was partially demonstrated at the Renaissance convention in California this summer where we used just one corner of a transmitter and placed all the coils we had for the Resonance kits there. Coils placed further out from the transmitter powered their loads a lot more when identical coils with loads were placed between them and the transmitter. Many of these were placed around one corner of the transmitter parallel to it. Three or four repeater coils were placed in series from the transmitter. Hundreds or thousands could be placed around the transmitter without causing the transmitter to draw more energy from the source. When this was done I then added a loaded coil with a ferrite rod to the transmitter and the input source lowered in half to 20ma (at somewhere between 9 and 18V). The receiver coils did not put out less as that load was applied, and these all developed a sympathetic relationship with the transmitter where it was hard to disturb the strong lock. The same thing applies to crystal radios with transmitters. These, as we have shown with the kits, also show more gains when adding the ground (which allows the system to draw in more electrons).

The multiplied possible outputs from magnetic flux is more importantly seen in each turn of the L2 coils where this can be easily measured as increasing from bottom to top of the coil (by holding neon bulb and touching other end of bulb to different points, or by measuring flux).

The **fourth point** is obvious once stated, but is not acutally realized by most people. Only magnetic flux passes through an isolation transformer so that electrons on either side are not the same. Magnetic fluxing is more efficient than electrical fluxing.

The **fifth point** is a reason for making this book. People disregard radiated magnetic flux in designing of electrical appliances, motors, etc. By the end of this book you should *pick up* on this point enough to figure out how to tap this wasted energy from anything producing significant radiation. Remember, once you are away from the inner near field of a transmitter, what you collect from it is not going to change what it draws from its source to run it. Anything that produces fields, waves, or rays can be a free source of energy completely disregarded by those holding to conventional theory (except for solar panels and radios).

We covered the **sixth point** in the Resonance book in detail. With resonance a tank circuit has the energy multiplied. Notice the trumpet wave. This is not denied but is almost completely ignored. While this fact is used to "amplify" energy for other applications (like signaling), it is never admitted to be used for amplifying energy for electrical power generation. Only some technicians and others figure this contradiction out, even though it is implied in all of their books. Because no one makes the obvious application in school no one realizes what this could mean. In stead the Q in a resonant circuit is made out to be merely a tuning matter in relation to radio stations. The bigger coils provided at my last conventions had Qs (quality factor) of around 1000, which means I was able with 5V input at 20 ma to produce 5000V at 20 ma. You can make coils having Q in the thousands. And the amperage can equal the voltage with proper grounding...

The **seventh point** pertains to the square laws. Don mentions these laws many times and adds on page 18 that they are actually cubed because they are 3D. This pertains to both voltage and frequency at the same time. The gain can be enormous very quickly as seen with these systems and Marx generators.

"Basically electrical energy is achieved when an electron density at point A becomes greater than electron density at point B. This is your potential difference between two points. If that doesn't exist you don't have any energy, regardless of how many electrons are there. It's just like the high pressure and low pressure areas of the weather system. Things are going to move away from the high pressure area into the low pressure area. The electrons density acts the same way."

This foundational point should be clear now.

The second homework assignment:

What is fundamentally wrong with conventional generators?

What is Fundamentally Wrong with Conventional Generators?

Homework assignment to figure out as you read this book.

Typical Generators

Only magnetic flux induces electrical charges at a distance.

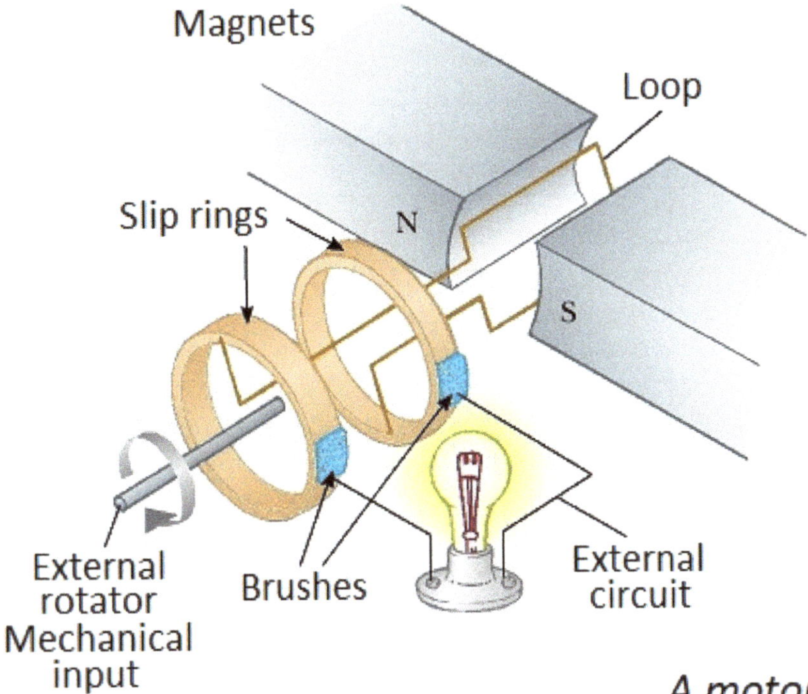

Electrical circuit replaces permanent magnet to create field excitation.

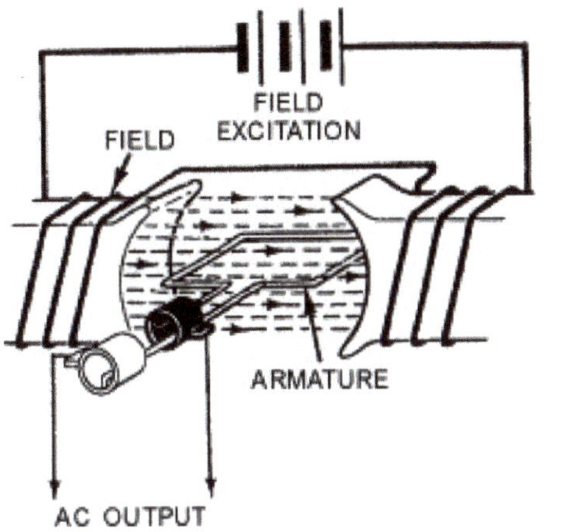

Rotating Armature Alternator

A motor/generator set

Electrons do not pass from fuel or motor to output.

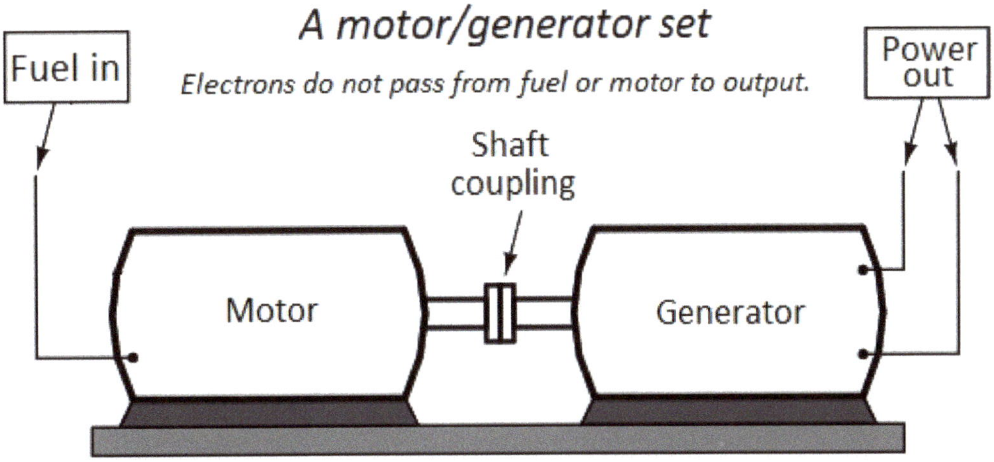

Input can be from motors powered by fuel, wind, hydro or steam, where mechanical force creates a dipole separation of charges in the generator to extract energy from.

Significant Electrical, Magnetic, and Mechanical Losses from Conventional Power Generation

Turning your available horsepower into usable wattage for your power needs *is a little more complicated than you may think*. This is mainly because *there are two types of kilowatt (kW) ratings*, kWe (electrical equivalent) and kWm (mechanical or work equivalent). Even though both kWe and kWw have the same result, roughly:

1 horsepower = about 745.5 watts.

Real world example. Diesel engine turning 1 horsepower into 745 watts. This isn't going to happen because of *friction*. Doing the opposite, turning 745 watts from an electrical motor into 1 horsepower won't happen either.

Here's why:
1. Copper loss from copper resistance.
2. Magnetic loss from the magnetic field being applied to the Iron in the stator core.
3. Stray loss from stray currents or harmonics when the motor is under load.
4. Mechanical losses from friction in the motor bearings and the fan for air cooling.

Considering off-the-shelf generators, for a 10 kw generator you would need roughly 18 hp to drive it after factoring in efficiency losses. ***Besides these losses, what is it in the design of these that results in so little output?***

Differences between Under and Over Unity Generators

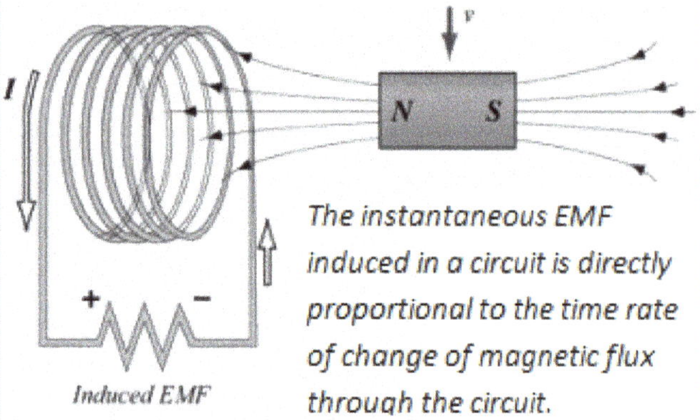

The instantaneous EMF induced in a circuit is directly proportional to the time rate of change of magnetic flux through the circuit.

Faraday Shake Flashlight Generator

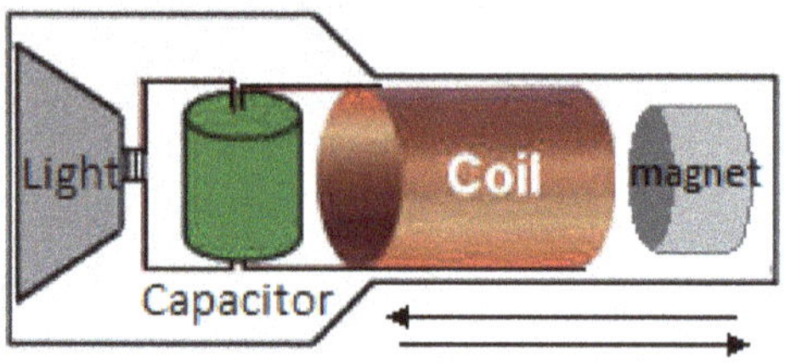

Magnet moves back and forth through coil, inducing EMF and charging capacitor. Rate of charging related to speed of shake.

Conventional generators require much effort to create magnetic flux to indirectly cause electrical charging. Non-moving magnets (or armatures) do not normally create useable flux. The moving magnet or electromagnet is resisted by the coil so other types of energy are needed to overcome drag. There is a better way however.

Bloch Wall Interrupting Magnet Generator

Pumping Useful Magnets without Movement or Effort

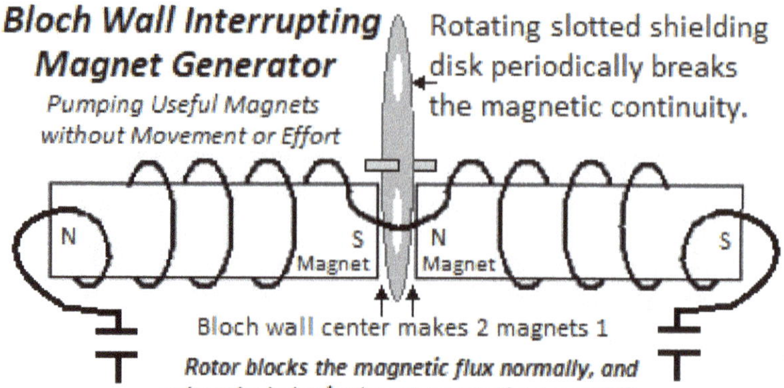

Rotor blocks the magnetic flux normally, and when the holes in the rotor pass the magnet the magnetic flux comes back causing an energy surge.

Instead of moving magnets or armatures, *the bloch wall is switched between two permanent magnets*, which has the same result upon the coil as if they were moving, but without creating a resistive force or drag. The magnet always has its so-called fields* around it, so moving its lines of magnetic flux *easily* in this way generates the same electricity. The faster the switching the higher the EMF in the coil.

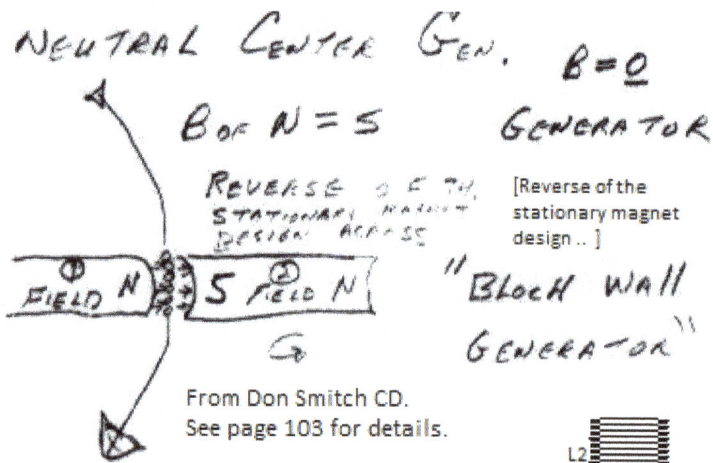

From Don Smitch CD.
See page 103 for details.

Resonant Tesla Coil Generators

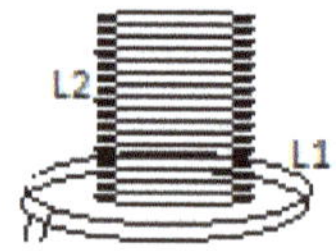

Resonate Electrical Coil Systems (regular Tesla) are analogous to the system observed in the Bar Magnet (shown). The Bloch Wall Area is Located at the base of the L-2 Coil. The Left Spin portion (Voltage Only) part of the Coil predominates. The right spin portion (magnetic-Amperage) portion is mostly absent.

BLOCH WALL (Space)
(Area of Electron Spin Separation)

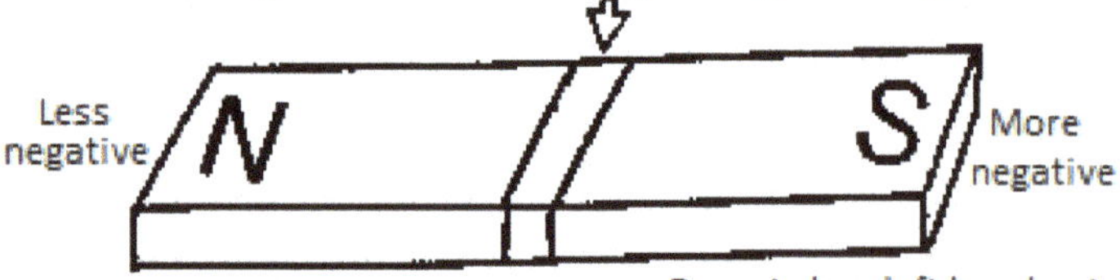

Domain has right hand spin (= Source of Amperes)

Domain has left hand spin (= Source of Volts)

Electrons* are commonly obtained from Earth and Air Groundings. They exist in Doublet Pairs, one being more negative than the other. The more negative one has a Left Hand Spin. The less negative one has a Right Hand Spin.

Don Smith Center-Tap Resonant Coil Generators

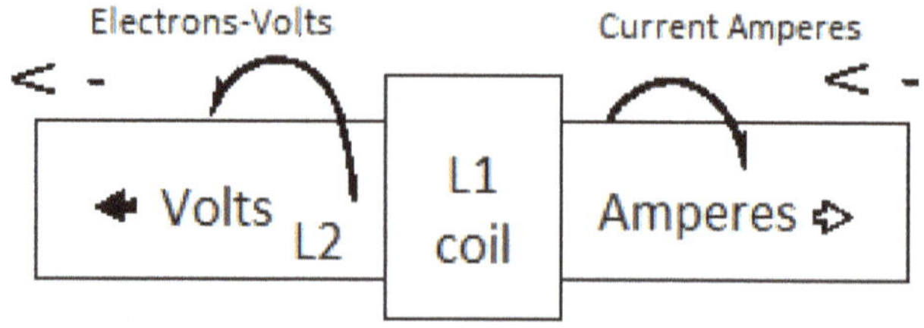

Moving L1 of Tesla coil to center balances out A with V, and center tap L2 allows grounding to supply more Amps. L1 replaces magnet switching with HF, HV Resonant Amplification for COPs in the thousands.

* The author does not hold to the meaning of words like "fields" and "electron" in the same way as conventional physics, and does not necessarily agree with Don Smith's views on them or some other parts of his theory. These diagrams, and this book, present Don's words *and* basic truths while revealing inconsistencies in conventional theory.

The ***Typical Generators*** page shows the basic principles and operation of conventional generators. They require external input to force a charge. These do *not* actually convert mechanical energy into electrical. The mechanical energy is needed to overcome the needless drag caused by opposing magnetic fields. The source of energy is the same with both the conventional and the unconventional over unity systems. The conventional system just adds needless work so that you have to pay for it. One practical thing to notice in the last section is that you have significant losses so that it takes a lot more than 1 horsepower (easily 30% more) to generate 745.5 watts with conventional generators. Also, 745.5 watts of electric does not give you 1HP of power from a motor.

The next page shows the difference between the two types of energy generation systems. The conventional example illustrates the shake flashlight with moving magnet. We will come back later to Faraday's law specifically. We need to consider that a magnet is more than something with a north and south pole. Most people are not aware of the *Bloch Wall* area between the more and less negative sides of a magnet. This is a very important area and is to be understood in the same way as in Don's coils. This page has a lot of information briefly stated so it will require some reading of the book to fully realize.

Don's model 9 system is the ***Bloch Wall Interrupting Magnet Generator***. This is the name I have given it for lack of title. This needs to be understood as one long magnet that is switched back and forth from one to two magnets. Since the center part of a magnet is the Bloch Wall, therefore this area is being switched with magnetic shielding that interrupts the magnetic continuity. Each time this happens the magnetic fields significantly move, and according to Faraday, the changing magnetic field gives you the goods. If you consider the loop of wire to be like the secondary of a transformer you can understand how that applies. At least ***five significant advantages*** are seen in this basic illustration:

1. The magnetic flux-blocking disk creates no drag with the magnets and only deals with air flow resistance. It takes less than 1 watt to power a motor that can produce 50 kilowatts of output by *letting the magnets work for you, rather than forcing fields against each other.*
2. The *faster the switching the more the output* due to the fields reversing and resulting in electron spin.
3. When the *frequency is at the resonant point* of the coils the output is amplified.
4. The *stronger the magnets the more output is produced* without adding any more input.
5. The *coils are wound* so that they produce *maximum amperage and voltage*.

The reason I have never wanted to recommend generators to my customers is because of this system. I could never recommend doing anything less. There are several different ways to do this.

The coil systems that follow are far easier. One example is the *regular Tesla coil,* followed by the *centered primary with center-tapped secondary*.

Conventional Power Grid System Simplified *

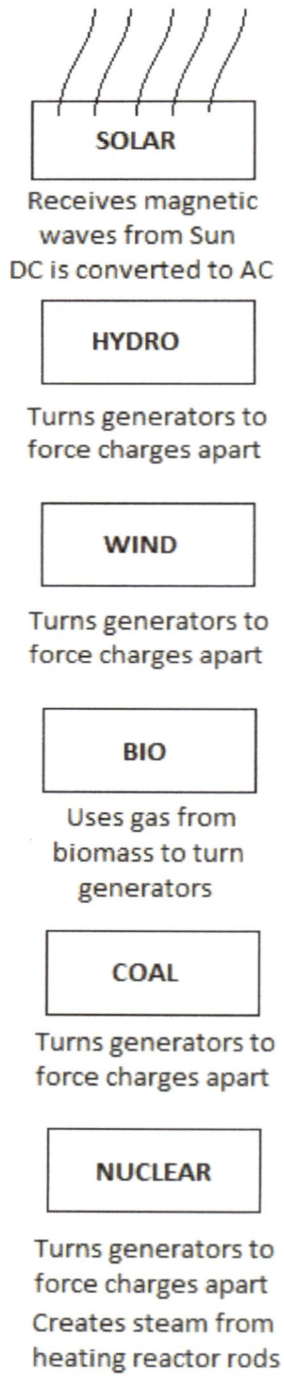

Expensive ways to harvest energy from environment to force charges needlessly without resonance. Creates pollution, and keeps people dependent.

Electrons from plant stay at plant and do not pass through transformers. Magnetic flux passes through transformers only. Steps up voltage to reduce cost of wire and losses.

Transmission through wires at low frequencies rather than through ground (Tesla) at high frequencies is very expensive and inefficient. Enormous amount of energy radiates off power lines as wasted magnetic energy causing cancer, etc.

Low voltage and low frequency power results in low efficiency, high costs, and unhealthy radiation. Closed loop circuits with 2 wires and appliances that waste most of the energy prevent over unity and energy independence.

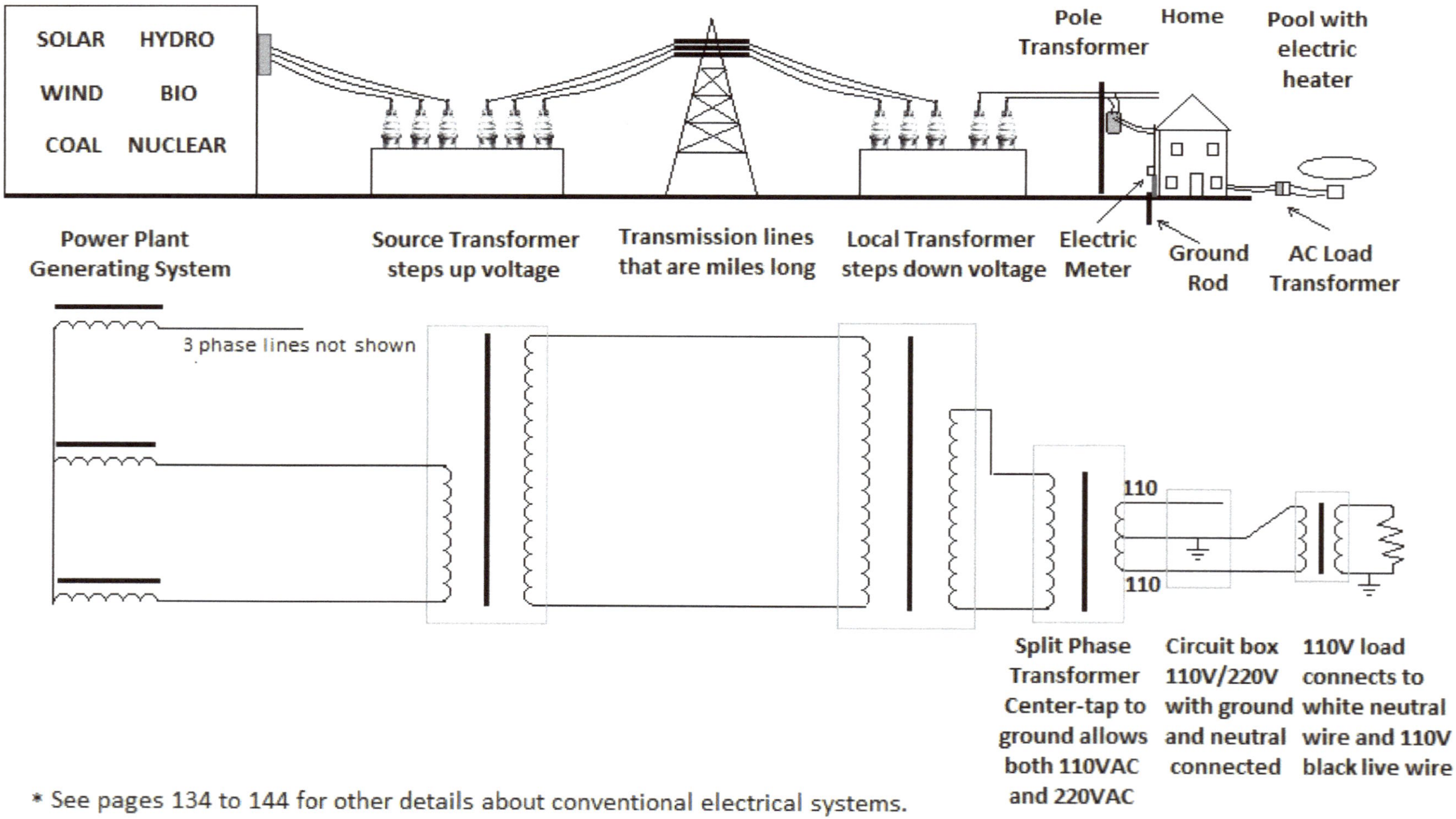

* See pages 134 to 144 for other details about conventional electrical systems.

Don Smith Power System Outline

EES II, use of naturally occuring electrical differences in the Earth's crust and Radio Device for Transmitting the Energy.

Motor-Generator
Interruptible Magnet Field Switching Bloch Wall with alternating magnetic shield.

Gaseous Tube, Resonant electrons excited: Tesla Coil, Plasma Tube and Laser Systems.

Propulsion-Star
Spin System like Tornado's: Rail Guns, Gyroscopes, and Negative Gravitational Systems.

Power System, Piezoelectric, Solid State Marx Generator.

Power System, Supermagnet Battery relating to Magnetic Waves as in Solar Power. Domains aligned/resonance.

Energy Acquired by Magnetic and Electrical Impulsing

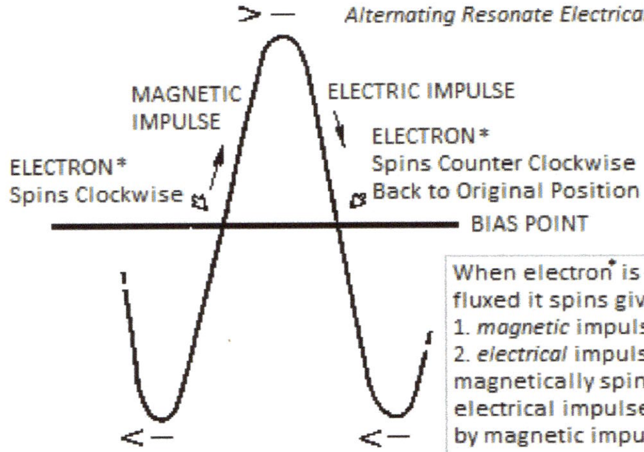

When electron* is electrically fluxed it spins giving off
1. *magnetic* impulse then
2. *electrical* impulse. If fluxed magnetically spin results in electrical impulse followed by magnetic impulse.

Inexpensive, local, non-polluting, fueless, safe and powerful energy systems can replace or supplement grid power. Uses unique switching methods with varying potentials from ambient to create electron* spin for energy harvesting. Source of electrons* is ambient background as earth ground, or chassis frame, or negative battery terminal.

Table top electrical power systems capable of powering loads as big as needed. Energy provided freely from the local environment through processes found in nature and sometimes used in conventional systems like radio, radar, optics, etc.

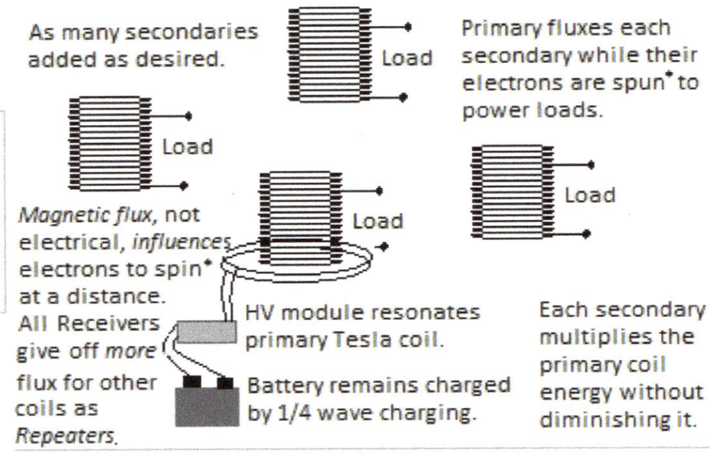

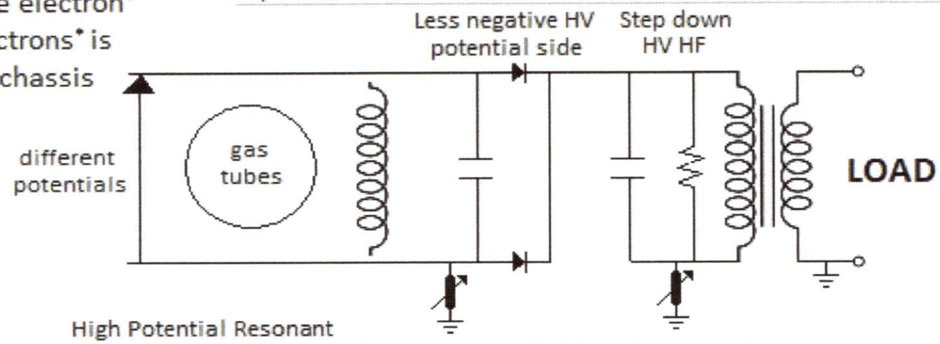

High Potential Resonant Fluxing Systems create imbalance from local ambient background.

Electrons are pulled in from ambient ground, trapped, and spun.

Electrons alternately impulse load through transformer while returning to ambient.

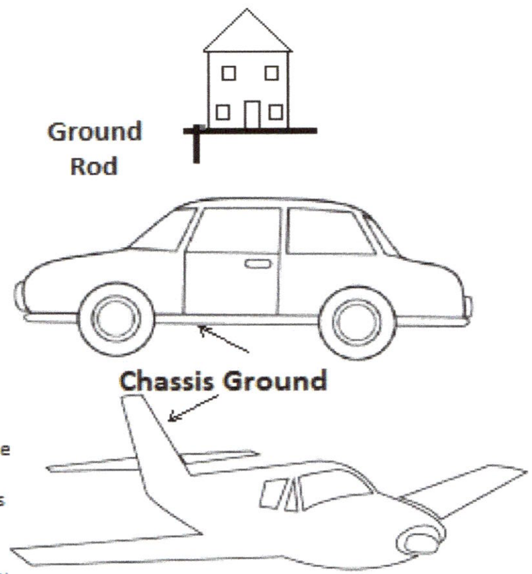

Don Smith Effect (DSE)

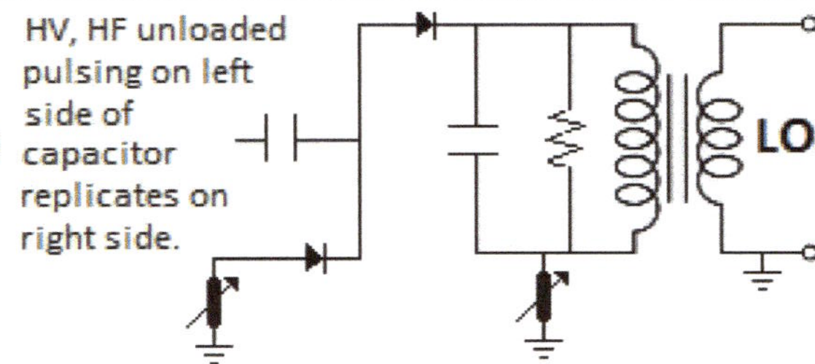

* The author does not hold to the meaning of words like "electron" in the same way as conventional physics, and does not necessarily agree with Don Smith's views on that or some parts of his theory. These diagrams, and this book, present Don's words *and* basic truths while revealing inconsistencies in conventional theory.

The ***Conventional Power Grid*** page is overly simplistic and straighforward. Besides making this a contrast to the Don Smith Systems I wish to point out that electrons from the source power station do not transfer to the destination load. At many points only magnetic flux is transferring between sections.

The ***Don Smith Power System*** page is a basic comparison. Both systems use grounding for similar reasons. Don's systems are thousands of times smaller than conventional means of power generation. They are environmentally friendly.

The left side of the page shows several different kinds of energy harvesting systems that Don developed and attempted to patent in the United States. The right side of the page shows the first Don Smith system (model 2). The bottom schematics are important variations of the basic process as conceptualized on the *Electron Cycle* page. Notice the details of the grounding. The last schematic refers to pages 76 and 77.

"Free Energy Harvesting Requirements,

1. Electron acceleration or deceleration must vary from ambient.
2. Spin rate CPS fluxing provides electrical with magnetic energy.
3. At Resonance (negative resistance), all electrons move in the same direction without interfering with each other.
4. Electrons must Avalanche, pulsing magnetic waves provide three dimensional and nonlinear energy at each cycle, electrons presently available disturb and bring along many more electrons, resulting in gain."

Summary of Magnetic Resonant Stages

1. One way or another magnetic Flux, in the form of fields, Waves, or rays:
   a) are engineered,
   b) or are supplied freely from the local environment.
2. A media resonantly tuned (to the incoming signal) as means of intercepting the magnetic communication:
   a) is connected to a source of electrons through a one-way valve, and
   b) these electrons get pulled into the collector media transducer, and
   c) are spun (influenced) and bunched (pressurized) as they are magnetically flipped,
   d) which causes the electrons to alternately produce electrical fluxing, followed by magnetic fluxing of their own,
   e) which results in charges in a collection capacitor and/or inductor after being rectified.
3. This high pressure charge
   a) is shunted to a lower frequency and transformed to a lower voltage,
   b) while the accelerated electrons decelerate through the load on their way back to ambient ground.
4. The load isolation transformer only supplies magnetic flux to the secondary which powers the loads at the desired frequency, voltage and amperage.

This is a similar summary as found on pages 403 to 413 of the *Systematic Index* pertaining to five stages. The *Sectional Processes in Over Unity Systems* is an organization of Don's words about these processes.

Inspired by videos from MIT Physics professor Walter Lewin

Kirchhoff's Law is for the Birds!

*Non-conservative Fields**

Faraday's Law

$$\varepsilon = -\frac{\Delta\Phi_m}{\Delta t}$$

induced emf — rate of change of magnetic flux through the circuit

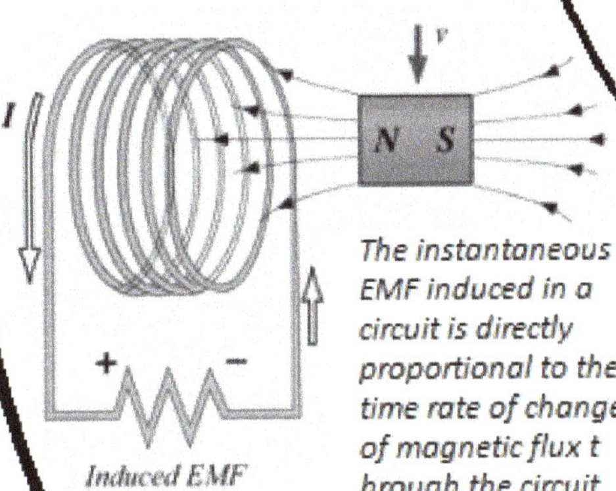

The instantaneous EMF induced in a circuit is directly proportional to the time rate of change of magnetic flux t hrough the circuit.

Kirchhoff's second law

based upon the law of the conservation of energy

The total voltage across a circuit loop is equal to the sum of the voltage drops across the devices in that loop.

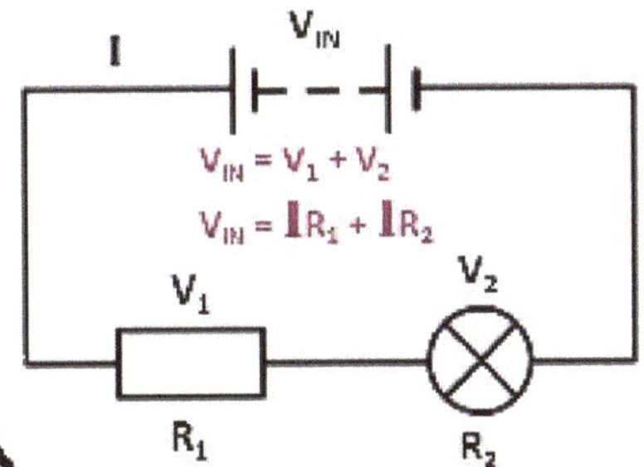

Conservative Fields Special case of Faraday's Law*

The energy you put into the circuit equals the energy you get out of each circuit loop.

"If you're in a closed system, and you're boxed in, Ohm's law works. That's on your electrical side.

But when you move over onto the counterpart of the magnetic side, none of those laws apply."

Resonant tank as transmitter amplified

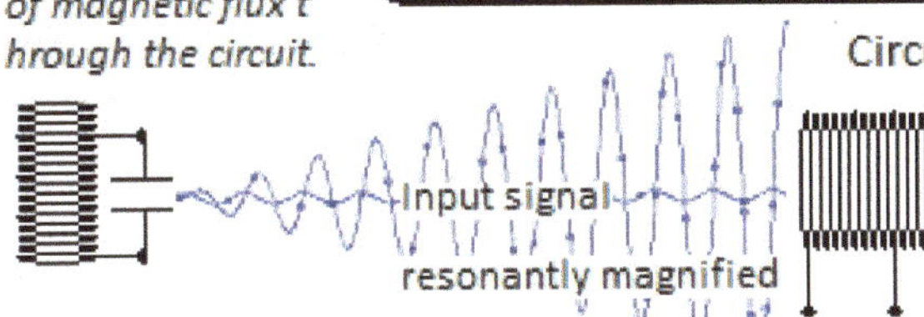

Circulating current in high Q transmitter can produce thousands of times more flux powering receiver than needed to power input.

In resonant tank circuits the circulating current or voltage is multiplied from the input current with the result that output fluxing is equally multiplied and available for energy harvesting.

*The author does not hold to the meaning of words like "field" in the same way as conventional physics, and does not necessarily agree with Don Smith's views on that or some parts of his theory. These diagrams, and this book, present Don's words *and* basic truths while revealing inconsistencies in conventional theory.

The establishment leaves the student and consumer *practically* believe that Kirchhoff's rule is a *universal law* when it is just a *convenient mathematical model for closed systems. It is a special case of Faraday's Law,* which also for non-conservative fields* in an *open system*. As soon as an unfamiliar non-conservative system is presented to such people they cry *violation of the conservation of energy* because they really believe this to be a universal law. Yet they have little problem with other systems that receive energy from external sources. And their textbooks also show that high Q resonant tank circuits have circulating current far greater than input current which thus produces enormously high rate of change fluxing that can result in EMF through man other coils exceeding total input. A similar deception passed onto students is the use of the word *transfer* as applied to magnetic flux as somehow turning into electric flux and *vice versa*, when magnetic flux actually spins electrons causing both electric fluxing and addiontal magnetic fluxing while the original magnetic flux is not absorbed or transferred but continues on.

Corrected Schematic and Avramenko Plug

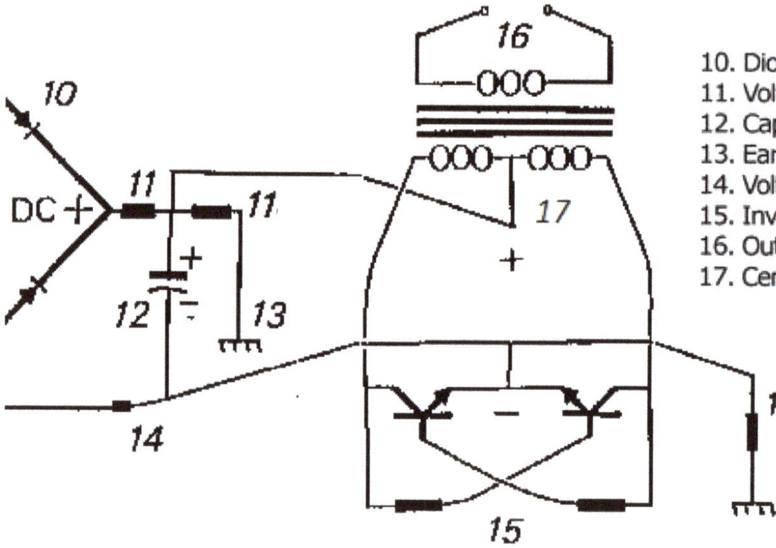

10. Diode Bridge, 200 Nanosecond, R.F., > 100 KV.
11. Voltage Divider Circuit, corrects voltage for next stage.
12. Capacitor, electrolytic, smoothes out DC + ripple effect.
13. Earth Ground (Adjustable).
14. Voltage Divider Circuit, corrects voltage for Transformer.
15. Inverter Circuit. DC + in and 60 CPS to Transformer.
16. Output from Transformer to Load (Work).
17. Center Tap.

This shows the corrected second half of an early Don Smith circuit shown on pages 68 and 69 where the positive and negative on the inverter was reversed. That mistake was completely independent from the special Smith process.

Avramenko

U.S. Patent 6,104,107
Aug. 15, 2000

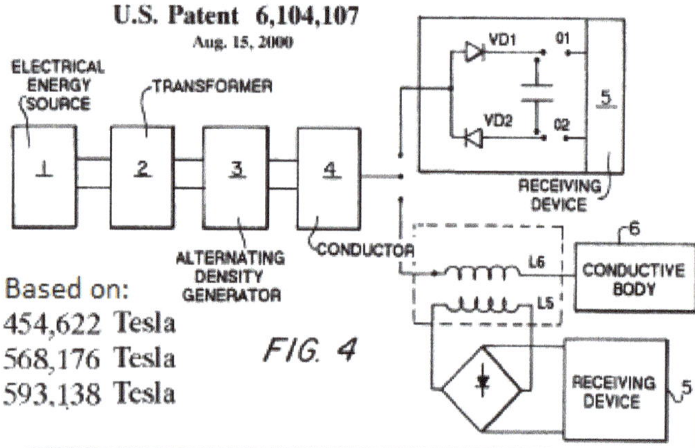

FIG. 4

Based on:
454,622 Tesla
568,176 Tesla
593,138 Tesla

"An object of the present invention is to provide an increase of the coefficient of efficiency to transmission from an initial source to a consuming device in electrical technology"

Practical Avramenko Plug

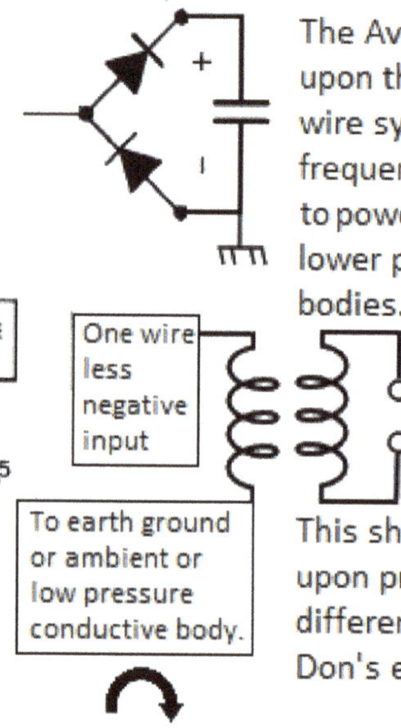

The Avarmenko Plug is based upon this patent and Tesla one wire system that uses a high frequency and high potential to power loads in series with lower pressure or ambient bodies. Attempting to power these loads without the terminal body results in low or no power conditions in the load. This shows that output depends upon pressure potential differences and accords with Don's electron spin theory.

Don Smith Circuit Modifications

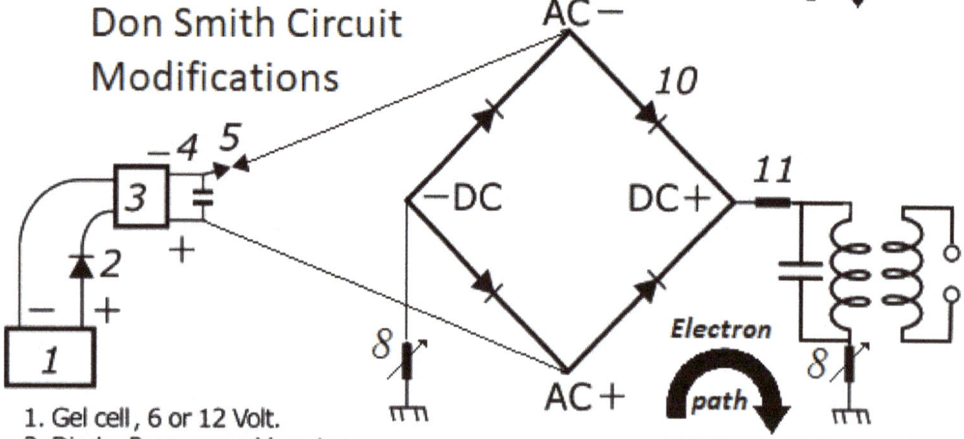

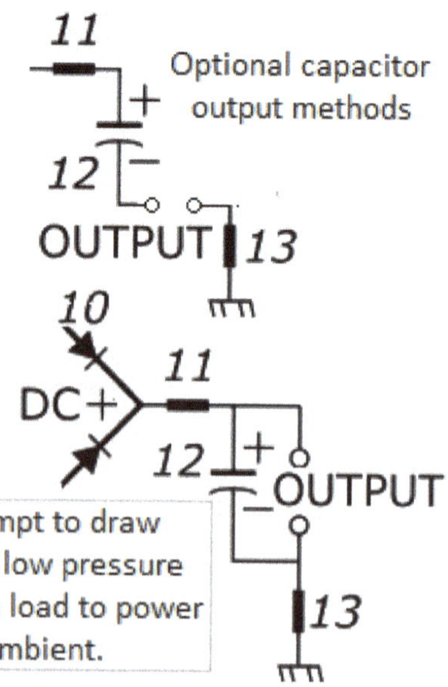

1. Gel cell , 6 or 12 Volt.
2. Diode, Poss. use a Varactor.
3. High Voltage Module, Constituting the L-1 and L-2 Coils.
4. Capacitor, TDK 10.9 Pf., 30 KV.
5. Spark Gap, Small Engine Spark Plug, Gap = .0025in.
8. Control Shunt.

High pressure pump: Attempt to draw electrons into circuit from low pressure ambient, spinning through load to power it on their return back to ambient.

Kirchhoff's Law is for the Bird's page was inspired by an old video from MIT Physics professor Walter Lewin. It is included at this point as a shift in this book from stated theory towards clearing up some of the misunderstandings people have who believe in conventional theory and reject over unity possibilities because they assume that would violate the conservation laws. The illustration is meant to show that *Kirchhoff's Law is actually only a Rule, a special case of Faraday's Law.* At some point this rule became practically considered a law because people generally only work with conservative fields, and people are uncomfortable with math and ideas that are not the same in all situations. I have extensively covered these kinds of mistakes in my first Resonance book in relation to one wire transfer and multi-body networks. The nonlinear circuits allow for a multibody network and amplification of energy that goes beyond the conservative experience most people only think of. *Kirchhoff's Rule does not pertain to a changing magnetic field* which results in additional energy outputs even if they are not used. An example has already been given where my motors produce a changing magnetic field in the motor coils for the purpose of bucking magnets. These take the same amount of energy to produce the torque of regular motors. So far all agree that we get out the energy we put in with some losses. But unlike everyone else, for 14 years I have run hundreds of motors like this while taking the second form of energy produced by these coils, or rather electron spins, and have also charged batteries. The amount of charging such batteries has been equal or more than what it takes to run the motor. This has been demonstrated thousands of times all over the world. I even demonstrated a 26' boat using this system running off the same batteries being rotated for three years. In order to continually rotate the batteries, considering all the losses, I would have to produce more charging in the charge batteries than the input battery was using. Yet in many cases I charged up battery banks many times larger in capacity and/or voltage than the input battery was so that the total amount of usable energy in the bigger charging bank was up to 250 times more. Even 10% charging of another battery bank of equal size would demonstrate that kirchhoff's rule is not a universal applying to all situations. In fact it rarely is the norm as most environments have pulsing circuits. Since the excess energy is not used people do not think about it. As soon as you start using the extra magnetic or electrical flux you can easily see, like a solar panel, that extra energy can be freely collected to power all your needs.

The next page shows the second half of a *Corrected Circuit* that only one person noticed over the years. It was a simple mistake that had nothing to do with the Don Smith System. More importantly, below that are portions of the Avramenko patent along with two forms of the ***Avramenko Plug***. This is like *Tesla's one wire system* and works as a *series load to ground* just as the *decompression negative resistance side* of Don's circuits. This is provided with the *Resonance Inductive Coupler Kit*.

The Simple Klystron Amplifier

https://youtu.be/Fvud81pYGOg

Summing up the whole action of the klystron. The electrons coming from the cathode enter the first resonator with *high kinetic energy*. This first resonator is set into oscillation by the input signal so that the *beam is modulated* and *bunches are formed* in the drift space. These bunches in turn *set up an electromagnetic field* of the same frequency in the second resonator. *From this field we can extract* ***a larger output signal***.

The essentials of this simple klystron amplifier are the election beam, and two metal boxes known as cavity resonators. The cavity resonators have a potential that is positive with respect to the cathode. Thus *the electrons emitted by the cathode are accelerated towards the first resonator*. Because the resonators have apertures in the form of grids the electrons will continue until they reach the collector which is at the same potential as the resonators. Therefore, in the region between the first resonator and the collector, the electrons would have a constant speed it if were not for the influence of the cavity resonators.

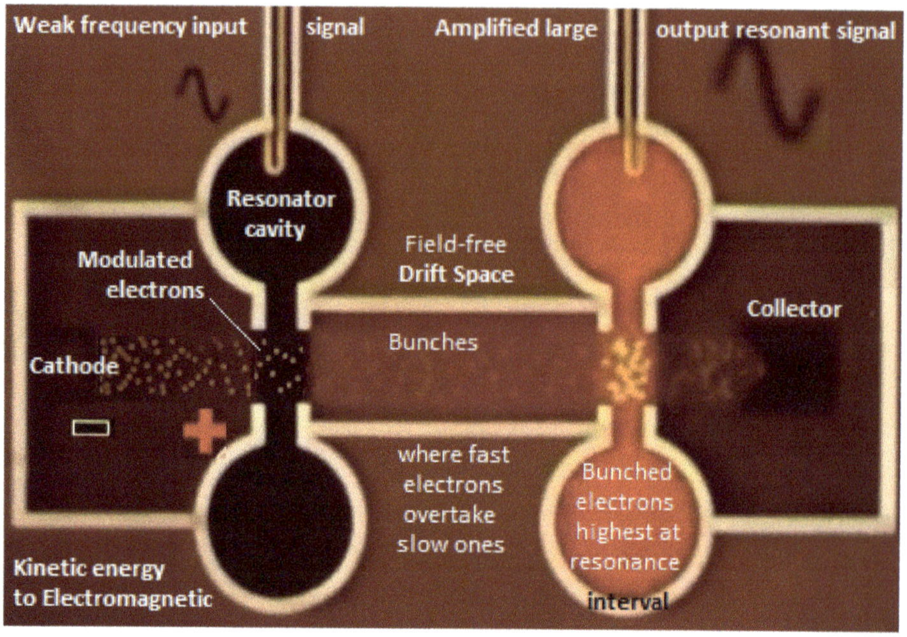

In reality the first resonator is supplied with the *weak frequency signal which is to be amplified*. This signal will initiate an electromagnetic field inside the cavity. Due to this field, the velocities of these electrons passing through the first resonator are modulated. On arriving at the second resonator the electrons will induce there another electromagnetic field of the same frequency. This field is so strong that it is possible to extract a *large signal* from the second cavity. *There is a certain similarity between a cavity resonator and a conventional resonant circuit. The* ***displacement of electrons*** *in the loop constitutes an* ***alternating electron current****. The* ***alternating current gives rise to an alternating magnetic field***.

The first resonator of the klystron is excited by *the external signal which is to be amplified*. This signal must have *a frequency which the cavity can resonate*. The current in the coaxial cable sets up a magnetic field, which in turn brings the resonator into operation. That is to say, that also an electric field is originated. It is sufficient to only consider the electric field in order to understand the modulation of the electron beam entering this first cavity resonator. The electric field lines indicate the direction of the force acting on the positive charge. The electrons, being negatively charged particles, *will therefore be* ***accelerated*** *if they move against the direction of the electric field*. On the other hand, *the electrons will be* ***decelerated*** *if they move in the same direction as the electric field*. With the alternating electric field present *the electrons will thus be* ***alternately*** *accelerated and decelerated*. Consequently they will enter the field-free space between the two resonators *at varying speeds*. In this space, known as the drift space, *the fast electrons will overtake the slow electrons*, and ***bunches*** *will be formed*. The second resonator is located in the position *where the bunches contain the* ***greatest number*** *of electrons. The bunches arrive at the second resonator* ***at intervals*** *corresponding to the frequency which the first resonator oscillates*. As both cavities are identical, *these intervals also correspond to a resonant frequency of the second cavity*. Therefore the later starts to oscillate.

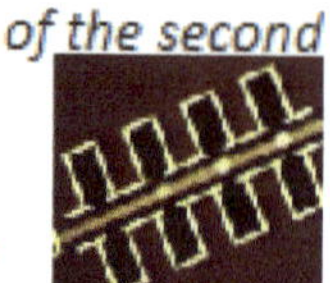

Almost ***no energy*** *is taken from the cavity*. At the second resonator however, *the majority of the electrons are in* ***bunches****, and these are* ***decelerated***. The result is that *almost all the* ***kinetic energy is converted into electromagnetic energy***. If amplification of such a two cavity klystron is too low then multi-cavity klystrons may be used with a very small signal applied to the first cavity. The bunches arriving at the second cavity are very weak. But they will set up a field there which is strong enough to produce small concentrated bunches at the third cavity, and so on, ***enabling very large amplification factors to be obtained***.

The Klystron Compared with the Don Smith Resonant Energy System

This is part of my project to show how old and existing technology sold to the public used these over-energy processes. In the previous page we see details about the Klystron Amplifier which modulates electrons through resonant cavities. This is very similar to what Don Smith did in his systems. But, instead of amplifying signals, Don amplified electrical energy.

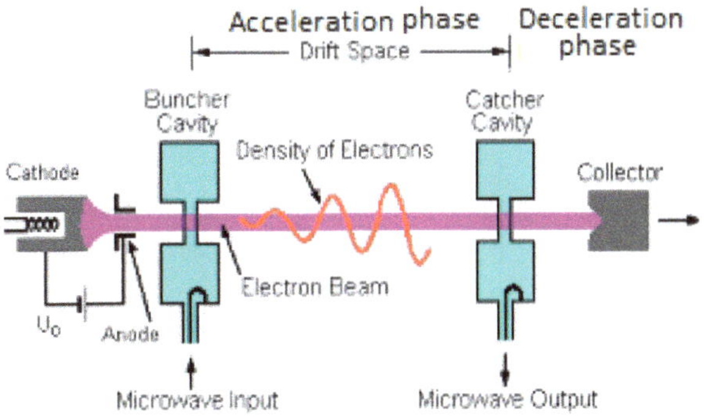

Klystron

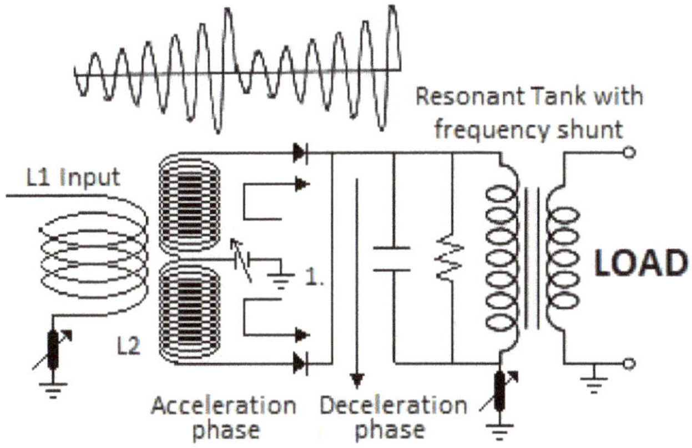

Don Smith System

1. The Klystron starts with an electron beam that pushes electrons forward to a collector.

2. The Klystron beams the electrons so that they pass through at a steady state before a resonant frequency is produced in the first cavity.

3. The Klystron sends an AC resonant frequency signal that triggers the first resonator cavity to create a magnetic modulation acceleration and deceleration of the electrons passing through it.

4. The first resonator's modulation results in electron bunching down the line that are maximized at the location of the next cavity that is tuned to the same resonant frequency.

5. At the point of maximum bunching the electrons decelerate as they influence the second resonant cavity and produce an amplified signal load.

6. Decelerating electrons arrive at collector.

1. In Don's system the electrons are provided from the earth ground or a good source of electrons from the local ambient. These cycle through the system and return back to ground.

2. The Don Smith ground system remains the same potential until the system is turned on.

3. Don's system sends the same sort of resonant signal through the L1 coil to draw in electrons through the resonant L2 coil in order to also modulate and avalanche them.

4. Don's L1 coil provides the same bunching in the L2 coil which is also made and tuned to a matching wavelength frequency.

5. The electrons bunch at the diodes and/or capacitor and decelerate as the frequency adjustment is made and as they pass through the primary of the output transformer. Models vary.

6. Electrons return to ambient.

The major difference between the two, besides the cavities and coils, is that the Klystron pushes electrons with a beam, whereas in Don's system the electrons are drawn in for free like a magnet. Then he taps their momentum and runs loads by them as they seek to return back to ambient through a different path from which they came up from. Similar comparisons could be made with sonar, radar, microwaves, multipactors, lasers, and many other commercially available devices over the last 100 years. By studying the various similar stages in these devices we can better understand what Don was doing in his.

This comparison was chosen because Don included the trumpet wave on the right (pages 98-99) in the book, and which was referring to the klystron as found in the 1948 book called: *Vacuum Tubes* tubebooks.org/Books/Spangenberg_vacuum_tubes.pdf

Resonant Tank Circuit with resistor shunt

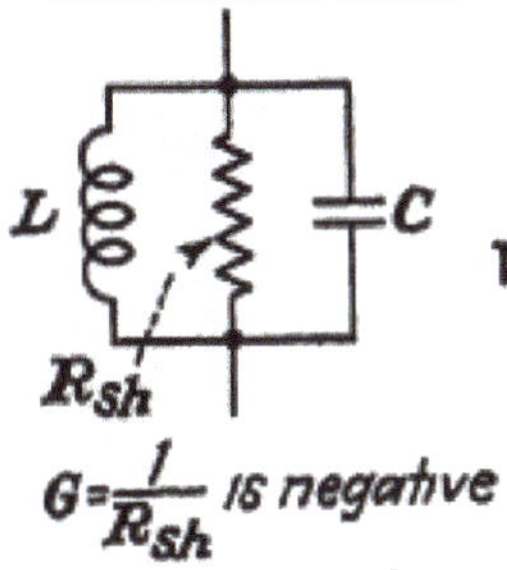

Trumpet Negative Resistance Wave

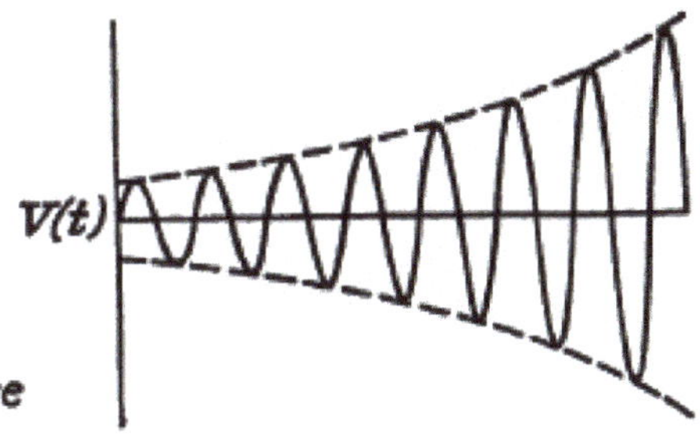

The **Klystron Amplifier** page is my transcription of an old video shown on YouTube that breaks down the Klystron in a simple way, and which shows it to be clearly an over energy system. The word amplifier in this context clearly means *resulting in a total gain* just as much as Don Smith used the same words. The processes were very much the same except that Don used more processes for higher gains. He also used the ground, not an electron beam, as a source of electrons. Notice also at the end where more resonant cavities allow for "very large amplification" just as we have considered with multiplying many receivers to be influenced to charge from the same waves.

On the next page we have **The Klystron Compared with the Don Smith Resonant Energy System**. My purpose in showing this kind of device is to help people understand that unconventional devices producing more output than input have been used for a very long time. *The limitation of all devices known by the public is that their amplification only pertained to purposes other than energy generation.* In this case we have signal generation. The same applies with sonar and radar equipment. In old televisions the gain was in the display. In microwaves food is cooked. In these examples an enormous amount of energy is not used, which could be used for energy generation to power the running device, and many others could be freely powered from the magnetic flux emissions. In regard to anything available or known to the general public we do not see systems that do the same sort of things to merely amplify energy.

The second reason why I chose to bring up this example is because the two pages [98-99] in the book, which are also displayed in several of Don's videos, are actually about the klystron. *The trumpet waveforms shown on both pages is the negative resistance experience we are looking for.* Don mentions in the video that as you shunt the resistance you *pull out* the wave (which is lowering the frequency) and *the resistor becomes a negative resistor.* This is what Lewin would call *counter-intuitive*. It is hard to get used to doing things that you would expect to destroy the process. But it is more than just a frequency converter, it becomes an amplifier of sorts.

You can see the same processes in both systems. Electrons are bunched through resonance, and at an ideal point they influence a later cavity to have an amplified output. The same kinds of explanations are given which help people to see that Don was not making things up, and that his theory accords with existing technology from the earliest days.

The subject of electron pairs has not come up yet. The coil system with the center-tap L2 allows for electron pairs, according to Don's electron theory, to easily separate and travel down opposite coils and meet up again after the diodes (and that out of phase due to different coil lengths). This theory seems very plausable and is demonstrable. It is not my intention to have this Introduction cover every point in the book, but only highlight the most important things you should not miss.

Frank Wyatt Prentice Electrical Power Accumulator

Patent # CA253765 1925-09-15

My invention relates to improvements in Electrical power Accumulators and the like, wherein the earth acting as rotor and the surrounding air as stator, collects the energy thus generated by the earth rotating on its axis, and utilizes the same for power and other purposes.

In the development of my Wireless Train Control System for railways, covered by my United States Letters Patent Number 843,550, I discovered that with an antenna consisting of one wire of suitable diameter supported by insulating means 3 to 6 inches above the ground and extending one half mile, more or less in length, the said antenna being grounded at one end through a spark gap and energized at the other end by a high frequency generator of 500 watts input and having a secondary frequency of 500,000 would produce in said antenna oscillatory frequency the same as that of the earth currents, and thus electrical power from the surrounding media was accumulated along the length of the transmission antenna and with a closed oscillatory loop antenna 18 feet ion length run parallel with the transmission antenna at a distance of approximately 20 feet it was possible to obtain by tuning the loop antenna, sufficient power to light to full candle power a series bank of 50 60-watt carbon lamps [3 kw]. Lowering or raising the frequency of 500,000 resulted in diminishing the amount of power received in the 18-foot antenna. Likewise, the raising of the transmission antenna resulted in a proportionate decrease of power picked up on the receiving antenna and at 6 feet above the earth no power whatever was obtainable without a change of potential and frequency.

It is the objective of my generic invention to utilize the power generated by the earth by means herein described and illustrated in the drawings. The two figures in the drawings illustrate simple and preferred forms of this invention, but I wish it understood that no limitation is necessarily made as to the exact and precise circuits, shapes, positions, and structural details therein shown, exhibited and herein described in combination or otherwise and that changes, alterations and modifications may be made when desired within the scope of my invention and as specifically pointed out in the claim.

Referring particularly to Figure 1; 1 and 2 are alternating current feed wires supplying 110 volts 60 cycles to a high frequency generator. 3 is a switch and 4 and 5 are poles of same. 6 and 7 are connections of high frequency transformer 8 for stepping up the frequency to 500,000 and the voltage to say 100,000. 9 is an inductance coil. 10 is the spark gap, and 11 is the adjustable condensor, while 12 is the primary winding of transformer 8 and 13 is the secondary winding of transformer 8. Secondary winding 8 goes to ground through lead wire 15 to adjustable condensor 16 and lead wire 17 and 18. 1`4 is lead wire from the other side of secondary winding of transformer 8 to main transmission antenna 19 supported by insulating means 20. 21 is spark gap from transmission antenna 19 to ground through lead wire 22, adjustable condensor 23, lead wire 24 to ground 24'. Transmission antenna 19 may be of any length desired.

In Figure 2: 25 is a closed oscillating loop antenna of any desired length and for greatest efficiency is run parallel with transmission antenna 19 of Figure 1. 26 is lead to step down transformer 27, of which 27' is the secondary, 28 is lead to adjustable condensor 29, lead 20 to ground 31. 32 is primary winding of transformer 27, 33 is adjustable condensor, and 34 and 35 are windings of frequency transformers supplying current through leads 36 and 37 to motor 38, or other power devices.

Having described the drawings, I will now describe the operation of my invention. Throw switch 3 connecting feed wires 1 and 2 with transformer leads 6 and 7, adjust spark gap 10 and condensor 11 so that a frequency of 500,000 and 100,000 volts is delivered from secondary leads 14 and 15 of step up transformer 8 of Figure 1. Next adjust spark gap 21 of transmission antenna 14 so that all nodes and peaks are eliminated in the transmission of the 100,000 volts and 500,000 frequency along said antenna 14, by the surges occurring pass over the gap 21 to lead 22 to adjustable condensor 16 to lead 15 to secondary winding 13 of transformer 8. The oscillatory current of 100,000 and frequency of 500,000 being of the same frequency as the earth generated currents and thus in tune with same, it naturally follows that accumulation of the earth currents will assimilate with the same tuned currents and frequency with those of the output from transformer 8 along wires 14, affording a reservoir of high frequency currents to be drawn upon by a tuned circuit having the same characteristics of 500,000 frequency capacity which is shown in Figure 2; the antenna 25 is tuned fundamentally to receive a frequency of 500,000which current passes to lead 26 through winding 27' of transformer 27, thence to lead wire 28 through adjustable condensor 29 to lead wire 30 to ground 31. The high frequency currents of 500,000 and voltage of 100,000 passes through to winding 32 and by adjustable condensor 33 and windings 34 and 35 of the frequency transformer 27 is stepped down to a voltage and frequency suitable to operate motor 38 receiving current from leads 36 and 37. This makes available a current supply for any purpose whatever, such as operation of railway trains and current for industrial plants, lighting, heating, etc.

The return of current through the earth from transmission antenna 14 is preferable to a metallic return as a higher percentage of accumulation of earth currents is noticeable on receiving antenna of Figure 2 than from a metallic return, accountable for because of the condensor effect the grounded circuit affords.

I also prefer under certain conditions to use a single antenna receiving wire in place of the closed loop shown in Figure 2. Under certain operation requirements I have found it expedient to have the transmission antenna elevated and carried on poles many feet above the earth, and in that case a different voltage and frequency was necessary to accumulate earth currents along the transmission antenna 14.

Similarity between Don Smith Ground System and Frank Wyatt Prentice Electrical Power Accumulator

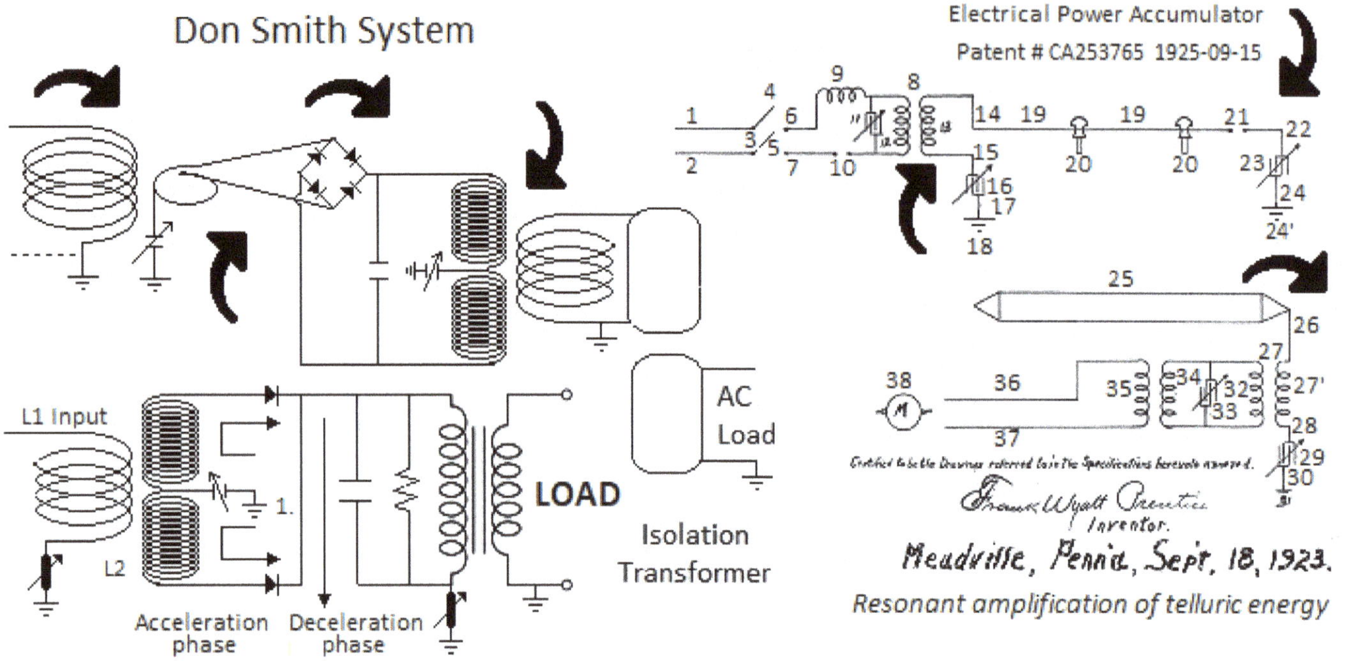

"Frank Wyatt Prentice's Electrical Power Accumulator, this Device works as stated. When properly constructed it works exactly as stated. The 3 Kwatts is not to be sneezed at. My devices are a vast improvement Prentice's. If and when You get Your Device up and running I will supply You with additional information which will move You in as a big winner. Regards, Don Smith"

"Howard, magnetic resonance flux which is every where present in huge amounts having a Gauss Meter is very important. One milli Gauss is 100 Volts of resonace flux. If PRENTICE or MORAY had known about this rest would have been HISTORY. Using a magnetometer in placing the magnetic dipoles, instead of 3 Kilowatts for PRENTICE would have been MEGAWATTS. This very correct as I use Gauss Meters Professionally for more than 60 years. Don Smith"

While there are significant differences between these two systems, Don did use all of the processes in various different models. Here we just want to notice the particular aspects of the Prentice system that relate to the themes of this book. 1. It is a tuned resonant system. 2. The wavelengths are matched. 3. The grounding has a definite beginning and termination. 4. There is similar tuning in series to groundings. 5. The loading is powered by magnetic flux.

1. What are the different ways these do the same things?

2. What did Don know which Prentice didn't?

3. Does Prentice understand electron spin in a way to make full use of it?

4. Does Prentice use 1/4 wavelengths?

5. What is the 21 spark gap in Don's system? What is it's function?

The Prentice Electric Power Accumulator is a 1925 Canadian patent. Not all the words of the patent could be included on the first page. The first words and claims were removed. The idea is that a resonant tank circuit was used to create an amplification of energy in the transformer 8. The secondary was part of a transmitter circuit that was outputting to a paralleled receiver antenna 25, which was in series to ground. The primary of the second transformer is in series to ground. The process results in 6 times the energy output more or less.

I mention this patent because one of the last things Don wrote to someone on the forums was to consider this patent, and he was going to privately show this person how to improve upon it. From what we have shown so far, it should be easy to see what could be improved upon and what is missing. On the second page showing **Similarity between Don Smith Ground System and Prentice Accumulator** we see the schematic compared with Don's related systems. The arrows indicate the direction of electron flow. Notice the second part of the circuit would have the electrons coming from the air (point 25) rather than the ground. The air is a grounding, but this creates problems and is limited. The length of wire is rather problematic. Don mentions later that you can see that Prentice didn't understand the magnetic amplification potential. He worked with relatively low frequency which limited his output and required long lengths of wire. The main reason I showed this was because it shows much similarity in the grounding beginning and ending paths with variable capacitors at both groundings. It also shows the loads in series on the way to earth ground. The transformers are also arranged in the same sort of ways. This gives the reader more examples to work with as well as more context.

One additional point about existing technology using actual Don Smith devices:
"The Devices (many) are presently hidden inside other Devices such [as] refrigerators and other appliances, other power Systems which are considered as commonplace... The object is to make My Devices commonplace before most of the World finds that It has been using them all along."

The table of context provides information about the various pages and sections of the book without requiring me to repeat that information here. The following explains some of the handwriting on pages 119-122:
"Inadvertently the book which they have in the bookstore right now has some of those original notes in there. What happened was I left my briefcase in the bookstore and they headed out copying everything in the briefcase. So the book that they have in there, now has maybe 10 or 15 pages or more of actual original notes and things in my own handwriting, in the back of it."

There are several more schematics that I have made and placed after several videos that were not included in the book. The videos and their wording sufficiently explains them. The reader is encouraged to carefully read and understand the full context before attempting to replicate these kinds of systems. There will be a learning curve. We have tried to make that process easier in providing the Resonance Kit as a safe and affordable (it is not sold at profit) option (more accesories coming). We can see below that Don recommends for people to start where he started (model 2), but this is only after the Resonance Kit. We are making the Kit to be able to do the same thing as model 2 in a way that is more affordable and safer.

"The best is to plan the construction, so no tuning is required, if so, very little, then the pica and nano farad capacitors can be used. I think where the confusion is surfacing is that capacitors are also used as temporary storage prior to continuing on into the load. This is the tricky part because it must supply ample coulombs (volt amperes) to the requirements. Best bet is to build the separate basic ciruits used seperate boards so they can be switched in and out until You settle on the particular Device that matches Your changing requirements. There are several circuits involved which with time and knowledge may change. keep the high voltage module separate, keep the temporary storage and inverter modules separate. as well as the final usage module. Set them up such that a particular board can be used in many ways for latter experiments. Saves buying a lot of possibly expensive components many times. The high voltage modules are already made in many and varied outputs. Do not go for the highest or biggest or fancyist. One to two thousand volts works just fine and preferably with a low amperage, 12 volt DC is just fine. The Idea being with a known input and latter a known output makes it easy to assess the results. I would recommend the Device photo # 2 on My Website with the small Tesla Coil output with the 3 separate Coils which duplicate the energy field from the center output. The ready made Coils shown B and W Coils (They are on the Internet). Each of the separate receiving Coils have a tunable mica capacitor and a tunable resistor in the circuit Neon 2 lights with the resistor on one leg are from Radio Shack and are placed with the resistor leg attached to the circuit of each of the receiving coils. This simplifies seeing the energy present when the coils are tuned to the sending Coil in the center. Start out with something that's manageable until You get the concept and then get complicated if You wish. Otherwise frustration will win.
Take Care, Don Smith"

The above quotation, and others like the following, gave me the idea to *organize the various processes and modules into sections that are systematically arranged*. Since it was difficult to organize with so many variations, I opted to do a *mixture* of modules and processes in the five sections of the Systematic Index (pages 403 to 413).

"Now this has all of the different components on it, which sometimes there's more over here, and this part, and this part, will be on different blocks. Because I can use them interchangeably with other things I'm doing."

IMPORTANT STATEMENTS BY DON SMITH

Chances are you have a fair chance of getting there the first time if you, you know, if you're lucky.

* * * * *

Think about it, the physics books are wrong.

University is there for the system, and it's there to make a rubber stamp out of you. If you don't like being a rubber stamp, you're in big trouble. Apparently there are enough people with that renegade attitude that it's beginning to change.

The Field of Science proflgates by notation, that which does not, is not.

Those with closed Circuit (Minds) conveniently put the world in a box, what goes in, comes out, end of question !

The university people like things that you could do mathematically, and you always end up with the same answer regardless. That way they look good if they're competing with something that's not going to give the answers that they want to be repeated every time. Then from university standpoint they're in danger of needing another job. The University is not interested in people that can't repeat what they're doing.

Most of what is presented here has been in place since mid 1800's. Putting it in words so that it becomes useful has not.

Now since last year [1994] there have been hundreds of people that have tried to duplicate what I did. There's about 12, I guess, out of several hundred that were successful. The main reason most of them failed was the impedance problem. Because when they started putting capacitors and resistors and you know various other things in there, which controlled the timing circuit, so that the different coils could talk to each other, but basically they blew themselves out of the water. And none of them, or very few of them ever bothered to check to see if there was something there other than resistance or capacitance or whatever. Or even in your L1 and your L2 coils.

In Eastern Europe there are probably four or five hundred people that have duplicated my devices with real good results [said in 2005].

I realized that there are a lot of people that had trouble understanding things unless you could go turn the wall switch on and that was the extent of it.

In chemistry physical change, from one form to another results in one of three actions: 1. Mixing only, physical only, no reaction. 2. Endothermic, loss of energy by absorption. 3. Exothermic, additional energy not previously available. This is referred to in all chemistry Texts as Free Energy or (Δ G). These changes occur at elemental and or molecular level. At particle size, as in Electrons there is an overlap into Physics. In Physics all Matter relates too, and is Energy. Electrical Energy at its denominator is the acceleration or deceleration of electrons. Ambient or background being Earth Ground. Variations from Earth Ground provides useful electrical energy. By accelerating (disturbing) electrons as in Resonate Coils and Magnet Systems, or Converting Mechanical (dynamo) into other than ambient, results in useful energy. The faster the cycling of electrons the more energy present. Energy greater than ambient, based on the established definitions from the above Gas Laws is then referred to as Free Energy.

Most people think that lots of volts and no amps, and that's the thing that they'll try to get you with every time. However volts are basically the water that's behind the dam. Until you poke a hole in the dam, you don't get any flow rate or amperage. Whatever your voltage is it's whatever your potential is, and how you go about letting that thing through the dam, is what's going to give you your useful energy.

* * * * *

You're not getting something for nothing because it was all there to begin with.

You're not getting something for nothing, and you're not creating anything. You're just simply changing it from one form to another, and in that change process from one form to another you have useful energy. If the energy is static and not moving, or it's mass, or other forms of energy, you don't really see anything useful. It's only when it's going from one form to another that it becomes useful, and you capture it just like the water running down from the mountain back to the sea level. In between you put a paddle wheel there, and you've captured the energy. This is true of all forms of energy, whether it's electrical or whatever.

Electromagetic stiulation disturbs electrons by avalanche, dislogdging adjacent electrons, resulting in more energy out than in.

Electrons, when disturbed, first produce magnetic flux and then electrical flux when they spin back to their normal position. Therefore, any electron movement produces above ambient energy, being over unity.

Electrons as a source of energy have been moving and yielding magnetic and electrical flux since the beginning of time. They are never diminished and are an excellent example of an over-unity source of energy.

The first appearance of radio glass tubes as a diode having a cathode (electron out put) and an anode (collector of electrons) for electromagnetic stimulation that dislodges electrons which avalanche in dislodging many-many more electrons. The excess of electrons at the Cathode magnify and react with the electrons present at the anode, resulting in more energy out than input.

* * * * *

Useful Energy is Defined as that which is other than Ambient.

Only that electrical energy above or below ambient is useful.

Once you get the energy balance working for you then you're on your way.

The idea is to make a magnetic field work for you.

*Anywhere there's a large amount of magnetic flux
I consider a source of usable energy.*

Resonant-fluxing of electrons activates potential, which is present everywhere.

Flux lines result both from Induction-henrys-amperage and capacitance-coulombs-volts and define electrical energy.

All electrical systems, natural or man made work exactly the same, by fluxing and-or resonanting magnetic waves.

Radiated magnetic flux from the Dipole when intercepted by capacitor plates at right angle changes to useful electrical energy.

Useful electrical energy is achieved when the electron density at point "A" becomes greater than at point "B", (being the more negative moving to the less negative concept).

All electrical Power Systems have only the magnetic flux move from point A to point B.

In Electrical Systems, Electrons active at point "A" are not the same Electrons active at point "B". The Electons activated at the Electrical Station are not the ones used at your house. When you ground your system by flipping the wall switch, you use your own electrons. In closed energy systems, electrons communicate with and replicate the activity of the overbalanced potential, when provided with Earth and or Air Groundings.

With alternating electrical current, electrons do not move from point "A" to point "B" as commonly envisioned! Electrical potential (oscillating electrons) at point "A" results in harmonic electron activity at point "B", when the grounding switch (circuit) is closed. Point "B" supplies it's own electrons and mirrors the activity of point "A". Impulsing (turbulence) by magnetic induction causes electrons to be pulled into the system, which then oscillates. When the magnetic field collapses (becomes absent) the electrical potential returns to it's natural background.

In nonlinear systems, interaction of magnetic waves amplify (conjugate) energy, providing greater output than input.

Any coil system when fluxed causes electrons to spin and produce useful energy, once converted to the style required.

Let us start backwards, with a conventional output transformer. Select one with desired physical characteristics, being also an isolation type... A power correction factor system, being a capacitor bank maintains an even flow of flux... With the down stream system in place, all that is now needed is a potential system. Any flux system will be OK... Any input system wherein a lessor amount of electrons disturbs a larger amount of electrons, resulting in the output being greater than the input is desirable.

Free Energy Harvesting Requirements, 1. Electron acceleration or deceleration must vary from ambient. 2. Spin rate CPS fluxing provides electrical with magnetic energy. 3. At Resonance (negative resistance), all electrons move in the same direction without interfering with each other. 4. Electrons must Avalanche, pulsing magnetic waves provide three dimensional and nonlinear energy at each cycle, electrons presently available disturb and bring alone many more electrons, resulting in gain.

The basic unit of electricity (the electron) upon encountering a moving magnetic field (or wave) spins, giving off an electric impulse. When this impulse collapses, it spins back to it's natural position, giving off a magnetic impulse. Therefore, magnetic and electric are two sides of the same coin. When the magnetic side is pulsed, it yields electricity and conversely, pulsing of the electrical side yields a magnetic field. Moving one in relation to the other produces useful energy. When done consecutively, each cycle pushes (current) forward, while pulling electrons into the system...in much the same way as a water pump moves water. These electrons are obtained from Earth and air grounding.

The flipping device (reactor L-1 Coil) is pulsed which then provides a resonant induction pulse, in turn this flips the electrons present at the (reactant L-2) Coil. The energy input in L-1 is divided by the number of turns present. The induced magnetic pulsing in turn flipps the electrons in each turn of L-2. If more turns are present in L-2 than L-1, there is a net gain in Energy present, as demonstrated by the dominos.

** * * * **

This whole device is like a pump. All it's doing is like the dominoes we were talking about, its tripping the electrons here and that electrons trip in the next one, and this whole bunch of electrons in all directions are being tripped. Each time one of those would be tripped it gives off a magnetic impulse first. Then, when it comes back on that infinity loop, it gives off an electrical impulse.

Electromagnetic Energy Harvesting methods include a catalyst, a collector, and a pump. Catalysts include such things as Air and Earth grounding, sensitization through doping with certian elements. Collectors include temporary storage as in capacitors, coils, and transformers. The pump system includes movement onward to point of use. The rotating coils ang magnet systems activate electrons present such that action at a distance can occur and therefore is an energy activation pump.

When electrons are spun or pushed apart ambient is disturbed, and yield magnetic and electrical energy. The rate of disturbance (cycling) determines the energy level achieved.

* * * * *

Of course in order for it to work you have to have more of something, and there has to be less somewhere else. The leg in between the more and the less is the active system which you're going to be using to get your useful energy.

Basically electrical energy is achieved when an electron density at point A becomes greater than electron density at point B. This is your potential difference between two points. If that doesn't exist you don't have any energy, regardless of how how many electrons are there. It's just like the high pressure and low pressure areas of the weather system. Things are going to move away from the high pressure area into the low pressure area. The electrons density acts the same way.

An earth ground is where your amperage comes in.

In order to turn this into useful energy it's the way in which you ground it will determine whether it becomes a practical source of useful energy.

I can tell you this, you can have hundreds of thousands of volts out of a tiny little matchbox and at high amperage.

I noticed that Joe Newman over there has already got his book finished pretty much. It looks real impressive. He's doing some of the things which I'm involved in. I noticed he's using the 200KV capacitor there which is one of the tricks to getting useful energy out of one of these devices.

Useful Circuits from Nikola Tesla

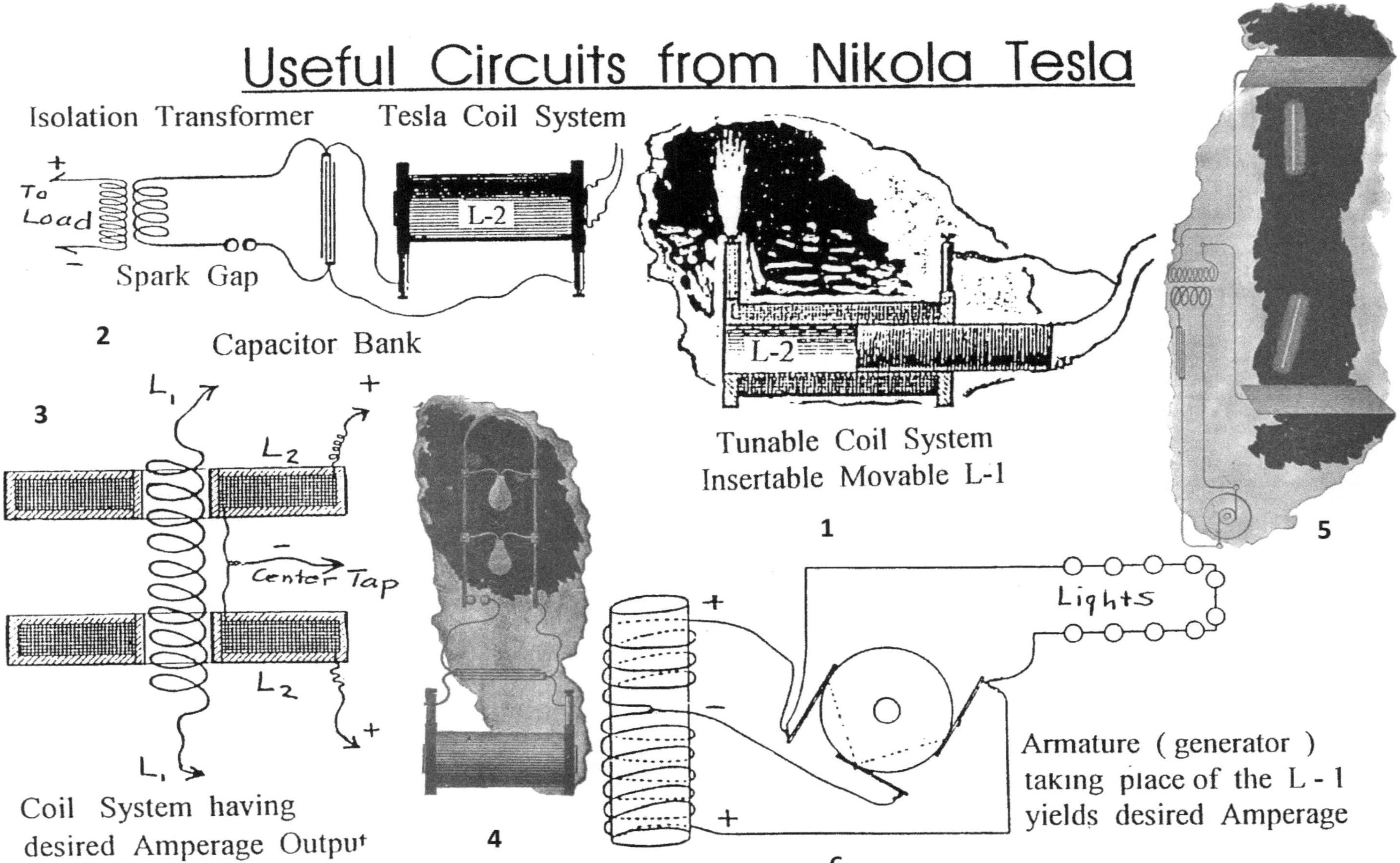

[Editor: Added the right image to the page as image 5. This was in the video but not in the book. Several other images shown in video 2 can be found on pages 170 to 197 in *The Inventions Researches and Writings of Nikola Tesla*: http://www.gutenberg.org/files/39272/39272-h/39272-h.htm]

Resonate Electrical Power System

Don L. Smith

Potential Energy is everywhere at all times, becoming useful when converted into a more practical form. There is no energy shortage, only gray matter. This energy potential is observed indirectly through the manifestation of electromagnetic phenomenon, when intercepted and converted, becomes useful. In nonlinear systems, interaction of magnetic waves amplify (conjugate) energy, providing greater output than input. In simple form, in the piano where three strings are struck by the hammer, the center one is impacted and resonance activates the side strings. Resonance between the three strings provide a sound level greater than the input energy. Sound is a part of the electromagnetic spectrum and is subject to all that is applicable.

Useful Energy is Defined as that which is other than Ambient. Electrical Potential relates to mass and it's acceleration. Therefore, the Earth's Mass and Speed through space gives it an enormous electrical potential. Humans are like the bird sitting unaware on a high voltage line. In nature turbulence upsets ambient and we see electrical displays. Tampering with ambient allows humans to convert magnetic waves into useful electricity.

Putting the above in focus requires a look at the Earth in general. Each minute of each day (1,440 minutes) more than 4,000 displays of lightening occurs. Each display yields greater than 10 million volts at greater than 200,000 amperes in equivalent electromagnetic flux. This exceeds 57,600,000 million volts and 1,152,000 million amperes of electromagnetic flux each 24 hour period. This has been going on for more then 4 Billion Years. The USPTO insist that the Earth's electrical field is insignificant and useless, and that converting this energy violates the laws of nature. At the same time they issue patents wherein electromagnetic flux incoming from the Sun is converted by Solar Cells to DC Energy. This is further converted to the required usage. Aeromagnetic flux (in gammas) Maps-World Wide, includes those provided by the US Department of Interior-Geological Survey , clearly shows a spread of 1,900 gamma, above Ambient present, reading instruments flown 1,000 feet above the (surface) source. Coulomb's Law requires the squaring of the distance of the remote reading, times the reading for the corrected amount. Therefore, 1,000 X 1,000 equals One Million X 1,900 gamma, being 1,900 million gammas.

There is a tendency to confuse gamma ray with gamma. Gamma is ordinary every day magnetic flux. Gamma Ray is high impact energy, not flux. One gamma of magnetic flux is equal to that of 100 Volts RMS. To see this take a Plasma Globe emitting 40,000 volts. A gamma meter when properly used will read nearby, 400 gammas The 1,900 million gamma previously mentioned then becomes the magnetic ambient equivalent of 190,000 million volts of electricity. This is on a Solar Quite day. On Solar Active days it may exceed five times that amount. The Establishment's idea that the Earth's electrical field is insignificant goes the way of their other great ideas.

There are two kinds of Electricity, potential and useful, until converted all electricity is potential. Resonate-fluxing of electrons activates potential, which is present everywhere. The Intensity/CPS of the resonate-frequency-flux rate, sets the available energy. This must then be converted into the required physical dimensions of the equipment in service. For example, energy arriving from the Sun is magnetic flux, which Solar Cells convert to DC Electricity, being further converted to required usage. Only the magnetic flux moves from point "A" (Sun) to point "B" Earth. All electrical Power Systems work exactly the same. Movement of Coils and Magnets at point "A" (generator) fluxes electrons, which in turn excites electrons at point "B" your house. None of the electrons at point "A" are ever transmitted to point "B". The electrons in both cases remain forever in tact for further fluxing. The above is allowed by Newtonian Physics (electrodynamics and the laws of conservation). Clearly these laws are all screwed up and inadequate.

In modern physics, USPTO Style, all the aforementioned can not exist since it opens a door to overunity. The Good News is that the PTO has already issued hundreds of Patents related to Light Amplification, all of which are overunity. The Dynode used to adjust the self powered shutter in your camera receives magnetic flux from light which dislodges electrons from the cathode, reflecting electrons through the dynode bridge to the anode, resulting in billions of more electrons out than in. There are currently 297 direct patents issued for this system and thousands of peripheral patents. all supporting overunity. More than one thousand other Patents which have been issued, to the discerning eye are overunity devices. What does the indicate about Intellectual Honesty ?

Any coil system when fluxed causes electrons to spin and produce useful energy, once converted to the style required. Now that we have qualified the method required, let us see how this concerns us ?

The entire System already exist and all we need to do is hook it up in a way, that it is useful in the required manor. Let us start backwards, with a conventional output transformer. Select one with the desired physical characteristics, being also an isolation type. Only the magnetic flux passes from the input side to the output side. No electrons pass through from the input to the output side. Therefore, we only need to flux the transformer to have output. Bad design by the establishment, allowing hysteresis of the metal plates, limit's the load which can be applied . Up to this point only potential is a consideration. Heat (energy loss) limits the output/ amperage. Correctly designed composite cores run cool, not hot.

A power correction factor system , being a capacitor bank maintains an even flow of flux. These same capacitors, when inserted with a coil system (transformer) become a frequency-timing system. therefor the inductance of the input side of the transformer, when combined with the capacitor bank, provides the required fluxing in activating the proper style of electrical energy (cycles per second).

With the down stream system in place, all that is now needed is a potential system. Any flux system will be OK. Preferably any amplification over-unity output type is desirable. The input system is point "A" and the output system is point "B". Any input system wherein a lessor amount of electrons disturbs a larger amount of electrons, resulting in the output being greater than the input is desirable.

At this point it is necessary to defecate in the punch bowel by substituting updated information relating to electrons, and laws of physics . A large part of this is original, from me, so don't expect peer review to be kind, rats will play.

Non - Ionic Electrons

As a source of Electrical Energy, non-ionic electron doublets exist in immense quantities throughout the universe. Their origin is from the emanation of Solar Plasma. When (spun or pushed apart) ambient is disturbed, they yield magnetic and electrical energy. The rate of disturbance (cycling) determines the energy level achieved. Practical methods of disturbing them includes, moving coils past magnets, or vice versa. A better way is the pulsing (resonate induction) with magnetic fields and waves near coils.

In coils systems, magnetic and amperage are one package. This suggest, that electrons in a natural non-ionic state exist as doublets. When pushed apart by agitation one spins left, yielding (Volts-potential) electricity and the other spins right, yielding (Amperage-magnetic) energy. One being more negative than the other. This further suggest that when they reunite, we have (Volts X Amperes = Watts) useful electrical energy. The above idea, until now, has been totally absent from the knowledge base. Amperage as previously defined is then flawed.

Electron Related Energy

		Energy Available	Method of Storage	Common Unit	Units of Measure
		Electrical	Capacitor/Coulombs	Volts	Flux Units
	Spin	Gravity	Momentum	Torque	Ergs
		Magnetic	Coils/Amp. turns	Amperes	Flux Units
Electrons					Teslas, Gauss, Gammas, Oersteds
		Light	Laser	Lux	Photons
	Impact	resistance			
		Heat	Various	Fahrenheit/Celsius	Temp

Left hand spin of Electrons results in Electrical Energy and right hand spin results in Magnetic Energy. Impacted Electrons emit visible Light and heat

Electrons as a source of energy have been moving and yielding magnetic and electrical flux since the beginning of time and will continue to do so, till the end of time. They are however, never diminished. Therefore, an electron is an excellent example of an over-unity source of energy.

The sea of Continuum "Ether", goverened by the Law of Squares is from it's electrical and magnetic character composed largely of right and left handed electrons.

Useful Circuits, Suggestions for Building an Operational Unit

1. Substitute a Plasma Globe such as Radio Shack's Illumna-Storm for the Source-resonate induction system. It will have about 400 milligauss of magnetic induction. One milligauss is equal to 100 volts worth of magnetic induction.

2. Construct a coil using a 5 - 7" diameter, piece of PVC for the core winding.

3. Get about 30 feet of Jumbo-Speaker Cable and separate into two pieces. A carpet knife stuck in a piece of card board or wood and pull carefully the cable past the blade.

4. Wind the Coil with 10 - 15 turns and leave about three feet of tail at each end. Use glue gun with a drop to hold start and finish of the winding.

5. This will become the L -2 as shown in the Circuits page.

6. When sitting on the top of the Plasma Globe (Crown - Like), you have a first class resonate air core coil system.

7. Now substitute 2 or more 5,000 plus voltage capacitors for the capacitor Bank shown on the circuits page. I use 2 plus , 34 microfarad capacitors.

8. Finish out the circuit as shown, upper left hand corner of the accompanying Useful Circuits page. You are now in business !

9. Voltage - Amperage Limiting resistors are required across the output side of the Load transformer used for adjusting out put level and the desired cycles per second.

SUGGESTIONS:

Obtain copy of "Handbook of Electronic Tables and Formulas", Published by Sam's, ISBN 0-672-22469-0, also an LCR meter is required. Chapter One in the Book has important time constant (frequency) information and a set of reactance charts in nomograph style which makes working, and approximating of the required three variables, capacitance, inductance and resistance, much easier. If two of the variables are known, the required missing one is obtained from the nomograph.

For example the input side of the isolation transformer needs to operate at 60 CPS. that's 60 up and 60 down, being 120 cycles. Obtain the Inductance in henries with the LCR meter from the input side of the isolation transformer. Plot this value on the proper reactance chart above mentioned. Plot the needed 120 Hertz and put a straight line between the two known points.. Where the line crosses the Farads line and the Ohms line yields two readings. Chose one (resistor) and insert between the two leads of the input side transformer winding.

The Power Correction Factor Capacitor (or Bank /more than one), now needs adjusting. The following formula is helpful in obtaining the required missing information. The capacitance is known, as is the desired potential to pulse the output transformer. One Farad of capacitance is one volt for one second (one Coulomb). Therefore if we want to keep the bucket full with a certain amount, how many dippers full are required? Should the bucket require 120 volts, how many coulombs are required..

$$\frac{\text{Desired Potential In Volts}}{\text{Capacitance in Microfarads}} = \text{Required CPS}$$

$$\frac{\text{Say 120 Volts Potential}}{\text{Say .004.000 (4,000 Microfarads)}} = \text{Yields 30,000 CPS}$$

Now go to the Reactance Chart above mentioned and obtain the required resistor jumper to place between the poles of the correction factor capacitor.

An earth grounding is desirable as a voltage limiter and transient spike control. Two are necessary, one at the power factor capacitor and one at the input side of the isolation transformer. Off the shelf surge arresters/ spark gaps and varistors, having the desired voltage/potential and amperage control are commonly available. Siemans, Citel America and others, make a full range of surge arrestors, etc. varesistors resemble in appearance, the coin sized flat capacitors. Voltage Limiters are herein combined as V - 1.

It should be obvious that several separate closed circuits are present in the suggested configuration.. Firstly, power input source, the high voltage module, a power factor capacitor / bank combined with the input side of the isolation transformer. Lastly would be the output side of the isolation transformer and it's Load. None of the electrons active at the power source (battery) are passed through the System to downstream usage. At any point should the magnetic flux rate vary, the amount of electrons active also varies. Therefore controlling the flux rate controls the electron (potential) activity. Electrons active at point "A" are not the same electrons active at point "B", "C" and so on. Should the magnetic flux rate (frequency- CPS) vary, a different amount of electrons will be disturbed. This does not violate any Natural Law and does produce more energy out than in, should that be desirable.

A optional convenient high voltage module would be a 12 volt DC neon tube transformer. The power factor, correction capacitors should be as many microfareds as possible, this allows a lower operating frequency. The 12 volt neon tube transformer oscillates at about 30, 000 cycles per second.. At the power correction factor capacitor / bank we lower the frequency to match the input side of the isolation transformer.

Other input optional high voltage sources would be automobile ignition coils, television flyback transformers, laser printer modules and possibly others. Always lower the frequency at the power correction factor capacitor and correct if needed at the input side of the isolation transformer. The isolation transformer comes alive when pulsed. Amperage becomes a part of the consideration only at the isolation transformer. Faulty design, resulting in histerisis creates heat which self distructs the transformer if overloaded. Transformers having the new composite core in place of the pandered many layered thin sheets of soft iron, run cool and tolerate much higher amperage.

Electrical Energy Related - Plasma thru Solids

	Status	Subatomic	Atomic	Molecular
None Resonate	Disorder-Trashed Ohm's Law Is King	Resistive	Resistive	Resistive
Resonate	Ordered - Useful Negative Resistance At Resonance ** Impedance = Zero	Nonionic Electrons *	Radioactive	NMR

* Solar Plasma is 65 % None Ionic Electrons .
Earth's Atmosphere and surface has massive amounts of None Ionic (Free) Electrons. When Resonate they become an Unending source of Electromagnetic Energy. These electrons are In no way diminished by this activity .

** When Resonate, adjacent electrons, not a part of the original are pulled into the composite, becoming greater energy out than in. This allows for a self sustainable energy source if desired.

8 - 2

RESONATE ELECTROMAGNETIC POWER SYSTEM

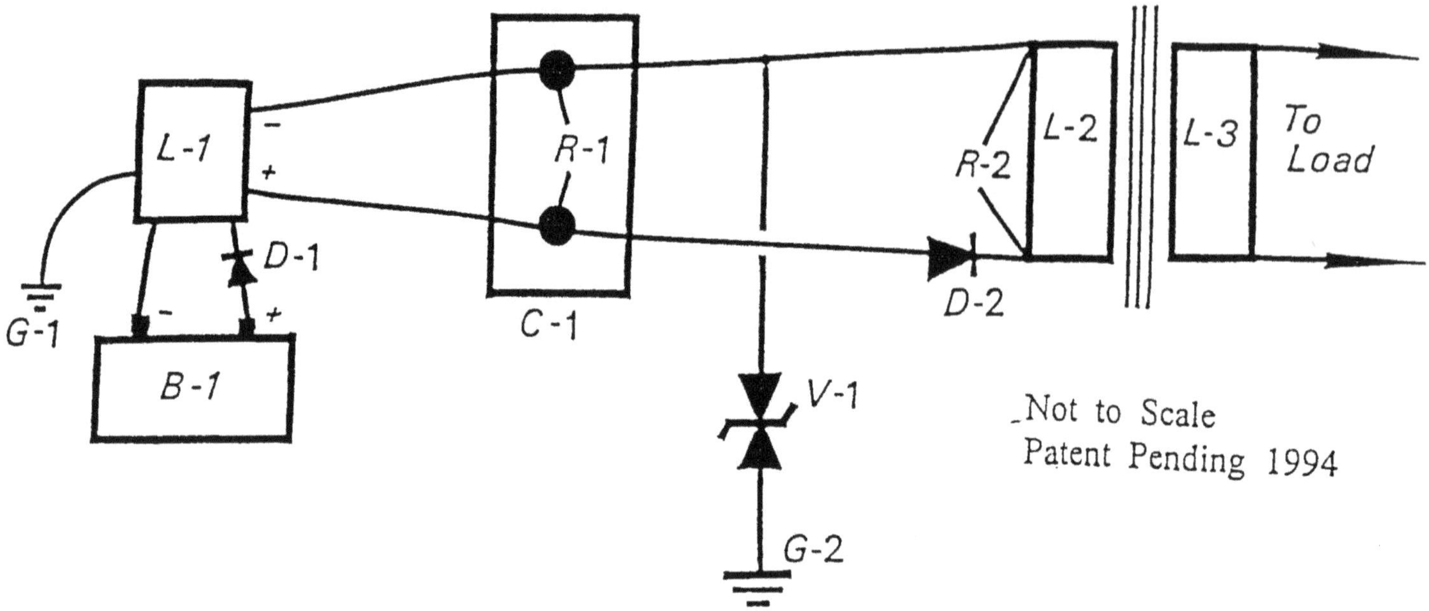

Power Source: B - 1 Gelcell, 12 Volt, 7 Amp Hour
D - 1 Kick back protection for L - 1
L - 1 Bertonee, NPS - 12D8, constant burn Neon Tube transformer, Bertonee, Boston, MS

Power Conditioner: C - 1, Capacitor or Capacitor Bank, 8,000 microfarads for 480 volts DC. R - 1, Resister used to set electron pump rate, frequency of the capacitor. Maintains the desired voltage level required to operate the system.

Voltage Control: V-1, Varistor, limits the voltage as required for the Output Transformer L -2. (480 V @ 60 Amps)

Output Transformer: Isolation Type, (L - 2 / L-3) with R - 2 (resistor) correcting the output frequency to 60 CPS, being 60 UP and 60 DN (120 total). (28.8 KVA)

Useful Timing Formulas:

T = frequency in cycles per second
C = capacitance in microfarads
L = Inductance in milliheneries
R = resistance in ohms

Therefore: $T = RC$ and $T = \frac{L}{R}$

9

Comments

The information here given reflects the small Suitcase Model demonstrated at the Tesla Convention (1996), presented on VCR as Don Smith's Workshop. This unit was a very primitive version. Newer versions have atomic batteries and Power Output ranges into the Giga Watts. The battery requirement is low level and is no more harmful than the radium on the dial of a clock. Commercial Units (Boulder Dam Size) are currently being installed at Several Major Locations throughout the World. For reasons of Personal Security and Contract Obligations the information herein given is incomplete.

Ultimate Energy Sources

Don L. Smith

Recent history of the Late Great Earth evidences a considerable amount of convenient ignorance regarding Sources of Energy. The purpose of this Preface is to call attention to and put the spotlight on the establishments approach to energy.

In order to understand where from and where to, one must focus on that which is obvious. For example, the wind must constantly and dependably blow and the sun must constantly and dependably shine for the System to Work. Energy sourced from these are at the fancy of nature. Massive amounts of energy required, subject to the caprices of nature are an explosive mixture.

Atomic Energy is a throwback to the age of steam engines. It has been demonstrated as an expensive Rube Goldberg approach to solving the energy problem. This type of energy is politically correct because it manipulates massive amounts of tax payer dollars, economy and practicality are not important considerations. Conspiracy is the name of the Game. Major Benefactors are the Industrial, Governmental Agencies, University R & D Programs and Military Complexes.

Fringe Sources of Energy along with Coal and Petroleum are precursors of a horrible fate and extinction of all air breathing animals. Frogs absorb oxygen through their skin are showing considerable distress with large numbers becoming extinct. This conventionally and intentionally muddled ignorance relating to the human condition and this precursor places the establishment in default.

The importance of Environmentally Friendly Energy Sources are the only answer to survival of the Human Masses. For example the Oxygen Ratio obtained from a bubble in a piece of Amber (2000 years old) is 39 %. Same information of record from Air Test dated 1900 shows 30 %. Carl Sagen's Book " Blue Planet " shows 18 % for 1984. Today the best case world wide is 15 %. Larger cities such as Tokyo, Jakarta, Mexico City and numerous others World Wide have 7 % or below. For Health reasons 15 % is critical. All Disease mechanism's, germs, virus's, etc. are anaerobic. That is they thrive best in oxygen deficient situations. The acknowledged and recorded 0.5 degree average per year temperature increase equals 50 degrees plus during the next 100 years, propelling the Earth into the same Fate as Venus.

DECLINE OF OXYGEN IN THE EARTH'S ATMOSPHERE RATIO OF OXYGEN TO NITROGEN

The absolute minimum required for human survival is 15 %
The ratio in 1900 AD was 30 %, today [1996 AD] it is 18 %

This ratio indicates the approximate time beyond which humans will not survive. Disappearance of the World's Forest, Chemical Combination of Oxygen from domestic and industrial usage. Burning of Coal for electrical power generation, petroleum usage by internal combustion engines, automobiles, airplanes, etc., are the major contributors to the decline.

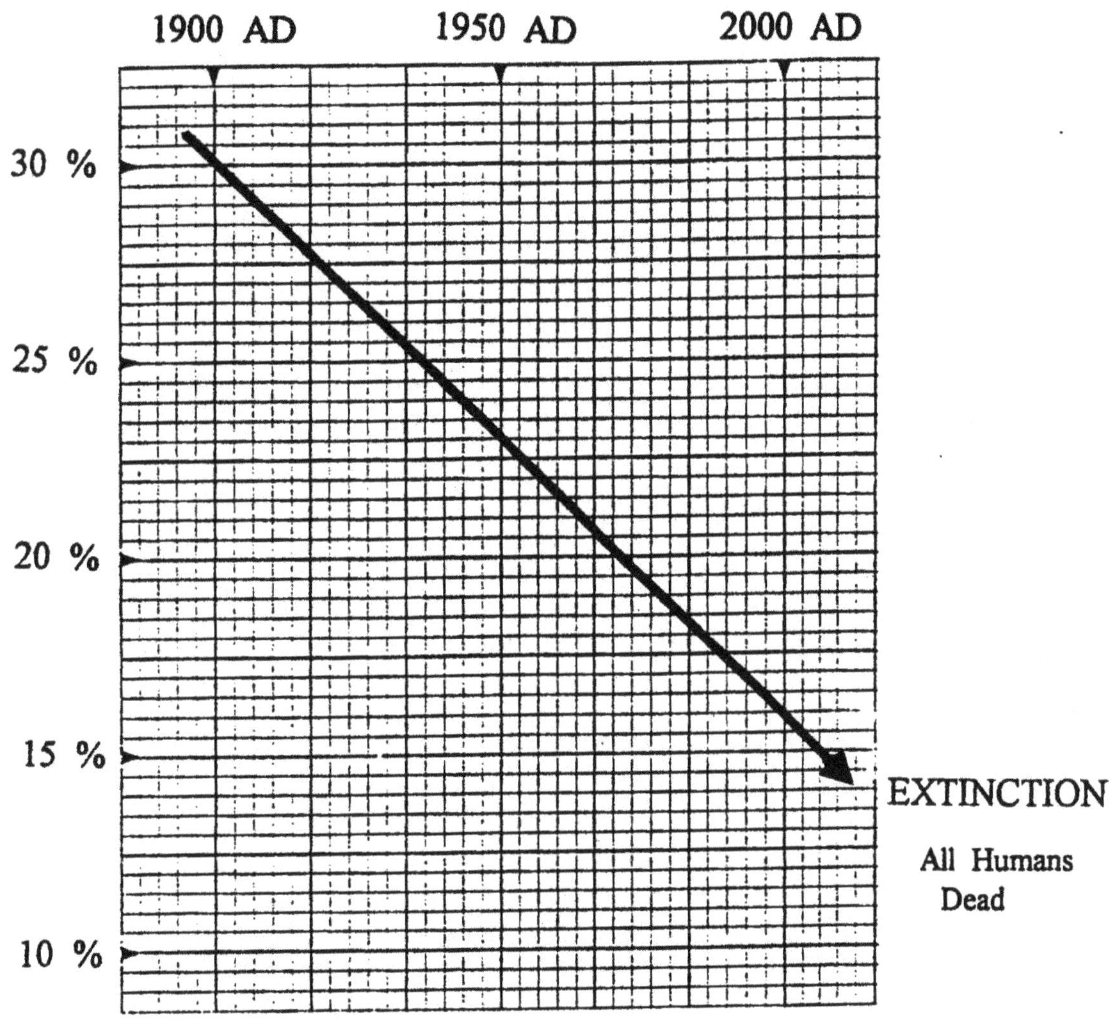

For further Reference look up related material on the Oxygen and Carbon Dioxide Cycles in advanced Biology Text Books.

Major Causes of this sad situation are now presented. Each gallon of gasoline chemically combines 200 pounds of Oxygen, such that it is out of reach by Humans. Based on the Oxygen Nitrogen Ratio a much larger amount of Nitrous Oxide (pollution) is produced. Ozone, natures house keeper is consumed at horrible rates in an effort to counter the pollution. Depletion of the limited Ozone allows deeper infrared penetration into the Earth's Atmosphere, causing temperatures to rise, plus disrupting the health system world wide.

More than four million barrels of gasoline is used each day, one barrel being 44 gallons. Therefore, 44 gallons x 200 pounds of Oxygen x 4 million indicates 35.2 Billion Pounds of Oxygen is extinguished and removed from Human accessibility each day. In One Year, 365 x 35.2 Billion, equals 12.9 Trillion pounds of Oxygen removed from the System each year, in one hundred years, that's 1,290 Trillion Pounds of Oxygen and many times that amount of Nitrous Oxide Pollution Emission from Internal Combustion Type Engines.

This Horrendous Fate with Earth Changes can be slowed down and temporarily avoided if the Inventions herein are permitted. The inventions now discussed are environmentally benign and are fully self renewable. Universally available Electron's are the Source , simply recycled and are in no way diminished by the process. The Systems are inexpensive to own and last through many generations. These Devices conform to all known Laws of Physics, easily recognizable as fully functional and simplistic to verification.

Inventions herein described eliminate the conflict between Energy and the Environment, by providing a none Malignant Source. They are also designed to level the Playing Field between the have's and have-not's such that future Conflicts and Wars are unnecessary. It provides a Safe, Fully Renewable, Un-ending Source of Energy. They are easily managed and work as required, being inexpensive to build and operate, only the Imagination limits their usage.

Pre Civil War electrical devices included successful usage of the Earth as a battery, use of kits with metal string and spark-gap (hand manipulated becomes a telegraphic key). If the length of wire (now called an antenna) was an equal multiplication or division of, when one Kite system sparked, the second one at distance, at same time also sparked. The Patent Office, Military and Other Governmental Agencies would not touch such stupid Ideas that violated all scientific laws and dictums, such as Radiotelegraphy.

Early Radio Type Devices could not be properly explained until after Thompson, about 1911 theorized that electrons exist. Related Devices began to seriously appear in the 20's and 30's. The coher of Tesla, precursor of the diode and solid state devices gave radio a push in the right direction. The first appearance of radio glass tubes as a diode having a cathode (electron out put) and an anode (collector of electrons). Simply put electromagnetic stimulation dislodges electrons which avalanche in dislodging many-many more electrons. The excess of electrons at the Cathode magnify and react with the electrons present at the anode, resulting in more energy out than input. Without this action, radio could not exist. Electrons at point "A" do not travel to point "B", only the equivalent in magnetic wave is transmitted which causes electrons at point "B" to become active.

Farnsworth and others built more complicated Devices related to the more primitive coher and radio tube. Today's television tube is related to the aforementioned. Devices that use the electron avalanche system include dynode bridges, multipactors, masers and lasers. Solid state devices presently exist which perform the electron spin-reversal and recovery required of Electrical Power Generation. The Aforementioned devices through the avalanche effect produces huge amounts of electrical energy not previously present and therefor if desired can become self sustaining once operational.

None Ionic Electrons constitute greater that 65 % of the Solar Plasma and very large percent of the Earth's Atmosphere. They are what support (backbone of) the atmospheric molecules (nitrogen, oxygen, etc.). The Earth's gravitational field saturates it's atmosphere with them beyond that occurring in the Solar Plasma. Enriched zones are clearly observable in the radiation belt (shear zones). Their enrichment also defines the upper limit of the Earth's Atmosphere by their visible stimulation resulting in daylight as we know it. Outside that limit only darkness exist allowing the observation of the Universe.

This Ocean of enhanced concentration of Electrons is fully accessible and needs only gentle effort is producing unlimited quantities of Electrical Power. The complicated, dangerous and expensive procedure of ripping apart atoms in order to get at the Electrons is in the Inventors Opinion, undignified and stupid. It is politically correct since it puts enormous amounts of money into the Political Black Hole controled by friends of the establishment.

The Ultimate Energy Source, is then a never ending source, a fully self renewable source and does not compromise the World's Oxygen Supply. The Only Universally available Source which meets these requirements is the Electron. It's energy accessed through Spin Technology seen in Coils and Magnet Systems. The required spin left (electricity) and recovery spin right (magnetic - amperage) provides any required amount of Electrical Energy. This Electromagnetic System and It's Derivatives provides the Devices required in accessing this energy.

A rather large number of individuals and companies world wide are presently engaged in New Energy Research. Their failure to focus on practicality, economic feasibility and simply reproducible methods limits their success. The gravitational effect of and the pontificating on what is and what is not, in an effort to preserve status quo, blows an ill wind. Peer Review being the principal weapon of the establishment in keeping out those considered as destabilizing to it's self and thought system. Their Intellectual Dishonesty being the Hula Hoop for all times is the norm, sadly tainting all true research and development. The only hope then would be privately funded Inventors, who have freedom in developing that which is scorned by the Establishment.

I would like to introduce you to some of my Inventions which address the foregoing . I will use anointed Patent Abstracts, in each case, with limited release information, further data is on hold until Patent's are Issued. No Filing numbers will be provided for obvious reasons.

1. Earth Electrical System II, use of naturally occurring electrical differences in the Earth's Crust and Radio Device in combination for Transmitting of the Energy to Points of Use.
2. Motor-Generator \ Improved , This Invention is a new version of the Coils and Magnet System. Each continuous coil is opened in the center with each magnet separated in the middle, such that a rotating magnetic shield with alternating cut-outs, alternatingly breaks the composite magnetic field. This provides the necessary off-on, electron spin as required for harvesting electrical energy. Back e.m.f. expected is not present in this configuration. This interruptable Bloch Wall allows a smaller motivating force to produce a larger output resulting in a Self-sustainable Power Source.

3. Generator-Gaseous Tube Electrical Power System, A System wherein resonate electrons in a gaseous mixture are excited, such that the resulting electromagnetic energy can be harvested. The Device relates to Tesla Coil, Plasma Tube and Laser Systems.

4. Propulsion-Star System, A System that uses the same Spin System seen in Star Systems, Solar Systems, which is observed on Earth in Tornado's, Cyclones and the Like. It is related to Rail Guns, Gyroscopes, and Negative Gravitational Systems.

5. Power System, Piezoelectric Solid State Marx Generator; First Use of Such Device, being a major improvement of previous Systems.

6. Power System, Supermagnet Battery System, Relates to Magnetic Waves as in Solar Power. It is based on the original definition of a Gauss (being the magnetic flux equivalent of 100,000. Volts RMS). Domains aligned equates with resonance and provides a long term energy source.

7. Several Others, not open to discussion at this time.

The aforementioned are the results of about ten years of independent Research and Development. No Third Party interference, verification or general meddling is solicited or required. I simply offer the above as an effort to show what can be done with a clear mind and simple approach to New Energy.

Energy as Relates, Chemistry Vs Physics

In chemistry physical change, from one form to another results in one of three actions:

1. Mixing only, physical only, no reaction.
2. Endothermic, loss of energy by adsorption.
3. Exothermic, additional energy not previously available. This is referred to in all chemistry Texts as ***Free Energy*** or (ΔG). These changes occur at elemental and or molecular level. At particle size, as in ***Electrons*** there is an overlap into Physics.

In Physics all ***Matter*** relates too, and ***is Energy***. Electrical Energy at it's denominator is the acceleration or deceleration of electrons. ***Ambient*** or background ***being Earth Ground***. Variations from Earth Ground provides useful electrical energy. By accelerating (disturbing) electrons as in ***Resonate Coils*** and ***Magnet Systems***, or ***Converting Mechanical*** (dynamo) into other than ambient, results in useful energy. The faster the cycling of electrons the more energy present. Energy greater than ambient, based on the established definitions from the above Gas Laws is then referred to as ***Free Energy***.

The Required Formulary Short Hand is then:

A = Ambient Electrons (mass) ***Electrical Background***
CPS = ***Acceleration Constant*** (cycles per second)
F = ***Free Energy*** ***(ΔG)*** Therefor:

F = A times CPS (squared),

(Being a three dimensional entity, it requires being cubed not squared).

Donald L. Smith
20 April, 1998

Free Energy Harvesting Requirements

1. Electron acceleration or deceleration must vary from (background) ambient.
2. Electron Spin rate (cycles per second) fluxing - mechanism provides the electrical (left spin) combined with magnetic (right spin) energy, such that harvesting becomes possible.
3. Electrons at Resonance (negative resistance), all electrons at each cycle move in same direction at the same time, without interfering with each other as in conventional ohms related systems.
4. Electrons must Avalanche, pulsing magnetic waves provide three dimensional and nonlinear energy at each cycle, electrons presently available disturb and bring along many more electrons, resulting in a greater amount of energy out than was input. Low energy input becomes high energy output.

Highly Obvious Devices demonstrating the above includes dynode bridges, multipactors, masers, klystrons and All other radio type devices. Once Avalanche-regeneration is started, it is self sustainable, if desired. This action occurs at all frequencies when resonate and is the Source of Free Electrical Energy. Electrons are always there, just as the fish in the ocean, once caught become useful. Only a capture device is required. If the value received is greater than the cost, free energy is evident.

In chemistry, free energy is caused by chemical action resulting in more energy out than in. Explosives are an obvious example. Brownian movement is a good example, a direct quote from Maxwell, Boltzmann and Clausius " the molecules of a gas carry out ceaseless transitional movement at all temperatures above absolute zero ". This should send those opposing perpetual motion to the woodshed. Zero Point Quantum Physics then takes over, allowing that, an enormous amount of energy is still present. Perpetual electron movement disturbs the molecules and provides the Brownian Movement.

Enormous amounts of electromagnetic energy overflows (is wasted) and returns into the background unused by conventional methods. Simply measure the electromagnetic radiation bleeding out from motors, transformers, power lines and any other electrical device and you will see much more energy is wasted than is ever used. Unnecessary friction resulting in heat is another way the establishment assures that mandated energy wasting will occur.

Observations of an Inventor

1. Two hundred psi tire pressure supports a half million pound Plus Aircraft.
2. Living creatures start as one cell, which becomes billions of cells.
3. Systems were in a smaller source controls a larger source, such as Levers, pulleys, gears and hydraulic systems.
4. Avalanches start small and increase enormously.
5. A bolt of lightning starts with one electron, becoming an uncountable number of electrons.
6. All exothermic chemical reactions and Explosions of any kind.
7. All none linear three dimensional sources exhibiting greater out put than input. No rational person fails to observe this or do they ?

Near the turn of the last century when inventors minds were working overtime the French Patent Office became overloaded. The Staff being underpaid, overworked, lazy and deeply resent full of anything which interfered with playtime, decided to undercut the troublemakers by limiting or stopping certain categories of inventions.

A very large number of inventors were expanding on knowledge related to items listed above. Simply put, when level of output exceeds input , over unity is obvious. Bureaucratic French logic permitted declaring all inventions having greater out put than input a violation of the laws of nature and therefore would not be considered.

Edison and his mentors considered the items above as a mortal offense and through grants and other favors managed to control any and all information dispersed by publications, text books and sources of higher learning. In effect, greater output than input was deviously bastardized and continues today as the only acceptable official creed of the Establishment Accepting the establishments logic would disallow anything related to the Seven Items, above listed and thousands of other related useful Inventions. It's all a matter of very simple logic !!!!!!

20-B

Ultimate Energy Systems
By Don Smith

Simply put electromagnetic stimulation disturbs electrons by avalanche, dislodging adjacent electrons, resulting in more energy out than in. This system, if desired, becomes self sustainable. Another characteristic of the system is demonstrated by electromagnetic waves incoming from point " A ", the Sun, reacting with point " B", a solar cell which spins the local electrons side ways, yielding electricity. The electrons at point "A " are not the same electrons at point "B". All electrical systems, natural or man made work exactly the same, by fluxing and-or resonating magnetic waves.

Resonate fluxing provides the avalanche effect as does electromagnetic waves when reacting with an electron rich source. Examples would be dynode bridges, multipactors, masers and lasers. Without this type of action, radio type devices can not exist.

The Ultimate Energy Source, is then, a never ending Source, a fully self renewable source and does not compromise the World's Oxygen Supply. The Only Universally available Source which meets these requirements is the Electron. It's energy is accessed through Spin Technology as seen in coils and magnet systems and Avalanche Technology as demonstrated herein.

21

Physics of Avalanche

Avalanch Phenomenon is universal, undeniable and has profound effects when related to various displays of energy found in nature.

Head in the sand Physics varies from a religious (Quantum) experience to intellectual (Orgasmic) Posturing, in a profane attempt at ignoring facts.

Those with open Circuit tendencies allow for the possibility of a first rain drop, first snow flake, first shot in a war, first grain of sand in bringing down a mountain by avalanche., ad infinitum.

Those with closed Circuit (Minds) conveniently put the world in a box, what goes in, comes out, end of question !

The very same laws of nature which allow for one cell to become a human, one small explosion to ignite a hydrogen bomb applies to all matter. All matter and energy have a shared expressions of being, composing the real world and the related natural laws defining it.

A closed circuit world is a necessity for preserving the status quo and keeping the special interest, special. We are all familiar with a time in history when only the establishments viewpoint was permitted, anything otherwise was immediately destroyed (Dark Ages).
Only a slight glance will tell the most stupid that the dark ages is still with us very much present as relates to vested interest. No where is it more obvious than in the US. Patent Office. How much difficulty do you think General Electric Company would have in getting a Patent verses an individual ?
The deck and field is stacked against the individual and profoundly tilted so as to maintain those already in power. Greed must protect it's self.

Just as a stupid person repeats the same mistake over and over, the intelligent person learns from his mistakes and a wise person learns from mistakes of others, the world goes on.

Which brings us to the task at hand. The Field of Science profligates by notation, that which does not, is not. As an accommodation to the closed circuit and dark age constituency, herein is an offering of a piece of best Roquefort presently available.

The closed circuit coffin already has several stakes driven through it. Isaac Newton (1642 - 1727) with the help of Charles Coulomb (1736 -1806) provide us with the square laws which are based on an open circuit. Oliver Heaveyside (1850 - 1925) provided us with Energy equals mass times the square of it's acceleration. Albert Einstein (being an insider) took credit for Heavyside's work and the rest is history.

Mathematically Leonardo Fibonacci (1170 - 1230) of pisa demonstrated very clearly that, which directly translates from laws of nature, an orderly form of progression.

This same ordering of nature occurs in all radio type circuits. No where is it more obvious than in air core induction transformers and radio tubes.

Leaning on that which is already a part of history, the writer will propose that which is of nature :

Energy (a unit of being - static)
Energy Out (plus self acceleration - kinetic) *

* = must be squared at each stage of increase.

Simply stated, the unit of being , biological, physical or chemical when accelerated relates to Fibonacci and the square laws mentioned.

Energy and mass being one, indicates a very small portion can , and does initiate vast changes.

Scientific Notation Formulas to Be Supplied Later

(Input is solicited)

Dedicated Fax Line 281 - 370 - 4911

No Ans

The establishment's carpet has some rather large lumps under it. Coulomb's and Newton's inverse square law is politely ignored and it's opposite is allowed only the most abstract status. Without opposites we have no definition.

The source value of a remote flux reading, requires the squaring of the distance, times the remote reading, to obtain the original value. The opposite of this, being the derivations relate to Energy equals Mass times the Velocity constant squared. The electrical equivalent, being Energy equal capacitance times voltage squared and Energy equals induction times amperes squared. Flux lines increase as the law of squares and then activate electrons (energy) not previously a part of the sum. The cumulative capacitance and inductance as the distal ends of a Tesla coil are approached, results in energy greater than the input being present. This Energy is real when properly understood. It can be safely measured by magnetic flux methods and electrostatic voltmeters, based on the inverse square law.

As seen above flux lines result both from Induction-henrys-amperage and capacitance-coulombs-volts and define electrical energy. The non-linearity of this system does not obey ohm's law, which is replaced with impedance and reactance for alternating current systems. Impedance is the sum of the system resistance, which becomes zero at resonance. In a resonate induction systems, cycles per seconds increases, invokes a second round for the law of squares.

The degree to which flux lines are present, disturbs an equal amount of electrons, upsetting ambient, resulting in useful electrical energy. The frequency at which the disturbance occurs, obeying the law of squares further accelerates away from ambient, increasing the useful energy available. Two square law entities, flux density and frequency are envoked. Enter resonance which cancels the resistive effect.

Only that electrical energy above or below ambient is useful. For the Central U.S. going east to west, ambient as approximated by electro-static voltmeters and flux methods is on a solar quite day 200,000 volts.

At night time ambient drops to about one half the daytime value. On a solar active days it may reach more than five times that of a quite day. Ambient background energy at the polar regions is approximately 500,000 volts on a solar quite day. The background varies as relates to the north-south component and the east-west continuum.

This leaves us with an interesting problem. Electrons, when disturbed, first produce magnetic flux and then electrical flux when they spin back to their normal position. Therefore any electron movement produces above ambient energy, being over unity.

AN ANSWER TO AMERICA'S ENERGY DEFICIT

1995 EDITION

Donald L. Smith

Energy Consultant
8110 Bent Oak Lane
Spring, Texas 77379
(713) 370-4547

Donald L. Smith
TransWorld Energy
Spring, Texas 77379
January, 1997
Fax 218 - 370 - 4547
281 - 370 4547

Dear Reader;

Welcome to the Fifth Edition of " An Answer to Americas Energy Deficit" It has been translated and republished in all major languages, including Japanese, Chinese, Russian, German, Dutch, Finnish and numerous others. More than 30,000 copies are in circulation at this time.

The style is intended to be accessible without the usual intellectual posturing , common in the scientific community. It is intended to establish an intuitive base for understanding Electric Energy. Therefore, opening a new panorama of opportunity for the inventors to be.

Scientific Notation showing overunity, is and has been available prior to Newton. It is conveniently ignored for various reasons. The formulas are simple, honest and basically not very sophisticated.

Working from the knowledge base here presented the extensions thereof would do Tesla proud. Most of what is presented here has been in place since mid 1800's. Putting it in words so that it becomes useful has not.

Now that a burr has been placed under the saddle, perhaps more interesting things will occur. I wish you well for your voyage into the future.

An Answer to America's Energy Deficit

Donald L. Smith
Energy Consultant

Energy, energy everywhere and not a Joule to Jounce. Conventional wisdom, when properly tuned will appreciate the nature of energy, as here presented. The basic unit of electricity (the electron) upon encountering a moving magnetic field (or wave) spins, giving off an electric impulse. When this impulse collapses, it spins back to it's natural position, giving off a magnetic impulse. Therefore, magnetic and electric are two sides of the same coin. When the magnetic side is pulsed, it yields electricity and conversely, pulsing of the electrical side yields a magnetic field. Moving one in relation to the other produces useful energy. When done consecutively, each cycle pushes (current) forward, while pulling electrons into the system...in much the same way as a water pump moves water. These electrons are obtained from Earth and air grounding.

The word "electric" comes from the Latin word electron "amber". When rubbed, amber develops an electrical charge, which can be transferred to a dissimilar substance. During the seventeenth and eighteenth centuries, a great deal of attention was centered on this attribute of amber. Amber was used to differentiate the non-metals. Carbon-related substances and other non-metals, when subjected to friction, give up negative electrical charges. On the other hand, metals when subjected to friction, simply conduct the charge. It is important to note approximately 70% of the Earth's exposed crustal portions (surface) consist of silicone related non-metals (electron donors) and become a direct source of electrical energy when properly agitated.

Useful electrical energy can be obtained by grounding into the Earth's non-metal crust and into it's atmosphere as a natural source of electrons. These electrons have accumulated from the solar plasma during the aging of the Earth for more than 4.5 billion years, at a rate exceeding 3.9 exajoules per year. This indicates that the Earth's electrical field contains in excess of 17.6 X 10 to the 18th power of cumulative exajoules of energy. One Exajoule is the approximate energy equivalent of 125 million barrels of oil. The electrical energy in one display of lightning is approximately ten trillion joules. Each 24 hours, the land portions of Earth's surface yields in excess of 200,000 emissions, which involves more than 2,000 quadrillion watt seconds of active energy in the display. This physical phenomenon indicates the Earth's crust is an unending source of electrical energy. The surface area involved is a very small portion of the Earth's crust.

J.C. Maxwell (1891) suggested an active electron field gives rise to an associated magnetic field. Therefore, both are present with pulsating current. Early studies, involving observations of compass needles by microscopy, revealed the needle vibrates as with alternating current. More recent studies by A. Nishida and others, confirm alternating current is common in the Earth's crust.

In the early 1700's Benjamin Franklin demonstrated electricity flows from one point to another. This report by Franklin set forth the concept of positive and negative charges. The electron (negative) charge is the one which moves to the positive area where it is balanced out by the proton (positive) charge. This concept is flawed in several respects. The author's concept of the forces of nature differs from conventional. The concept is that of weak force and strong force, each being additionally composed of electrical, magnetic and gravitational (fields & waves). Any two of the three constitute the third member. Gravity "B" of the weak force competes with humans on a daily basis. Gravity "A" of the strong force is the force that holds the solar system and universe in place. Energy from electrons represents the weak force. Energy inside the atom represents the strong force "A". Controlled resonate induction of any two of the three, changes into the third and is the motor that runs the Universe. We see this in the electrically induced magnetic thrust against gravity in electric motors. Weak force is required to dislodge electrons and strong force (atomic) to dislodge protons. Unless dislodged, these particles are of little value in producing electrical power.

Therefore, the particle of importance is the negative electron. Electrons have a "grudging" relationship with other electrons. This is to say they like each other, but especially at arms length. Therefore, we see charges of the same potential repelling charges of the same gender...like charges repel and unlike charges attract. This can be demonstrated by taking two batteries of the same type but at different charge levels (unequal potentials) and put plus and minus ends facing the same direction. Then, with a volt meter, measure the electrical potential between the two negative ends and the two positive ends. It is obvious the "more negative" moves to the "less negative" is the correct concept for electrical energy generation.

C.F. Gauss (1777-1855) and H.C. Oersted (1777-1851) both were separately trying to define the Earth's electrical field with external influences removed. These external influences being solar quiet periods and being remote from the land's surface. The air electricity background which they measured varies with latitude. Their European measurements correspond to approximately the latitude of Washington, D.C. They were measuring magnetic field flux as an indicator of negative electron energy active and present. A related family of measurement are now presented. Units of measure used to define flux fields include Gauss (one unit = 100,000 volts), Oersted (one unit = 50,000 volts), Tesla (one unit = 10,000 Gauss) and Gamma (one unit = 1/10,000 of a Gauss). Much confusion exists in electrical related publications about these units. As presented here they are correct with values taken from their original definitions.

The entire surface of the Earth has been surveyed by aerial magnetometer, in most cases using gamma or nano teslas. One gamma is the magnetic flux equivalent of 10 active volts of electricity. When the data is corrected for flight height it becomes obvious that there are numerous areas where the gamma readings exceed one trillion gammas. Lightning strikes from the ground up are in that energy range. With knowledge of these electron enriched areas, the quality of Earth grounding, becomes enhanced. The correction necessary for land surface data when acquired from aerial magnetometer maps (using Coulomb's law) requires that the remote distance be squared and then multiplied times the remote reading. As an example, the remote reading is 1,600 gammas and the flight height being 1,000 feet. Take 1,000 X 1,000 = 1,000,000 X 1,600 gammas = 1.6 trillion gammas X 10 volts = 16 trillion volts equivalent for land surface data. Present day methodology requires mechanical energy in exchange for electrical energy. Once obtained, this energy is subject to Ohm's Law. Present Methodology obtains it's electrical energy from it's non-metal and air groundings.

This same energy can be obtained without the wasteful mechanical approach and at a much, much lower cost. Any required amount of electricity is available by resonate induction transfer from the Earth's magnetic and electrical fields. The major difference is in the functioning of Ohm's Law in relation to resonate circuits. In the resonate induction system here suggested, system resistance (Z) becomes zero at resonance.

Therefore, volts and Amperes are equal (V.A.R.) until work (load) is introduced.

Each cycling of this resonate induction system pulls in additional electrons from the Earth's electrical field, generating electrical energy in any required amount. In this system, a small amount of electrical energy is used to activate and pull into the system a much larger amount of energy.

This electrical advantage corresponds to the pulley and lever of the mechanical world. The electrical system here presented is extremely efficient. Using present methodology as a basis for comparison, with it's 60 Hz. per second system. The resonate induction system, cycling at 60 million times per second produces one million times the energy produced by the present energy systems. A single small size unit of the resonate induction system has more usable electrical output than a major conventional unit. The radio frequency energy here produced is easily changed to direct current, then to the present 60 Hz. per second system in preparation for commercial usage.

---

*Patent Pending # 08/100,074, Electrical Energy Generating System, 4 February, 1992.

Definitions: Joule is one watt for one second
One watt is one volt ampere
V.A.R. is Volt Amperes Reactive

Additional Reading:

Electricity and Magnetism
By B.I. Bleany and B. Bleany
Oxford University Press 1991
I.S.B.N. 0-19-851172-8

Engineering Electromagnetics
By W. H. Hayt, Jr.
McGraw-Hill 1989
I.S.B.N. 0-07-027406-1

Speech presented the evening of 23 July, 1994 at the International Tesla Society Convention at Colorado Springs, Colorado

DONALD L. SMITH
ENERGY CONSULTANT
8110 BENT OAK LANE
SPRING, TEXAS 77379
(713) 370-4547, (713) 251-5114 Fax

ELECTRICAL ENERGY REFERENCE POINTS
Electrical Energy Generating System
Patent Pending # 08/100,074, 2/4/92

The word "electric" comes from the Latin word electron "amber". When rubbed, amber develops an electrical charge, which can be transfered to a dissimilar substance. During the seventeenth and eighteenth centuries, a great deal of attention was centered on this attribute of amber. Amber was used to different the none-metals. Carbon-related substances and other none-metals, when subjected to friction, give up negative electrical charges. On the other hand, metals when subjected to friction, simply conduct the charge.
It is important to note that approximately 70% of the Earth's exposed crustal portions (surface) consist of silicone related none-metals (electron doners) and therefore becomes a direct source of electrical energy when properly agitated.

Useful electrical energy is obtained by grounding into the Earth's none-metal crust and into it's atmosphere as a natural source of electrons. These electrons have accumulated from the solar plasma during the aging of the Earth for more than 4.5 billion years, at a rate exceeding 3.9 exajoules per year. This indicates that the Earth's electrical field contains in excess of 17.6 X 10 to the 18 th power of cummulative exajoules of energy. One Exajoule is the approximate energy equivalent of 125 million barrels of oil. The electrical energy in one display of lightning is approximately ten trillion joules.
Each 24 hours, the land portions of the Earth's surface yields in excess of 200,000 emissions, which involves more than 2,000 quadrillion watt seconds of active energy on display.

Page 1

[Editor's Note: The first two pages here are almost identical to the previous later section which was expanded upon and then cut off and this continues for several more pages.]

ELECTRICAL ENERGY REFERENCE POINTS (Continued)

This physical phenomenon indicates the Earth's crust is an unending source of electrical energy. The surface area involved is a very small portion of the Earth's crust.

J.C. Maxwell (1891) suggested an active electron field gives rise to an associated magnetic field. Therefore, both are present with pulsating current. Early studies, involving observation of compass needles by microscopy, revealed that the needle vibrates as with alternating current. More recent studies by A. Nishida and others, confirm alternating current is common in the Earth's crust.

C.F. Gauss (1777-1855) and H.C. Oersted (1777-1851), both were separately trying to define the Earth's electrical field with external influences removed. These external influences being solar quite periods and being remote from the land's surface. The air electricity background which which they measured varies with latitude. Their European measurements correspod to approximately the latitude of Wasington, D.C. They were measuring magnetic field flux as an indicator of negative electron energy active and present.

A related family of measurement is now presented. Units of measure used to define flux fields include Gauss (one unit = 100,000 volts), Oersted (one unit = 50,000 volts), Tesla (one unit = 10,000 Gauss) and Gamma (one unit = 1/10,000 th of a Gauss). Much confusion exist in electrical related publications about these units. As presented here they are correct with values taken from their original definitions.

The entire surface of the Earth has been surveyed by aerial magnetometer, in most cases using gamma or nano teslas. One gamma is the magnetic flux equivalent of 10 active volts of electricity. When this data is corrected for fleight height it becomes obvious that there are numerous areas where the gamma readings exceed one trillion gammas. Lightning strikes from the ground up are in that energy range. With knowledge of these electron enriched areas, the quality of Earth grounding, becomes enhanced.

ELECTRICAL ENERGY REFERENCE POINTS (Continued)

The correction necessary for land surface data when acquired from aerial magnetometer maps (using the inverse square law) requires that the remote distance be squared and then multiplied times the remote reading. As an example, the reading is 1,600 gammas and the flight height being 1,000 feet. Take 1,000 X 1,000 = 1,000,000 X 1,600 gammas = 1.6 trillion gammas X 10 volts = 16 trillion volts equivalent for land surface data.

INSERT AND DISCUSS, Aerial Magnetometer Map GP-948 East-Central United States

Present day methodology requires mechanical energy in exchange for electrical energy. Any required amount of electricity is available by resonate induction transfer from the Earth's magnetic and electrical fields. Each cycling of this resonate induction system pulls in additional electrons , generating energy in any required amount. A small amount of electrical energy is used to activate and pull into the system a much larger amount of energy.

ENERGY VERSES MASS

INSERT AND DISCUSS ELECTRICAL ENERGY

Steady State	Unsteady State
Static "Pre-Energy "	Kinetic "Energy"
Mass attracts Mass, Gravity Dominates	Expanding, Magnetic Energy Dominates
	Electrons moving apart
	Pressure decreasing
	Cooling effect dominates
	Less scattering of Energy
	Negative resistance
	Contraction, Electrical Energy Dominates
	Electrons moving together
	Pressure increasing
	Heating effect dominates
	Scattering of Energy
	Positive resistance

Functions of active Electrons

Electrons become active when placed inside the critical distance allowed by their negativity.

Active Electrons provide:

1. Electricity
2. Magnetics
3. Gravitational thrust as in Electric Motors
4. The source of Visible Light
5. It's charge is Negative

They move in a closed loop as seen in the Icon for infinity, not in a circle as shown in many books.

One half of the loop consist of a magnetic impulse and the return half consist of the electrical impulse. This is seen as the classic sine wave of alternating electrical energy.

A flash of light occures when two electrons suddenly find they are too close to gather. Daylight results from the impengement of Electrons in the Earth's atmosphere with the Electrons of the Solar Plasma.

My Concept of the Forces of Nature differs from the conventional. It consist of a weak and a strong force, each being additionally composed of electrical, magnetic and gravitational (fields & waves). Any two of the three constitute the third member. Gravity "B" of the weak force competes with humans on a daily basis. Gravity "A" of the strong force is the force that holds the Solar System and the Universe in place. Energy from the Electrons represent the weak force. Energy inside the Atom represents the strong force "A". Controlled resonate induction of any two of the three, changes into the third and is the motor that runs the Universe. We see this in the electrically induced magnetic thrust against gravity in electric motors.

Weak force is required to dislodge electrons and strong force (atomic) to dislodge protons.

Unless dislodged, these particles are of little value in producing Conventional Electrical Energy.

Functions of active Electrons (Continued)

INSERT AND DISCUSS:

1. Electrical Energy with Associated Phenomena
2. Energy Acquired by Magnetic and Electrical Impusing
3. Electrical Energy

Therefore, in conventional electrical energy production, the particle of importance is the negative electron. Electrons have a "grudging" relationship with other electrons. They like each other, especially at arms. length. Like potentials repel each other and unlike potentials attract. To demonstrate this, take two batteries of the same type, but of a different charge level (unequal potentials). Put the plus and minus ends facing the same direction. Then with a volt meter, measure the electrical potential between the two negative ends and then the two positive ends. It is obvious that the " more negative " moves to the " less negative ", is the correct concept for electrical energy generation. Electrical Energy flow consist of a higher concentration of electrons moving to an area of lesser concentration.

OHM'S LAW WITH CORRECTIONS:

A major obstruction in reference to the correct function of electrical energy is the establishment's incorrect interpretation of Ohm's Law. The corrected version is:

Volts = Energy Available (Potential)

Ohm = Scattering, dissipation of Energy (Load)

Ampere = the rate of, dissipation / scattering of energy

It is important to note that Ohm and Ampere are after the fact, and are not decisive except for the dissipation factor. High Voltage at low amperage simply means that the High Voltage is still intact for future usage. In no way is the potential diminished by low amperage.

EXAMPLES OF OVERUNITY

Dominos did not exist in England when the Laws of Conservation were originally put in place. Otherwise they might have been very different. For example let us take a long row of upright dominos, (many thousands) and flip number one. The Energy required to flip # 1 must now be added with that of thousands more in order to have a correct assessment.

The Electron it's self is an excellent example of overunity. The electron provides various forms of energy continuously throughout eternity and is in no way dimenished. It simply cycles through the system and is available thereafter.

In Electrical Systems, Electrons active at point "A" are not the same Electrons active at point "B". That is to say, the Electrons activated at the Central Electrical Energy Station are not the ones used at your house. When you ground your system by flipping the wall switch, you use your own electrons. In closed energy systems, electrons communicate with and and replicate the activity of the overbalanced potential, when provided with Earth and or Air Groundings.

The number of Radios and Televisions running at any one time do not diminish, in any way the electrical output of the source station.

For example, let now use use an Air Coil Resonate Induction System for the purpose of flipping some electrons.
The flipping device (reactor L-1 Coil) is pulsed which then provides a resonate induction pulse, in turn this flips the electrons present at the (reactant L-2) Coil. The energy input in L-1 is devided by the number of turns present. The induced magnetic pulsing in turn flipps the electrons in each turn of L-2. If more turns are present in L-2 than L-1, there is a net gain in the Energy present, as demonstrated by the dominos above. The farads and henrys of the resonate system provide the resonate frequency when pulsed by an external energy system. A system shunt in the resonate circuit sets the containment level for energy potential.

EXAMPLES OF OVERUNITY

Dominos did not exist in England when the Laws of Conservation were originally put in place. Otherwise they might have been very different. For example let us take a long row of upright dominos, (many thousands) and flip number one. The Energy required to flip # 1 must now be added with that of thousands more in order to have a correct assessment.

The Electron it's self is an excellent example of overunity. The electron provides various forms of energy continuously throughout eternity and is in no way dimenished. It simply cycles through the system and is available thereafter.

In Electrical Systems, Electrons active at point "A" are not the same Electrons active at point "B". That is to say, the Electrons activated at the Central Electrical Energy Station are not the ones used at your house. When you ground your system by flipping the wall switch, you use your own electrons. In closed energy systems, electrons communicate with and and replicate the activity of the overbalanced potential, when provided with Earth and or Air Groundings.

The number of Radios and Televisions running at any one time do not diminish, in any way the electrical output of the source station.

For example, let now use use an Air Coil Resonate Induction System for the purpose of flipping some electrons.
The flipping device (reactor L-1 Coil) is pulsed which then provides a resonate induction pulse, in turn this flips the electrons present at the (reactant L-2) Coil. The energy input in L-1 is devided by the number of turns present. The induced magnetic pulsing in turn flipps the electrons in each turn of L-2. If more turns are present in L-2 than L-1, there is a net gain in the Energy present, as demonstrated by the dominos above. The farads and henrys of the resonate system provide the resonate frequency when pulsed by an external energy system. A system shunt in the resonate circuit sets the containment level for energy potential.

EXAMPLES OF OVERUNITY DEVICES (Continued)

Let us take a closer look at resonate induction. As an example, let a room full of ping pong balls randomly bouncing at a high speed represent the Conventional method of underunity energy generation.

Suppose that by resonate induction the balls all move in the same direction at the same time. When this occures a huge amount of energy not previously available is present. The resonate air core coil system lines up the electrons in such a manor that the energy factor is near 100 % , not 2 and 3 % as in Conventional underunity devices sanction by the establishment.

Some other devices where overunity is common would be resonate induction circuits present in conventional radio tubes (high plate voltage), negative feedback systems found in Op-Amps and possibly others.

SUMMARY

Useful electrical energy is achieved when the electron density at point "A" becomes greater than at point "B", (being the more negative moving to the less negative concept). Coils moving through a magnetic field or vice versa causes this imbalance.

The mind set of the professional electrical engineer is restricted to none-resonate and iron core coil resonate systems. Ohm's Law when applied to resonate air core induction systems, becomes, system resistivity (impedence, Z). "Z" at resonance becomes zero. Therefore, in this system, volts and amperes are equal until load (resistivity) is introduced. This is called the Volt Ampere Reactive (V.A.R.) System. With impedence being zero, the System grounding is coupled directly into the Earth's immense electrical potential.

Efficency of induction relates to the square of the cycles per second. Compare the ratio of the conventional 60 c.p.s. System and the 220 million plus cycles of my Earth Electrical System II.

SUMMARY (Continued)

Electrons which cycle through this system, after being used, are returned in tact to their former state for future usage.

Electron spin causes electrical current and magnetic lines of force.

The effect of current results from the unequal distribution of negativity (electrons).

Magnetic imbalance causes the gravitational effect. This is evidenced in electric motors by magnet-gravitational displacement of mass which causes the motor to rotate.

The System is an extension of present technology

The System and it's source utilizes magnetometer studies.

This System (Earth Electrical System II, E.E.S. II) utilizes a fully renewable energy source.

This System utilizes a non-poluting energy source.

This System utilizes an universally available energy source.

Endorsement and Certification of The System can be anticipated by States with pollution problems.

ELECTRICAL ENERGY WITH ASSOCIATED PHENOMENA

1. The effect of current results from the unequal distribution of negativity (electrons).
2. Electron spin causes electrical current and magnetic lines of force.
3. Magnetic imbalance causes the gravitational effect. This is evidenced in electric motors by magneto-gravitational displacement of mass which causes the motor to rotate.

ENERGY LINES OF FORCE FIELDS & WAVES *

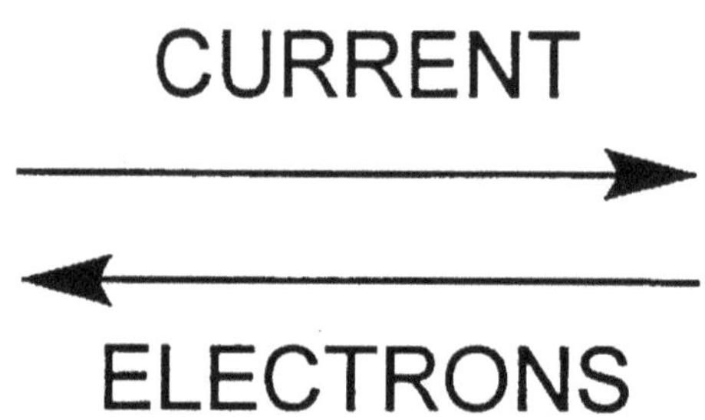

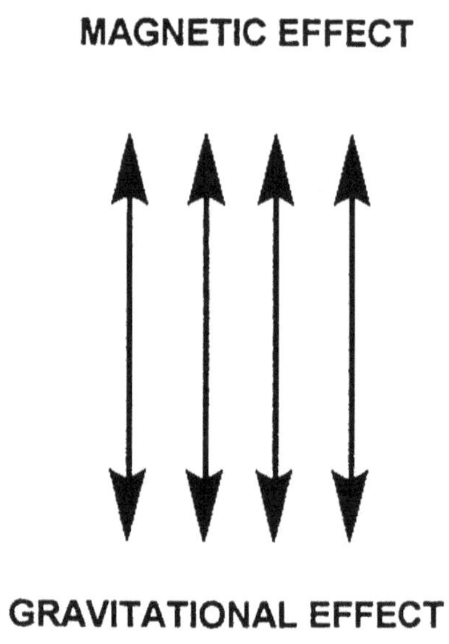

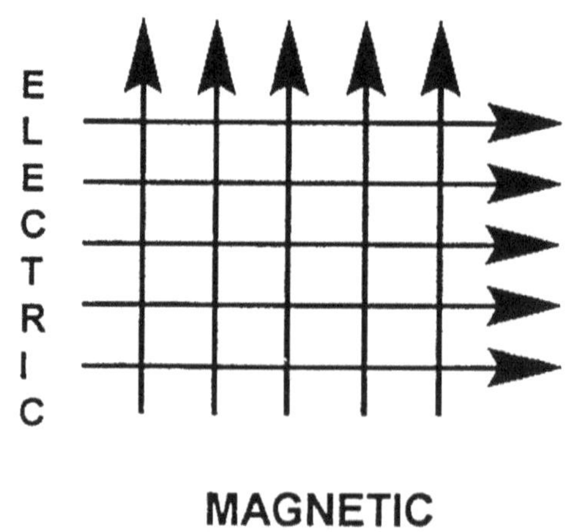

* Below 20,000 Hertz per second = Fields
Greater than 20,000 Hertz per second = Waves (radio frequency).

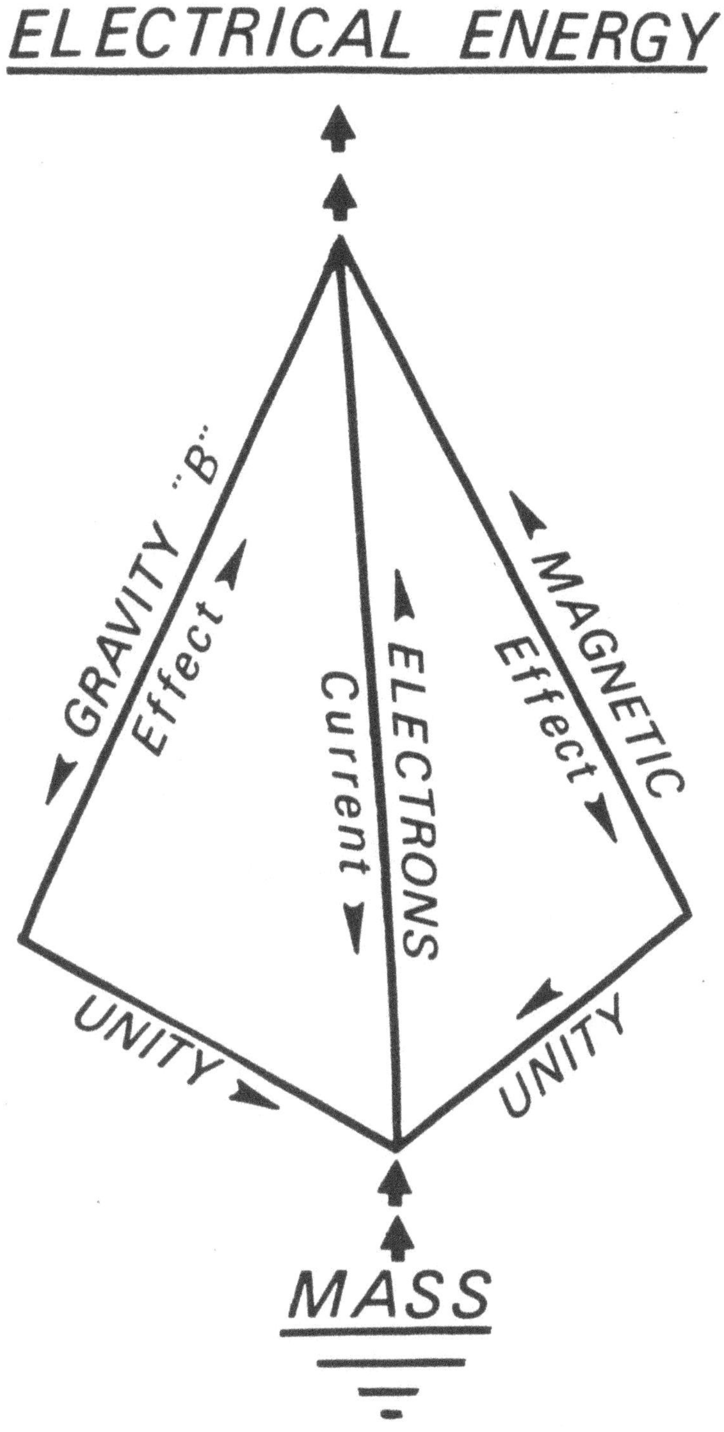

$$\mathcal{M}C^2 = \epsilon$$

D.L.S., 93

9

ENERGY ACQUIRED BY MAGNETIC AND ELECTRICAL IMPULSING

(ALTERNATING RESONATE ELECTRICAL POWER)

["As it turns out there's good evidence that some electrons have more negative charge than others. And being more negative or less negative that that's the same as being plus or minus. So therefore the electron with a weaker charge will attract one with a stronger charge, and they will be orbiting in each other. They're relatively easy to pull apart. So the electrical and magnetic charge of the electron, of course, when you spin them, spin them right you've got a magnetic field. When you spin them left you get an electrical field. So when they spin apart you're getting a benefit of both of those. And then whenever they come spinning back together you're getting the volt's and the magnetic field coupling again, which means your amps, amps times volts equal watts, or you've got useful energy."]

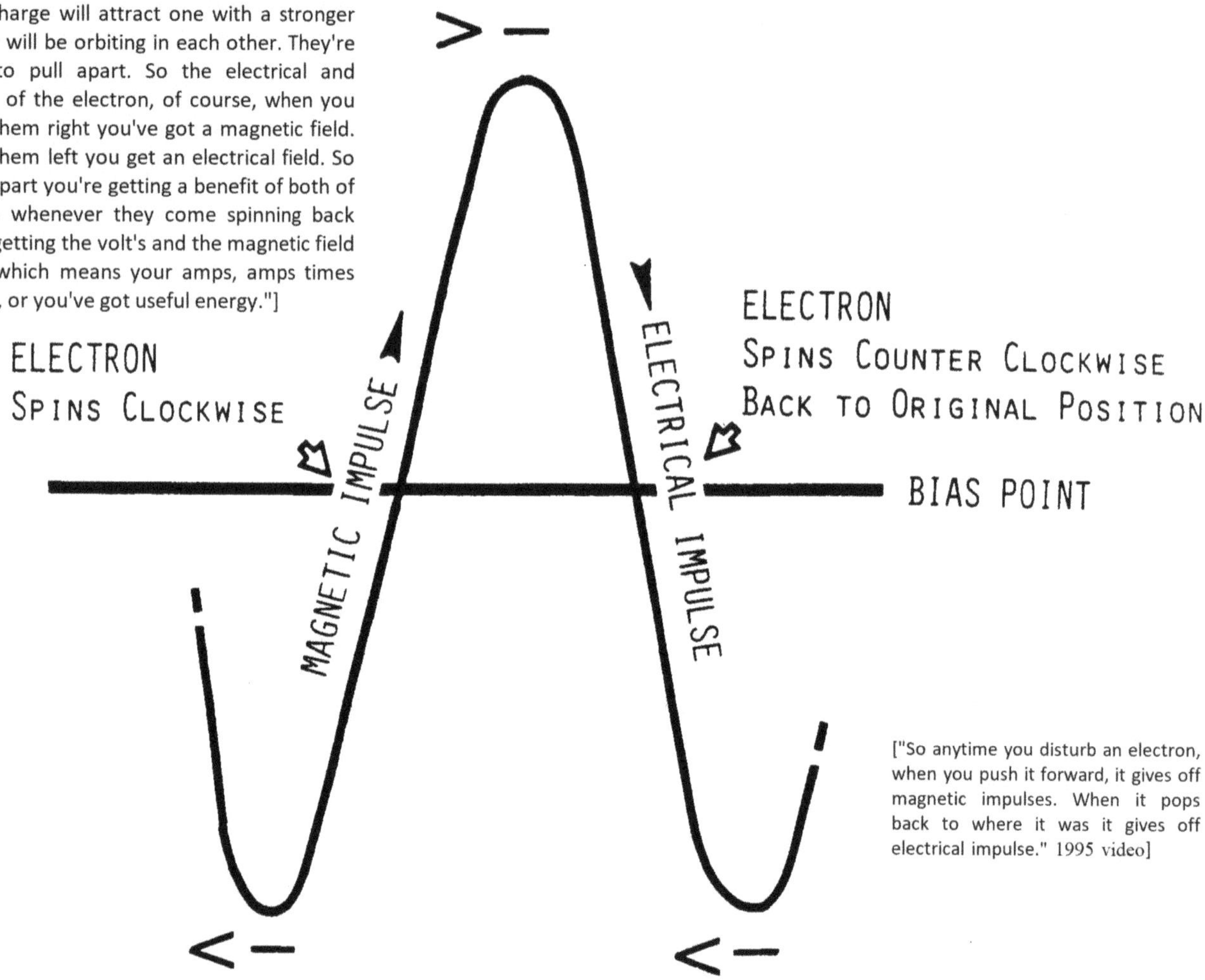

["So anytime you disturb an electron, when you push it forward, it gives off magnetic impulses. When it pops back to where it was it gives off electrical impulse." 1995 video]

EXPECTED ENERGY

$$\frac{\text{Volts}}{\text{Ohms}} = \text{Amperes}$$

RESONATE ENERGY

$$\frac{\text{Volts}}{\text{System Ohms}} = \text{Amperes}$$

SYSTEM OHMS BECOMES ZERO WHEN RESONATE.
THEREFORE VOLTS AND AMPERES BECOME EQUAL, BEING VOLT AMPERES REACTIVE (V.A.R.).

10

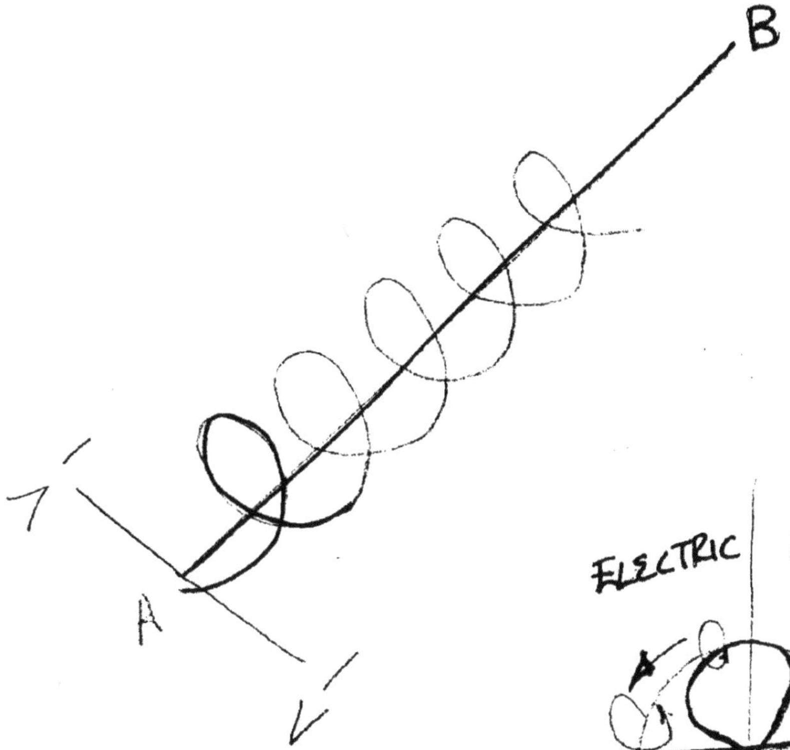

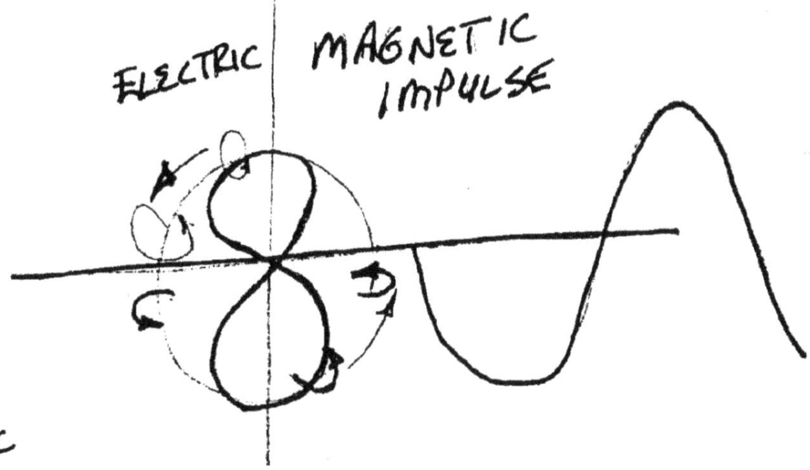

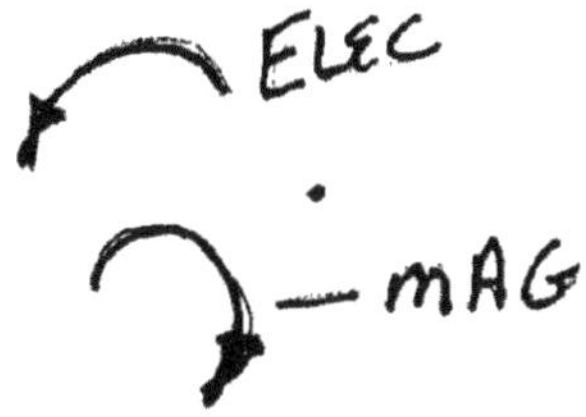

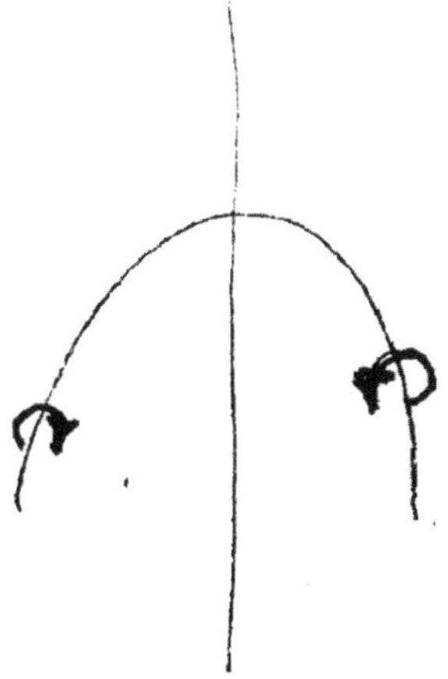

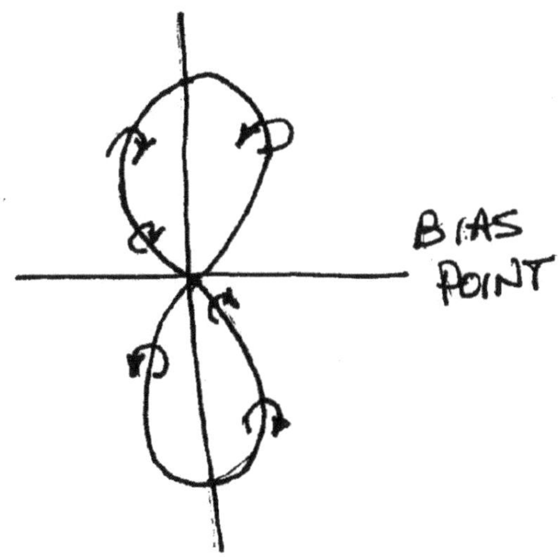

13 SEPT 98

QPX

Earth Electrical System I I

Information & Concept

Donald L. Smith

ENERGY CONSULTANT
8110 BENT OAK LN.
SPRING, TEXAS 77379
PH: (713)370-4547

ABSTRACT:

Earth Electrical System II, Power Generation

Useful electrical energy is achieved when the electron density at point "A" becomes greater than at point "B", (being the more negative moving to the less negative concept).

Coils moving through a magnetic field or vice versa causes this imbalance.

The mind set of the Professional Electrical Engineer is restricted to none resonate and iron core coil resonate systems. Ohm's Law when applied to resonate air core induction systems, becomes system resistivity (impedence, Z). "Z" at resonance becomes zero. Therefore, in this system, volts and amperes are equal until load (resistivity) is introduced. This is called the V.A.R. (Volt Amperes Reactive) System.

With Impedence being Zero, the System is then Coupled directly into the Earth's immense electrical potential.

Efficiency of induction relates to the square of the cycles per second. Compare the ratio of 60 cycles per second with the 200 million plus cycles per second for the system here presented.

Electrons which cycle through this System, after being used are returned in tact to their former state for future usage.

This System utilizes a fully renewable energy source.

This System utilizes a none poluting energy source.

The Evidence Against Under Unity

1. Use of Logarithmic Scales on electrical measurement instruments. Linear measurement works fine where Ohm's Law applies (direct current). In alternating current ohms are replaced by impedence and the measurements become non linear.
2. Infinite "Q" at resonance confirms that voltage and amperage is squared, as in the kinetic energy formula. See the formulas of this report.
3. Square waves are clipped infinite "Q"s.
4. Maxwell and others show that magnetic-inductance-amperage and electrical-capacitance-voltage are two sides of the same coin, Magnetic-inductance is directly equal to amperage. Both obey the Law of Squares, which has over unity built in.
5. Magnetic and electrical flux are present in enormous amounts at the distal ends of an operating Tesla Coil.
6. Ignorance in how to measure and relate magnetic and electrical flux, is the chief weapon of the under unity gaggle.
7. The Cumulative inductance and capacitance of the Tesla Coil grounds it's self out if not properly utilized. See this report for the temporary energy storage accessible, if properly managed.
8. The Patent Office refers devices related to over unity to their metering group, which is a sure indication that they are aware and accept the logarithmic measuring devices.

This is direct and absolute evidence that they accept the square law as relates to kinetic energy. This also indicates they are aware that over unity exist. Since their bureaucratic brain is improperly motivated they continue to badger inventors who are working in the over unity arena. Their level of intellectual dishonesty is sanctioned by, and is a real part of doing business with a government which prides it's self in being a hooliganistic bureaucracy.

E.E.S. II, BACKGROUND INFORMATION & CONCEPT

With alternating electrical current, electrons do not move from point "A" to point "B" as commonly envisioned! Electrical potential (oscillating electrons) at point "A" results in harmonic electron activity at point "B", when the grounding switch (circuit) is closed. That is to say, point "B" supplies it's own electrons and mirrors the activity of point "A". Impulsing (turbulence) by magnetic induction causes electrons to be pulled into the system, which then oscillates. When the magnetic field collapses (becomes absent) the electrical potential returns to it's natural background.

Several major flaws are present in the conventional 60 Hertz's per second method of electrical power generation and it's iron core transformer system. ***This system is handcuffed by the inverse relationship of volts to amperes.*** This represents a stodgy inflexible inheritance, courtesy of Mr. T.A. Edison and his concept of electrical power generation.

Nikola Tesla stood, almost alone, against Edison and managed to prevail with his alternating current system. Without the alternating current system, electronic things in the modern sense would not exist.

This report will be concerned with some of the extensions and benefits of the alternating current electrical system. ***This study will limit it's scope to air core coil transformers at radio frequency and upward. The electrical power produced by this method is inverted to direct current and then to alternating current as required for popular usage.*** There are several important advantages of this system over conventional power generation.

Start with two coils (separate-apart), one being a reactor coil (L-1) and a second coil (L-2), being the reactant coil. Magnetic field fluxing (off-on of the electrical source) causes inductive reactance of L-1 which replicates by induction in L-2. Pulsing of the magnetic field (from L-1) in the presence of L-2, generates electrical potential. For example, should the L-1 coil have ten turns, with an imposed A.C. potential of 1,200 volts. This results in each turn of L-1 acquiring 120 volts of potential. This induced magnetic field, then replicates itself in each turn of the L-2 coil. The L-2 coil may have one or many hundreds of turns. Modern encapsulation techniques makes high frequency and high energy controllable.

Let's take another important step in this air core transformer process. For purpose of discussion, let the value of inductive reactance at 60 Hertz's per second equal one. Each time the Hz's. are doubled, the effectiveness of induction is squared. At about 20,000 Hz., ***when radio frequency is achieved, the electrons begin spinning free, outside of the inductor. They become increasingly free of the inverse relationship of volt-amperes.*** From this point on, they replicate by the inductive

process as V.A.R.. ***That is to say, volts and amperes are equal, until resistance (work) is introduced.***
Therefore, additional, not previously available electrons become incorporated for a very large net gain in potential. This gain is real!

The quality of the grounding system determines the effectiveness of this method of producing electricity. A handy reference to locate the negative grounding areas for power generation can be found in the Aeromagnetic Map Studies of the US. Geological Survey. They provide an excellent method for locating the best sites for optimum negative grounding areas.

When this method is combined with the induction coil system, already described, it provides an electrical power generating system millions of times more efficient than any known conventional method.

This new system (E.E.S. II) is uncomplicated, small in physical size and inexpensive to build. Technology required already exists. Maintenance is near zero, as there are no moving parts. Once operating, this system could last forever.

Small mobile E.E.S. II units are presently available as replacements for the batteries used in electric automobiles. Larger E.E.S. II units can be provided as a replacement source of power for hotels, office buildings, subdivisions, electric trains, manufacturing, heavy equipment, ships, and; generally speaking; in any present day application of electrical power.

Earth Electrical System II, Modular Units

The system consists of three separate modules. Reverse engineering is used in matching the modules to the desired usage.

HIGH VOLTAGE INDUCTION TRANSFORMER MODULE:

1. Preferably an off-the-shelf-unit similar to a TV flyback and/or automobile ignition type related coil (transformer).
2. Ratio of input to output may be from less than 1/100 to greater than 1/1,000. A voltage tripler may then be used.
3. A connection allowing the high voltage output to pass onward through the induction coil L-1 and then to it's grounding.

AN AIR CORE INDUCTION COIL TRANSFORMER MODULE:

1. Two coils, the reactor coil L-1 and the reactant coil L-2. L-1 has a high voltage radio frequency capacitor between it and it's grounding.
2. Input into the L-1 inductor is divided by the number of turns therein. The magnetic flux field provided from each turn of L-1 replicates itself as an electrical potential in each turn of L-2.
3. L-2 may have one turn or many hundreds of turns. The net gain depends upon the number of turns in L-2. Output from L-2 is in V.A.R. ***With this type of output, volts and amperes are the same until work(resistivity) is introduced***

THE INVERTER MODULE:

1. Inverts to direct current (D.C.)
2. Inverts to alternating current (A.C.), as desired.
3. Provides customized output of electrical power ready for designated usage.

Alternating Impact Energy Transfer Analogy

Energy Transfer by Swinging Steel Balls

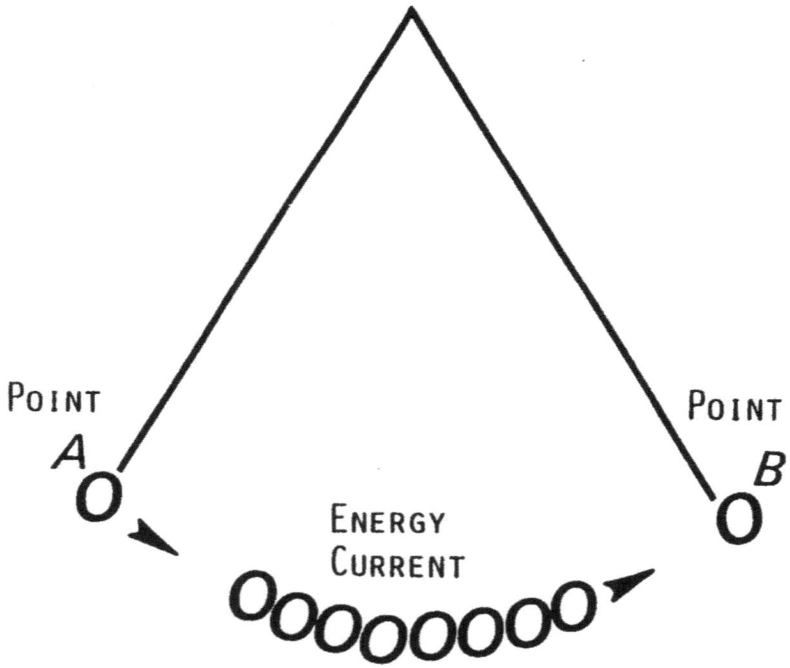

Electrical Energy Transfer by Induction Pulsed (Impacted) Electrons

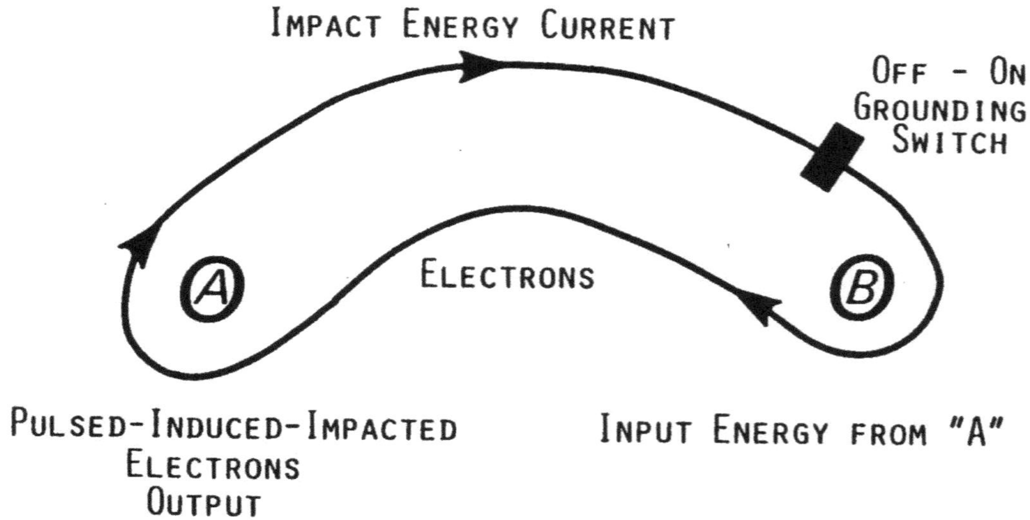

In Alternating Electrical Energy Systems, Only the Potential-Impact Level Replicates It's Self at Point "B". Electrons do not Travel From Point "A" to Point "B".

Earth Electrical System II, Domestic Use Range Module
Up to Two Million Volt-Amperes-Reactive Output *
Plan "A", With Variable Controls
Not to Scale

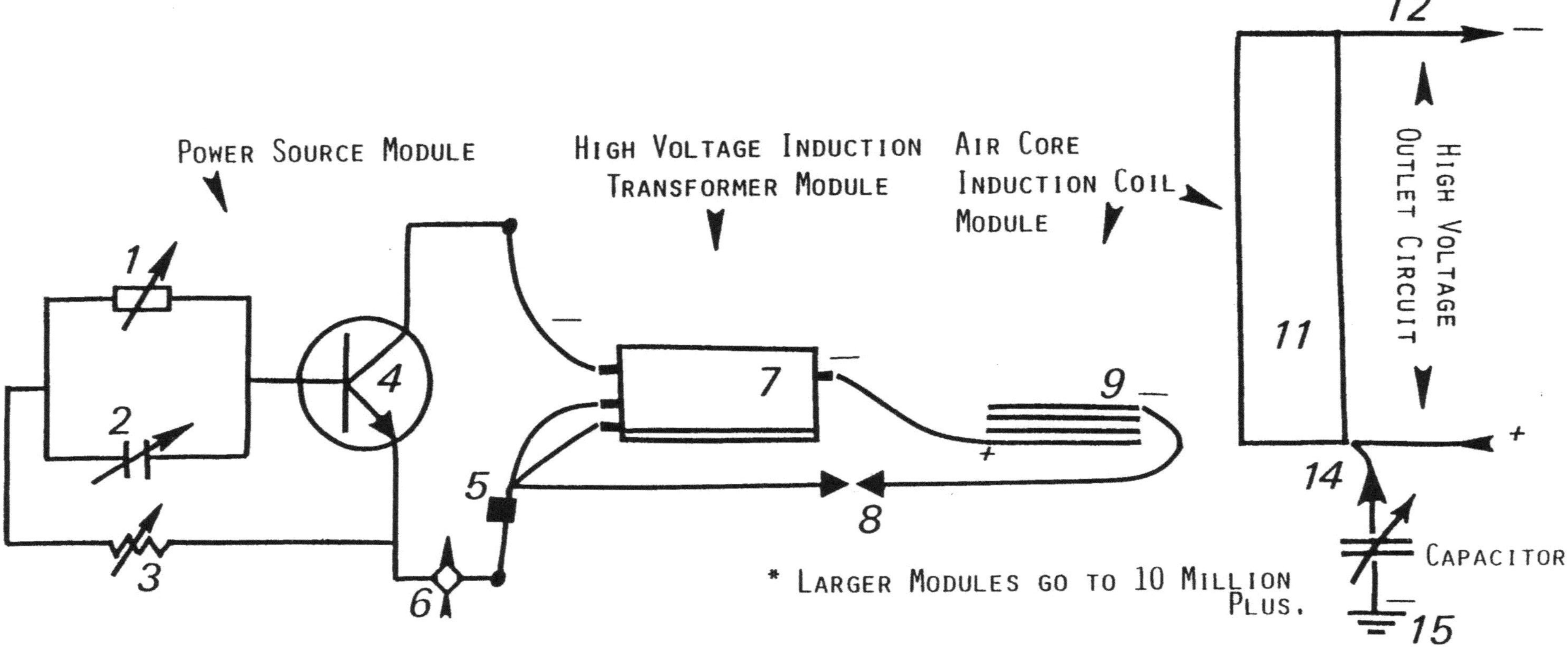

Parts: 1. Coil, Variable, 2. Capacitor, Variable, 3. Resistor, Variable, 4. Transistor, R.F., 5. Battery, Rechargeable, 6. Off-On Switch, Variable, 7. High Voltage Transformer, 8. Feed Back with Spark Gap, 9. Reactor, Induction Coil, 10. Feed Back with Spark Gap, 11. Reactant Coil, 12. Output for # 11, 13. Input for eleven, 14. Grounding for eleven.

5

Earth Electrical System II, Domestic Use Range Module
Up to Two Million Volt-Amperes-Reactive Output *
Plan "B", Electrical Automobile Energy Source

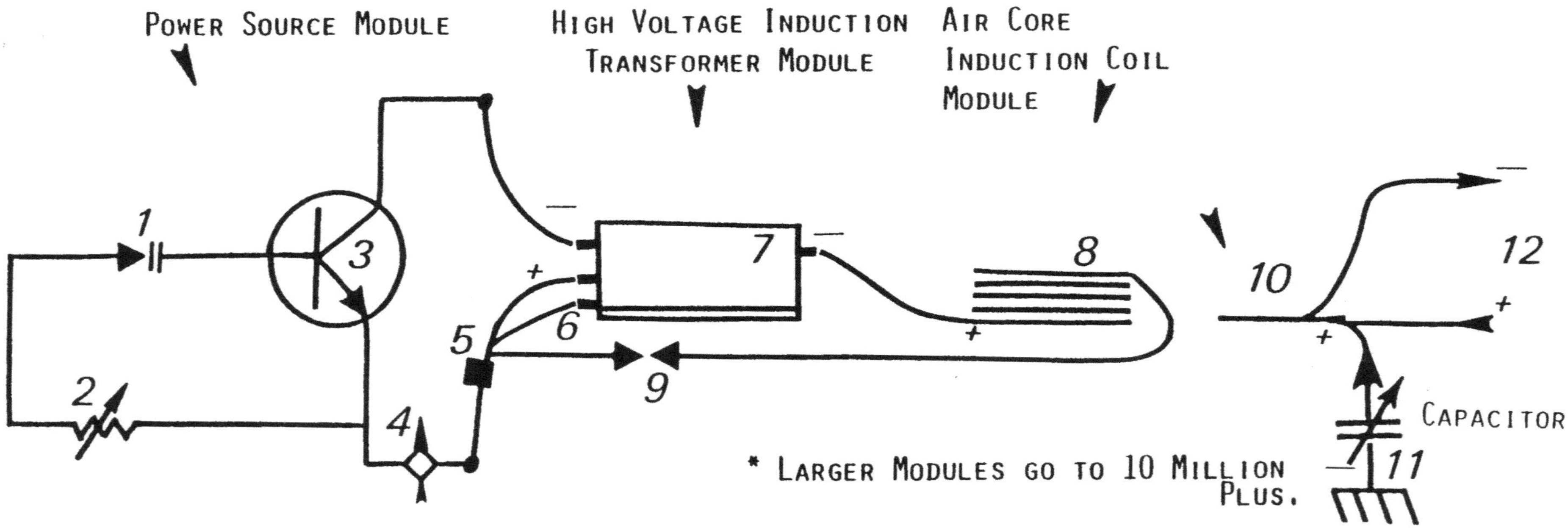

* Larger Modules go to 10 Million Plus.

Parts: 1. Varactor, Radio Frequency, 2. Resistor, 3. Transistor, Radio Frequency, 4. Off-On Switch, Multi-Position, 5. Battery, Rechargeable, 6. Transformer Grounding, 7. High Voltage Induction Transformer, 8. Reactor, Induction Coil, 9. Feed Back with Spark Gap, 10. Reactant, Induction Coil, 11. Grounding for # 10. Out-put Circuit, In Volt-Amperes-Reactive.

Memo

UNIQUE ASPECTS OF E.E.S. II

1. THE SYSTEM UTILIZES A FULLY RENEWABLE ENERGY SOURCE.

2. THE SYSTEM UTILIZES A NON POLLUTING ENERGY SOURCE.

3. THE SYSTEM UTILIZES AN UNIVERSALLY AVAILABLE ENERGY SOURCE.

4. THE SYSTEM IS SAFE. PARTS OF TESLA COIL SYSTEMS ARE LIFE THREATNING.

5. IT IS AN EXTENSION OF PRESENT TECHNOLOGY.

6. THE DISRUPTIVE DISCHARGE OF THE TESLA COIL IS ABSENT.

7. IT IS BASED ON THE MORE NEGATIVE TO LESS NEGATIVE CONCEPT.

8. ELECTRONS WHICH CYCLE THROUGH THE SYSTEM, AFTER BEING USED ARE RETURNED IN TACT TO THEIR FORMER STATE FOR FUTURE USAGE.

9. THE VOLT AMPERES REACTIVE CONCEPT APPLIES.

10. E.E.S. II IS LESS COMPLICATED THAN TESLA.

11. IT USES A DIFFERENT CONCEPT OF GROUNDING THAN TESLA.

12. THE PHYSICAL SYSTEM IS MUCH SMALLER THAN TESLA.

13. IT'S RATE OF ENERGY PRODUCTION AND EFFICIENCY IS MUCH HIGHER THAN TESLA.

14. EFFICIENCY OF INDUCTION RELATES TO THE SQUARE OF THE CYCLES PER SECOND. TESLA COILS OPERATE NORMALLY BELOW 200,000 C.P.S., IN THE 40,000 TO 50,000 C.P.S. RANGE. FOR A DIRECT COMPARISON SQUARE THE TESLA C.P.S. AND THEN THE MEGAHERTZ (100 PLUS) OF E.E.S. II . THE RATIO OF THE TWO DEMONSTRATES THE ACQUIRED IMPROVEMENT.

15. THE SYSTEM AND IT'S SOURCE UTILIZES MAGNETOMETER STUDIES.

Electron Related Energy

	Energy Available	Method of Storage	Comon Unit	Units of Measure
Electrons	Electrical	Capacitor	Volts & Amps.	Coulombs
	Spin	Momentum		Ergs
	Magnetic	Coils	Units of Flux	Teslas, Gauss, Oersteds & Gammas
	Light	Laser	Lux	Photons
	Heat	Various		Temperature

Useful Energy occurs when the Energy availale differs from the surrounding [ambient]. Therefore when any variation above or below the ambient results in movement towards the ambient.

When Electrons are disturbed by turbulence-movement of any sort they become active and try to regain equalibrum. This movement towards equalibrum results in magnetic displacement, which causes the motor to rotate.

Foreword thrust of Electrons results in Magnetic Energy and backword thrust results in an Electrical Energy. Impacted Electrons result in Visible Light.

Pumping of Electrons by Resonate magnetic induction results in the accumulation of Electrons exceeding the ambient, therefore usefull Electrical Energy Occurs. Electrons entering an active energy system are obtained from it's earth and air groundings. The energy system may be activated by external methods. These methods may be mechanical and-or electrical, as desired.

Let the electrons accessed by conventional methods be represented by a room full of ping pong balls randomly bouncing. It is obvious that most of the energy present is cancled out by the electrons going in oposing directions. This is the sanctiond method of under unity energy production.

In the Resonate Induction System here presented, the electrons are all going in the same direction at the same time, resulting in a huge amount of energy which is not available in the establishments idea of Energy.

TransWorld Energy
8110 Bent Oak Lane
Spring, Texas 77379

16 May, 2001

Phone/Fax 281-370-4547

Web Site: altenergy-pro.com
E-mail donsm1@earthlink.net

Dear Reader,

TransWorld Energy is dedicated to improving the Human Condition in the Field of Energy which at the same time makes possible Healthy Water and increases the Food Supply. A never ending source of energy found throughout the universe is easily accessed with a minium of effort and cost. The technology for doing this has been around since the 1820's. Selfish special interest has made sure that the technology remains discredited. People who control the Energy Sources control the World.

Extensive research and development by TransWorld and Associates has been progressing for more than 15 years. Numerous successful Energy Producing Devices have been produced and demonstrated throughout the World. Some of these can be viewed by the Web Site located at all major search engines (Such as Lycos, Yahoo, Altavista, NorthernLight and more than 2,000 others throughout the World) .

The Book which You are viewing has more than 40,000. Copies in circulation. It has been translated and distributed in all major languages including Japanese, Arabic, Portuguese, French, Italian, Russian, Chinese, German, Spanish and many more. There are seven editions in circulation. An enormous interest is evident in the subject matter. An average of fifty e-mails per day from the ends of the Earth (about 1,500 per month).

Once the Web Site and the book are viewed, it will become evident that abundant, self sustainable energy is available every where for the taking. This is natural energy which does not harm the environment or those using it. The proper Device for Collecting is all that's required.

The Good News is that the problem is solved and with assistance an ultimate source of energy that is environmentally benign, in abundance throughout the universe and inexpensive to capture is there for the taking.

Thank You for Consideration,

Donald L. Smith, CEO

MEM

Electrical Energy Generating System

Description and Function:

Generation of Electrical Power requires the presence of electrons with various methods of stimulation, yielding magnetic and electrical impulses, collectively resulting in Electrical Energy (Power). In place of the mechanical - coils and magnet system, present in conventional electrical power generation, visible moving parts are replaced by resonate magnetic induction, using radio frequency. Transfer of energy by resonate induction is related to the ratio of the square of the cycles per second.

The Energy System, here presented, operates at millions of cycles per second verses the conventional 60 C.P.S.'s. This tells us that it has a size advantage over conventional methods. The same advantage applies to the amount of electrical energy output. Therefore the Device is small in size and produces large amounts of Electrical Energy. The Electrons acquired are from the surrounding Air and Earth Groundings, being the same source as in conventional methods. This is accomplished by magnetic resonate radio induction.

Applications:

This Electrical System adapts nicely to all Energy Requirements. It is a direct replacement to all now existing Energy Systems. This includes such things as Manufacturing, Agricultural, Home Usage, Office Complexes, Shopping Centers, Rail Transportation, Automobiles, Electrical Power Grids, Municipalities, Subdivisions, and Remote Areas. Briefly, only the imagination is the limiting factor.

Economic Possibilities:

No Historical Reference Point exist for a comparison of the Possibilities of this System. One can see from the impacted applications listed above that the magnitude exceeds any known invention, presently a part of the Human Experience.

Present and Future Plans:

The Energy System has been in the developmental stage during the past seven years. It is Patent Pending # 08/100,074 with the Patent Office. No prior art exist based on the Patent Office response. The System is presently being introduced into the World Market.

ELECTRICAL ENERGY SYSTEMS PREFACE

Useful Electrical Energy is obtained directly from electron spin induced by incoming magnetic waves or indirectly through mechanical exchange as in dynamo type devices. Simply put electron spin converts from magnetic to electrical energy and vice versa. Nature for free provides grand scale magnetic wave induction through out the universe. In Electrical Systems movement is at right angle to the direction of current movement. This explains the rotary movement of the Earth and other related Systems. The rate of Spin for the Earth is known as well as the mass (5.98 X 10 to the 24 th. kg)* therefore the amount of Electrical Energy incoming which produces this action can be calculated. It can be easily be seen that the incoming magnetic wave energy is Vast and Continuous. As an accretion mass, the Earth is an Energy Sink. getting it's energy from elsewhere, being Cosmic, Galactic and Solar.

Conversion of incoming magnetic waves into electrical energy provides an unending, inexpensive and environmentally friendly source available to all. Cosmic and Galactic Energy is available twenty four hours per day. Large amounts of this Energy accumulates in the Earth's radiation belts. The Giant Energy Storage, when properly understood provides a major source of free unending electrical energy. One of My Inventions plugs into this vast energy source.

A perverse Intentional Ignorance on the part of the Establishment prevents recognition of the importance of the Energy Systems herein. Any new system favorable towards the masses is considered as disruptive and therefore not allowed. Those who have the (Gold) Energy Rule (Golden rule) Mandated Destruction of all Humanity is not a consideration..

This Presenter will remove some of the Fog placed with intent of preventing the recognition of this unending, environmentally clean electrical energy Source, present throughout the Universe. The Cost of Harvesting and Using this Free Energy is a function of Human Stupidity.

* "Physics for Scientist and Engineers", Raymond A. Serway, Saunders College Publishing, Second Edition, page 288, Table 14.2

ELECTRICAL ENERGY SYSTEMS METHODS

1. **DIRECT** - Faraday's Action at a Distance from incoming magnetic wave conversion to useful electrical energy. Includes Cosmic, Galactic, Solar and Magnets. Technology Transfer from Solar Cell Technology.
2. **INDIRECT**-Electron Stimulation-Induced Electron Spin Systems
   Electron Avalanche Pumping Systems

   Primitive, Indirect Conversion from another form of energy.
   Coils and Magnet as in Dynamo Systems (Closed Systems).
   Chemical Systems, Atomic, Pons & Fleischman and etc.

   Advanced, Direct Conversion, Magnetic Wave (Open Systems).
   Ambient Sources
   Air Core Coil Systems
   Gaseous Tube Systems
   Solid State Marx Generator Avalanche Type Systems.
   Leyden Bottle Capacitor Types inserted in Lakes and other.

   Magnet Systems
   Electron Beam Antenna Systems
3. **TRANSFER MECHANISMS**

   Solids-as in metal conductors

   Gaseous as in radio wave transmission, a form of ionization.

   Sensitizing of Systems by use of Trace Doping with Radioactive elements, includes metal surfaces.

   Open Systems, none linear with external forces. Albert Einstein as a direct quote from his biography states that these are excluded from the conservation of energy laws.

   Closed Systems Maxwellian Type Systems. Mathematics are predictable requiring deductive reasoning. Ohm's Law is King and Establishment Intellectuals being comfortable with this brand all else as a violation of the Laws of Nature by obtaining something for nothing. This is Dishonesty grand mal.

AMBIENT ENERGY SOURCES

	RADIATION SYSTEM	*DIFFUSION METHOD*	*MAGNETIC WAVE ENERGY*
1.	**Cosmic**	*	**Ultraviolet**
2.	**Galactic**	*	**Infrared**
3.	**Solar**	*	**Visible Light**
4.	**Earth**	*, **, ***	**Earth's Electrical System**

* Reflection, Deflection and Absorption

** Faraday's action at a Distance from Magnetic Waves

*** A Composite of from All the Above

A deep fog pervades the entire Scientific Community with regards to the Significance of the Above Energy Sources. Magnetic Waves convert directly into Electrical Waves (useful electricity). Two sides of the electromagnetic system are always present and never separate. Local electron spin provides (action at a distance) the flip side of the incoming magnetic wave energy.

Enormous amounts of incoming magnetic wave energy becomes a part of the Ambient Background and as such can not be measured directly. Reconstruction from indirect information allows us to establish actual energy levels present. Instruments provided by the Scientific Community measure only point "A" to "B", when both are ambient no potential energy is shown. This is the bird sitting on the million volt power line and sensing nothing approach. The Earth's actual ambient back ground has as it's Energy level multi Billions of Volts, which are conveniently and obliviously ignored by the scientific community. When properly understood this enormous, never ending source of environmentally friendly energy becomes available.

Electrical Energy System

Don L. Smith, Energy Consultant

At a meeting between J.P. Morgan, Edison and Tesla, Tesla proposed an Electrical Energy System which could be connected into directly, with out using a meter. Tesla's Idea of " Free Energy " was not compatible with their thinking. Courtesy of Morgan and Edison, from that day foreword, a complete and total bastardization of the Idea has been in progress. Agents for Morgan and Friend include the U.S. Patent Office and Academia. Academia's bad habit of incestuous quoting of each other eliminates them as a possibility in cleaning up the mess. This selective ignorance permeates through out the study of electricity.

Many persons, otherwise as intellectuals, have a total blackout and become jabbering idiots when " free energy" is mentioned. The term has been amended to say, "something which was never there is being harvested and that this violates the laws of physics". For the selectively ignorant this seems the way to run. Those who choose Morgan's drum beat, have severely limited the possibilities built into electricity.

This paper will be an exercise in creative understanding, in placing updated knowledge at your disposal. Whether it becomes a useful tool or is selectively ignored is your choice.

Electrons are defined as the practical source of electrical and magnetic energy. The electron as a particle was postulated by professor J. Thompson in early 1900's. It is now universally accepted that, the electron exist and that it is the source of electricity. When the electron is agitated it produces magnetic and negative electrical energy. Physics as it exist, can not explain why the electron remains in tact and is not diminished by the energy it releases. This is a part of the built in ignorance provided by the Morgan and Edison Camp.

A volts worth of electrons, when cycled yields a volts worth of electricity. This can be repeated continuously forever and never deplete or diminish the electrons in question. They simply return to their air and, or earth source, waiting to do the whole thing, again and again. Therefore, electrical energy is available, any and every where humans go. Persons who intercede for profit, set the cost of electrical energy. Otherwise, all electrical energy is free, Morgan and Edison be dammed.

Improving upon Professor Thompson's postulation, other obvious character further defines the electron. It has both magnetic and electrical emanation resulting from a right and left hand spin. Since magnetic and amperage are one package this suggest, that electrons in a natural none ionic state exist as doublets. When pushed apart by agitation one spins and supplies electricity and the other spins and provides magnetic (amperage) energy. When they reunite, we have Volts X Amperage = Watts. This Idea, until now, has been totally absent from the knowledge base.

The times an electron is cycled sets the collective energy potential present. The electrical equivalent of E = M C squared is E = (Volts X Amperes) X Cycles Per Second squared. Those who choose, are now free to head for the bushes and make their usual contribution to humanity.

Prior to Tesla, there was a large group of persons in Europe, who were building resonate coil systems related to medical usage. Amperage was dangerous in their coil systems. The Tesla Coil is only the Voltage half of their coil system, as will be demonstrated herein.

A short list of those active (1860-1880 onward) in resonate high frequency coil systems include the Curies, Roentgen, Ruhmkoff, Oudin, Hertz, Levassor, Dumont, D'Arsonval and many others.

Peugeot, Panhard-Levassor, Bollee, Renault and others had successful electric automobiles in production using A.C. motors. Various electrically powered airships including the Dirigible France were in service.

D'Arsonval , Professor of Experimental Medicine at the College of France, invented the electrocardiograph, oscilloscope, amp and volt meters, thermography and numerous other medical applications of high frequency electricity. As early as 1860, he was building high frequency coil systems, used in his experimental work.

There is a strong connection between the work of Tesla and the above mentioned.

Electric vehicles of all sorts, dominated until the 1920's, when the electric starter motor made the internal combustion engine practical. Prior to that, upon cranking, it frequently would break the owner's arm. At that point the use of batteries as a source of power was replaced by oil..

The establishment's carpet has some rather large lumps under it. Coulomb's and Newton's inverse square law is politely ignored and it's opposite is allowed only the most abstract status. Without opposites we have no definition.

The source value of a remote flux reading, requires the squaring of the distance, times the remote reading, to obtain the original value. The opposite of this, being the derivations relate to Energy equals Mass times the Velocity constant squared. The electrical equivalent, being Energy equal capacitance times voltage squared and Energy equals induction times amperes squared. Flux lines increase as the law of squares and then activate electrons (energy) not previously a part of the sum. The cumulative capacitance and inductance as the distal ends of a Tesla coil are approached, results in energy greater than the input being present. This Energy is real when properly understood. It can be safely measured by magnetic flux methods and electrostatic voltmeters, based on the inverse square law.

As seen above flux lines result both from Induction-henrys-amperage and capacitance-coulombs-volts and define electrical energy. The non-linearity of this system does not obey ohm's law, which is replaced with impedance and reactance for alternating current systems. Impedance is the sum of the system resistance, which becomes zero at resonance. In a resonate induction systems, cycles per seconds increases, invokes a second round for the law of squares.

The degree to which flux lines are present, disturbs an equal amount of electrons, upsetting ambient, resulting in useful electrical energy. The frequency at which the disturbance occurs, obeying the law of squares further accelerates away from ambient, increasing the useful energy available. Two square law entities, flux density and frequency are envoked. Enter resonance which cancels the resistive effect.

Only that electrical energy above or below ambient is useful. For the Central U.S. going east to west, ambient as approximated by electro-static voltmeters and flux methods is on a solar quite day 200,000 volts.

At night time ambient drops to about one half the daytime value. On a solar active days it may reach more than five times that of a quite day. Ambient background energy at the polar regions is approximately 500,000 volts on a solar quite day. The background varies as relates to the north-south component and the east-west continuum.

This leaves us with an interesting problem. Electrons, when disturbed, first produce magnetic flux and then electrical flux when they spin back to their normal position. Therefore any electron movement produces above ambient energy, being over unity.

Derivation of Magnetic and Electrical Power

Analogous Relationships:

1. Potential Power is present in a bar magnet as shown.

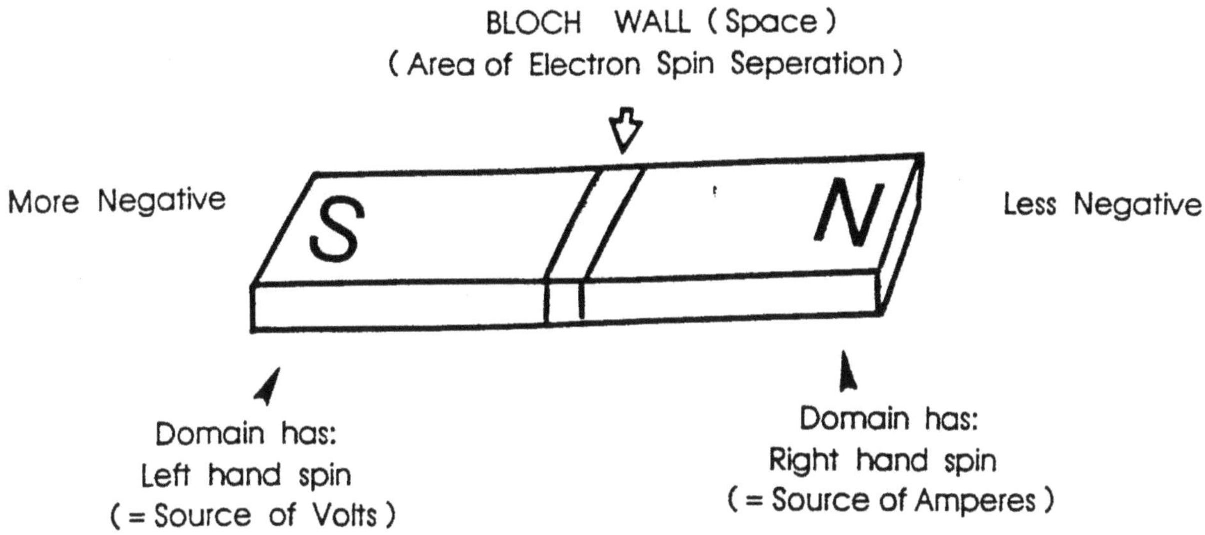

2. The Source of these Electrons being from the Solar Plasma, are none ionic and occupy all Free Space. They are commonly obtained from Earth and Air Groundings. They exist in Doublet Pairs, one being more negative than the other. The more negative one has a Left Hand Spin. The less negative one has a Right Hand Spin.

3. Resonate Electrical Coil Systems (Tesla) are Analogous to the System observed in the Bar Magnet (above). The Bloch Wall Area is Located at the base of the L - 2 Coil. The Left Spin portion (Voltage Only) part of the Coil predominates. The right hand spin portion (Magnetic-Amperage) portion is mostly absent.

4. The Electrical Circuit equivalent of the Bloch Wall is a Diode. Electron Doublets once seperated must recouple, resulting in useful energy, (Volts X Amperes = Watts) = Power.

I = Current = Amperes = Right Hand Spin ⇨

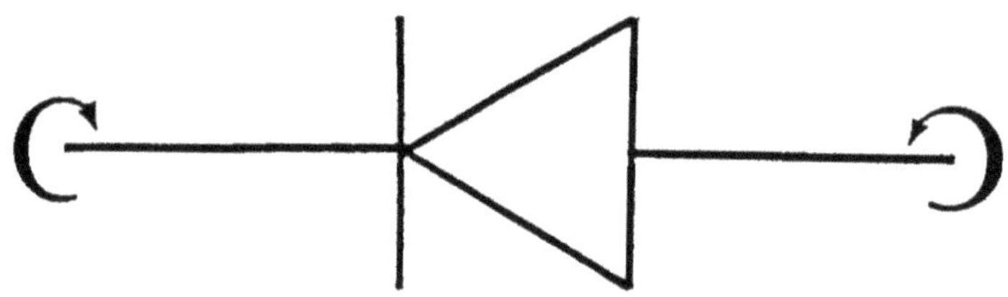

V = Electrons = Volts = Left Hand Spin ⬅

Tesla Coil Geometry *

Correct Method of Measurement
Use the Inverse square law with an Electrostatic Volt Meter to measure from a safe distance. they are frequency nonspecific.

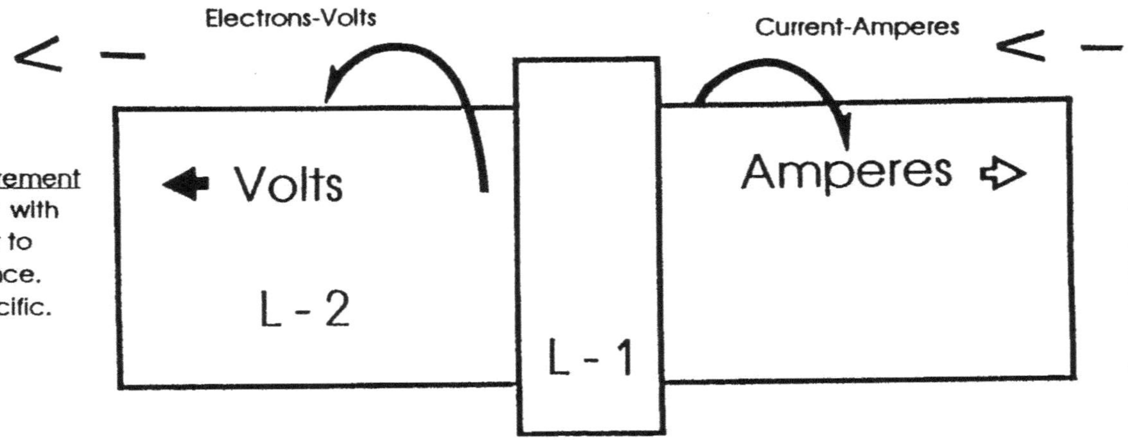

Correct Method of Measurement
Use the Inverse square law with Hall-Effect devices to measure from a safe distance, they are normally frequency nonspecific. The magnetic field they measure converts directly to Amperage

Volts dominate if the L-1 Coil is far right
Volts and Amperes about equal if L-1 Coil centered

This End Has Greater Voltage
Distributive Capacitance at Maximum
Capacitance-Coulombs-Voltage
Voltage has Electrons spinning to Left

This End Has Greater Amperage
Distributive Inductance at Maximum
Induction-Henrys-Amperage
Amperage has Electrons spinning to Right

Volts represent the more negative and Amperes the less negative. Therefore the electrical charge of more negative must seek the less negative to regain ambient.

* Contains proprietary information related to Patent Procedure
Geometry - properties of lines, surfaces and solids

Donald L. Smith
2 November, 1995

Power Triangle

Power System 1 | Power System 2 | L-1 Power System | L-2 Power System

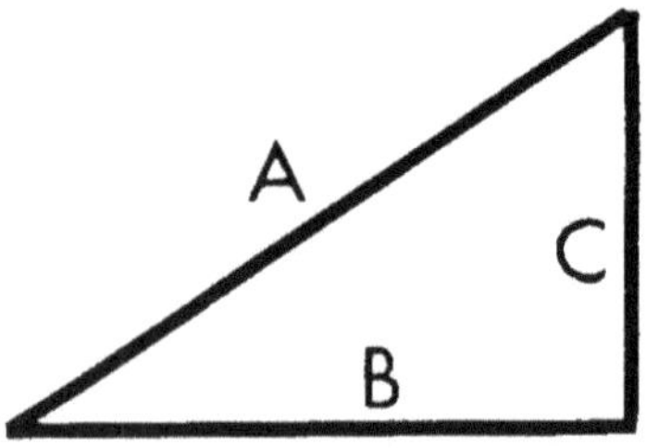

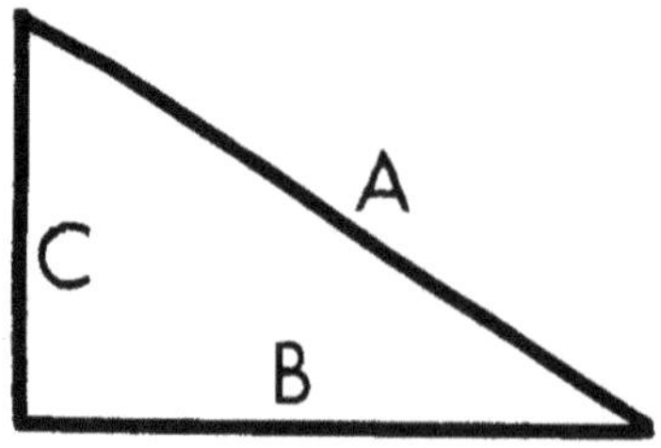

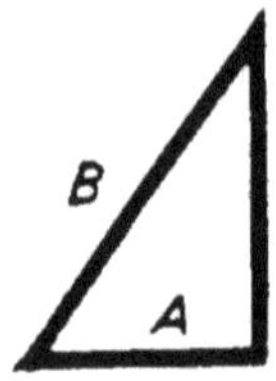

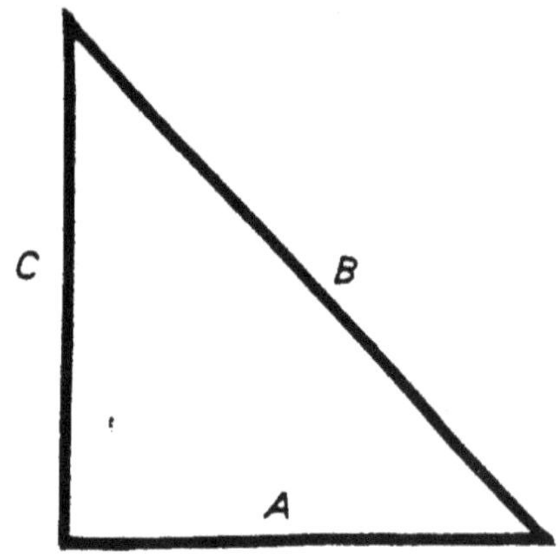

A: Volts x Amperes (Available Power).
B: Volts x Amperes x Time (Used Power).
C: Volts x Amperes x Reactive (Resonate Power).

1. "A" and "B" represent the establishments View of electrical power. Random movement of electrons in these systems, mostly cancel out each other. Therefore, always resulting in less energy out than in. This dampning, or wasteful concept of energy, is a source of much pleasure for the establishment.

2. "C" (Volt Amperes Reactive V. A. R.) , all the electrons move in the same direction at the same time. Therefore, near unity energy output by resonate induction transfer. This is the room temperature equivalent of super conductivity.

3. Resonate induction transfer from one power system allows, other resonate induction systems to duplicate the original source, which in no way diminishes the original source. Air core coils systems, when a part of a functioning transformer unit confirm this. A less perfect illustration would be, the number of radio devices or televisions, concurrently running in no way diminishes the output source.

4. Resonate induction transfer disturbs a large number of adjacent electrons which were not a part of the original source. The pulsating (pumping effect) then incorporates the newly available extra electrons into the ongoing energy generation source-system. A near unity energy system of resonate air core coils and the extra acquired electron-energy source constitutes an over unity system.

Don Smith
pwrt1.sam

[Editor's note: There were two pages almost idential with an added sentence on point 1 shown here, and different shapes with different lables so the images on the right were added.]

ELECTRICAL ENERGY GENERATING SYSTEM

Patent Pending 08 / 100,074

NOTE: IF THE PLUS CONNECTOR (2) WIRE'S LENGTH IS A WAVE LENGTH COMPONENT OF THE L - 6 OR L - 7's FREQUENCY, THE BATTERY WILL RECHARGE IT'S SELF AT THE SAME TIME IT IS BEING USED. TO SEE THIS SUBSTITUTE TWO OPPOSING LIGHT EMITTING DIODES. THEY SHOW THE DIRECTION THE ELECTRONS ARE MOVING. DC ELECTRONS OUT AND RADIO FREQUENCY ELECTRONS INTO THE BATTERY. [See page 275]

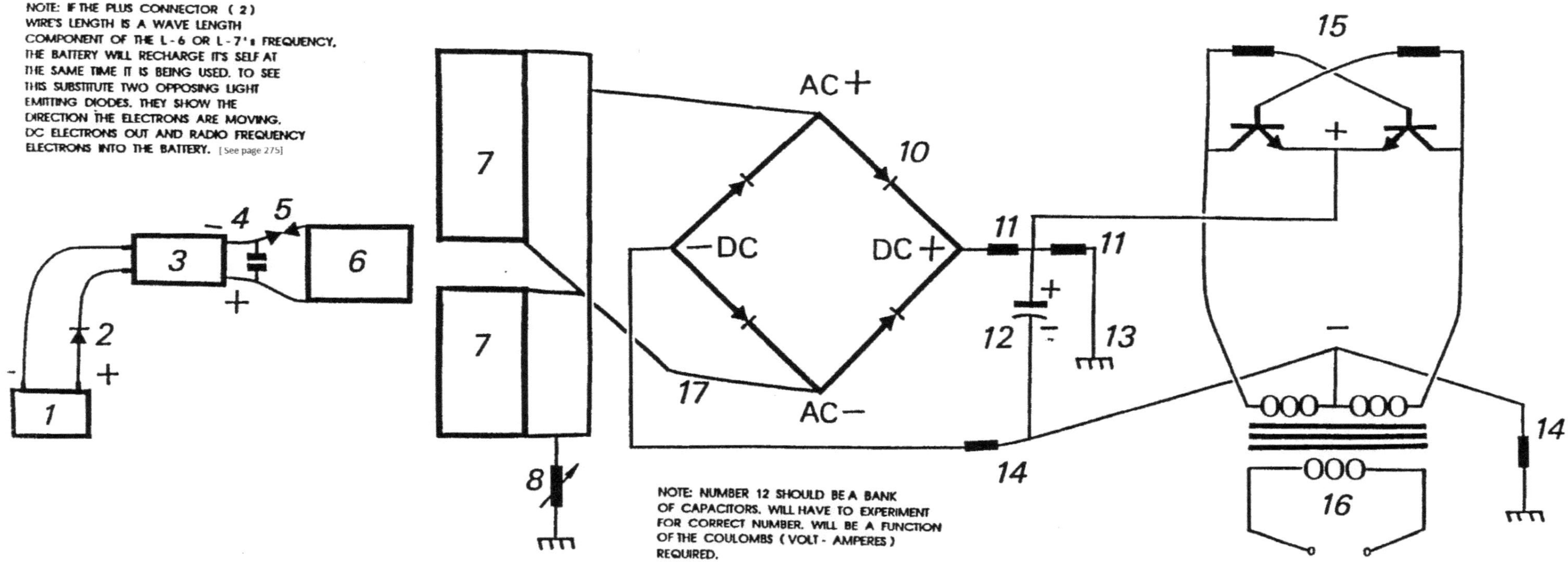

NOTE: NUMBER 12 SHOULD BE A BANK OF CAPACITORS. WILL HAVE TO EXPERIMENT FOR CORRECT NUMBER. WILL BE A FUNCTION OF THE COULOMBS (VOLT - AMPERES) REQUIRED.

1. Gelcel, 6 or 12 Volt.
2. Diode, Poss. use a Varactor.
3. High Voltage Module, Constituting the L-1 and L-2 Coils.
4. Capacitor, TDK 10.9 Pf., 30 KV.
5. Spark Gap, Small Engine Spark Plug, Gap = .0025 in.
6. Induction Transfer Coil L-3.
7. Induction Receiving Coil L-4.
8. Voltage Control Shunt.
9. Frequency Adjustor, prevents derating by Diode Bridge
10. Diode Bridge, 200 Nanosecond, R.F., > 100 KV .
11. Voltage Divider Circuit, corrects voltage for next stage.
12. Capacitor, electrolytic, smooths out DC + ripple effect.
13. Earth Ground.
14. Voltage Divider Curcit, corrects voltage for Transformer
15. Inverter Circuit, DC + in and 60 CPS to Transformer
16. Output from Transformer to Load (Work).
17. Center Tap

17 JANUARY, 1997

[Editor's note: The positive and negative notation in the center of the inverter is backwards. This was a mistake copied from a schematic on page 464 of the book *Encyclopedia of Electronic Circuits*, which is found on page 110 of this book. However, page 298 of the same book has the correct input polarity, which Don also included in his book, which is found on page 109 of this book.]

"The device which you're looking at now is this particular device."
Video 2 minute 39.
[This page has been modified to fit the above quote and below picture.]

ELECTRICAL ENERGY

Patent Pendir

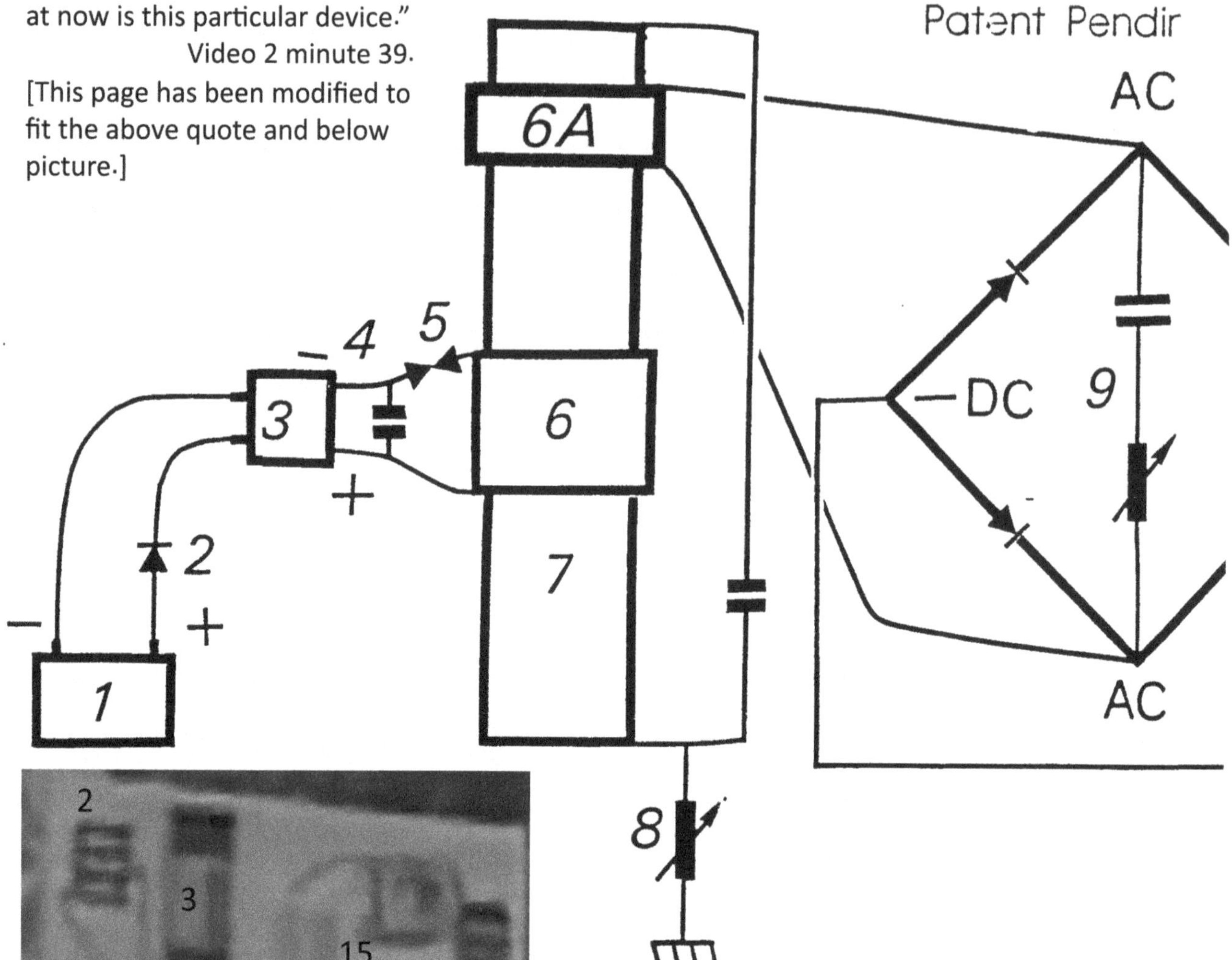

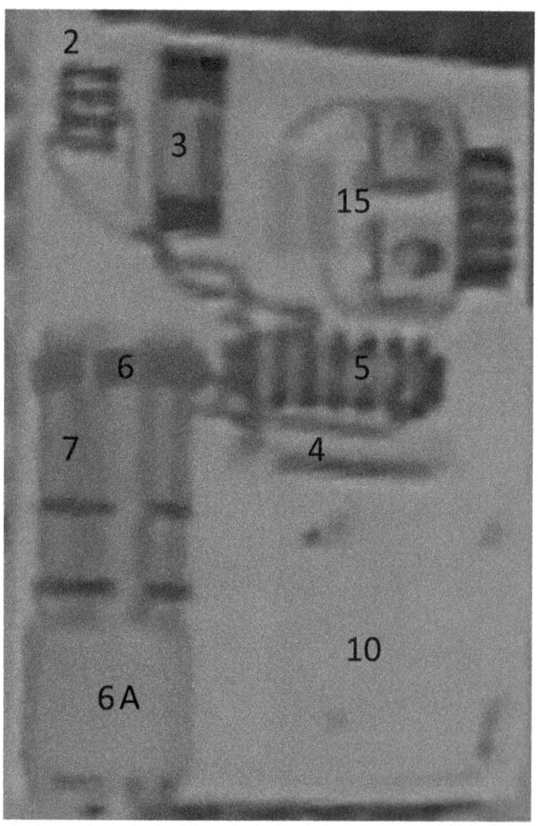

1. Gelcel, 6 or 12 Volt.
2. Diode, Poss. use a Varactor.
3. High Voltage Module, Constituting the L-1 and L-2 Coils.
4. Capacitor, TDK 10.9 Pf., 30 KV.
5. Spark Gap, Small Engine Spark Plug, Gap = .0025 in.
6. Induction Transfer Coil L-3., 6A = L-5
7. Induction Receiving Coil L-4. [B&W coil]
8. Voltage Control Shunt.
9. Frequency Adjustor, prevents derating by Diode Bridge

This picture is from video 2 minute 36.
9, 11, 12, 13, 14 not displayed.

GENERATING SYSTEM

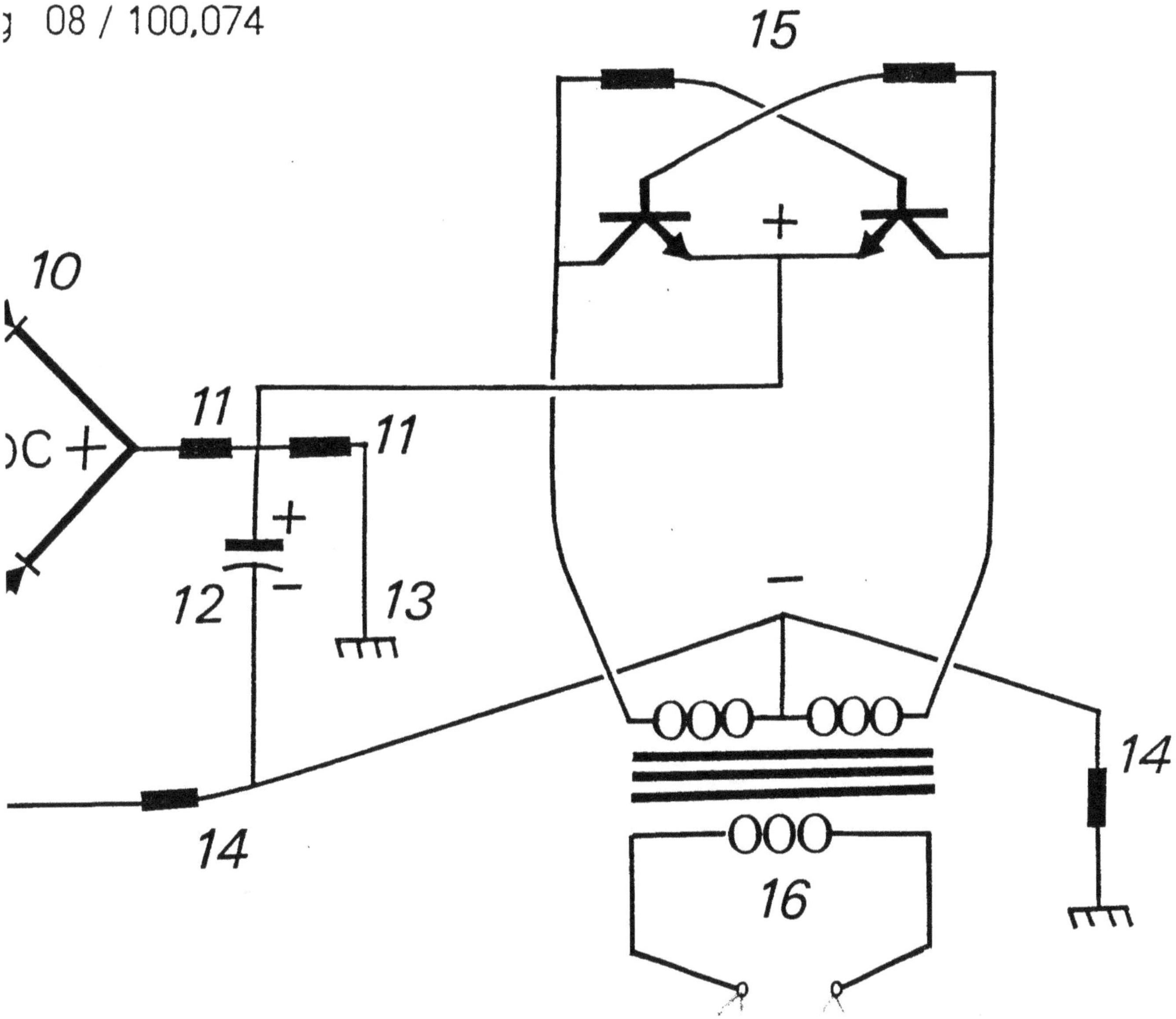

10. Diode Bridge, 200 Nanosecond, R.F., > 100 KV .
11. Voltage Divider Circuit, corrects voltage for next stage.
12. Capacitor, electrolytic, smooths out DC + ripple effect.
13. Earth Ground.
14. Voltage Divider Curcit, corrects voltage for Transformer
15. Inverter Circuit, DC + in and 60 CPS to Transformer
16. Output from Transformer to Load (Work).

20 Dec., 1994

[Editor's note: The positive and negative notation in the center of the inverter is backwards. This was a mistake copied from a schematic on page 464 of the book *Encyclopedia of Electronic Circuits*, which is found on page 110 of this book. However, page 298 of the same book has the correct input polarity, which Don also included in his book, which is found on page 109 of this book.]

Induced Electrical Energy System

Collection and transfer of energy requires temporary storage, which occures as capacitors and coils of a resonate circuit are cycled, off and on. The frequency at which the capacitors and coils are pumped determines the amount of electrical energy that moves onward.

The amount of Energy transfered relates directly to the density of lines of flux present. The Kenetic Energy Formula is helpful in establishing the amount of energy present. This formula squares the velocity times mass. In the case of electrical energy, intensity of voltage and amperes times cycles per second replace velocity.

Note the "acceleration" of the Voltage "E" and Amperage "I", which increase as none linear, then obeys the Law of Squares.

Each unit of increase causes a squareing of the flux lines present The amount of energy transfer caused by this increase in flux lines is demonstrated below.

Increase in Flux Lines Present Symbolized

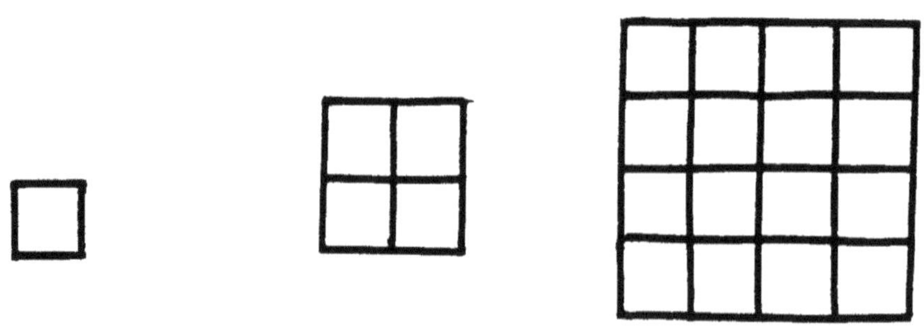

One Volt   Two Volts = Four   Four Volts = *Sixteen*

In resonate air core coil energy transfer, the increase in flux lines present disturbes more electrons than previous, resulting in over unity energy being present and available.

Energy stored, times the cycles per second, represents that being pumped by the system. Capacitors and inductors temporarly store electrons.

Capacitor formula: $W = .5 \times CE^2$

- W = energy in Joules (Watt Seconds)
- C = capacitance in farads
- E = applied potential in volts squared

Inductor (Coil) formula: $W = .5 \times LI^2$

- W = energy in Joules (Watt Seconds)
- L = inductance in henrys
- I = current in amperes squared

Both one henry and one farad equal one volt. The higher the cycles per second, including the squareing of the flux lines cause a large increase in the amount of energy being produced.

The above combined with a resonate energy induction system (all electrons moving at the same time in the same direction), make the next move into overunity practical.

The dampning process of conventional electrical power generation has all the available electrons randomly bouncing, mostly canceling out each other. In this System, useful energy available is a very small percent of that present.

In the resonate induction system a very high percent of the energy present is useful. When resonate, (ohms-Impedence-Z) becomes zero and all energy present is available, undegraded. Ohms is load or wasted energy and amperes is the rate of wasting.

Using the previous information, now apply it to an air core coil, resonate transformer energy system. L-one and L-two coils are now present. L-one has a smaller number of turns and is several times the diameter of L-two. Input from a 12 volt gelcel souce* produces 8,000 volts with low (wasted energy) amperage into 4 turns of coil L-one. Each turn of L-one then acquires 2,000 volts of resonate potential. Each turn of L-two is then exposed to the electric flux of 2,000 volts. Each turn at the bottom end of L-two acquires 2,000 volts. The flux lines are squared and are additive as the voltage progresses towards the top end of L-two's many turns.

A huge amount of flux lines not previously present occur at the top end of L-two. These flux lines excite the electrons nearby in it's earth and air and groundings. This high level of excitment above the ambient causes a large amount of electrons not previously a part of the energy present to become available. At this point overunity is present in large amounts.

The bubble gum between the ears response to this is, lots of volts but no amperes. Please recall that amperage is wasted energy, and that until wasting occurs there are no amperes.
A good way to demonstrate this is let the bubble gum crowd, put their hand on the high voltage end of the device while standing on wet ground (a people zapper).

This overunity device produces energy at radio frequency, ranging into the megahertz range. This allows the device to be small in size, and produces of very large amounts of energy.
A megawatt sized unit will sit comfortably on a breakfast table.
This energy is changed to direct current and then to the desired working frequency.

* A high voltage laser module.

Electrical Power Generation / Points of Reference

Useful Electrical Power is Generated when Electrons from Earth and Air Groundings are disturbed by the movement of coils and magnets with reference to each other. The resulting electrical and magnetic energy is then changed to joules [watt seconds, Volts x Amps x Seconds]. Each forword electron movement results in a magnetic impulse and each return movement causes an electrical impulse. The composit of the electrical energy impulsed from these electrons yields useful energy [Power].

Let the above electron movement be represented by a room full of ping pong balls randomly bouncing. Most of the energy present cancels out by random impaction. This is the Classic Under Unity approach to Electrical Power Generation, sanctioned by the Establishment.

In the Electrical Energy Generation System here presented, the resonate Electrons are all moving in the same direction at the same time. This allows Near Unity Electrical Power to Develop. This is the room temperature equivalent of super conductivity.

The Energy System here presented consists of a properly adjusted and functional resonate air core coil tank . The Electrical Energy is stored in capacitors and magnetic energy in the coil system. From Maxwell and others, we know that electrical related energy has an equal amount of magnetic energy associated with it..

" The formula which establishes the Useful Energy of the Inventors System is : "

Joules = [0.5 C x V's squared] x C.P.S. squared

Joules { Volts x Amperes x Seconds } Watt Seconds

C = Capacitance in microfarades
V = Potential in Volts
C.P.S. = Cycles per second

Transfer of Electrical Power by Resonate Induction is a direct function of the squaring of the cycles per second. For example square 60 C.P.S. and then square the radio frequency C.P.S.'s of the System here presented. Obviously One Million Cycles per Second transferes more energy than Sixty Cycles per second. The Sanctioned Method of Electrical Power Generation uses the 60 C.P.S. Method. Usage of the 60 C.P.S.. and the random scattering of the Electrons System assures the Establishment of it's desired Under Unity Goal.

This random bouncing of the Electrons is the Ohms of Ohm's Law and is used to establish the rate of dissipation and or Load [Work].

In the Resonate Tank Induction Energy Transfer System here presented Impedence [system resistance] replaces the conventional ohm's usage. At Resonance impedence becomes Zero and the full force and effect of the Energy Transfer occurs. This is superconducter conditions at room temperature. At radio frequency the Electrons do not pass through the conductor as at lower frequencies. These Electrons encircle the conductor and are free of the conductor's resistance.

Let the Establishments Power Generation System be "A" and the Inventors System here presented be "B".

"A". Given 60 C.P.S. at 120 Volts using 10 microfarad Capacitor

Joules = [0.5 x .000,010 x 120 squared] x C.P.S.'s squared

[120 x's 120 = 14,400.]
[.000,010 x 14,400. = .144]
[.144 x 0.5 = .072]
[.072 x 3, 600 = 259.2]

If using the Inventor's Resonate Induction System the Electrical Power available would then be 259.2 Joules [Watt Seconds]. Using the Establishment's method only permits less than 10 Watt Seconds of Useful Electrical Energy.

"B". Given One Million Cycles per second at 100,000. Volts, using a 10 microfarad Capacitor.

Joules = [0.5 x .000,010 x 100,000. squared] x C.P.S.'s squared

[100,000 x 100,000. = 10,000,000,000.]
[.000,010 x 10,000,000,000. = 1,000]
[1,000 x 0.5 = 500]
[500 x One Miilion squared =]

The usefull Electrical Energy available is more than 500 Mega Joules [Watts] plus., Since the Resonate Electrons are nonimpacting, all the Energy is available for direct usage.

Benefits of the Inventors System are summarized:

1. Induction Energy transfer is enhanced by the squareing of the cycles second of the System.
2. Induction Energy transfer is enhanced by the squareing the input voltage.
3. Squareing of the flux lines occurs from the above, disturbing more electrons, therefore more electrical energy becomes available.
4. Resonate Induction has all the Electrons moving. unimpeded, resulting in superconductor conditions at room temperature.
5. A smaller amount of energy is used to disturb a larger number of Electrons. Electrons not originally a part of the System then contribute their energy, resulting in a net gain in available usable power.
6. The physical size of the System [Device] is small. The Device describe in "B" sits comfortable on a breakfast table.
7. A small energy source is used to start the device and remains fully charged at all times from the System .

Abundant Energy Every Where Present

USERS GUIDE

AMBIENT ENERGY GENERATOR

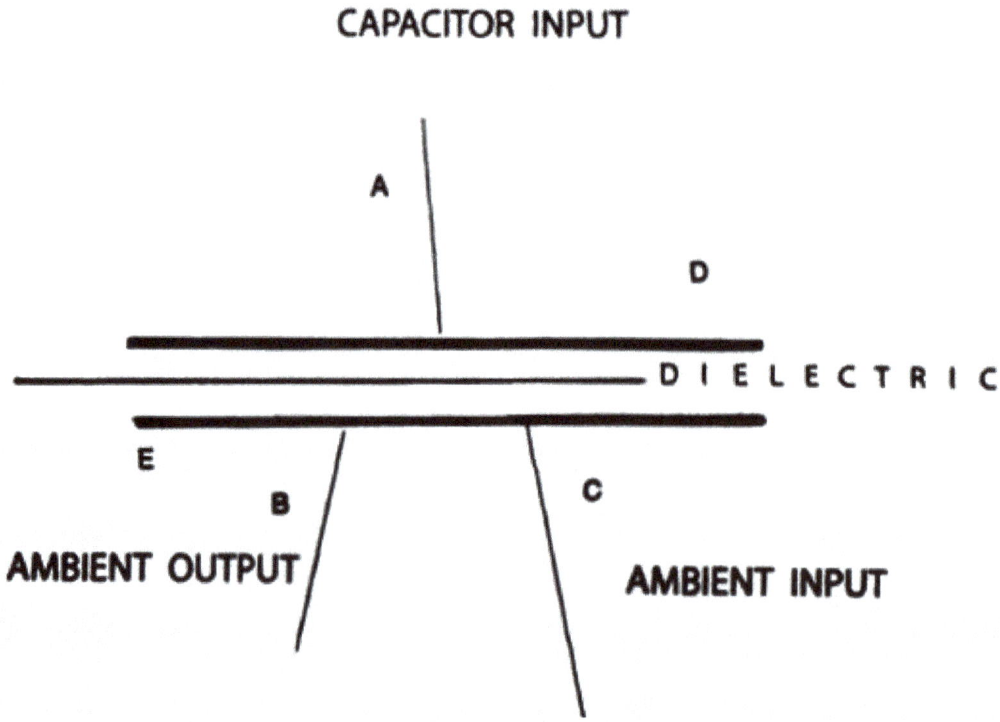

CAPACITOR [A] WITH A PREDETERMINED VOLTAGE PLACED UPON IT, CAUSES CAPACITOR [E] TO DUPLICATE FROM THE AMBIENT - EARTH GROUNDING THE ENERGY PRESENT AT [A] WITH AN INPUT DIODE PLACED [C] AND AN OUTPUT DIODE PLACED AT [B] THE ENERGY PRESENT ON [E] FLOWS THROUGH A TRANSFORMER INTO AN EARTH GROUNDING . USEFUL ENERGY IS OBTAINED FROM THE TRANSFORMER . DONALD L. SMITH , 14 FEBRUARY , 2004 .

CAPACITOR FUNCTIONS

Two main types being regular and electrolytic . Also Super Capacitor versions .

A Capacitor is a blocking device , the charge is placed on one side only The other side pulls and equal amount of Energy from the Ambient Background . The Capacitor is an Energy Pump . This second side captures ambient energy and will on it's own maintain it's charge even tho there is an energy is passing through the outlet . The beauty of this is that when used with a diode and or a diode bridge which are one way devices You have an Energy Pump with unlimited output is present . The second side feeds from the Ambient . By arranging the diodes You can get negative or positive output . Positive output will die a heat death and the negative will run cool . The only limit on the output is the capacitor size New shoe box seized super capacitors as single units greater than 400 Farads has 400,000. Watts of energy storage .

The Device does not deplete if only the output side is used . Since this action has not been named , I get to name it . I herewith name it the [DSE] DON SMITH EFFECT .

Further it is visually observed if one leg of a light emitting diode is attached to the output side capacitor plate . The main output side should be diode passing through to the point of use or load . Another phenomn is observable , namely a cooling effect as seen in the heat exchange modules

AMBIENT ENERGY GENERATOR [AEG - DLS]

My initials = DLS

ADVANTAGES OF THE GENERATOR

1. Simple to build with long trouble free life .

2. Few parts and all are easily available .

3. Moron level of comprehension required .

4. Once installed Energy is Cost Free .

5. The energy source is universally present .

6. Full fills any and all energy requirements . Automobiles and Airplanes are a couple of examples .

AIR CORE INDUCTION COIL BUILDERS GUIDE

DONALD L. SMITH

Energy Consultant

1. Decide frequency. Considerations are: (economy of size)
   a. Use radio frequency upward (above 20,000 Hz's).
   b. Use natural frequency (coils have both capacitance and inductance), that is match the wire length of the wire in the coil to the desired frequency.
   c. Wire length is either one quarter, one half or full wave length.
   d. obtain wire length (in feet) use the following: If using one quarter wave length divide 247 by the desired frequency (megahertz range is desirable). If using one half wave length divide 494 by the desired frequency. If using full wave length divide 998 by the desired frequency.

2. Decide number of turns, ratio of increase in number of turns sets the function. In the case of L-1 coil each turn divides the input voltage by the number of turns. In the case of L-2 the resulting voltage by division from L-1 is induced into each turn of L-2, resulting in an additive process. For example if the input into L-1 from a high voltage, low amperage module is 2,400 volts. L-1, for example has 10 turns. Then each turn of L-1 will have 240 volts of magnetic induction which transfers 240 volts of electricity to each turn of L-2. L-2 may be one turn or many, such as 100 to 500 plus turns. At 100 turns, 24,000 volts is produced. At 500 turns, 120,000 volts is produced.

3. Decide the height and diameter of the coil system. The larger the diameter of the coil, the fewer number of turns required, and shorter (lowering of) in height. In the case of L-2 this results in lowering the amplification of the induced voltage from L-1.

4. For example, 24.7 MHz is the desired frequency output from L-2. One quarter wave length would be 247 divided by 24.7 equals 10 feet of wire. The number of turns will be the amplification factor. The coil may be wound on standard size P.V.C. or purchased from a supplier. The supplier is normally a ham radio supply source. Once the length is determined and the number of turns decided, move the next step. For example let each turn of L-1 have 24 volts and desired output of L-2 being 640 volts. Therefore L-2 needs 26.67 turns. It has been determined that the wire length for one quarter wave length is 10 feet. The number of inches in 10 feet is 120. Using Chart "A" supplied look for next higher number of turns showing (being between 20 and 30 turns with a 2" diameter coil). This tells us

to use a 2" coil. If ready made as in the case of Barker and Williamson, 10 Canal Street, Bristol, Penna., 215-788-5581, they come in standard sizes of 4, 6 and 10 turns per inch. For higher "Q" use wider spacing of the turns. These coils come in a ready made length of 10 inches. Select from the coil 30 turns and put input clamps on the base of the coil and at 30 turns. For exact determination of the correct position of the output clamp, use an externally grounded voltage probe. The node of maximum intensity, being the natural resonate point. Off the shelf multimeters are not radio frequency responsive. The easiest way to accomplish the above is to get from the hardware store or Radio Shack a voltage detector having a neon bulb system (Radio Shack Cat. No. 272-1100b, NE2-Neon Lamps) will work. With your hand as a ground, move the wire extension of the neon lamp along the coil surface until bulb is brightest.
This is the desired point of resonance and connection.

5. The input power now needs consideration. A 2,400 H.V. module has been previously selected. This module can be made from a diode bridge or any combination of voltage amplifiers. The one used here is an off-the-shelf type, similar to those used for laser technology.
6. Construction of the input L-1 coil. For purposes already determined there will be 10 turns. Length of the wire here is not critical. Since L-2 is 2" in diameter, the next off-the-shelf larger may be used for L-1. Use a 3" diameter off the shelf coil having 10 turns to the inch. Remove (cut) a 10 turn portion from the larger coil. Use a L.C.R. meter and get the natural farads and henry's reading from L-2. Now do the same for L-1. It will be necessary to put a capacitor for matching L-1 to L-2 across the voltage input of L-1. Also a spark gap in parallel is required on the return voltage from L-1. A tunable capacitor of the pad type for L-1 is desirable.
7. L-2 can be further enhance by having an Earth grounding from the base of the coil. The maximum voltage output will be between the base and top of L-2. Lesser voltage can be obtained at intermediate points from L-2.

SUPPLY SOURCES

1. HAM RADIO SUPPLY STORES
2. COILS, AIR INDUCTOR IN HOUSTON

   BAKER AND WILLIAMSON (READY MADE), BRISTOL, PENNA.
   ALSO R.F. DUMMY LOADS AND WATTMETERS.

   CITY ELECTRONIC SUPPLY PH. 663-6066,
   7015 ATWELL M-F 9-5
   HOUSTON, TEXAS 77081 SAT 9-12
   LOCATED OFF OF BELLAIRE BLVD., JUST WEST OF WHERE BELLAIRE AND BISSONNET CROSS, ABOUT 1/2 MILE).
3. OTHER COMPONENTS AS REQUIRED:

   ACE ELECTRONICS PH. 713-688-8114 NOT OPEN SAT.
   3210 ANTOINE MICA PAD CAPACITORS
   HOUSTON, TEXAS

   ELECTRONIC PARTS OUTLET PH. 713-784-0140 OPEN SAT.
   3753 FONDREN CAPACITORS, ETC.
   HOUSTON, TEXAS

[Editor's Note: added third page to the bottom of this page as it is outdated references.]

Magnetic Resonance Power System
Suggestions for Construction

This is the Basic Sonar Power System which permits submarines to see approximately 50 miles distance. What is not commonly known is that it works better at higher frequencies into the Gigahertz. Any Device that can radiate 50 miles plus is producing an enormous electromagnetic disturbance from a small input into a rod of magnetostrictive material. Disturbing the Earth's Ambient Background plus the strong dipole being produced turns the magnetostrictive rod into a combination receiving antenna and a vastly superior output transformer.

The Drawing is only the Key Unit. A power input module and an output inverter circuit (diode bridge plus output transformer) is required. The metal core and wire size of the output transformer plus adjusting the Earth Grounding of the Load will determine the Amperage.

The Ideal rod material is Terfenol-D (check the internet). However a 1 ½" D and 10" L cost over $5,000. Each. Less expensive alternates are obvious. In construction use PVC tubing with removable caps. Wind the coils on it and insert the experimental rod. Use magneto-strictiv material only.

When You get it right You will have exactly what the Doctor ordered.

Magnetic Resonance Power System for Water Systems

Donald L. Smith

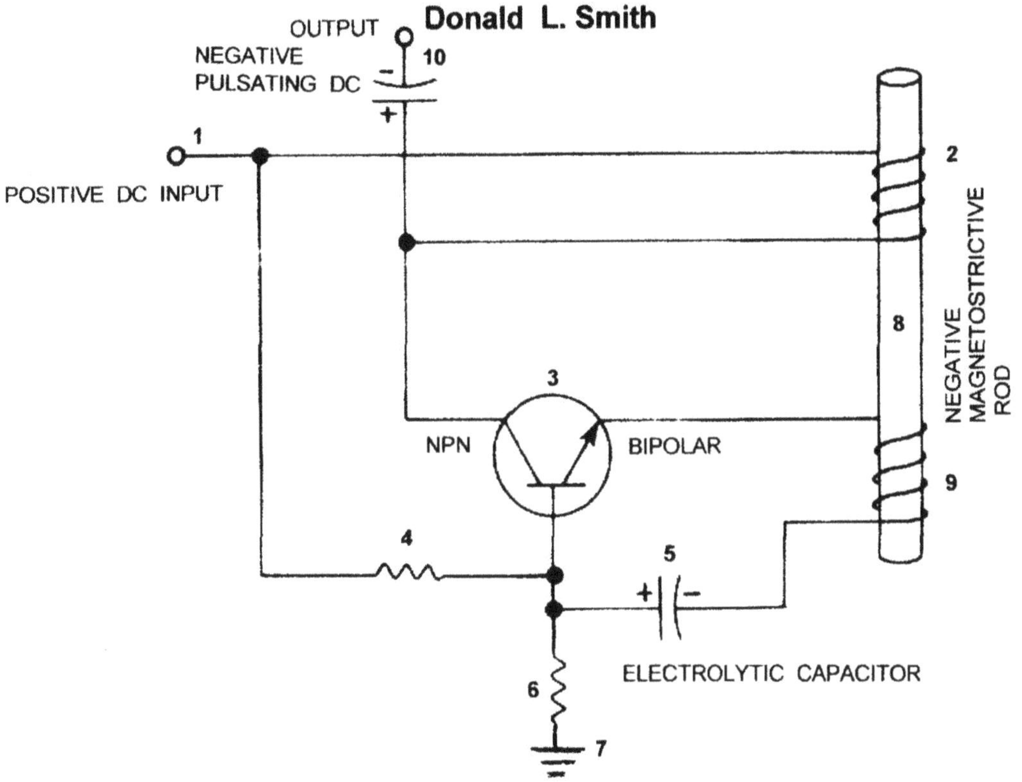

Magnetostriction oscillators work by magnetic resonance in a rod of magnetostriction material*. This rod serves two purposes, It vibrates at frequency of resonance oscillation, and becomes the feedback transformer. Frequency is determined by items 4, 5, 6 and 8. The diameter, length and volume of the rod and output windings determines the output. Item # 2 provides feed back into the system. Negative magnetic character of item # 8 plus the windings # 2 in reaction to the magnetic flux field provided by # 9, increases (amplifies - magnifies) the output. Magnetic permeability is the counterpart of negative resistance. Resonate with negative magnetic resistance it pumps energy from the Earth's ambient background. Magnetic permeability is the ratio of flux density (Earth's B field) to the magnetizing force (H) in oersteds.

* Magnetostrictive materials are piezoelectric in character, have very high resistivity to electrical current flow. Examples are:

1. Permealloy	Negative	Magnetic	Permeability	>	80,000
2. Sendust	"	"	"		30,000 - 120,000
3. Metglas	"	"	"	>	200,000
4. Iron with (34%) Cobalt			"		13,000
5. New Technology		"	"	>	1,000,000

<u>Unity (Ambient) Vs Greater Than Ambient (Over unity)</u>

1. Useful Energy is defined as anything other than Ambient.

2. Other than Ambient invokes the Kinetic Laws as expressed in Energy Equals Mass times Acceleration Constant.

3. Acceleration Constant invokes the Square Laws * which with increasing rapidity move away from Ambient.

5. Cumulative Energy present as an End Product is Greater than the Input. Any Energy **<u>greater than Ambient is Over Unity.</u>** The intellectually dishonest will have great difficulty with this, since in their **<u>Peanut Brain Logic</u>** it violates the Laws of Physics.

* Present Day Physics acknowledges that cubed not squared is the correct rate for the acceleration constant.

DIPOLE TRANSFORMER GENERATOR

DESCRIPTION

TECHNICAL FIELD:

The Invention relates to loaded Dipole Antenna Systems and their Electromagnetic radiation. When used as a transformer with an appropriate energy collector system it becomes a transformer generator. The invention collects and converts energy which, with conventional devices is radiated and wasted.

BACKGROUND ART:

An International search of Patent Data Base's for closely related methods did not reveal any prior Art with an Interest in conserving radiated and wasted magnetic waves as useful energy.

DISCLOSURE OF INVENTION:

The Invention is a new and useful departure from transformer generator construction, such that radiated and wasted magnetic energy changes into useful electrical energy. Gauss Meters show that much energy from conventional electromagnetic devices is radiated back into the ambient background as wasted. In the case of conventional transformer generators, a radical change in the physical construction allows better access to the energy available. It is found that creating a dipole and inserting capacitor plates at right angle to the current flow allows magnetic waves to change back to useful electrical (coulombs) energy. Magnetic waves passing through the capacitor plates do not degrade and the full impact of the available energy is accessed. One or many sets of capacitor plates as desired may be used. Each set makes an exact copy of the full force and effect of the energy present in the magnetic waves. The originating source is not depleted or degraded as is common in conventional transformers.

BRIEF DESCRIPTION OF THE DRAWINGS:

The Dipole at right angle allows the magnetic flux surrounding it to intercept the capacitor plate or plates at right angle. The electrons present are spun such that the electrical component of electron is collected by the capacitor plates. Essential parts are the South and North component of an active Dipole. Examples here presented exist as fully functional prototypes and were engineer constructed and fully tested for utility by the Inventor. Corresponding parts are utilized in each of the three examples as shown in the Drawings.

DRAWING 1 OF 4 : VIEW OF THE METHOD

N = North and S = South of the Dipole

1. North and South component of the Dipole.
2. Resonate High Voltage induction coil.
3. Dipoles electromagnetic wave emission.
4. Heaviside current component.
5. Dielectric separator for the capacitor plates
6. For purposes of the drawing, a virtual limit of the electromagnetic wave energy.
7. Capacitor plates, with dielectric in between.

DRAWING 2 OF 4 : COMPONENTS, 2A and 2B

2-A

1. Hole for mounting Dipole B-1.

2. Resonate high voltage induction coil.

5. Dielectric separator, a thin sheet of plastic separating the capacitor sheets.

7. Capacitor sheets, upper is aluminum and lower is copper.

8. Battery system, deep cycle.

9. Inverter input Direct Current, output is 120 Volts at 60 cycles.

10. Connector wires.

12. Output to point of use being the load.

2-B N = North and S = South component of the Dipole

1. Metal rod, being soft magnetic metal such as iron.

2. Resonate high voltage induction coil.

10. Connector wires.

11. High Voltage input energy source such as a neon tube transformer

DRAWING 3 OF 4 : Proof of Principal Device using a Plasma Tube as an active Dipole.

N = North and S = South Components of the active Dipole.

5. Dielectric separator of the capacitor plates.

7. Upper capacitor plate, upper being aluminum and lower being copper.

10. Connector wires.

15. Plasma Tube, 4 feet Length and 6 inches diameter.

16. High Voltage Energy source for the active Plasma Dipole.

17. Connector block, to outlet for testing and use.

DRAWING 4 OF 4 : Manufactures Prototype, Constructed and fully tested.

1. Metal Dipole rod.

2. Resonate High Voltage induction coil.

10. Connector wires.

17. Connector block for input from high voltage energy source.

18. Clamps for upper edge of capacitor packet.

19. Support Device for The Dipole Transformer Generator.

20. Packet of Capacitor Plates.

21. Output connectors of the capacitor, poducing energy into a deep cycle battery which then powers the inverter.

BEST METHOD OF CARRYING OUT THE INVENTION:

The Invention is applicable to any and all electrical energy requirements. The small size and it's high efficiency makes it an attractive option. It is particularly attractive for remote areas, homes, office buildings, factories, shopping centers, public places, transportation, water systems, electric trains, boats, ships and all things small or great. Construction materials are commonly available and the skill level required is moderate.

CLAIMS:

1. Radiated magnetic flux from the Dipole when intercepted by capacitor plates at right angle, changes to useful electrical energy.
2. A Device and method for converting for use, normally wasted electromagnetic energy.
3. The Dipole of the Invention is any resonating substance such as Metal Rods, Coils and Plasma Tubes which have interacting Postive and Negative Componets.
4. The Resulting Heavyside current component is changed to useful electrical energy.

ABSTRACT

A Electromagnetic Dipole Device and Method, wherein radiated and wasted energy is transformed into useful energy. A Dipole as seen in Antenna Systems is adapted for use with capacitor plates such that the Heavyside Current Componet becomes a useful source of electrical energy.

Dipole Transformer Generator

DRAWING 1 of 4

Not to Scale View of Method

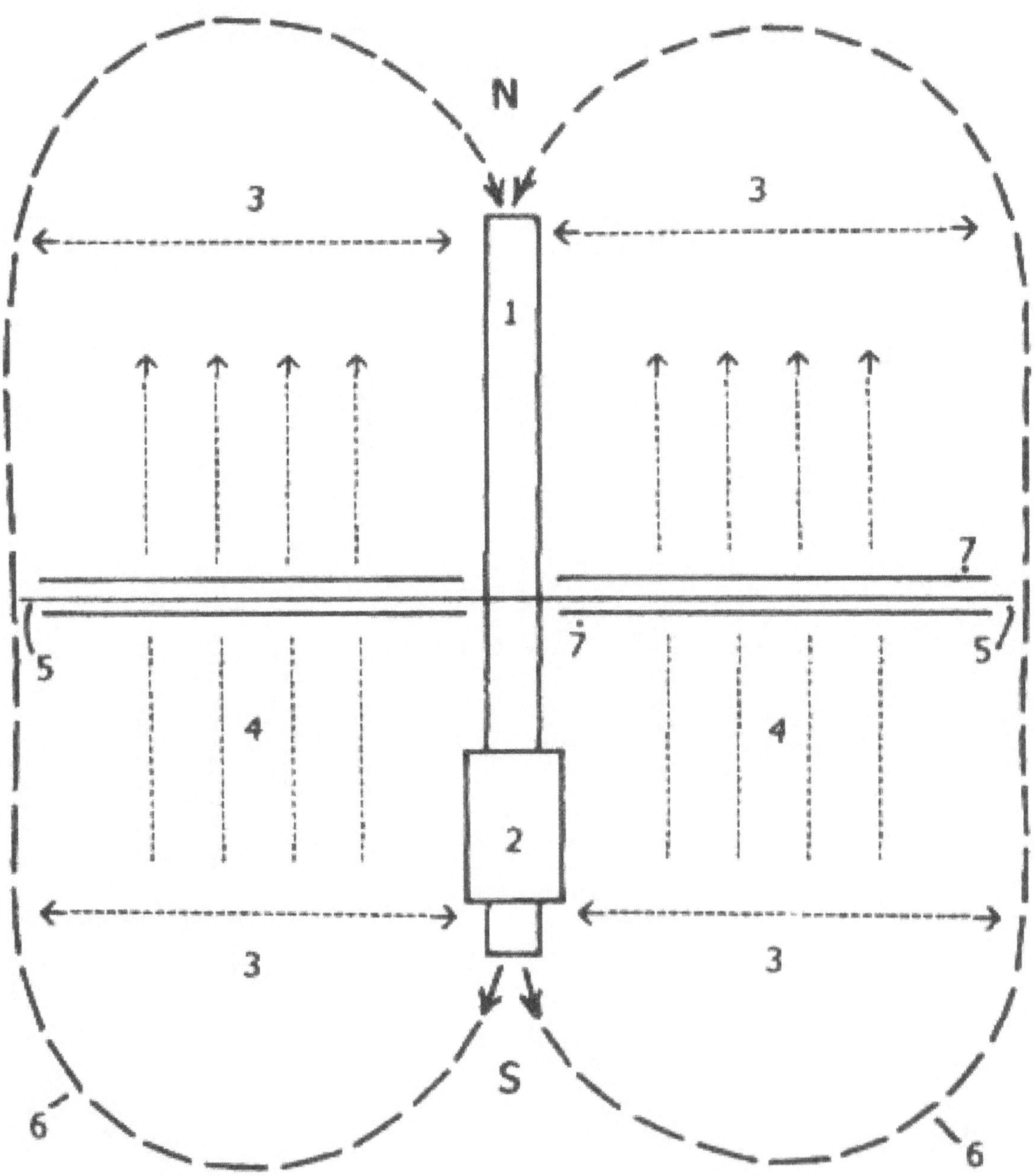

Dipole Transformer Generator

DRAWING 2 of 4

Not to Scale Components

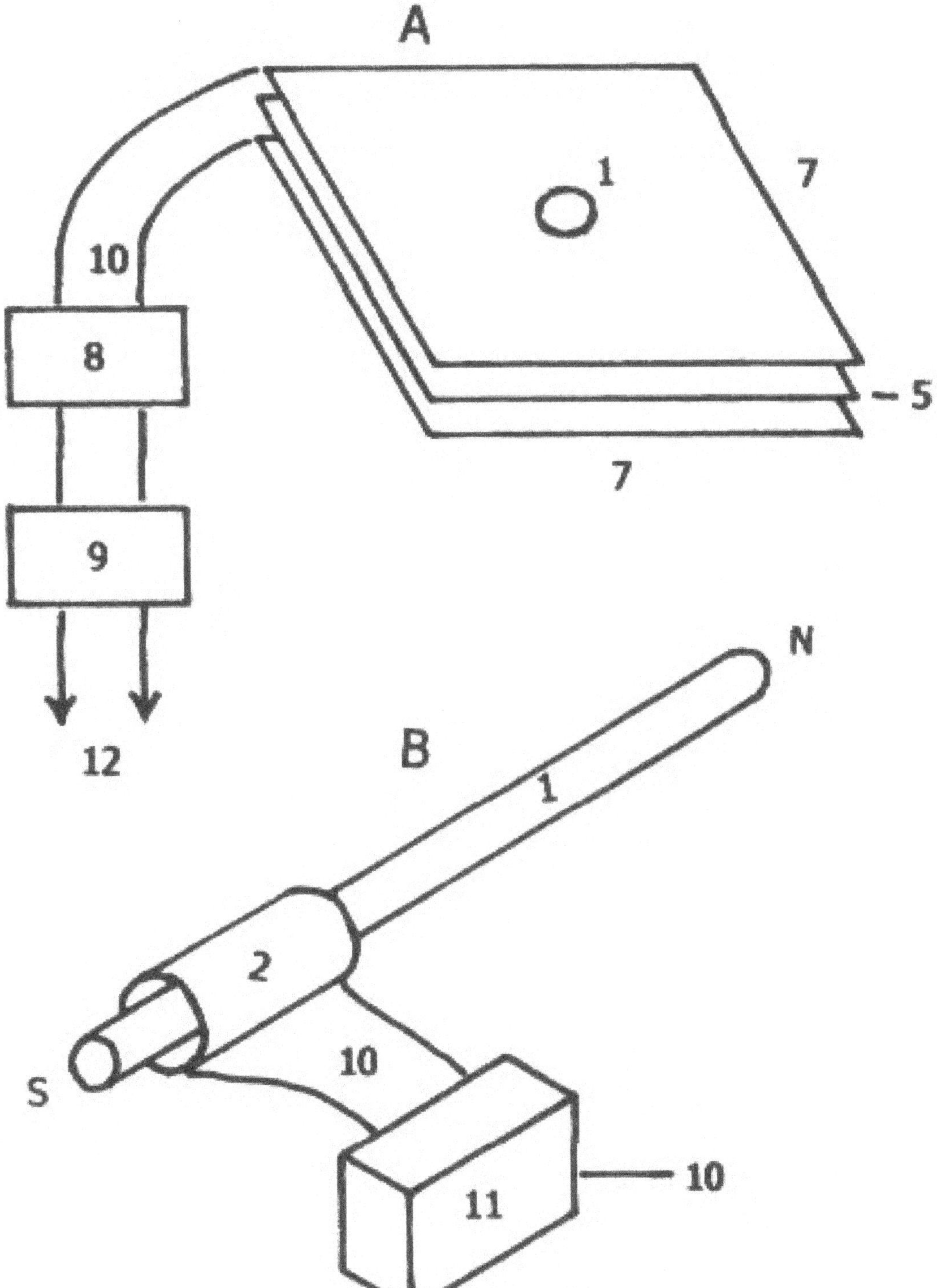

Dipole Transformer Generator

DRAWING 3 of 4

Not to Scale Proof of Principle Device

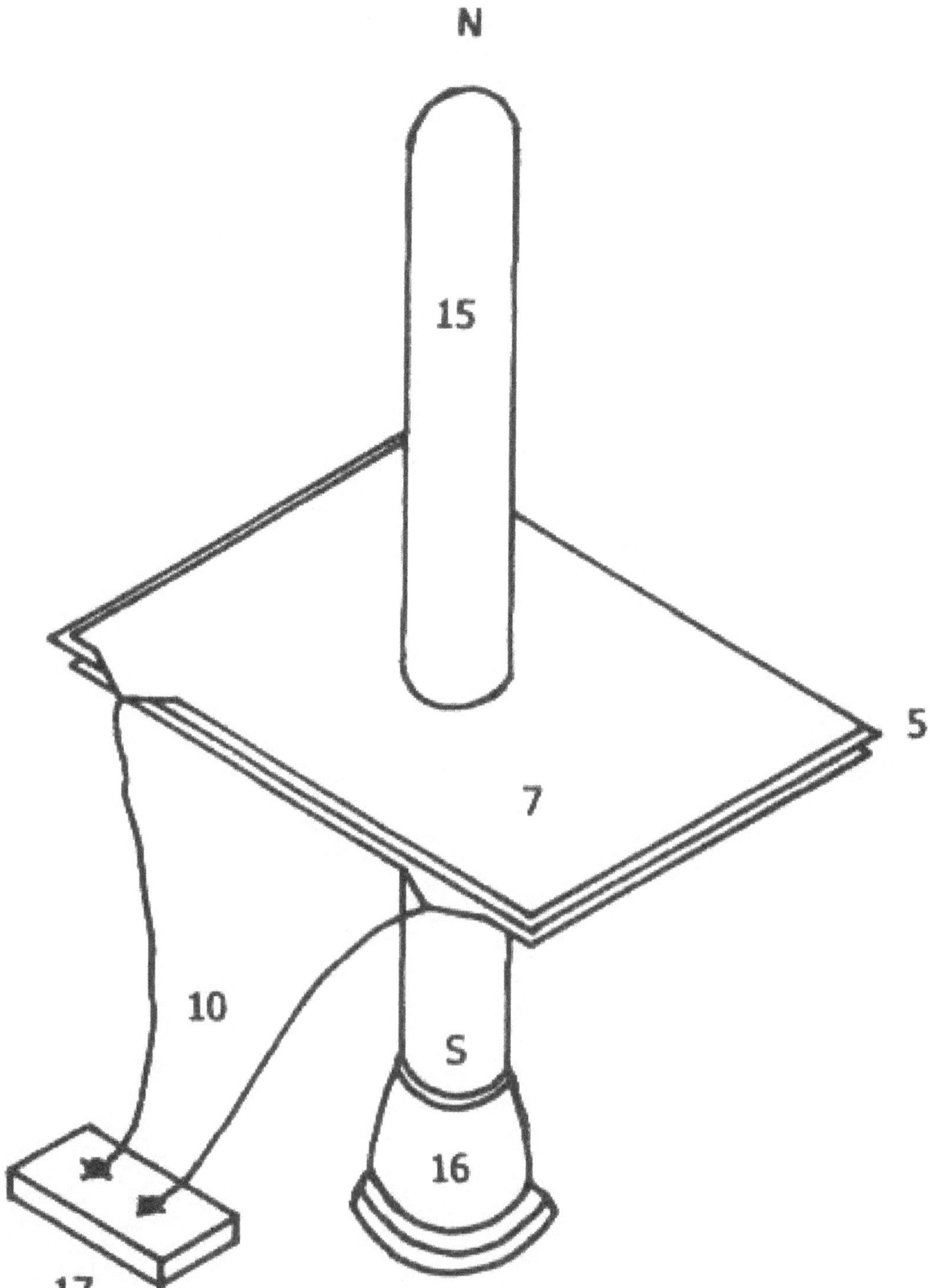

Dipole Transformer Generator

DRAWING 4 of 4

Not to Scale   Manufacture's Prototype

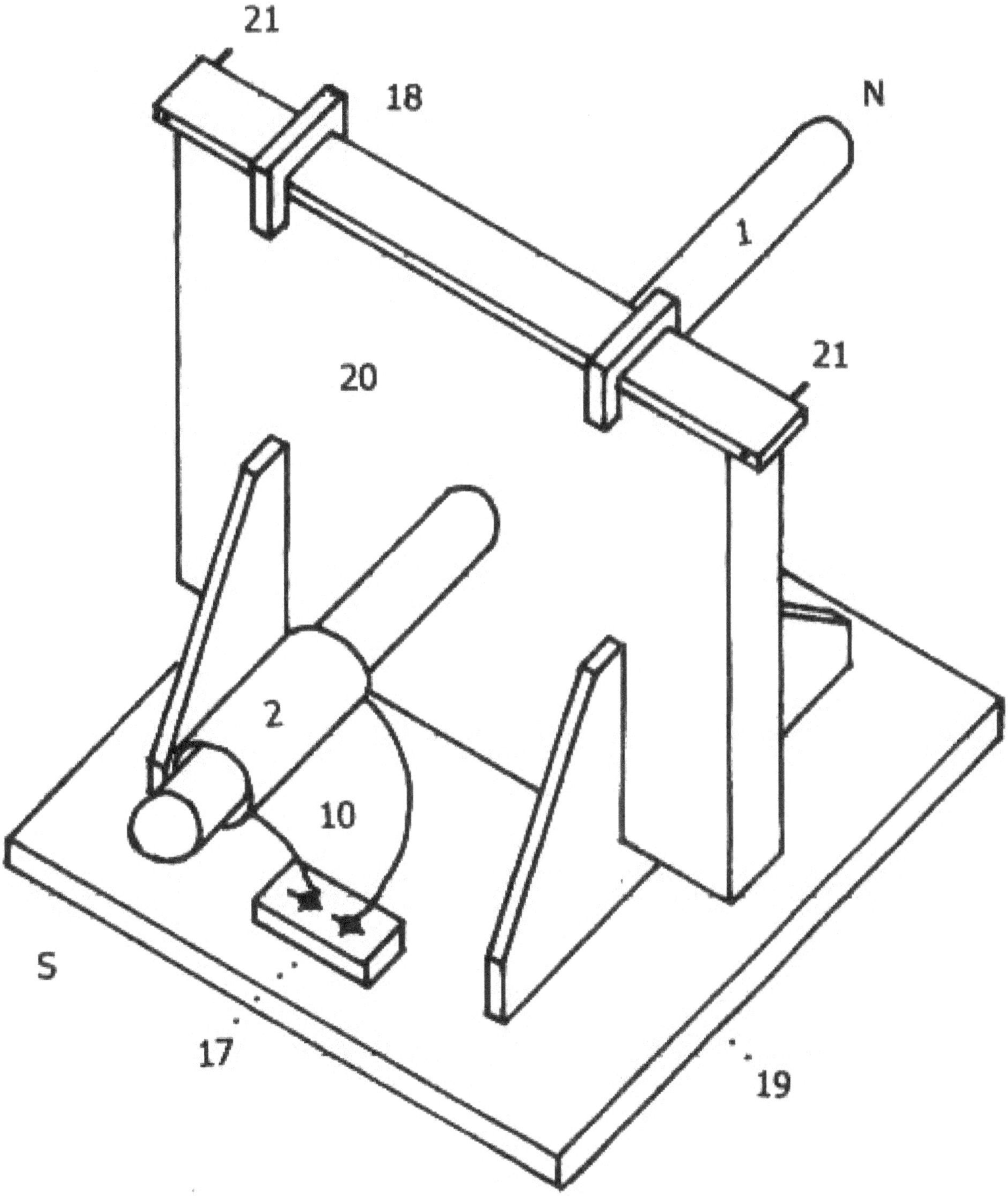

Ultimate Energy Sources

by Don Smith

Increasing the tolerance level for intellectual awakening opens doors of reality. These doors blink into, out of existence, and upon recognition benefit mankind. Opening some of these doors, presently seen only through a deep fog is our purpose. Exploring unrecognized energy sources, being a part of the ambient background is another goal. As such, they are ignored, much as the unaware bird sitting on a high voltage power line. Available Instruments do not use reference planes, which allow recognition, as we shall see, the vast energy sources that totally surround and which are available through technology transfer. These energy sources are here. They are inexpensive, fully self renewable and environmentally benign. Incoming electromagnetic wave energy with Faraday's action at a distance will be looked at closely. Particle Physics will be left for the Astrophysicists. Excited electrons at Point "A" the Sun do not travel to point "B" the Earth, however corresponding action occurrences at point "B." The electrons being disturbed at the central power plant in the same manner excite the electrons at your house, upon switching into an earth ground (flipping the switch). Correspondingly, there are four major power sources providing enormous amounts of magnetic wave energy. They are Cosmic, Galactic, Solar, and the Earth's electromagnetic background. The Earth's Field comes from reflection, deflection, and absorption, from action at a distance by the other sources. Prescription physics mandates that the Earth's background is of little interest. When you have considered the evidence herein, it will become obvious that special interest's effort at keeping the people ignorant has until now largely succeeded. From maps published by the US Geological survey out of Boulder Colorado, information for the entire world is available regarding the magnetic background of the Earth's surface. When examined properly and understood, these maps yield information regarding reflection, deflection, and absorption of these incoming magnetic waves, plus action at a distance. Partial reflection of these magnetic waves, when properly interpreted yields an ambient electromagnetic flux energy equivalent of 7.2 Billion Volts. This is a part of the Earth's ambient background as relates to the bird on the high voltage line. When deflected magnetic flux from electrons changes to electrical flux, they constitute a motor system that rotates/spins the Earth. In electrical systems, physical movement is in direction of inflow current movement. What is the electromagnetic current level required to accomplish this? Absorbed microwave flux energy heats from the inside out, therefore a hot interior for the Earth. Water is diamagnetic and ocean waves present on windless days provide visible proof of the overhead incoming magnetic flux. Using the weight of the Earth and it's rate of spin allows us to calculate the amount of incoming ambient background energy required to do this. As you can see, it not inconsequential as prescription physics mandates. Astrophysicists are concerned with charged particles, which do not move from point "A" to point "B". Associated action at a distance is highly active and translates into energy at point "B". The Galaxy is alive with energy, being millions of times greater than that of the Sun. Visible light is only a part of the electromagnetic spectrum. Other frequencies present in the Galaxy allow radio telescope photographs of their existence. One such photograph in the 408 MHz energy spectrum here presented shows the energy levels present. The Earth is only a tiny spec of dust (at the left end of the light central area) in this enormous ocean of energy.

The Light Center Is Our Galaxy, Colors Equals Ambient Energy Levels

A Radio Telescope Photograph of our Galaxy at 408 MHz All-Sky Survey as viewed looking sideways through the disk, showing Energy Levels.

http:/hp148.mpifr.bonn.mpg.de/survey.html

The Light Center Is Our Galaxy, Earth Is Near the Left End

This energy extends in all directions and totally surrounds the Earth. Energy sinks cause accretion and formation of Planets, Suns and Galaxies, through variable sized black holes. Mass retains heat and is cooked from the inside out by the microwave background provided by the Universe. Flux movement into energy sinks provides frictional force known as gravity. Spinning mass in the presence of incoming energy amplifies the gravitational effect. The key to harvesting this energy is recognition and direct conversion for useful purposes. At present, only Solar Energy is recognized. It is inconsistent, flaky and a very small part of the magnetic wave energy present. Technology transfer from solar power related provides uncomplicated and inexpensive direct access to the other greater energy sources. All Electromagnetic Energy harvesting methods include a catalyst, a collector, and a pump. Catalysts include such things as Air and Earth grounding, sensitization through doping with certain elements. Collectors include temporary storage as in capacitors, coils, and transformers. The pump system includes movement onward to point of use. The rotating coils and magnet systems activate electrons present such that action at a distance can occur and therefore is an energy activation pump. In direct access systems as in Solar Cells, the same occurs with out mechanical action. Direct access occurs when magnetic waves impact a catalyst, spinning the local electrons sideways, producing useful electrical energy. Indirect acquisition of electrical energy by mechanical means is inefficient, troublesome, expensive, and degrades the environment. The dynamo is a combination collector and pump of energy, acquired from the ambient background.

Generators do not make electricity; they only collect it from the Earth's Background and foreword it, as in Faraday's action at a distance. Einstein excludes the following from the Conservation of Energy Laws:

1. Existence of external forces,
2. Nonlinear systems, and Open systems.

This includes the entire Universe and all aspects of ambient energy sources, therefore all electrical energy systems. Patent Examiners in the US Patent Office become arrogant and conveniently ignorant when confronted. As the Agent for special interest, they are the judge, jury, and executioner. They in no way represent the best interest of humanity. Use of the knowledge base you have just viewed provides a direct understanding of the requirements for harvesting of unending electrical energy sources. Nine separate and very different patents pending of this writer demonstrate the simplicity of the direct access methods, unlike the Rube Goldberg approach sanctioned by the establishment. Related patents pending and working models of the Energy Systems related to the Sources here presented were personally funded, developed, with Patents Pending. A short List of Patents Pending is presented:

1. **Magnetic Radiant Energy Direct Conversion System**, use of energy present at or near the Earth's surface. It is Self-sustainable.
2. **Motor Generator**, a rotating magnetic shield, alternately breaks the composite magnetic field providing the off-on electron spin reversal for harvesting electrical energy. This i**nterruptible bloch wal**l allows a smaller motivating force to produce a larger output, resulting in a Self-sustainable Power Source.
3. **Generator Gaseous Tube Electrical Power System**, resonate electrons in a gaseous mixture are excited, such that the resulting electromagnetic energy can be harvested. It is a Self-sustainable energy source.
4. **Propulsion System, Magnetic**, provides floating to Tachyon Speeds and Space Travel. An uncomplicated, inexpensive and highly obvious System that does everything the Flying Saucers are said to do. No more Rube Goldberg approaches.
5. **Power System, Piezoelectric Solid State Marx Generator**, major improvements of previous Systems.
6. **Power System, Super magnet**, direct conversion power.

7. **Power System, Direct Energy Conversion** of incoming magnetic waves from the Galaxy and the Cosmos, technology transfer from Solar Power Systems. It's a resource without limits and does not damage the environment.

8. Others not open to discussion at this time. The inventions are the results of about ten years of independent Research and Development. These inventions are not for sale and third party meddling is discouraged. Any deal making will be on the inventors terms only. Any licensee issued requires commercial viability to validate the contract and any game playing for advantage is an immediate invalidation of all prior agreements. Verification of substantial financial capability before any effort at contracting can be considered.

TransWorld Energy
e-mail: donsm1@ibm.net

Reactance charts *continued*

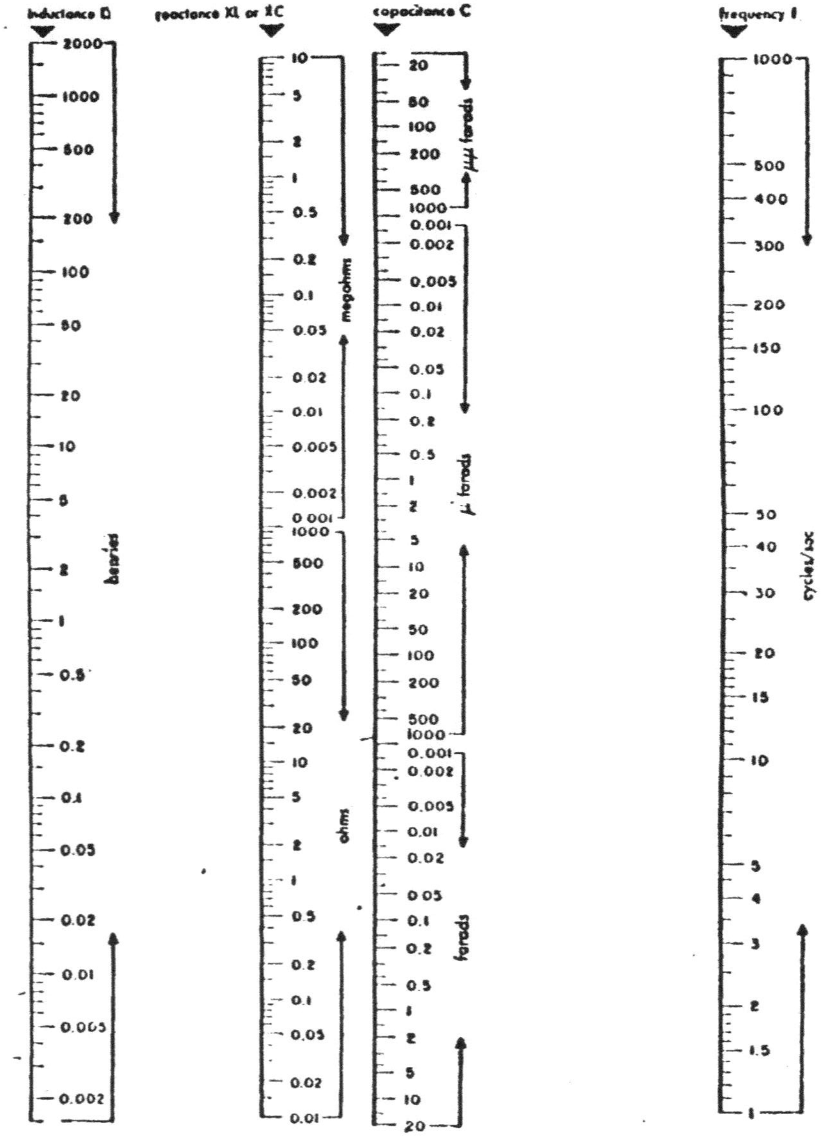

Fig. 4—Chart covering 1 cycle to 1000 cycles.

Reactance charts *continued*

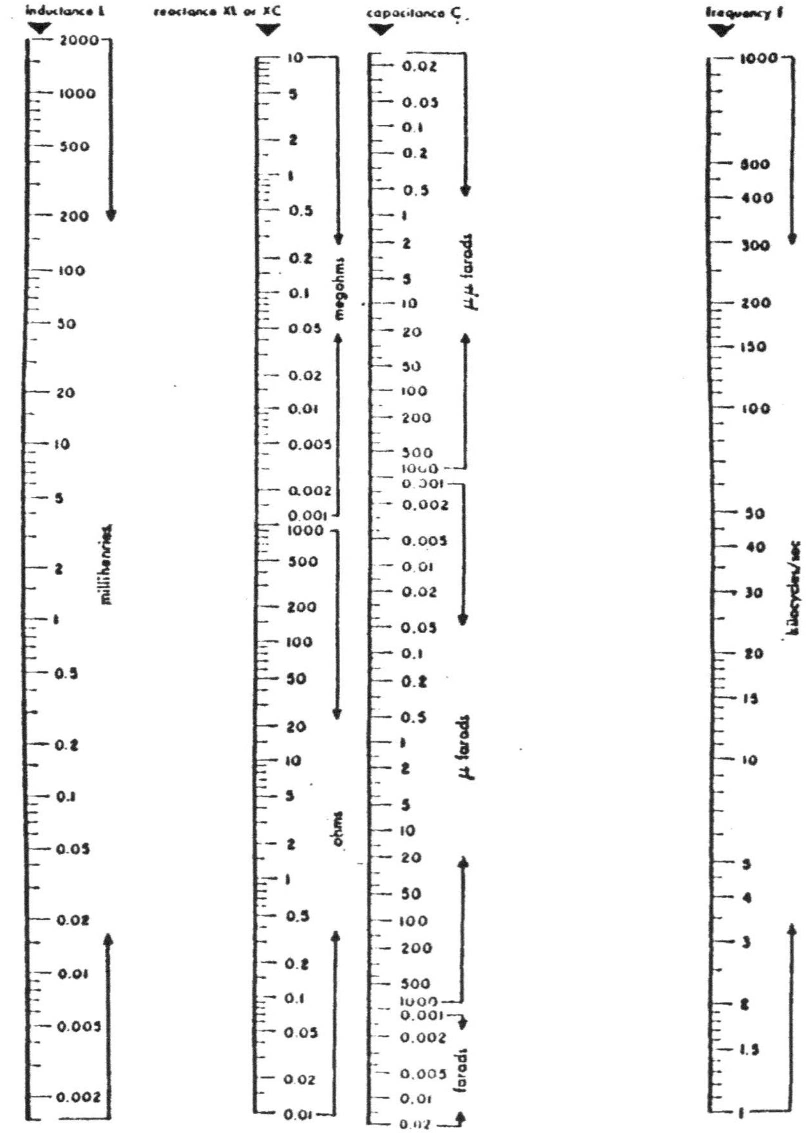

Fig. 5—Chart covering 1 kilocycle to 1000 kilocycles.

Ohm's Law for Reactance

Only ac circuits containing capacitance or inductance (or both) have reactance. Despite the fact that the voltage in such circuits is 90° out of phase with the current, circuit reactance does limit current in a manner that corresponds to resistance. Therefore, the Ohm's Law equations relating voltage, current and resistance apply to purely reactive circuits:

Fig 6.46 charts the reactances of capacitors from 1 pF to 100 µF, and the reactances of inductors from 0.1 µH to 10 H, for frequencies between 100 Hz and 100 MHz. Approximate values of reactance can be read or interpolated from the chart. The formulas will produce more exact values, however.

Although both inductive and capacitive reactance limit current, the two types of reactance differ. With capacitive reactance, the current *leads* the voltage by 90°, whereas with inductive reactance, the current *lags* the voltage by 90°. The convention for charting the two types of reactance appears in **Fig 6.47**. On this graph, inductive reactance is plotted along the +90° vertical line, while capacitive reactance is plotted along the –90° vertical line. This convention of assigning a positive value to inductive reactance and a negative value to capacitive reactance results from the mathematics involved in impedance calculations.

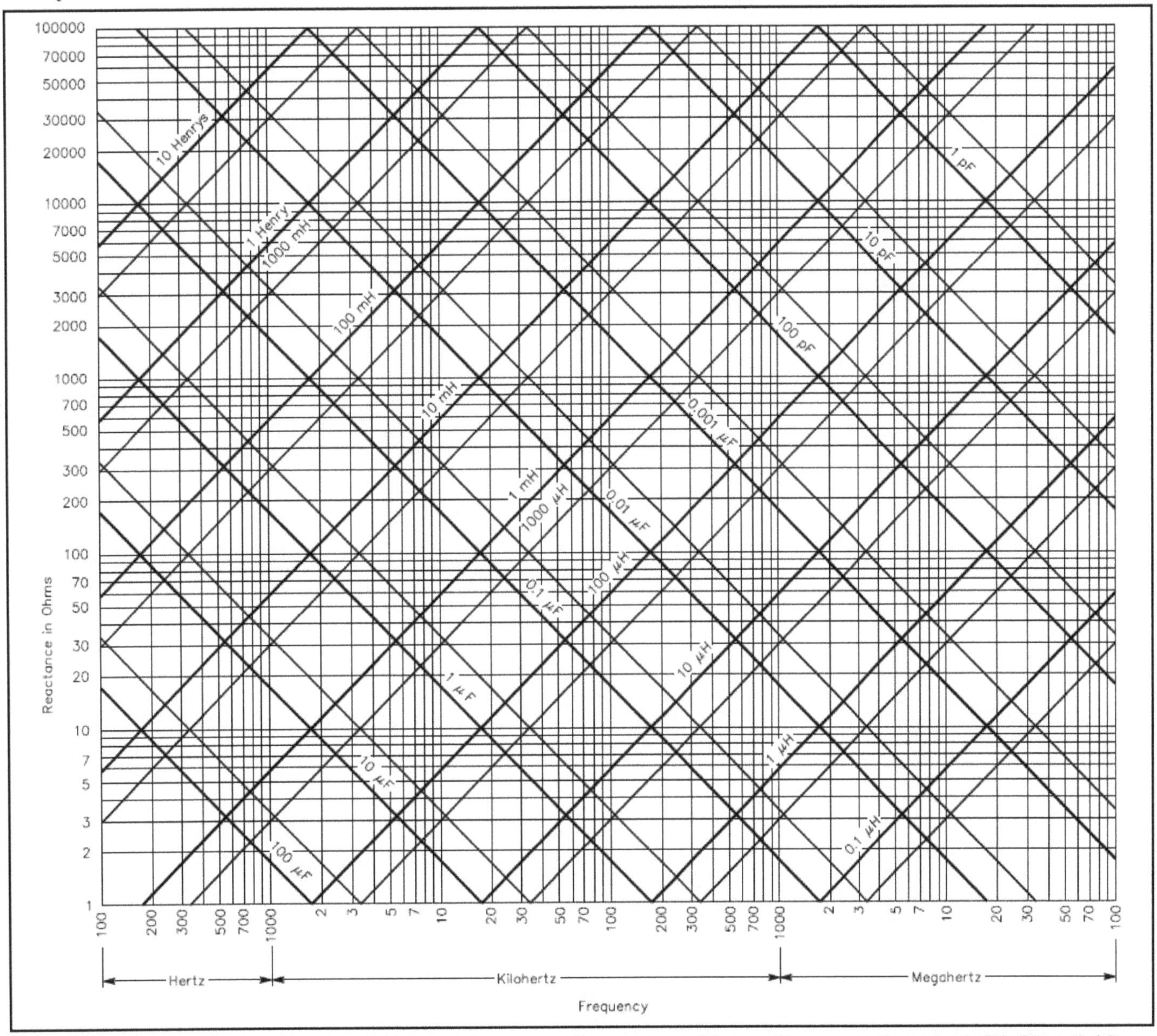

Fig 6.46 — Inductive and capacitive reactance vs frequency. Heavy lines represent multiples of 10, intermediate lines multiples of 5. For example, the light line between 10 µH and 100 µH represents 50 µH; the light line between 0.1 µF and 1 µF represents 0.5 µF, and so on. Other values can be extrapolated from the chart. For example, the reactance of 10 H at 60 Hz can be found by taking the reactance of 10 H at 600 Hz and dividing by 10 for the 10 times decrease in frequency.

The electronic admittance of the tube is seen to be a function of the d-c transit angle and the bunching parameter k, which also involves the transit angle Likewise it is seen that the electronic admittance has both a conductive and a susceptive component depending upon the value of the d-c transit angle Let it be assumed first that the value of k is zero, corresponding to zero r-f gap voltage. The factor $\frac{J_1(k)}{k}$ then has a value of ½ Accordingly, the conductance and susceptance of the beam have the form shown in Fig 17 31. The zero signal value of beam

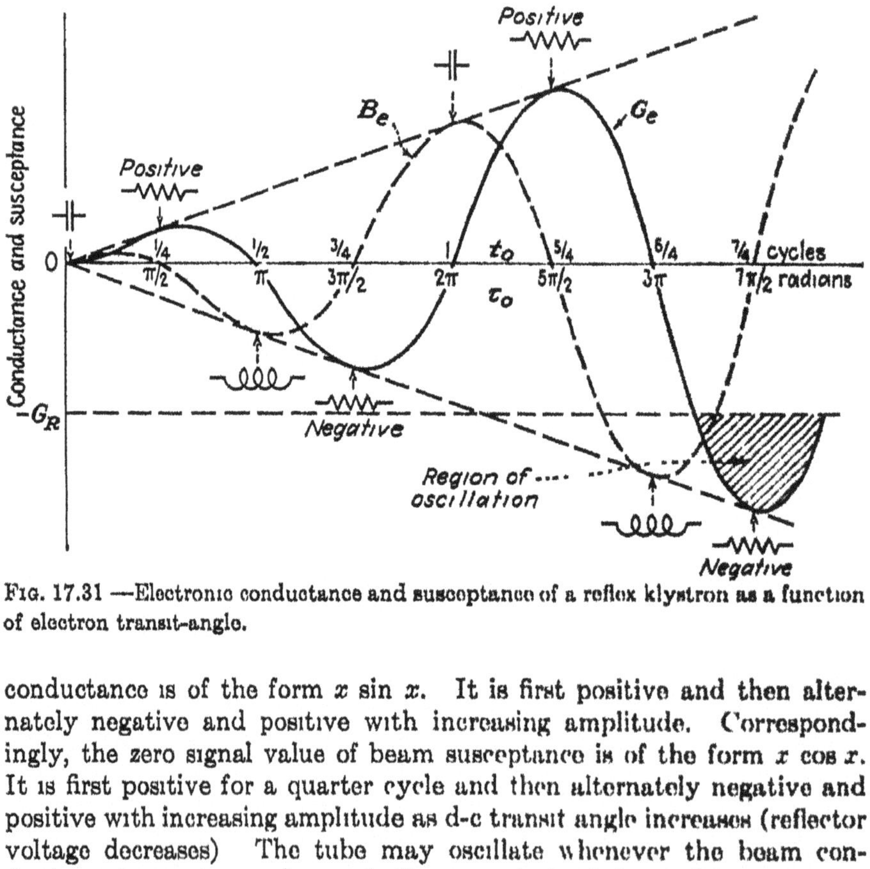

FIG. 17.31 —Electronic conductance and susceptance of a reflex klystron as a function of electron transit-angle.

conductance is of the form $x \sin x$. It is first positive and then alternately negative and positive with increasing amplitude. Correspondingly, the zero signal value of beam susceptance is of the form $x \cos x$. It is first positive for a quarter cycle and then alternately negative and positive with increasing amplitude as d-c transit angle increases (reflector voltage decreases) The tube may oscillate whenever the beam conductance is negative and exceeds the magnitude of the positive resonator conductance. The negative of the resonator conductance is shown in Fig 17.31. For the value shown, oscillations will not occur the first time the beam conductance is negative, for its magnitude is not large enough Oscillations will occur the second time and subsequent times the beam conductance is negative as d-c transit angle is increased, for the magnitude is then greater than the resonator conductance

Mechanism by Which Oscillations Start The above statements about conditions for oscillation are readily demonstrated by reference to the

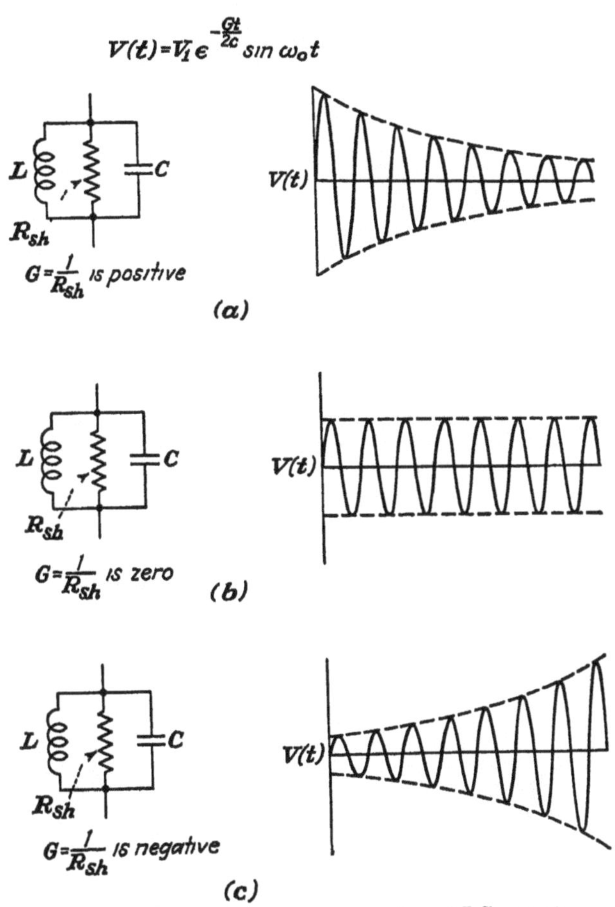

FIG 17 32—Transient response of an RLC circuit

transient response of a parallel combination of a resistance, inductance, and a condenser Let the circuit be as shown in Fig 17 32 The voltage transient across such a circuit that has been shock-excited by some disturbance has the form

$$V(t) = V_1\epsilon^{-\frac{Gt}{2C}} \sin \omega_0 t \qquad (17\ 84)$$

where G is the value of the shunting conductance, C is the capacity, and

DONALD L. SMITH
ENERGY CONSULTANT

HOUSTON

810 BENT OAK LANE
SPRING, TX 77379

(713) 370-4547

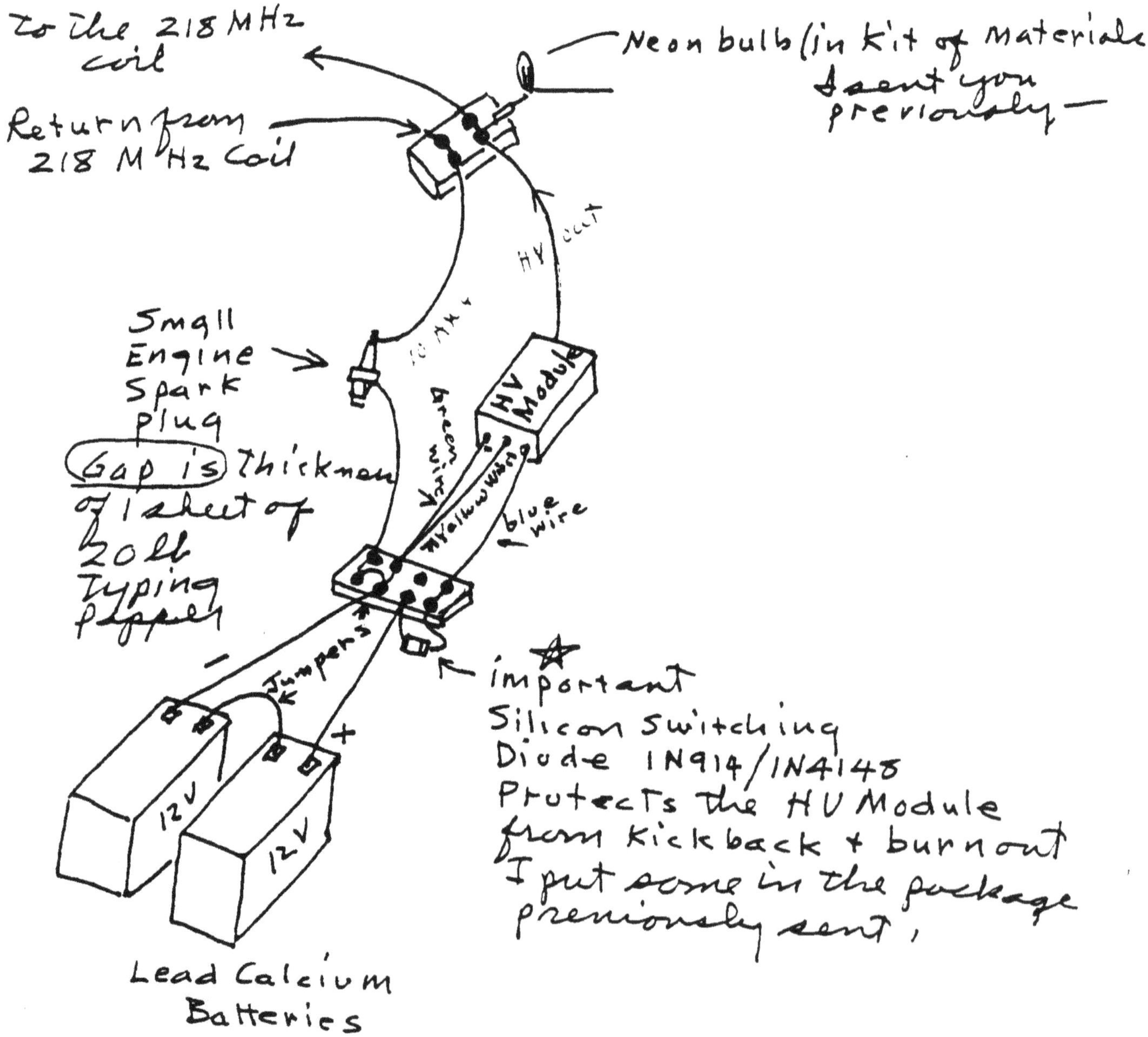

DONALD L. SMITH
ENERGY CONSULTANT
—
HOUSTON

(713) 370-4547

8110 BENT OAK LANE
SPRING, TX 77379

218 MHz Coil

Flat wound on out side edge of 50 MHz Coil

High Freq. Diode Bridge

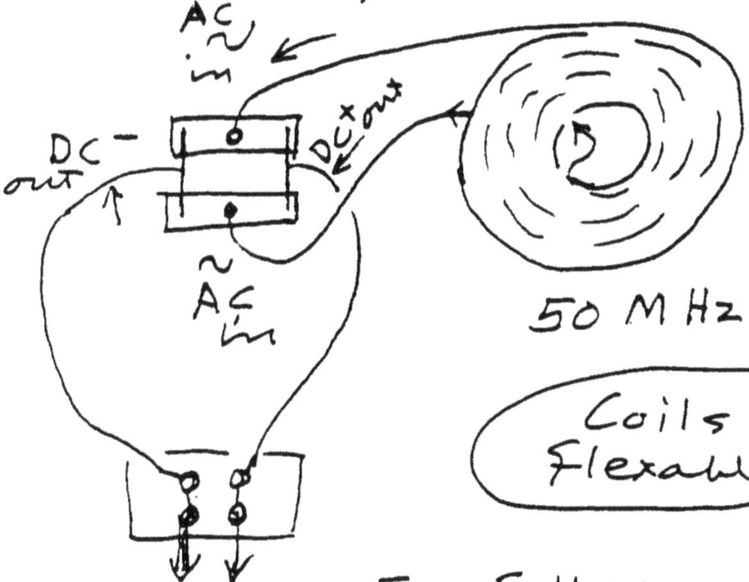

Flat wind on 3½" Centers

50 MHz Coil

Coils are # 6 Covered Flexable Litz type wire

For Full Wave Length Coil at MHz use Desired frequency into 990 for Length of wire having that natural frequency

990 ÷ 50 MHz = 19.8 feet (Make 20' Long)

990 ÷ 218 MHz = 4.54 feet (Make 4 3/4 feet Long

Has to be slightly Longer Than desired wave Length

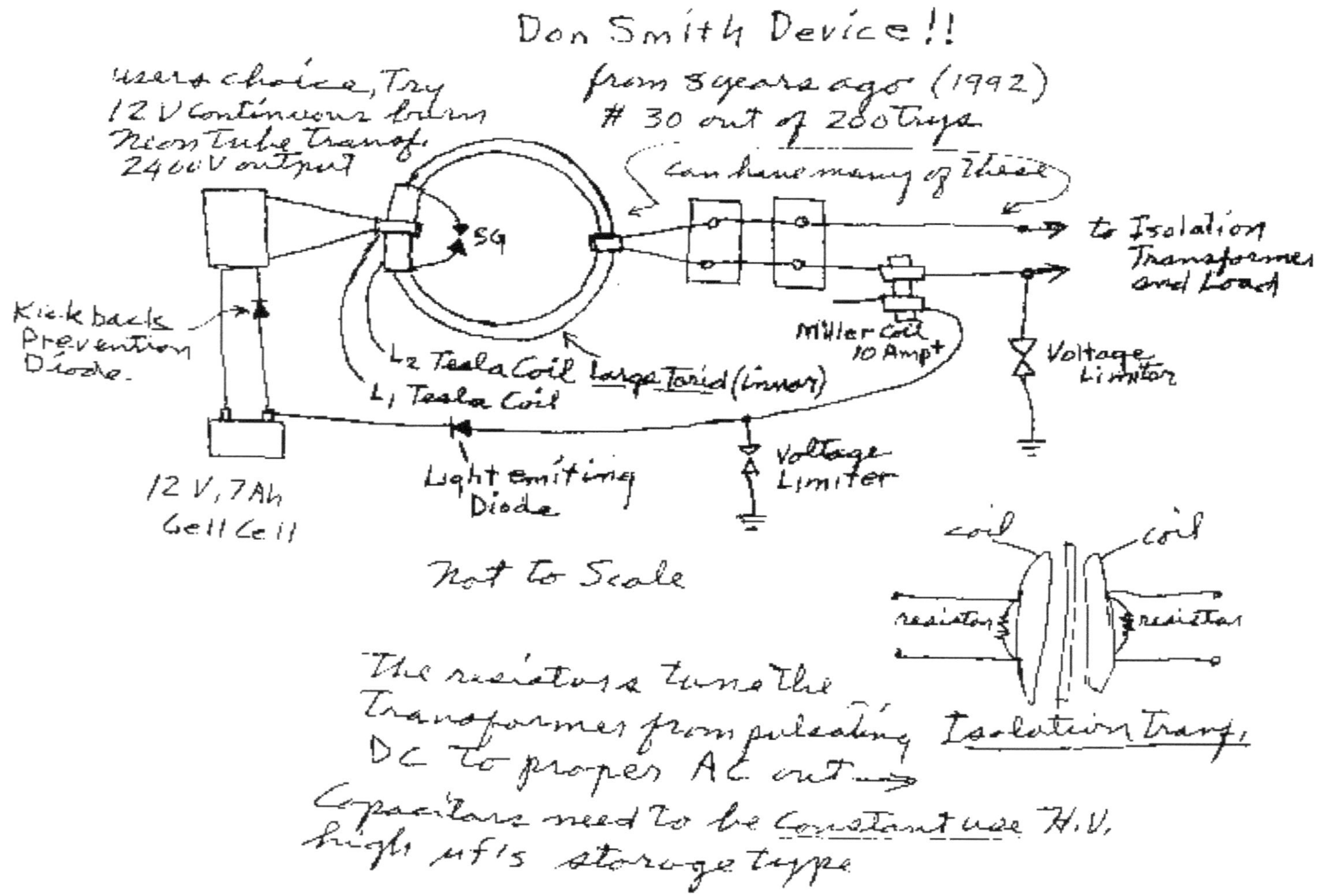
Don Smith Device!!
from 8 years ago (1992)
30 out of 200 trys
users choice, Try
12 V Continuous burn
Neon Tube transf.
2400 V output
can have many of these
SG
to Isolation
Transformer
and Load
Kickback
Prevention
Diode.
Miller Coil
10 Amp+
Voltage
Limiter
L2 Tesla Coil
Large Torid (inner)
L1 Tesla Coil
12 V, 7Ah
Gell Cell
Light emiting
Diode
Voltage
Limiter
coil
coil
resistor
resistor
Not to Scale
The resistors tune the
Transformer from pulsating
DC to proper AC out
Isolation Trans.
Capacitors need to be constant use H.V.
high uf's storage type

R.L. Clark, Ph.D.
4015 Crown Point Dr. P
San Diego, CA 9210[illegible]

1st Generation: Shaped Current Force Field Generators. (High Gain Generator Systems)

(*Other Embodiments of the Concept are Possible)

Dual Field Generator

{#1 Field magnet/coil can be used for D.C. Repulsation Motor/Transformer Drive Rotator; see Alexander Patent # 3,913,004 (1975) by attaching to rotor & central hub.}

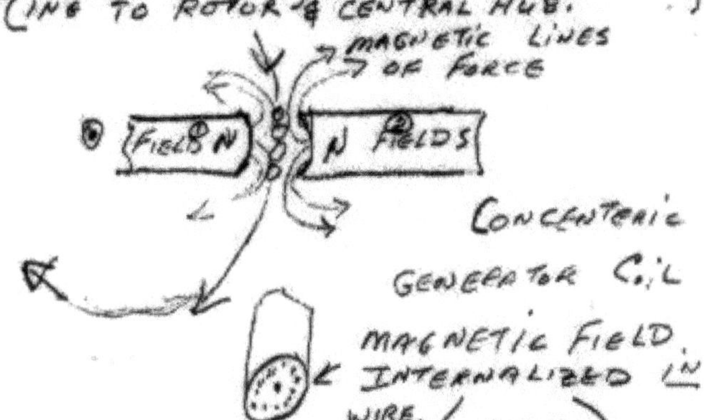

Concenteric Generator Coil

magnetic field internalized in wire. (NEW)

$E = BVdl$

Single Sided Field ("Push")

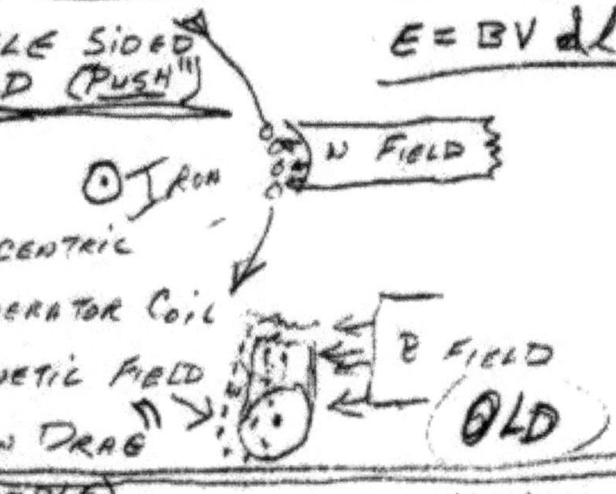

Excentric Generator Coil

magnetic field "Iron Drag"

(Example)

4 Pole Design

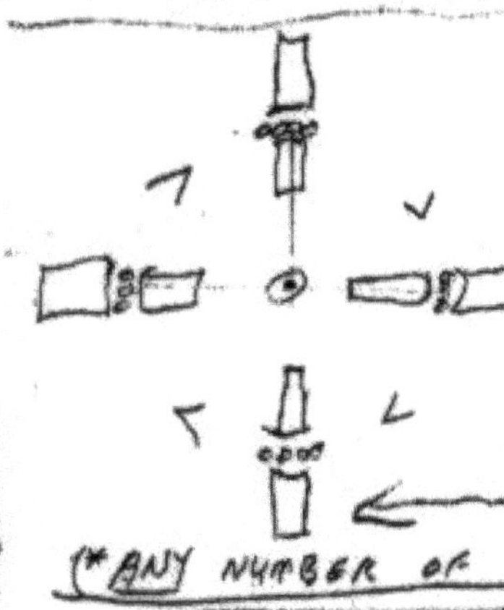

Hollow Gen. Coils Rotating Cylindar

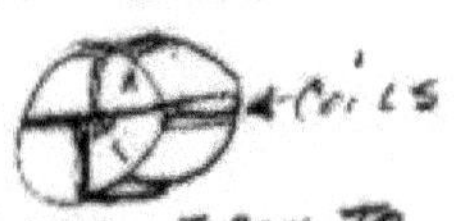

*Add iron to interior if version requires such.

End View

(*Any number of Poles are useable)

Note:

Try Generator Coils both with & without Iron Cores.
*Self Rotating design requires Iron under Coil for transformer action & repeller.

Neutral Center Gen.

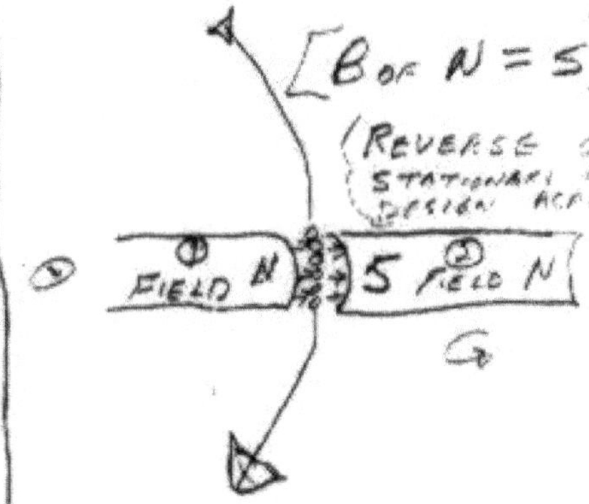

B=0 Generator
"Bloch Wall Generator"
(See, Hooper & Smith Data)

Witness: [signature]
Date: 2/1/2001

Witness: Rebekah Kent
Date: 2-1-01

Designed by: Richard L. Clark, Ph.D.
2/1/01

Magnets can be Permanent or Electro-magnets.

Dr. Richard L. Clark
4015 Crown Point Dr. Apt. P5
San Diego, CA 92109-0274

SEPT 7, 1998 DON SMITH WORKSHOP NOTES.

8 COIL DESIGN

MAGNET SHIELDING – DIO-MAGNET.

MAGNETIC PERMEABILITY.

MAGNETIC SHIELDING INDUSTRY –

THOMAS REGISTRY

SUBSTANCES SHIELD BTWN 5% → 90%.

MAGNETIC SPIN ↓ → GETS DEFLECTED 90°

350 GAIN FINE POWDER.

1 METAL KNOCKS SIDEWAYS

2ND METAL ABSORBS ENERGY.

GAUSSMETER 1 milligauss = 100 V RMS EQUIVALENT.

EVERYTHING ELECTRICAL HAS MAGNET FIELD. (MICROWAVE.)

TAKE FREQ METER + BUILD COIL HAS SAME FREQ AS MAG FLUX FIELD.

THEN MUST CONDITION FREQ'S TO LOWER LEVEL.

CAPACITOR BANK –

1 GAUSS = 100,000 VRMS

RADIO TOWER

↳ BUILD COIL TO PICKUP TOWER MAG ENERGY.

ELECTRONS FLIP – ELECTRIC / MAGNETIC

SPEED YOU FLIP = AMOUNT OF ENERGY AVAILABLE.

THEN YOU CAN HARNESS THIS ENERGY.

FREE ENERGY TERM – ACCEPTED IN CHEMISTRY.

$E = mc^2$ AS APPLIED TO ELECTRICS (SHOULD BE c^3)

M = ELECTRON MASS

TO MEASURE ELECTRIC POTENTIAL – $E = mc^{3}$

YOU CAN USE GAUSSMETER TO MEASURE

FROM A DISTANCE MUST DIVIDE DISTANCE BY C^3 (CUBE DISTANCE)

ANYTIME YOU HAVE GREATER THAN 50% EFFICIENCY

TUNE COIL SYS AROUND ANY MOTOR OR GENERATOR AND PLACE IT IN VICINITY TO PICK UP MAGNETIC FIELD.

MAGNETIC WAVES EXCEED THE SPEED OF LIGHT.

MAGNETIC NEUTRAL AREA

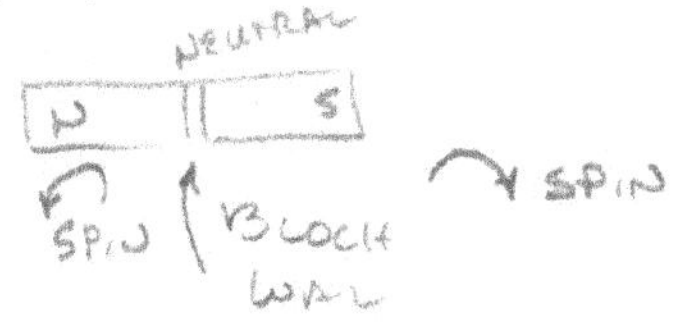

FLIPPING → ELECTRIC TO MAGNETIC

MAGNET – RESONANT SYSTEM IS OF THE METAL OF THE MAGNET

SUPERMAGNET BATTERY –

GROUND ELECTRICAL PICKUP IN THE GIGAHERTZ RANGE

PROPULSION / STAR SYSTEM

DIO MAGNETIC – AGAINST MAGNET UP TO 10,000 X MAGNETIC

PARA MAGNETIC –

RAIL GUN ANALOGY

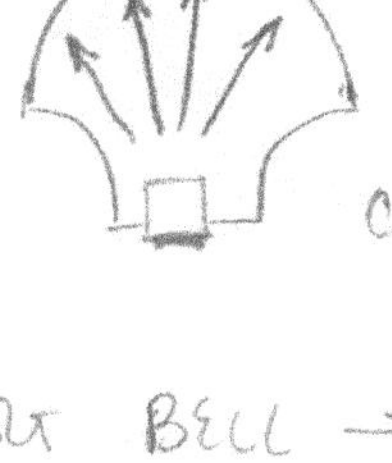

CREATES INTENSE MAGNETIC FIELD.

ART BELL → INTERNET

13½ DISC –

HOLES ½ DIAMETER OF MAGNET.

100 gauss

COMP 200 gauss MAGNETIC FIELD

CATAGORY 4 MAGNETS.

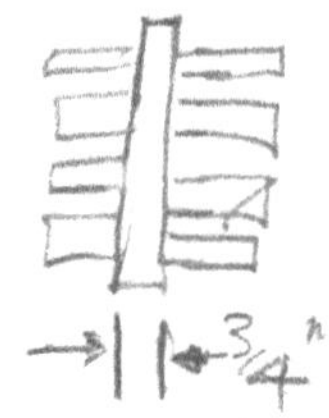

1000V x 50 AMPS x 8

GATEWAY ELECTRONICS –
MULTISTRAND WIRE (COPPER) #10
FLEX DC WIRE –

CYLINDERS FILLED WITH SUPER MAGNETS

ROTATE MAG

1½ DIAMETER MAGNETS.

WIND IN SAME DIRECTION ON OTHER SIDE.

CAPACITOR BANK & BLOCKED OFF
BACK TO PULSATING DC FEED THROUGH
TRANSFORMER. TAKE LC READINGS
120V

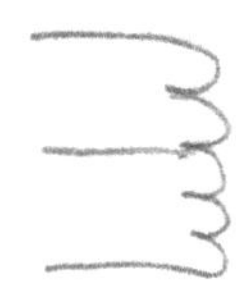

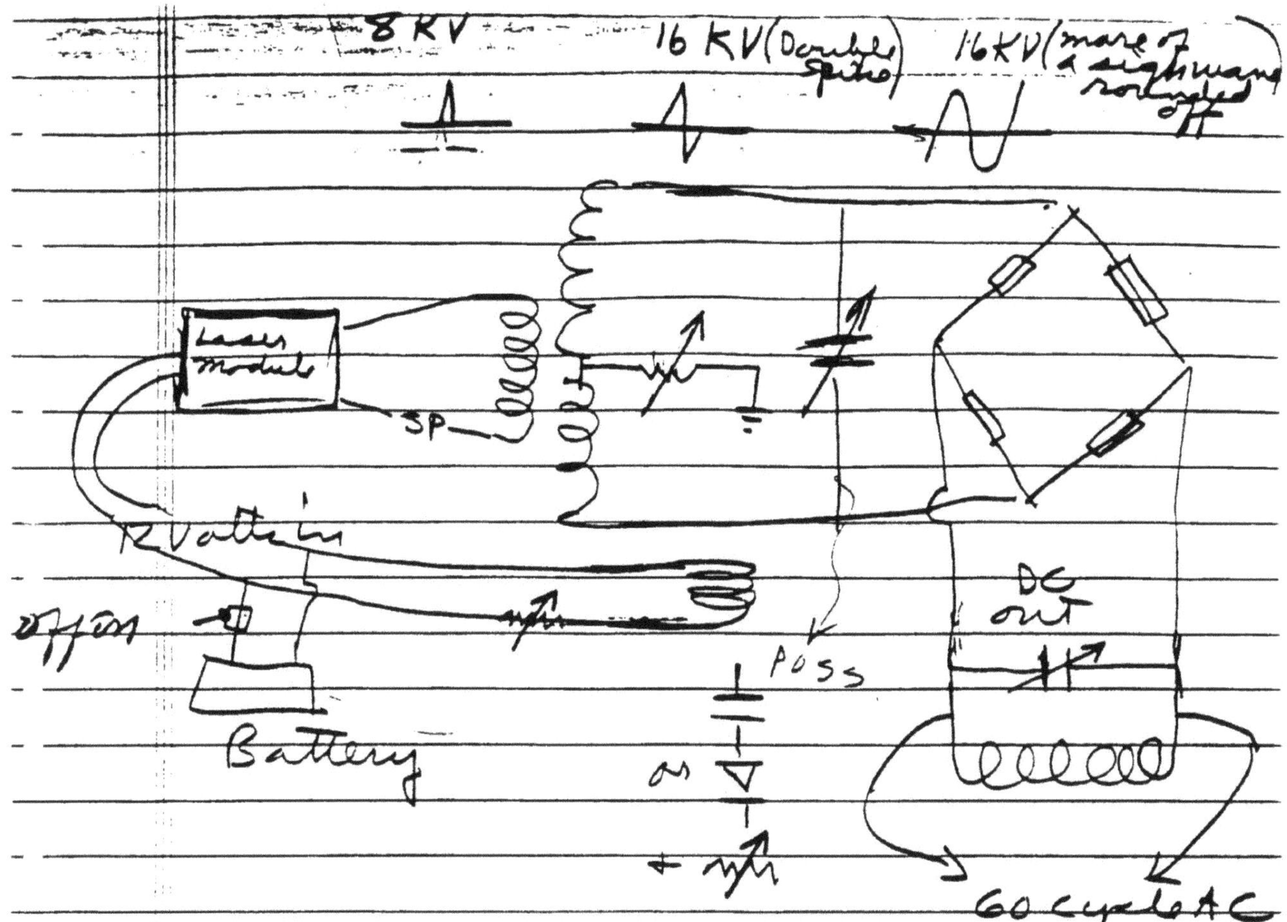

What Every One Knows, and What if! (Joe Gargola)

Then Compair Ions, atoms + molecules, all have negative and positive charactiristics which make them usefull –

For example the electrons, (charged parts) of an atom can be rearranged or pulled loose from the atom by slight agitation (weak force), Then while the positive part of the atom (protons) must be obtained

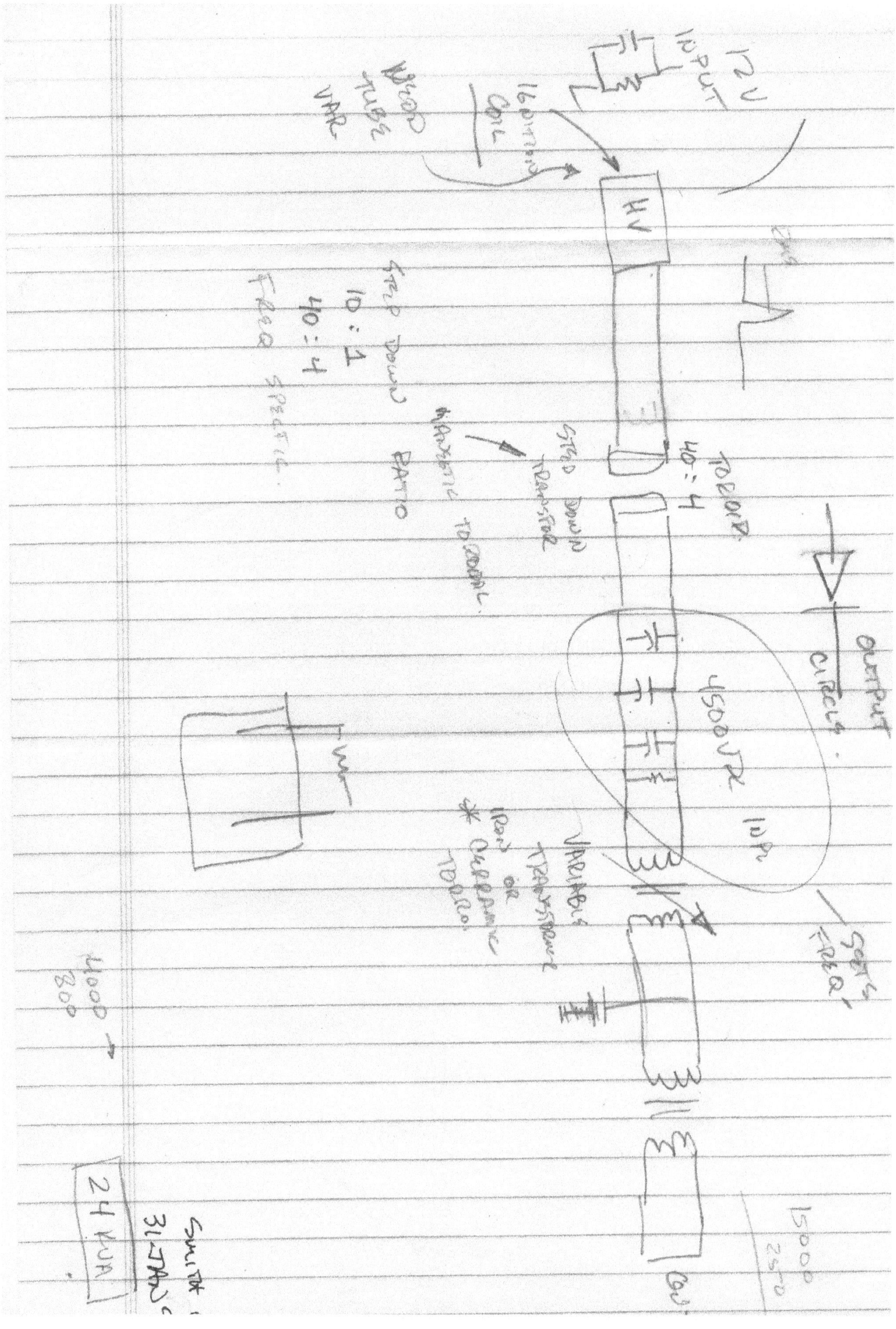
12 V INPUT
HV
IGNITION COIL
TUBE VAR
TOROID
40 : 4
STEP DOWN TRANSFOR
MAGNETIC TOROIDAL
STEP DOWN RATIO
10 : 1
40 : 4
FREQ SPECIFIC
OUTPUT CIRCLE
4500VPK
1NPK
SETS FREQ
VARIABLE TRANSFORMER
IRON OR
* CERRAMIC TORRO
15000
250
4000
800
24 KVA

PRECISION VOLTAGE INVERTER

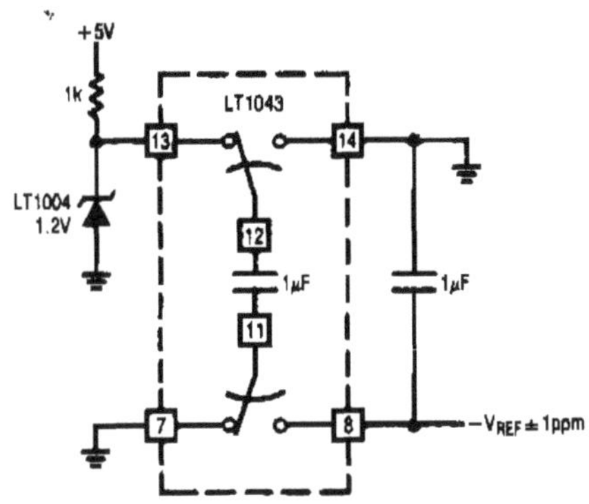

LINEAR TECHNOLOGY CORP. *Fig. 49-5*

This circuit allows a reference to be inverted with 1 ppm accuracy, features high input impedance, and requires no trimming.

POWER INVERTER

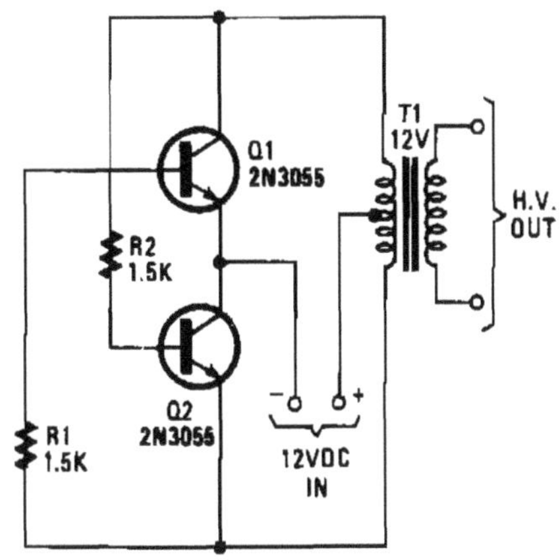

POPULAR ELECTRONICS *Fig. 49-6*

The transformer can be any 6.3 or 12.6 V type. Apply the 12-Vdc input so the positive goes to the transformer's center tap and the negative goes to the two transistor emitters. Any bridge-type rectifier and filter can be used at the output, if you need dc.

12-VDC BATTERY-OPERATED 120-VAC POWER SOURCE

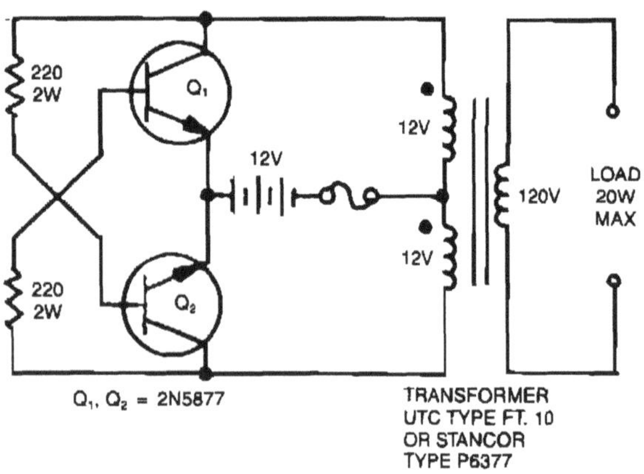

EDN

Fig. 73-10

A simple 120 V: 24 V, center-tapped control transformer and four additional components can do the job. This circuit outputs a clean 200 V pk-pk square wave at 60 Hz and can supply up to 20 W. The circuit is self-starting and free-running.

If Q1 is faster and has a higher gain than Q2, it will turn on first when you apply the input power and will hold Q2 off. Load current and transformer magnetizing current then flows in the upper half of the primary winding, and auto transformer action supplies the base drive until the transformer saturates. When that action occurs, Q1 loses its base drive. As it turns off, the transformer voltages reverse, turning Q2 on and repeating the cycle. The output frequency depends on the transformer iron and input voltage, but not on the load. The frequency will generally range between 50 to 60 Hz with a 60-Hz transformer and car battery or equivalent source. The output voltage depends on turns ratio and the difference between input voltage and transistor saturation voltage. For higher power, use larger transformers and transistors. This type of inverter normally is used in radios, phonographs, hand tools, shavers, and small fluorescent lamps. It will not work with reactive loads (motors) or loads with high inrush currents, such as coffee pots, frying pans, and heaters.

SIMPLE POWER SUPPLY

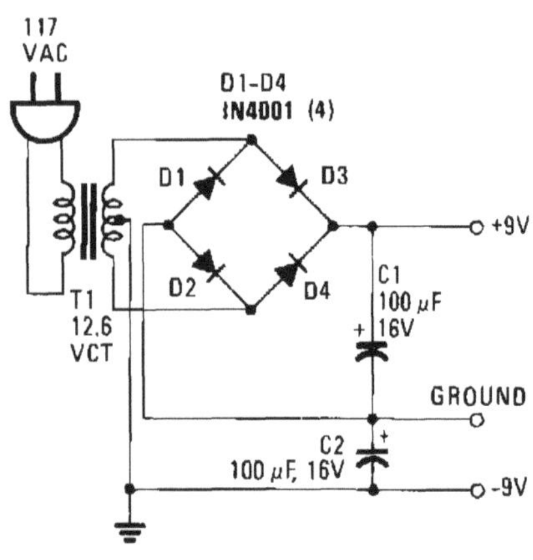

HANDS-ON ELECTRONICS

Fig. 73-11

This power supply delivers plus and minus 9 V to replace two 9-V batteries. The rectifier circuit is actually two separate full-wave rectifiers fed from the secondary of the transformer. One full-wave rectifier is composed of diodes D1 and D2, which develop +9 V, and the other is composed of D3 and D4, which develop −9 V.

Each diode from every pair rectifies 6.3 Vac, half the secondary voltage, and charges the associated filter capacitor to the peak value of the ac waveform, 6.3 × 1.414 = 8.9 V. Each diode should have a PIV, Peak Inverse Voltage, rating that is at least twice the peak voltage from the transformer, 2 × 8.9 = 18 V. The 1N4001 has a PIV of 50 V.

[Editor's note: The positive and negative notation in the center of the inverter is backwards. This is a mistake. On page 298 of this publication, which we have on our previous page 109, a similar schematic shows the correct polarity.]

Capacitor C5 and potentiometer R12 determine the frequency of the output signal at pin 3 of IC1, the 555 oscillator. The output signal is differentiated by C3 and C4 before it's input to the base of power transistors Q1 and Q2 via diodes D1 and D2, respectively. The signal from IC1 is adjusted to 120 Hz, because the flip-flop formed by transistors Q3 and Q4 divides the frequency by 2.

When Q3 is on, the base of Q1 is connected via R1 to the regulated 12-V supply. Then, when the flip-flop changes states, Q4 is turned on and the base of Q2 connected to the 12-V supply through R2. The 100 mA base current allows Q1 and Q2 to alternately conduct through their respective halves to the transformer's secondary winding.

To eliminate switching transients caused by the rapid switching of Q3 and Q4, capacitors C1 and C2 filter the inputs to the base of Q1 and Q2 respectively. Power for the unit comes from an automobile's 12-V system or from a storage battery. The power is regulated by IC2, a 7812 regulator. LED1, connected across the 12-V input, can be used to indicate whether power is being fed to the circuit. The neon pilot lamp, LMP1, shows a presence or absence of output power.

POWER MOSFET INVERTER

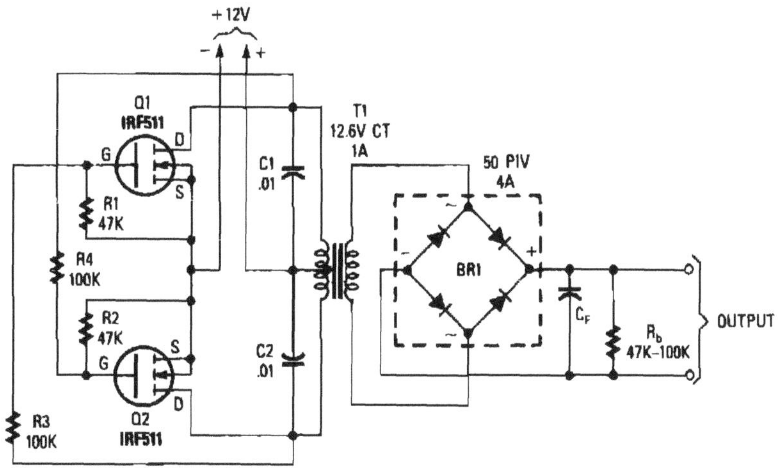

POPULAR ELECTRONICS *Fig. 49-2*

This inverter can deliver high-voltage ac or dc, with a rectifier and filter, up to several hundred volts. The secondary and primary of T1—a 12.6 to 440 V power transformer, respectively—are reversed; e.g., the primary becomes the secondary and the secondary becomes the primary. Transistors Q1 and Q2 can be any power FET. Be sure to heat sink Q1 and Q2. Capacitors C1 and C2 are used as spike suppressors.

high efficiency; lightweight, uniform construction; high power density; low manufacturing times and high frequencies of operation. Some disadvantages include high costs of design and tooling for printed circuit boards and special ferrite cores, inefficient means of terminating windings to the board, and thermal temperature rise of magnetics.

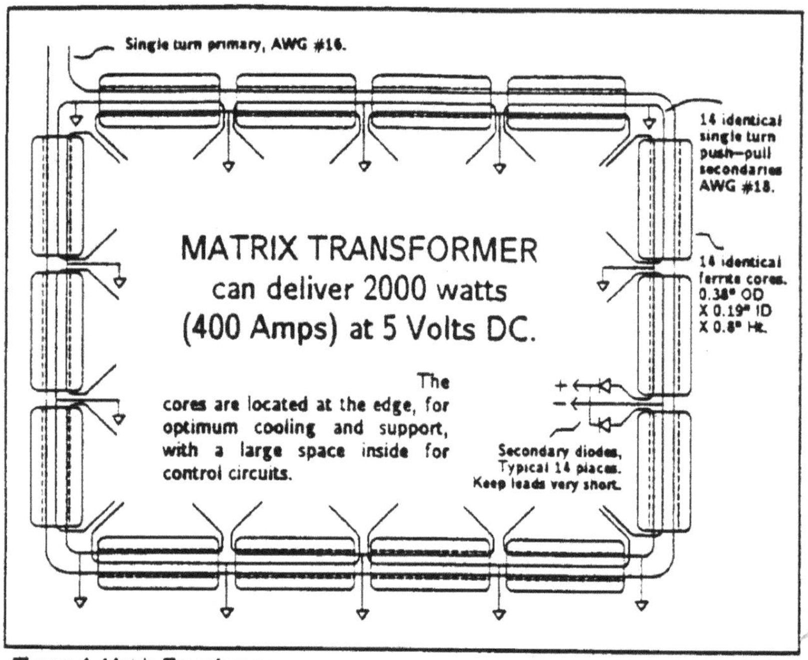

Figure 4. Matrix Transformer.

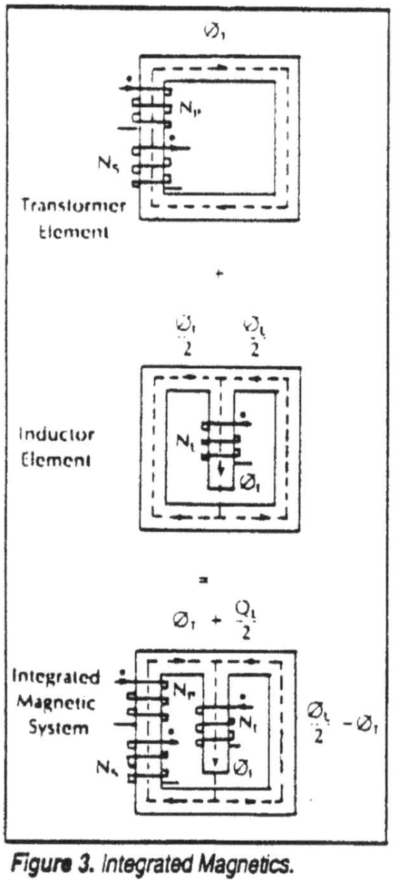

Figure 3. Integrated Magnetics.

A different approach to planar magnetics is the multilayer circuit board, a more condensed configuration. The primary and secondary windings are stacked vertically on different layers of a circuit board with a thin film of FR4 separating them. The multilayer circuit board offers all the advantages of the planar configuration and provides an even lower profile transformer. An additional advantage of the multilayer build is that heat is transferred out of the transformer more efficiently so that thermal rise is not as much of a problem as it is with planar magnetics.

Another magnetic breakthrough is the concept of integrated magnetics: combining the inductive and transformer functions within one magnetic core, shown in *Figure 3*[2]. This combination significantly reduces the total volume of an inductor/transformer combination without compromising electrical characteristics. The integration of the inductor and transformer create significant space reductions on the board. However, the optimum size requires custom cores and bobbins because of the unequal window and core area requirements of the transformer and inductor[3].

Matrix Transformers

The matrix transformer, created by FMTT Inc., is another recent development in transformer technology shown in *Figure 4*[4]. This new configuration uses many different cells or elements that consist of groups of two or more primary and secondary wire combinations. These combinations may be connected either in series or parallel, depending on the designer's needs. The wires pass through cores that are generally toroids or beads formed of high frequency ferrite material, or if necessary, a molyperm powdered iron material for energy storage.

A matrix transformer has several advantages over conventional transformers, such as a low profile construction, low leakage inductance, excellent winding-to-winding isolation, higher thermal dissipation due to increased surface area of cores and windings, highly accurate voltage ratios and inexpensive, readily available cores. Some disadvantages of this configuration include customized tooling for the new encapsulated core modules and a fairly large surface area that requires more board space. FMTT has made the matrix transformer technology available for license.

References

[1] Buerhker, D.J. Harper, "Improved Ferrite Materials for High-Frequency Power Supplies," Proceedings for the First High-Frequency Power Conversion Conference, 1986.

[2] Gordon Bloom, "Core Selection for Integrated-Magnetic Power Converters," e/j Bloom Associates Inc.

[3] G.E. Bloom, "New Integrated-Magnetic DC-DC Power Converter Circuits & Systems," 1987 Applied Power Electronics Conference Record, pp. 57-66 (IEEE Publication 87CH24C6-6).

[4] Edward Hebert, "Design and Application of Matrix Transformers and Symmetrical Converters," Fifth International High Frequency Power Conversion Conference 1990.

HOW TO SPECIFY POWER TRANSFORMER & FILTER RATINGS

The purpose of this section is to provide a practical guide for the selection of a power supply transformer and filter components. A number of basic assumptions are made to avoid an academic discussion of unnecessary material. For those interested in a rigorous theoretical analysis, there are a number of fine references available. Additionally, circuit analysis using appropriate analysis software (SPICE or its equivalent) is recommended in the rare circumstances where a better understanding of a particular situation is needed, or when it becomes necessary to optimize some aspect of the design. Computer analysis is particularly useful in understanding areas that are difficult to approach using traditional circuit analysis methods, areas such as capacitor RMS ripple current.

One of the more esoteric problems encountered by the circuit designer is the selection of power transformer ratings for a particular DC power supply. The designer is immediately confronted with a number of rectifier circuits and filter configurations. For the sake of simplicity, we will make some assumptions which should be valid for 99% of the average designer's applications and present some useful rules of thumb appropriate for conservative design.

Filters

We will immediately discard the consideration of choke input filters and confine our choice to capacitor input filters because of the following:

1. It is desirable to eliminate the weight and cost of chokes.
2. It can be assumed that if a regulator is used it will provide sufficient extra ripple reduction so that an L-C section is not required. In addition, the regulator will compensate for for the poor output voltage regulation with load, inherent in capacitor input systems.

The remaining disadvantages of the capacitive input filter system are caused by the discontinuous secondary current flow (high peak-to-average ratio of forward diode current). Current is drawn in short, high amplitude pulses to replace the charge of the filter capacitor which discharges into the load during diode off time. This results in higher effective RMS values of transformer secondary current. However, the transformer average VA rating is the same as the choke input filter because the higher DC output voltage obtained at the capacitor compensates for this effect. In addition, except perhaps for supplies handling very high currents, average semiconductor diodes will meet most of the peak or surge current requirements of capacitive filters.

Rectifier Circuit

The remaining choice is that of a rectifier circuit configuration. The most common single phase circuits are:

1. Half-Wave (single diode)
2. Full-Wave Center-Tapped (two diodes)
3. Full-Wave Bridge (four diodes)
4. Dual Complementary Supply – "Full-Wave Center Tap" (four diodes)

The only advantages of the half-wave rectifier are its simplicity and the savings in cost of one diode. Its disadvantages are many:

1. Extremely high current spikes are drawn during the capacitor charging interval (only one current surge per cycle). This current is limited only by the effective transformer and rectifier series impedance, but it must not be too high or it will result in rectifier damage. This short once-per-cycle current spike also results in very high secondary RMS currents.
2. The unidirectional DC current in the transformer secondary biases the transformer core with a component of DC flux density. As a result, more "iron" is needed to avoid core saturation.

About the only time it would pay to consider using the half-wave rectifier is for very low DC power levels of about ¼ watt or less. At these levels a power transformer cannot be reduced very much in size (at reasonable cost) and a small filter capacitor will be large enough for adequate DC smoothing.

The remaining single-phase rectifier circuits are of the "full-wave" type. Secondary current surges occur twice per cycle so that they are of smaller magnitude and the fundamental ripple frequency is double the supply frequency (i.e. 120Hz rather than the 60Hz of a half-wave system). All full-wave rectifiers also have the same basic rectified waveform applied to the filter capacitor.

Other Factors

- **Full-Wave Center-Tap**
  Uses ½ of secondary winding at a time
  Requires center-tap
  Uses 2 diodes
- **Full-Wave Bridge**
  Uses full secondary winding continuously
  No center-tap required
  Uses 4 diodes

As can be seen above, the choice between FWCT and Bridge configurations is a tradeoff. The bridge rectifier has the best transformer utilization but requires the use of 4 diodes. The extra diodes result in twice the diode voltage drop of a FWCT circuit so that the FWCT is usually preferable in low voltage supplies.

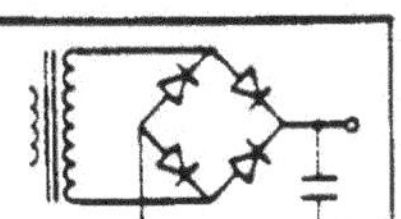

Full Wave Bridge

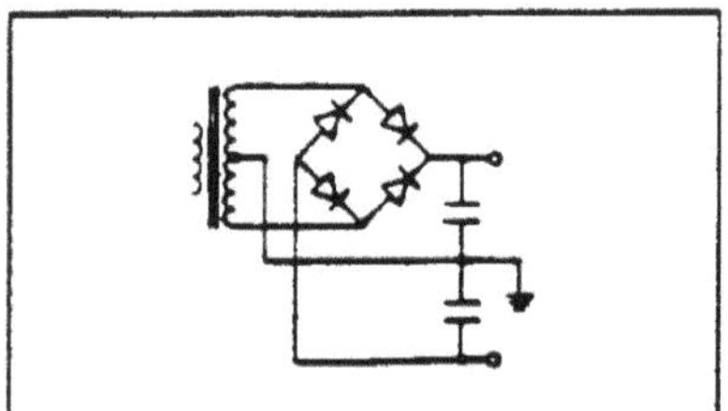

Dual Complementary Rectifier

The "dual complementary rectifier circuit" is the combination of two FWCT circuits and is a very efficient way of obtaining two identical outputs of reversed polarity sharing a common ground. It is also called a "center-tapped bridge rectifier."

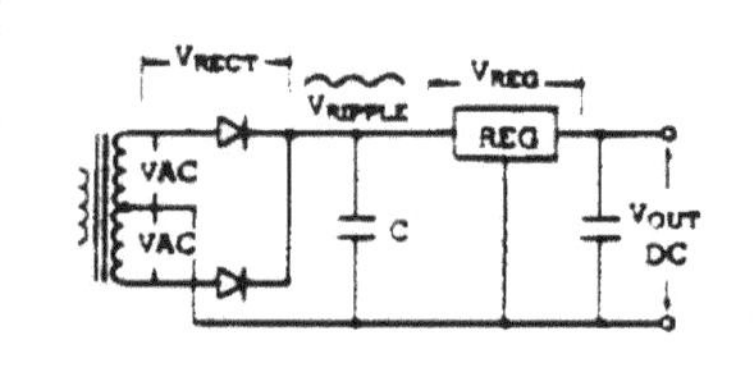

Full Wave Center Tap

The above diagram represents a full-wave center-tapped rectifier using a capacitive filter and is the most common selection for moderate power, regulated DC supplies. The following assumptions can safely be made:

1. V_{REG} is approximately 3 volts DC or greater.
2. V_{RECT} is about 1.25 volts DC.
3. V_{RIPPLE} is about 10% VDC peak.

The following formula may be used for determing the transformer secondary voltage:

$$V_{AC} = \frac{V_{OUT} + V_{REG} + V_{RECT} + V_{RIPPLE}}{0.9} \times \frac{V_{NOM}}{V_{LOWLINE}} \times \frac{1}{\sqrt{2}}$$

where: 0.9 = rectifier efficiency (typical)

$\frac{V_{NOM}}{V_{LOWLINE}}$ is the ratio of the nominal AC line voltage to the required low line conditions.

A sample illustration of the above will be shown for a supply requiring an output of 5V DC at 2A DC to operate down to an input voltage of 95V RMS.

$V_{OUT} = 5V$

$V_{REG} = 3V$

$V_{RECT} = 1.25V$

$V_{RIPPLE} = 0.5(1V\ p\text{-}p)$

$$V_{AC} = \frac{9.75}{0.9} \times \frac{115}{95} \times \frac{1}{\sqrt{2}} = 9.27V_{AC}$$

Therefore, the transformer secondary voltage can be specified as about 18 V CT. For a bridge rectifier of the same output requirements, the only change is that

$V_{RECT} = 2 \times 1.25 = 2.5V$

As a result V_{AC} will be reformulated as:

$$V_{AC} = \frac{11}{0.9} \times \frac{115}{95} \times \frac{1}{\sqrt{2}} = 10.46\ V_{AC}$$

so that the transformer secondary voltage now becomes about 10.5V.

Transformer Secondary Current

The remaining step is to determine the transformer RMS secondary circuit. This can be accurately determined only by complex analysis. However, for practical engineering purposes the chart below may be used.

Rectifier Type	Filter Type*	Required RMS Secondary Current Rating
Full-Wave Center-Tap	Choke Input	0.7 x DC Current
Full-Wave Center-Tap	Capacitor Input	1.2 x DC Current
Full-Wave Bridge	Choke input	DC Current
Full-Wave Bridge	Capacitor Input	1.8 x DC Current

*Even though we have dropped choke input filters from this discussion, they are included for reference.

For instance, in our particular example (5 V, 2A DC supply) the transformer RMS current would be:

for FWCT $1.2 \times 2 = 2.4$ A
for bridge $1.8 \times 2 = 3.6$ A

The total transformer specification would then be:

Circuit	Secondary Rating	Possible "Signal" Parts.
FWCT	18.5 CT @ 2.4 A RMS = 43.2 VA	241-7-20, 36-1
bridge	10.5 @ 3.6 A RMS = 36 VA	ST-7-10, 241-6 or 7-10, 10-4

Dual Complementary Supply

One more common example will be given, i.e., a dual complementary supply for ± 15 V @ 100 mA DC.

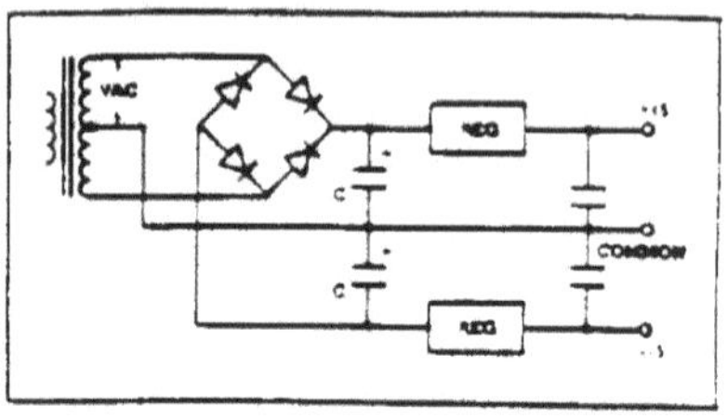

$V_{OUT} = \pm 15$ $V_{RECT} = 1.25$

$V_{REG} = 3$

$V_{RIPPLE} = 0.75$ (1.5 V p-p)

$V_{AC} =$

$$\frac{(15 + 3 + 1.25 + 0.75)}{0.9} \times \frac{115}{95} \times \frac{1}{\sqrt{2}} = 19\ V$$

$I_{AC} = 1.8 \times 100\ mA = 180\ mA$ RMS

The transformer secondary rating is 38 V CT @ 180 mA RMS. Possible Signal parts would be ST-4-36, PC-34-300, PC-40-250. A precautionary calculation remains to be made. That is, the increase in voltage at the filter capacitor (into the regulator) caused by a high line condition. If we assume our highest line voltage to be 130V AC then th transformer output (compared to low line) would rise by the ratio 130/95. In the 5V supply, for instance, the following would happen:

$$V_{AC} = \frac{130}{95} \times 9.27 = 12.7V$$

In the dual complementary +15V supply:

$$V_{AC} = \frac{130}{95} \times 19 = 26V$$

The increase in output must be absorbed by the regulator, which results in higher regulator power dissipation. The illustrated values are safe for the typical IC regulator but should be checked in any specific application.

Summary

This has been an attempt to provide a simple, practical method of determining transformer ratings. Certain basic assumptions have been made and this section is not meant as a rigorous academic analysis. However, such material is readily available in the literature (see below). This, we feel, may help bridge the gap for the working designer.

References

For more detailed theoretical analysis the following are recommended:

1. Reuben Lee, Electronic Transformers & Circuits, 1947, John Wiley & Sons
2. EE Staff – MIT, Magnetic Circuits & Transformers, 1943, John Wiley & Sons
3. O.H. Schade, Proc. IRE, vol 31, p. 356, 1943

Comments on capacitor & diode selection

Capacitor Selection

For low current supplies ($I_{OUT} < IA$) capacitor selection is relatively straightforward. Capacitance is found by the simple formula:

$$C = \frac{I_L \times 6 \times 10^{-3}}{\Delta V}$$

where: I_L = DC load current
ΔV = peak-to-peak ripple voltage
ripple frequency = 120 Hz.

This yields 2000μ F/amp for 3V p-p ripple. At DC currents below 1 amp, capacitor heating is usually not a problem and peak-to-peak ripple voltage is the determining factor in capacitor size.

At higher values of capacitance, where the ratio of capacitor outside surface area to volume is significantly lower, internal heating becomes a problem. Ripple current rating may be a determining factor in capacitor selection, rather than ripple voltage. In many cases, capacitor size will have to be increased to prevent excessive internal heating. Manufacturers' data sheets should be consulted (after an initial selection is made) to ensure that capacitor ripple current ratings are met. Remember that the RMS ripple current ratings shown on capacitor data sheets are not the same as DC load current. RMS ripple current in a capacitor input filter is 2 to 3 times the load current. In addition, the time-to-failure used to rate capacitors on data sheets is often 10,000 hours. For five-year life (40,000 hours), ambient temperature may have to be derated 30°C from the data sheet rating. Capacitor life roughly doubles for each 15°C reduction in operating temperature. The following calculations illustrate a typical design example:

assume $I_L = 3A$, $\Delta V = 4V$ p-p,
$V_{DC} = 12V$

$$C = \frac{(6 \times 10^{-3})\,(3A)}{4V} = 4{,}500\mu F$$

Manufacturer's rating on a 4,600μF/20V capacitor @ T_A = 65°C is 3.1 A RMS. Dividing by 2.5 to convert from RMS ripple current to output current yields a maximum DC load current of 1.24 amps. Obviously either a large capacitor is required or ambient temperature must be reduced.

As a final note, be sure to check whether the data sheet ratings are for still or moving air. Computer grade capacitors are often rated only for moving air. Other types may be rated for still air, and are therefore actually more conservatively rated.

Remember that capacitors are the number one cause of power supply failure. Don't let your supplies dominate the statistics column!

Diode Selection

The RMS value of the current flowing into a capacitor input filter is 2 – 3 times the DC output current because the current is delivered in short pulses. Assuming a full-wave center tap or bridge, this means that although each diode is conducting only on alternate half cycles, it should be rated for at least the full output current. To ensure adequate surge capability during turn-on, a diode rating of at least twice the output current is recommended, especially for higher current supplies where the ratio of filter capacitance to output current is somewhat higher. Keep in mind that axial lead diodes achieve most of their heat sinking through the leads. Short leads soldered to large area standoffs or printed circuit pads are definitely recommended.

For "short circuit proof" IC regulated supplies using three-terminal regulators, an additional diode derating may have to be used. Long-term output shorts do not harm the regulator, which goes into a current limit or thermal limit mode to protect itself. The diodes, however, may experience a substantial current increase during the short. Regulator data sheets should be consulted for current limit values, keeping in mind that current limit is a function of input-output voltage differential. At high input voltages, the short circuit current of IC regulators is often less than full load current, tending to alleviate this problem.

5 TO 900KVPK

B-1011-C

MATCHED PAIRS HIGH VOLTAGE VOLTAGE DIVIDERS

VD-60 B

Matched pairs high voltage dividers for high voltage differential measurements. Precision wide-band 1KV to 900KV 0.1%DC and 0.01%DC. 0.2% to 3% AC accuracies linear within 3%. DC up to 5 MHz.

Output difference as small as 0.1V at any input voltage 1KV to 900KV depending on model. For measuring small continuous, pulsed or transient voltage differences, point to point, or across resistive current shunts with both terminals at the high voltage level and instrumentation at ground level.

These Voltage Dividers can be adjustable to zero out any differential error or extraneous pickup and they are available in any ratio and can be made to match most loads. Matching differential fiber-optic transmission systems are also available for full isolation and minimum pickup for transmission of signals up to one kilometer, further with repeaters.

For more information on specifications and dimensions refer to Brochure B-1011 and B-1011-B or contact Ross Engineering Corporation.

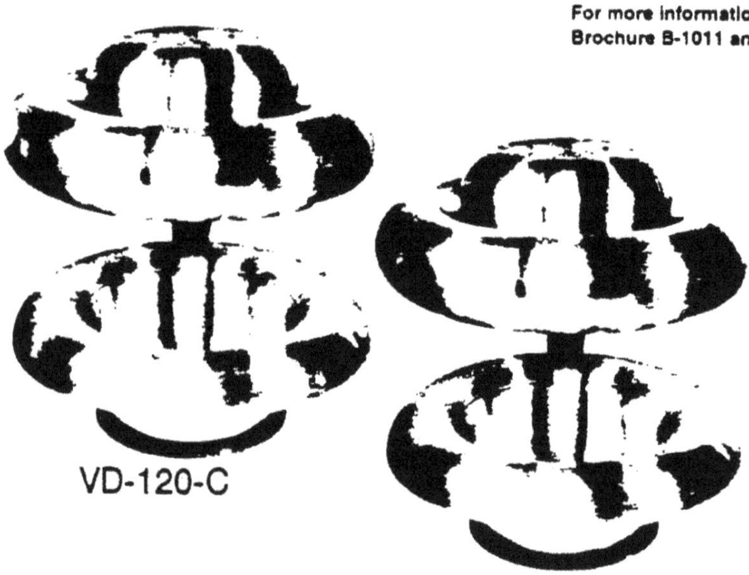

VD-120-C

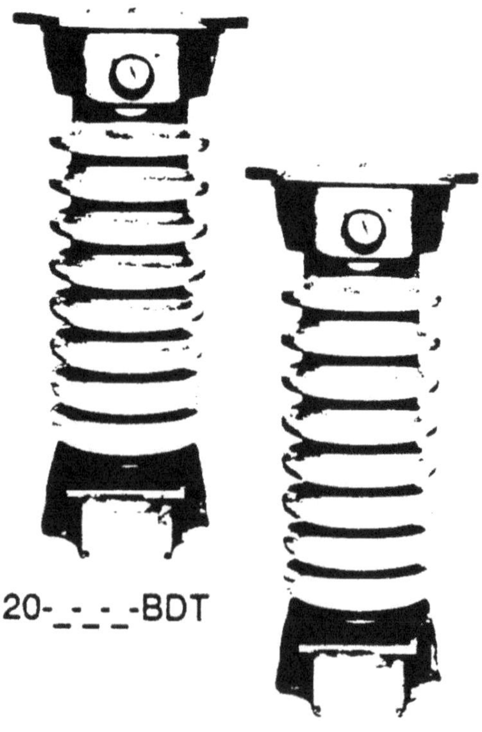

VD-120-_-_-_-BDT

VD-15-A

ROSS ENGINEERING CORP.

CAMPBELL, CA 95008 PH: (408) 377-4621 · FAX (408) 377-5182

1-94/5000

PRINTED IN U.S.A.

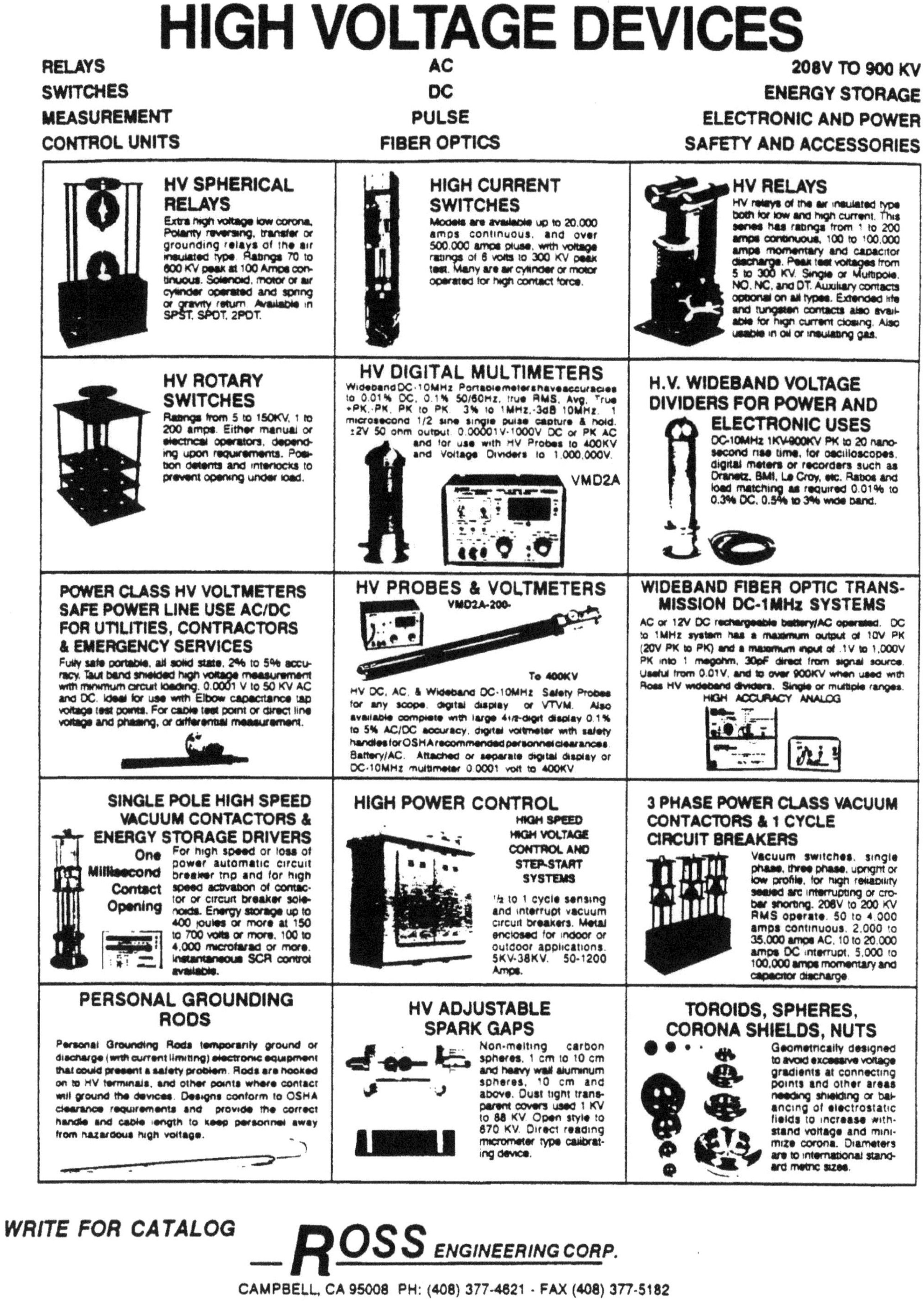
HIGH VOLTAGE DEVICES
RELAYS
SWITCHES
MEASUREMENT
CONTROL UNITS
AC
DC
PULSE
FIBER OPTICS
208V TO 900 KV
ENERGY STORAGE
ELECTRONIC AND POWER
SAFETY AND ACCESSORIES
HV SPHERICAL RELAYS
Extra high voltage low corona. Polarity reversing, transfer or grounding relays of the air insulated type. Ratings 70 to 600 KV peak at 100 Amps continuous. Solenoid, motor or air cylinder operated and spring or gravity return. Available in SPST, SPDT, 2PDT.
HIGH CURRENT SWITCHES
Models are available up to 20,000 amps continuous, and over 500,000 amps pluse, with voltage ratings of 6 volts to 300 KV peak test. Many are air cylinder or motor operated for high contact force.
HV RELAYS
HV relays of the air insulated type both for low and high current. This series has ratings from 1 to 200 amps continuous, 100 to 100,000 amps momentary and capacitor discharge. Peak test voltages from 5 to 300 KV. Single or Multipole. NO, NC, and DT. Auxiliary contacts optional on all types. Extended life and tungsten contacts also available for high current closing. Also usable in oil or insulating gas.
HV ROTARY SWITCHES
Ratings from 5 to 150KV, 1 to 200 amps. Either manual or electrical operators, depending upon requirements. Position detents and interlocks to prevent opening under load.
HV DIGITAL MULTIMETERS
Wideband DC-10MHz Portable meters have accuracies to 0.01% DC, 0.1% 50/60Hz, true RMS, Avg, True +PK, -PK, PK to PK. 3% to 1MHz, -3dB 10MHz. 1 microsecond 1/2 sine single pulse capture & hold. ±2V 50 ohm output. 0.00001V-1000V DC or PK AC and for use with HV Probes to 400KV and Voltage Dividers to 1,000,000V.
VMD2A
H.V. WIDEBAND VOLTAGE DIVIDERS FOR POWER AND ELECTRONIC USES
DC-10MHz 1KV-900KV PK to 20 nanosecond rise time, for oscilloscopes, digital meters or recorders such as Dranetz, BMI, Le Croy, etc. Ratios and load matching as required 0.01% to 0.3% DC, 0.5% to 3% wide band.
POWER CLASS HV VOLTMETERS SAFE POWER LINE USE AC/DC FOR UTILITIES, CONTRACTORS & EMERGENCY SERVICES
Fully safe portable, all solid state, 2% to 5% accuracy. Taut band shielded high voltage measurement with minimum circuit loading. 0.0001 V to 50 KV AC and DC. Ideal for use with Elbow capacitance tap voltage test points. For cable test point or direct line voltage and phasing, or differential measurement.
HV PROBES & VOLTMETERS
VMD2A-200-
To 400KV
HV DC, AC, & Wideband DC-10MHz Safety Probes for any scope, digital display or VTVM. Also available complete with large 4½-digit display 0.1% to 5% AC/DC accuracy, digital voltmeter with safety handles for OSHA recommended personnel clearances. Battery/AC. Attached or separate digital display or DC-10MHz multimeter 0.0001 volt to 400KV.
WIDEBAND FIBER OPTIC TRANSMISSION DC-1MHz SYSTEMS
AC or 12V DC rechargeable battery/AC operated. DC to 1MHz system has a maximum output of 10V PK (20V PK to PK) and a maximum input of .1V to 1,000V PK into 1 megohm, 30pF direct from signal source. Useful from 0.01V, and to over 900KV when used with Ross HV wideband dividers. Single or multiple ranges.
HIGH ACCURACY ANALOG
SINGLE POLE HIGH SPEED VACUUM CONTACTORS & ENERGY STORAGE DRIVERS
One Millisecond Contact Opening
For high speed or loss of power automatic circuit breaker trip and for high speed activation of contactor or circuit breaker solenoids. Energy storage up to 400 joules or more at 150 to 700 volts or more, 100 to 4,000 microfarad or more. Instantaneous SCR control available.
HIGH POWER CONTROL
HIGH SPEED HIGH VOLTAGE CONTROL AND STEP-START SYSTEMS
½ to 1 cycle sensing and interrupt vacuum circuit breakers. Metal enclosed for indoor or outdoor applications. 5KV-38KV. 50-1200 Amps.
3 PHASE POWER CLASS VACUUM CONTACTORS & 1 CYCLE CIRCUIT BREAKERS
Vacuum switches, single phase, three phase, upright or low profile, for high reliability sealed arc interrupting or crobar shorting. 208V to 200 KV RMS operate. 50 to 4,000 amps continuous. 2,000 to 35,000 amps AC, 10 to 20,000 amps DC interrupt. 5,000 to 100,000 amps momentary and capacitor discharge.
PERSONAL GROUNDING RODS
Personal Grounding Rods temporarily ground or discharge (with current limiting) electronic equipment that could present a safety problem. Rods are hooked on to HV terminals, and other points where contact will ground the devices. Designs conform to OSHA clearance requirements and provide the correct handle and cable length to keep personnel away from hazardous high voltage.
HV ADJUSTABLE SPARK GAPS
Non-melting carbon spheres, 1 cm to 10 cm and heavy wall aluminum spheres, 10 cm and above. Dust tight transparent covers used 1 KV to 88 KV. Open style to 870 KV. Direct reading micrometer type calibrating device.
TOROIDS, SPHERES, CORONA SHIELDS, NUTS
Geometrically designed to avoid excessive voltage gradients at connecting points and other areas needing shielding or balancing of electrostatic fields to increase withstand voltage and minimize corona. Diameters are to international standard metric sizes.
WRITE FOR CATALOG
ROSS ENGINEERING CORP.
CAMPBELL, CA 95008 PH: (408) 377-4621 · FAX (408) 377-5182

The curves at the right provide a comparative analysis of some of the key High-Beta parameters as compared to the conventional bipolar devices:

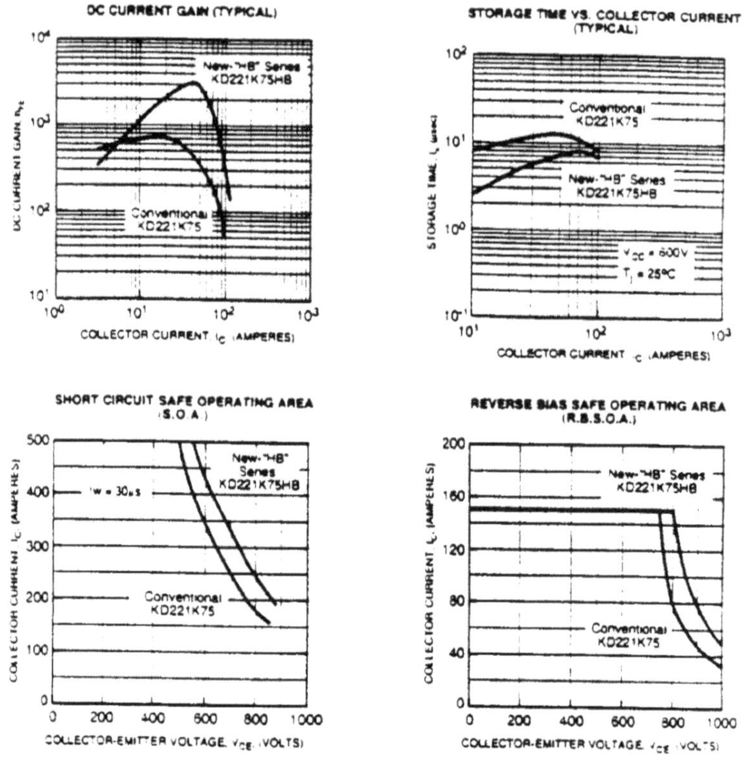

Current Product Range

Future Development Plan

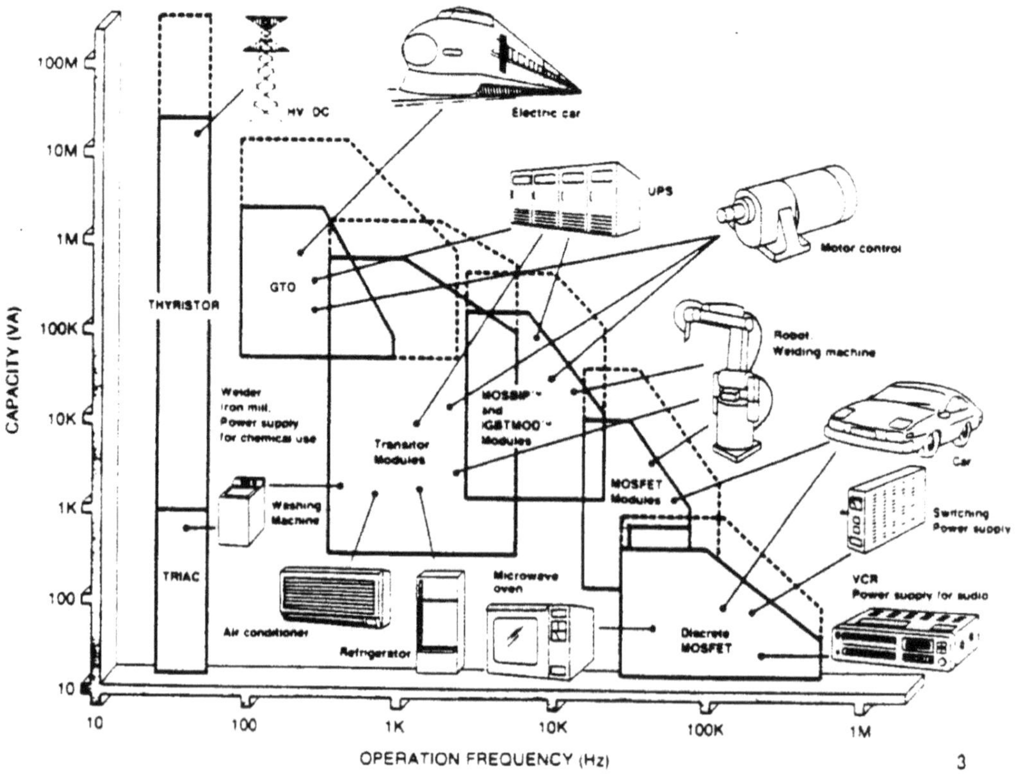

3

MARCONI ANTENNA

See ANTENNA DIRECTORY.

MARCONI EFFECT

A center-fed or end-fed antenna can show resonance at a frequency determined by the length of the feed line and radiator combined. This resonance usually occurs at the frequency occurring when the combined length of the feed line and radiator is ¼ wavelength, or some odd multiple of it. Marconi effect can result in unwanted radiation from an antenna feed line because the whole system behaves like a Marconi antenna. This can occur with antenna systems using coaxial feeders, as well as with systems using open wire line. The currents flow on the outside of the coaxial outer conductor or in phase along both conductors of open-wire line. *See also* MARCONI ANTENNA in the ANTENNA DIRECTORY.

MARITIME-MOBILE RADIO SERVICE

A two-way radio station in any private boat or commercial vessel not engaged in amateur communication is called a maritime-mobile station. Any fixed station used for nonamateur communication with the operator of any private or commercial vessel is also part of the maritime-mobile service.

The maritime-mobile radio service is allocated a number of frequency bands in the radio spectrum. The bands range from the very low frequencies through the ultrahigh frequencies; precise allocations vary from country to country.

MARX GENERATOR

A Marx generator is a form of high-voltage, direct-current impulse generator. Its principle of operation is analogous to a voltage-multiplier power supply. A source of direct current in connected to a set of capacitors that are arranged in parallel with series resistances, as shown in the illustration. A set of spark gaps is connected to the network of capacitors so that the gaps are subjected to the series combination of the capacitor voltages. The voltage appearing across the spark gaps is therefore much greater than the voltage across the individual capacitors.

When the capacitors reach a certain critical charge value, they discharge to create a spark across all of the gaps at once. This process occurs repeatedly as the capacitors alternately charge and discharge. The sparks produce short pulses of extremely high voltage at the output terminals. (AND HIGH AMPERAGE

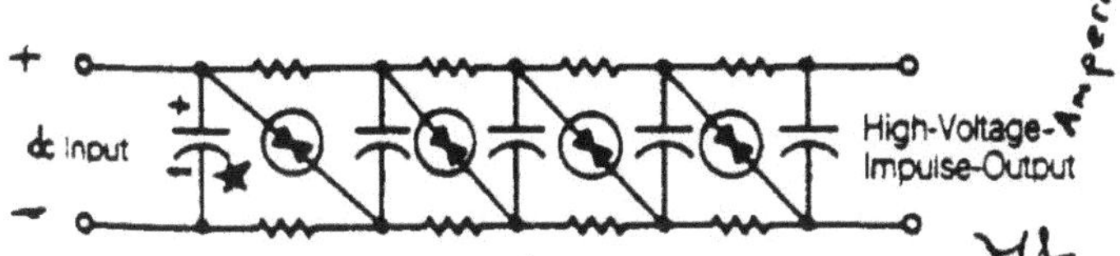

MARX GENERATOR: A high-voltage, pulsed output is produced from the low-voltage direct-current input to the Marx generator.

★ for D.C. uses Electrolytic Capacitors

MASER

The maser is an amplifier for microwave energy. The term is an acronym for *m*icrowave *a*mplification by stimulated *e*mission of *r*adiation. Maser output is the result of quantum resonances in various substances.

When an electron moves from a high-energy orbit to an orbit with less energy, a photon is emitted (*see* ELECTRON ORBIT). For a particular electron transition or quantum jump, the emitted photon always has the same amount of energy, and therefore the same frequency. By stimulating a substance to produce many quantum jumps from a given high-energy level to a given low-energy level, an extremely stable signal is produced. This is the principle of the maser oscillator. Ammonia and hydrogen gas are used in maser oscillators as frequency standards. Rubidium gas can also be used in masers. The output-frequency accuracy is within a few billionths of 1 percent in the gas maser.

Solid materials such as ruby can be used to obtain maser resonances. This kind of maser is called a solid-state maser, and it can be used as an amplifier or oscillator. When an external signal is applied at one of the quantum resonance frequencies, amplification is produced. The solid-state maser must be cooled to very low temperatures for proper operation; the optimum temperature is absolute zero. While satisfactory maser operation can take place at temperatures as high as that of dry ice, the most common method of cooling employs liquid helium or liquid nitrogen.

The traveling-wave maser is one form of the device for use at ultrahigh and microwave frequencies. The cavity maser is an alternate configuration. Both the traveling-wave and cavity masers use the resonant properties of materials made to precise dimensions. The traveling-wave and cavity masers are solid-state devices, and the temperature must be reduced to a few degrees Kelvin for operation. The gain of a properly operating traveling-wave or cavity type maser amplifier may be as great as 30 to 40 decibels.

Maser amplifiers find applications in radio astronomy, communications, and radar. Some masers operate in the infrared, visible-light, and even ultraviolet parts of the electromagnetic spectrum. Visible-light masers are called optical masers or, more commonly, lasers. *See also* LASER.

MASS

Mass is a measure of the amount of matter present in an object. Mass is expressed in kilograms (kg) in the standard international system of units. The kilogram is the amount of mass which, when subjected to a force of 1 newton in free space, undergoes an acceleration of 1 meter per second per second.

Mass is often confused with weight. However, weight depends on the intensity of a gravitational field in which the mass exists; mass is independent of the envi-

range at which iron-cored transformers can be used, the air-core transformer or transformers with powdered-iron slugs can be used effectively. However, in these cases the turns-ratio-squared impedance-transforming property is no longer true. Since the transformer is usually part of a tuned circuit, other factors influence the design of the transformer.

The impedance-transforming property of an iron-cored transformer is not always used to give maximum power transfer. For example, in the design of power-amplifier stages in audio amplifiers, the impedance presented to the transistor affects distortion. A study of a given circuit can often show that at a given output power level, usually the maximum expected, there is a value for the load resistance which will minimize a harmonic component in the harmonic distortion, such as the second or third harmonic. The transformer turns ratio is selected to present this resistance to the transistor. *See Transformer.*

Emitter follower. In electronic circuitry a signal source of large internal impedance must often be connected to a low-impedance load. If the source were connected directly to the load, attenuation of the signal would result. To reduce this attenuation, an emitter follower is connected between the source and the load. The input impedance of the emitter follower is high, more nearly matching the large source impedance, and the output impedance is low, more nearly matching the low load impedance. If the object were the delivery of maximum power to the load, it might be possible to design the emitter follower to have an output resistance equal to the load resistance, assuming that the load is a resistance. (Special audio amplifiers have been designed to use emitter followers, rather than a transformer, to connect the loudspeaker to the power amplifier.) In many cases, maximum power transfer is not the goal; the emitter follower is introduced primarily to reduce to a minimum the attenuation of the signal.

There exist a number of applications where the emitter follower is not useful as an impedance matching circuit. For example, if a very low-impedance source must be matched to a high-impedance load, then a transistor is used in the common base configuration. *See Emitter follower; Voltage amplifier.*

Christos C. Halkias

Bibliography. C. A. Schuler, *Electronics: Principles and Applications*, 2d ed., 1984; R. L. Thomas, *A Practical Introduction to Impedance Matching*, 1976; M. E. Van Valkenburg, *Network Analysis*, 3d ed., 1974.

Impsonite

A black, naturally occurring carbonaceous material having specific gravity 1.10–1.25 and fixed carbon 50–85%. Impsonite is infusible and is insoluble in carbon disulfide. A vein of impsonite 10 ft (3 m) wide occurs in La Flore County, Oklahoma, and there are other deposits in Arkansas, Nevada, Michigan, Peru, Argentina, Brazil, and Australia. The Peruvian impsonite may be derived from grahamite by weathering. The origin of impsonite is not well understood, but it appears to be derived from a fluid bitumen that polymerized after it filled the vein in which it is found. Impsonites are known to contain unusually high percentages of vanadium. *See Asphalt and asphaltite; Wurtzilite.*

Irving A. Breger

Impulse (mechanics)

The integral of a force over an interval of time. For a force **F**, the impulse **J** over the interval from t_0 to t_1 can be written as Eq. (1). The impulse thus represents

$$\mathbf{J} = \int_{t_0}^{t_1} \mathbf{F}\, dt \qquad (1)$$

the product of the time interval and the average force acting during the interval. Impulse is a vector quantity with the units of momentum.

The momentum-impulse relation states that the change in momentum over a given time interval equals the impulse of the resultant force acting during that interval. This relation can be proved by integration of Newton's second law over the time interval from t_0 to t_1. Let **P** represent the momentum at time t, with $\mathbf{P}_0$ and $\mathbf{P}_1$ being the values of **P** at times t_0 and t_1, respectively. Then Eq. (2) holds. If, as is ordinarily true, the mass m is constant, the momentum change can be expressed in terms of the velocities $\mathbf{v}_1$ and $\mathbf{v}_0$ at times t_1 and t_0, respectively, giving Eq. (3).

$$\mathbf{J} = \int_{t_0}^{t_1} \mathbf{F}\, dt = \int_{t_0}^{t_1} (d\mathbf{P}/dt)\, dt = \int_{\mathbf{P}_0}^{\mathbf{P}_1} d\mathbf{P} = \mathbf{P}_1 - \mathbf{P}_0 \qquad (2)$$

$$\mathbf{J} = m(\mathbf{v}_1 - \mathbf{v}_0) \qquad (3)$$

The concept of impulse is ordinarily most useful when the forces are large but act only for a short period. In most of these cases it is necessary to know only the momentum change, which is determined by the impulse. The relation between momentum and impulse thus has the advantage of eliminating the need for a detailed knowledge of how the forces, which can be very complicated, change with time. Forces which occur during collisions are of this type. *See Collision (physics).*

A further simplification of this type of system is obtained if the time interval is short enough for the system to be considered essentially stationary during the time of action of the force, so that the momentum change occurs almost instantaneously. Except for the changed momentum, the motion of the system is treated as it would have been if no impulse had occurred. The effect of the impulse can thus be considered to provide a set of initial conditions for the motion. In a ballistic galvanometer, for example, there is an electric current for only a short time, during which the galvanometer coil, originally at rest, is given an impulse by the force associated with the current. The only effect of the current on the later motion of the coil, which can be considered stationary during the time there is a current, is to provide an initial velocity for the subsequent motion. *See Galvanometer; Impact; Momentum.*

Paul W. Schmidt

Bibliography. F. Bueche, *Introduction to Physics for Scientists and Engineers*, 3d ed., 1980; R. Resnick and D. Halliday, *Physics*, 3d ed., 1977.

Impulse generator

An electrical apparatus which produces very short high-voltage, or high-current surges. Such devices can be classified into two types: impulse voltage genera-

A 1-MV impulse voltage generator. (Hipotronics, Inc.)

tors and impulse current generators. High impulse voltages are used to test the strength of electric power equipment against lightning and switching surges. Also, steep-front impulse voltages are sometimes used in nuclear physics experiments. High impulse currents are needed not only for tests on equipment such as lightning arresters and fuses but also for many other technical applications such as lasers, thermonuclear fusion, and plasma devices. *See Fuse (electricity); Laser; Lightning and surge protection; Nuclear fusion; Particle accelerator; Plasma physics.*

Impulse voltage generator. An impulse voltage generator (sometimes called a Marx generator, after E. Marx who first proposed it in 1923) consists of capacitors, resistors, and spark gaps. The capacitors are first charged in parallel through charging resistors by a high-voltage, direct-current source and then connected in series and discharged through a test object by a simultaneous spark-over of the spark gaps. The output voltage of such a generator is determined by the number of capacitors and the charging voltage. It is also determined by the efficiency of the generator. The waveform of the impulse voltage can be controlled by suitably adjusting the circuit components of such a generator.

The fundamental parameters of an impulse voltage generator are its rated voltage, energy, waveform T_f/T_t, and polarity. The wavefront time T_f, in microseconds, represents the steepness of the front of the impulse wave. The wave-tail time T_t, defined as the time, in microseconds, required for the tail of the wave to decline to half its peak value, represents the speed of attenuation of the impulse wave. The value of T_f/T_t for a standard lightning impulse wave is 1.2/50, and for a switching impulse wave it is 250/2500. The highest rated voltage of an impulse voltage generator is 10 MV.

The 1-MV impulse voltage generator shown in the **illustration** consists of 10 stages. Each stage has one capacitor. The maximum charging voltage of these capacitors is 100 kV.

Impulse current generator. The impulse current generator comprises many capacitors that are also charged in parallel by a high-voltage, low-current, direct-current source, but it is discharged in parallel through resistances, inductances, and a test object by a spark gap. It provides very high current. Its parameters are rated amperage, waveform T_f/T_t, and polarity. For standard impulse current waves, the values of T_f/T_t, in microseconds, are 4/10 and 8/20. The waveform can be controlled by suitably adjusting the value of the circuit resistances and inductances. The highest rated amperage of an impulse current generator is above several hundred kiloamperes.

Changchang Wang; T. C. Cheng

Bibliography. T. J. Gallagher and A. J. Pearmain, *High Voltage Measurement, Testing and Design*, 1983.

Impulse turbine

A prime mover in which fluid (water, steam, or hot gas) under pressure enters a stationary nozzle where its pressure (potential) energy is converted to velocity (kinetic) energy. The accelerated fluid then impinges on the blades of a rotor, imparting its energy to the blades to produce rotation and overcome the connected rotor resistance. The impulse principle is basic to many turbines. *See Prime mover.*

The impulse principle can be distinguished from the reaction principle by considering the flow of water from a hole near the bottom of a bucket (**Fig. 1**). The hole is a nozzle that serves to convert potential energy to the kinetic form ΔE; the impulse force F_i in the

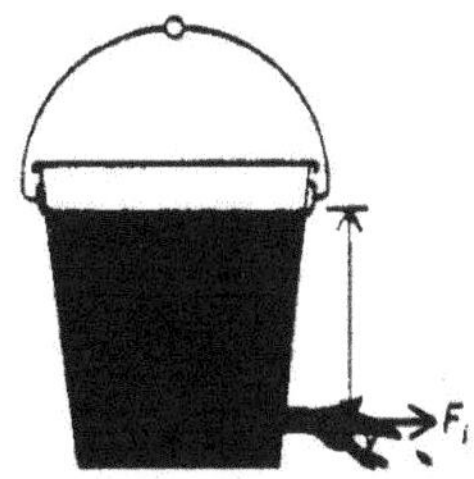

Fig. 1. Water escaping through a nozzle near the base of a bucket, free to swing, illustrating the impulse F_i and reaction F_r forces of the jet issuing with a velocity v.

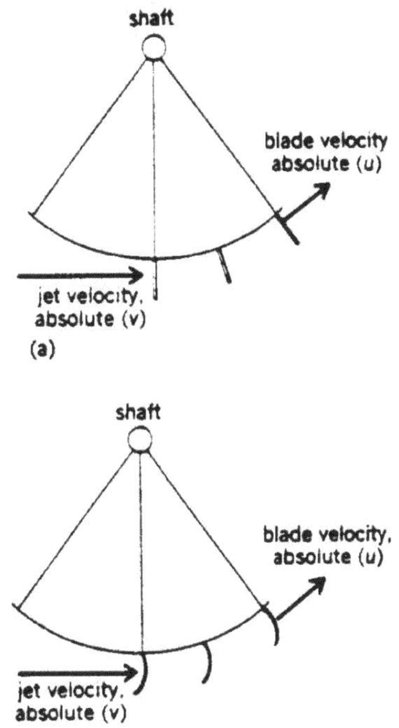

Fig. 2. Diagrams of a jet impinging on a series of (a) flat blades and (b) curved blades mounted on the periphery of a wheel.

mariner's compass *See* MAGNETIC COMPASS.

mark **1.** In telegraphy, the dot or dash portion of a character as opposed to the dead space between such portions. **2.** The intelligence part of a similar signal such as sound, light, and the like. **3.** The "on" or "one" state represented by a binary bit as opposed to the "off" or "zero" state. **4.** A character identifying the end of a data set. Also called MARKER.

marker A pip indicating a particular frequency on a response curve displayed on an oscilloscope screen.

marker beacons Individual coded-signal transmitters placed along a radio range and indicating features of the course marked by them.

marker frequency **1.** A known frequency which may be used to identify a spot-frequency harmonic of a frequency-standard signal. **2.** A known accurate signal employed to identify the limit of a radio band. **3.** The frequency at some point on a response curve as identified by a marker pip (*see* MARKER).

marker generator An oscillator that supplies a marker pip (*see* MARKER).

mark hold In telegraphy, an unmodulated signal meaning information is not being sent.

mark reading The reading by an optical scanning device of marks made in specific areas of a document; the process also includes the marks' conversion to digital signals for input to a computer.

mark scanning *See* MARK READING.

mark sensing A process similar to MARK READING except that the marks, say, penciled in blocks on a form, are sensed electrically.

marker trap A wavetrap that supplies a dip-type marker pip when used in conjunction with an RF test oscillator (*see* MARKER).

market scanner A device which scans a black-bar type of binary label printed on a carton or other package (or magazine) and indicates the corresponding price of the merchandise on the readout of the store's checkout register.

mark-to-space ratio In radiotelegraph code transmission, the ratio of the duration (mark) of a dot to the interval (space) between successive dots.

Marx generator An impulse-type high-voltage dc generator circuit in which several capacitors are charged in parallel through a high-resistance network and, when capacitor voltage reaches a critical high value, discharge in series through spark gaps, producing a high-voltage pulse for each discharge.

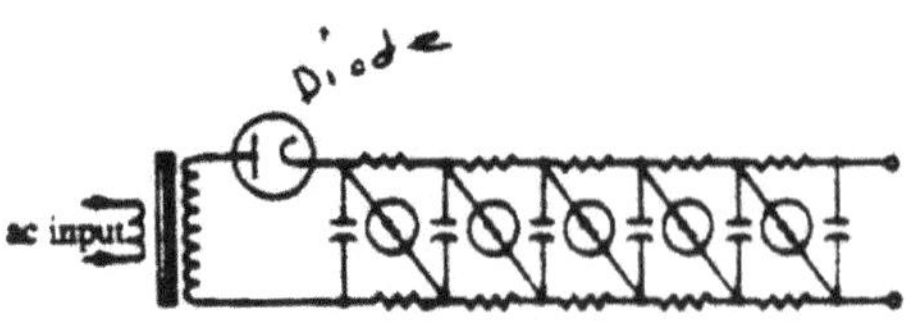

Marx generator

Mary Phonetic alphabet code word for letter M.

maser A low-noise microwave amplifying device in which a microwave input signal causes high-energy-state molecules of ammonia or ruby to fall to the low-energy state and, as a result, to emit large amounts of energy as an output signal. The name is an acronym for *m*icrowave *a*mplification by *s*timulated *e*mission of *r*adiation.

mask **1.** A kind of stencil through which plating, electrodepositing, or diffusion can be carried on. **2.** The viewing screen, or GRATICULE, of an oscilloscope. **3.** To provide with a mask. **4.** To obliterate a signal with a stronger one. **5.** A bit or character pattern used to change or extract bit positions in another pattern.

masking To use a mask (*see* MASK, 5).

Masonite Masonite Corporation's tough fiberboard used for panels and bases of some electronic equipment.

mass The quantity of matter in a body. Like weight, mass is expressed in kilograms in the metric (SI) system and in pounds in the English system. For a given piece of material, mass may be determined by dividing the weight by the acceleration due to gravity.

mass data Data in excess of the maximum amount that can be stored in the main (internal) storage unit of a digital computer, i.e., that which can only be accommodated by external media such as magnetic tape or disks.

mass-energy equation Energy, E, is the product of a given mass, m, and the square of the speed of light c^2, or $E = mc^2$. *See also* EINSTEIN EQUATION.

mass number **1.** Symbol, A. A number representing the total of neutrons and protons in the nucleus of an atom. The approximate mass of an atom is equal to $A \times m_p$ where m_p is the total proton (rest) mass. **2.** The number indicating the sum of nuclear protons and neutrons in an atom. It is usually written as a superscript to the right of the symbol for the atom: thus, U^{238} is uranium having 238 nucleons. An isotope of an element will have a different mass number than that of the normal atom.

mass of electron at rest Symbol, m_e. The amount of matter in an electron; $m_e = 9.109\,558 \times 10^{-31}$ kg.

mass of proton at rest Symbol, m_p. The mass of the proton in the nucleus of an atom; $m_p = 1.672\,614 \times 10^{-27}$ kg.

mass resistivity For wire, the resistance of a length 1 meter long having a mass of 1 gram, or of a mile-long piece having a mass of a pound.

mass spectograph An instrument employed to analyze chemical compounds and mixtures in terms of their distinctive mass spectra, which is exhibited by ionized samples of the materials in a magnetic field.

mass spectrometer Abbreviated MS. An instrument that permits rapid analysis of chemical compounds.

voltage crest *See* VOLTAGE PEAK.

voltage-current characteristic *See* VOLTAGE-CURRENT CURVE.

voltage-current crossover The point at which a voltage-regulated supply becomes a current-regulated supply.

voltage-current curve For a circuit or device, the plot of the interrelationship of current and voltage, with voltage as the independent variable.

voltage-current feedback *See* CURRENT-VOLTAGE FEEDBACK.

voltage decay The exponential decrease of voltage across a discharging capacitor. *Also see* EXPONENTIAL DECREASE.

voltage-dependent capacitor A capacitor (such as a varactor) whose capacitance varies with applied voltage.

voltage-dependent resistor A nonlinear resistor whose value varies inversely with the voltage drop across it. Also called VARISTOR.

voltage detector A circuit or device which delivers an output voltage only when the input voltage is of a prescribed value. *Compare* VOLTAGE DISCRIMINATOR.

voltage-determined property A property (capacitance, current, frequency, resistance) whose magnitude depends upon the value of an applied voltage. *See*, for example, VOLTAGE-DEPENDENT CAPACITOR; VOLTAGE-DEPENDENT RESISTOR; VOLTAGE-CONTROLLED OSCILLATOR.

voltage differential Symbol, dE or dV. An infinitesimal change in voltage represented by $E_2 - E_1$. *Compare* VOLTAGE INCREMENT.

voltage-directional relay **1.** A relay that is actuated when voltage exceeds a certain value in a given direction. **2.** A relay which closes only when the applied voltage is in a specific direction.

voltage discriminator A circuit or device whose output voltage is zero when the input voltage is of a prescribed value. When the input voltage is greater than this value, the output is positive; when it is less, the output is negative.

voltage distribution **1.** The delivery of operating voltage throughout a circuit, e.g., high and low dc voltages in the various stages of a control circuit. **2.** Sometimes, the distribution of electrical energy ("power" distribution).

voltage divider A resistive or capacitive potentiometer used to divide an applied voltage by a desired amount.

voltage doubler A circuit which supplies a dc output voltage that is about twice the peak value of the ac input voltage.

voltage drift *See* DRIFT VOLTAGE.

voltage drop Abbreviation, VD. The voltage (E) set up across a resistance (R) carrying a current (I); $E = IR$. For alternating current, reactance (X) and impedance (Z) may be substituted for resistance where applicable.

voltage-equalizing resistors In a power-supply filter, resistors connected across each capacitor in a string of electrolytics connected in series to withstand a high voltage. The resistors protect the capacitors by equalizing the voltage across them.

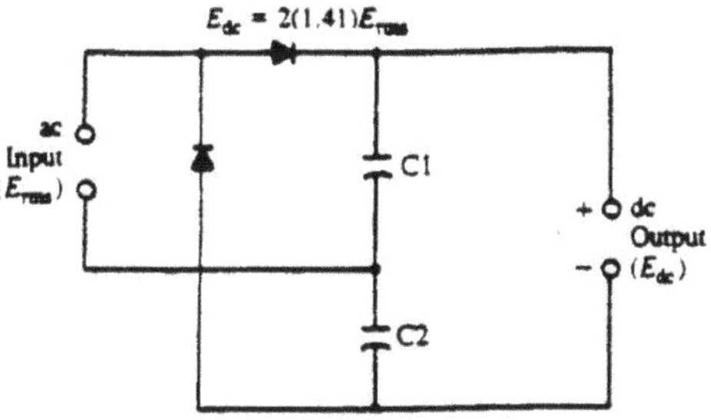

voltage doubler

voltage-fed antenna An antenna in which the feeder is attached to the radiator at a voltage loop (current node). *Compare* CURRENT-FED ANTENNA.

voltage feed The delivery of voltage to a device or circuit at a point where voltage, rather than current, is dominant. *Compare* CURRENT FEED.

voltage feedback A feedback signal consisting of voltage fed from the output to the input circuit of an amplifier or other device. *Compare* CURRENT FEEDBACK and CURRENT-VOLTAGE FEEDBACK.

voltage-frequency transducer **1.** A device whose output voltage is a function of the frequency of a signal at the input. **2.** A device whose output frequency is a function of the voltage at the input.

voltage gain *See* VOLTAGE AMPLIFICATION.

voltage generator *See* VOLTAGE SUPPLY.

voltage gradient The voltage per unit length along a path.

voltage increment Symbol, ΔE or ΔV. A change in voltage represented by $E_2 - E_1$, when the difference is finite. *Compare* VOLTAGE DIFFERENTIAL.

voltage input encoder An analog-to-digital encoder for which the input is an analog voltage. *See*, for example, VOLTAGE-TO-SHAFT-POSITION ENCODER.

voltage inverter A circuit or device whose voltage output has the sign opposite that of the input voltage. The device may or may not provide amplification. Thus, the inverter output E_o is AE_i where A is positive or negative gain, loss, or unity; E_i is the input voltage; and E_o is the output voltage.

voltage jump **1.** An upward transient in voltage. **2.** In a glow-discharge tube, a sudden break or increase in voltage drop across the tube.

voltage lag The condition in which changes in voltage follow corresponding changes in current. *Compare* CURRENT LAG and VOLTAGE LEAD.

voltage lead The condition in which changes in voltage precede corresponding changes in current. *Compare* CURRENT LEAD and VOLTAGE LAG.

voltage level 1. A prescribed reference value of voltage, e.g., the black level in a TV picture signal. 2. The discrete value of a steady voltage, or average value of a fluctuating voltage, as observed or measured in a circuit.

voltage limit The maximum or minimum level in a voltage range.

voltage loop In a standing-wave system, such as an antenna or transmission line, a maximum-voltage point. *Compare* VOLTAGE NODE.

voltage loss 1. Reduction of a voltage across a load, occurring because of a series resistance. 2. The ratio, in decibels, between the input voltage to a transmission line and the output voltage at the load end of the line.

voltage maximum *See* VOLTAGE PEAK.

voltage minimum *See* VOLTAGE TROUGH.

voltage-mode switching circuit A resistor-transistor-logic (RTL) NAND or NOR circuit in which (in the off state) the transistor is cut off by the V_{BB} bias voltage. The output is then approximately equal to the collector supply voltage V_{CC}. The proper combination of input pulses overrides the cutoff bias, and the transistor switches on. The output then drops to a level equal to V_{CC} minus the voltage drop across the external collector resistor.

voltage multiplier A special type of rectifier circuit which delivers a dc output voltage that is a multiple of the peak value of the ac input voltage, thus affording voltage step-up without a transformer. *See*, for example, VOLTAGE DOUBLER; VOLTAGE QUADRUPLER; QUINTUPLER, 1; VOLTAGE TRIPLER.

voltage node In a standing-wave system, such as an antenna or transmission line, a minimum-voltage point. *Compare* VOLTAGE LOOP.

voltage of self-induction The voltage drop across an inductor, resulting from the flow of alternating current through the inductor and due to self-induction.

voltage peak The highest value attained by a voltage during an excursion. Also called VOLTAGE CREST or VOLTAGE MAXIMUM. *Compare* VOLTAGE TROUGH.

voltage-power directional relay A relay system in which two circuits are connected when their voltage difference exceeds a predetermined value in one direction, and are disconnected when the power in them in the opposite direction exceeds a predetermined level.

voltage quadrupler A special rectifier circuit which, without a transformer, supplies a dc output voltage that is approximately quadruple the peak value of the ac input voltage.

voltage quintupler *See* QUINTUPLER, 1.

voltage-range multiplier 1. A multiplier resistor (*see* VOLTMETER MULTIPLIER) connected in series with a voltmeter that has an internal multiplier, to increase the range of the instrument. 2. For an ac voltmeter, an input amplifier used to increase the sensitivity of the instrument.

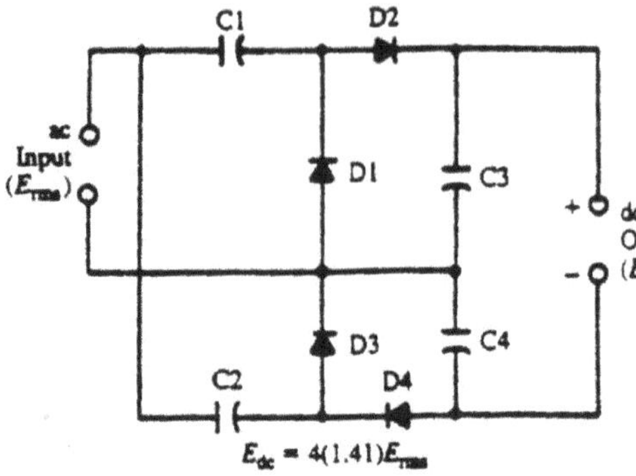

$E_{dc} = 4(1.41)E_{rms}$

voltage quadrupler

voltage rating 1. For a circuit or device, the recommended maximum voltage which may be applied, or the recommended working voltage, as specified. 2. For a generator, the specified output voltage.

voltage ratio The ratio of voltage at two points. These may be input and output voltage, primary and secondary voltage, and so on.

voltage-ratio box *See* VOLT BOX.

voltage reference *See* STANDARD CELL and ZENER-DIODE VOLTAGE REFERENCE.

voltage-reference cell *See* STANDARD CELL.

voltage-reference diode A zener diode biased into its zener region. The voltage drop across the diode is comparatively constant and under proper conditions may be used for reference purposes. *Also see* ZENER-DIODE VOLTAGE REFERENCE.

voltage-reference tube A special gas tube providing a voltage drop of such constancy that it may be used as a standard in voltage comparisons.

voltage reflection coefficient In a reflected-wave situation, the ratio E_r/E_i, where E_i is the field-strength voltage of the incident wave, and E_r the field-strength voltage of the reflected wave.

voltage-regulated supply *See* CONSTANT-VOLTAGE SOURCE.

voltage-regulating transformer A special transformer in which a resonant circuit and core saturation (*see* SATURATED OPERATION, 1) are employed to provide a constant ac output voltage.

voltage regulation The stabilization of a voltage against fluctuations in source or load.

voltage-regulation constant Symbol, K_v. For a voltage-regulated power supply, the ratio $\Delta E_L/\Delta E_i$, where ΔE_i is a change in input voltage, and ΔE_L is the corresponding change in load voltage.

voltage regulator A circuit or device which holds an output voltage constant during variations in the output load or input voltage.

voltage-regulator diode *See* ZENER-DIODE VOLTAGE REGULATOR.

electrical noise from gas tubes and mercury-vapor tubes. *Also see* HASH.

hat 1. A capacitance hat. 2. A procedure for randomizing data.

hash total *See* CHECKSUM.

hatchdot pattern A TV test pattern consisting of a crosshatch pattern with dots around its outer edges and one dot at its center.

hatted code A form of code in which randomization is used to maximize the difficulty of breaking the code.

hav Abbreviation of HAVERSINE.

haversine Abbreviation, hav. A trigonometric function equal to half the VERSED SINE.

Hay bridge An ac bridge for measuring the inductance and Q of an inductor in terms of resistance, frequency, and a standard capacitance.

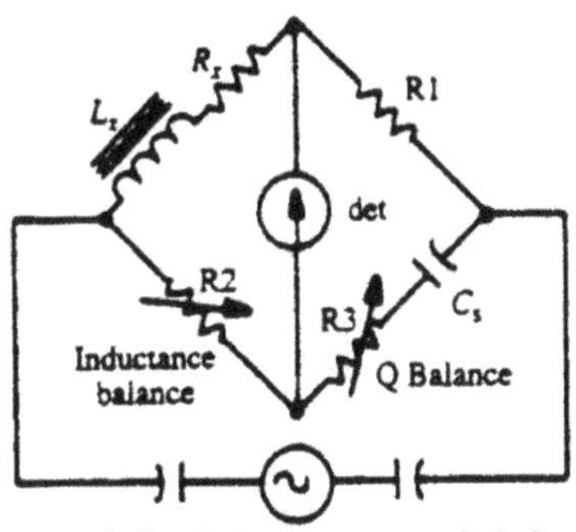

$$R_x = \frac{W^2C^2R_1R_2R_3}{1+(W^2C^2R_3^2)} \qquad L_x = \frac{C_SR_1R_2}{1+(R_3^2W^2R_3^2)}$$

$$Q_x = (WL_x/R_x)$$

Hay bridge

haywire Loose, disorderly, or apparently careless wiring.

haz Abbreviation of HAZARD.

hazard Abbreviation, haz. A dangerous or potentially dangerous circuit, device, material, method, situation, or system, e.g., electric-shock hazard.

Hazeltine circuit *See* NEUTRODYNE, 1, 2.

Hazeltine neutralization Plate neutralization (*see* PLATE-NEUTRALIZED AMPLIFIER).

H beacon A form of homing beacon with an omnidirectional radiation pattern and a radio-frequency output of between 50 W and 2 kW.

H bend *See* H-PLANE BEND.

HCD Abbreviation of *hard-copy device.*

HCM Abbreviation of HALF-CYCLE MAGNETIZER.

HCT Abbreviation of HEATER CENTER TAP.

HDB-3 code Abbreviation of HIGH-DENSITY BIPOLAR-3 CODE.

HDF Abbreviation of HIGH-FREQUENCY DIRECTION FINDER.

H-display *See* H-SCAN.

He Symbol for HELIUM.

head 1. The top or operating portion of a device, e.g., microphone head, dynamic-speaker head. 2. In magnetic recording and reproduction, the magnetic device (transducer) that delivers or picks up recorded impulses.

head alignment 1. Positioning the cone of a dynamic speaker so that the voice coil moves freely, i.e., without rubbing against the core. 2. Positioning a magnetic-recorder head so that a proper relationship to the moving tape is maintained.

head amplifier A self-contained amplifier or preamplifier in the head of a microphone or sound-on-film pickup.

head degausser A device employed for the purpose of demagnetizing the head of a tape recorder. Unwanted magnetization can build up because of direct-current components in the driving signal.

head demagnetizer A head degausser.

header 1. A (usually glass) disk or wafer through which one or more leads pass and to which they are fully sealed. A header can be used as the bottom of the bulb in a miniature tube, or as the terminal base of an enclosed plug-in unit, such as a miniature coil, filter, or similar components. *Also see* GLASS-TO-METAL SEAL. 2. A data set placed before other sets as a means of identifying them and, possibly, including control data pertinent to the sets so identified.

header card In a batch of data cards, the first card which contains data about the following set of cards.

header capacitance Capacitance between the leads in a header.

header label A header recorded on a magnetic tape file.

head gap 1. In data processing equipment for reading from or writing to magnetic tape or disks, the distance between the head and the medium. 2. In audio practice, the spacing between tape-unit head electrodes; gap width.

heading The direction taken by a vehicle with reference to some point such as a radio beacon, true north, or magnetic north.

headlight In radar practice, a small rotating antenna.

headphone A transducer (miniature speaker) worn against the ear for (1) listening to music without disturbing others or (2) for monitoring live or recorded material without being disturbed by noise in the environment. *Also see* RECEIVER, 2.

headphone amplifier An audio frequency amplifier designed and operated primarily to supply a signal to headphones.

headphone receiver A portable radio receiver, usually for AM/FM broadcast, consisting of a pair of headphones or a headset with the radio built into it.

head room In tape recording, the region between the maximum recording level specified by the manu-

SIMPLE TESLA COIL

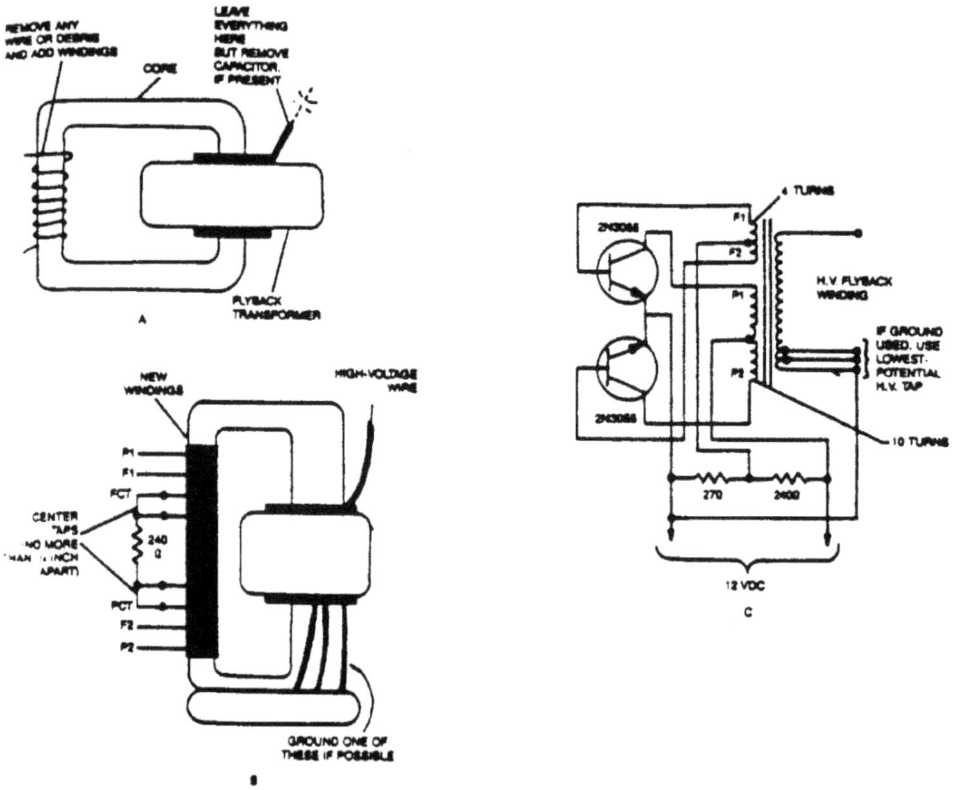

GERNSBACK PUBLICATIONS INC.

Fig. 106-1

The Tesla coil described here can generate 25,000 V. So, even though the output current is low, be very careful! The main component is a flyback transformer from a discarded TV.

A new primary winding is needed. Begin by winding 5 turns of #18 wire on the core. Then, twist a loop in the wire, and finish by winding five more turns. Wrap with electrical tape, but leave the loop exposed.

A four-turn winding must be wound over the ten-turn winding that you've just finished. That is done the same way. First wind two turns of # 18 wire, then make a loop, and finish by winding two more turns. Again, wrap the new winding with electrical tape, leaving the loop exposed.

When the windings are finished, the two loops shouldn't be more than 1/4-inch apart, but take care that they do not touch. Connect a 240-Ω resistor between the two loops. The modified transformer now should look like the one shown. Connect the transformer as shown. The 27-Ω resistor and two transistors should be mounted on a heatsink and must be insulated from it.

The output of the high-voltage winding should begin to oscillate as soon as the circuit is connected to a 12-Vdc power supply. If it does not, reverse the connections to the base leads of the transistors. In normal operation, you should be able to draw 1-inch sparks from the high-voltage lead using an insulated screwdriver.

AUTOMOBILE IGNITION SUBSTITUTE

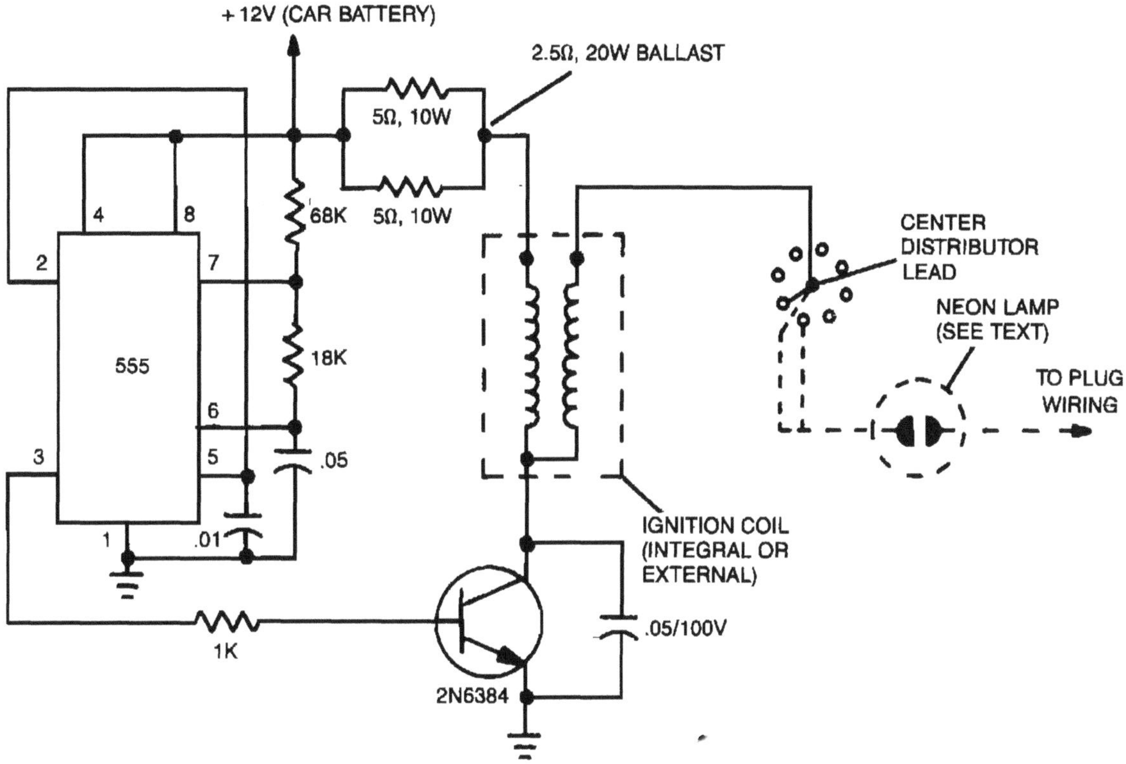

GERNSBACK PUBLICATIONS INC.

Fig. 8-1

The ignition substitute provides a constant power source for the ignition coil. Its frequency, 0.5 – 1.0 kHz, is that used by an 8-cylinder engine with an idling speed of 650 RPM, and the unit provides a rapid spark at a 17% duty cycle, while staying within the power dissipation limits of the components.

The circuit consists of a 555 timer IC configured as an astable free-running multivibrator that is used to drive a high-current npn transistor, such as a 2N6384. The transistor should be heavily heatsinked because it might be drawing several amps over quite a long period of time.

The coil ballast can be from 0.68 to 6.5 Ω, depending on what's available. The 2.5-Ω, 20-W ballast shown works well. All the other resistors can be either 1/4- or 1/2-W devices, and the capacitor, between pins 1 and 5 of the 555, can range from 0.01 to 0.05 μF. Do not omit the 100-V, 0.05-μF capacitor across the transistor; it prevents voltage spikes from damaging the device.

Although designed for an 8-cylinder engine, this device can be used with other types. In addition, a neon bulb can be added to the circuit to verify the presence of a spark.

HIGH-VOLTAGE BUCKING REGULATOR

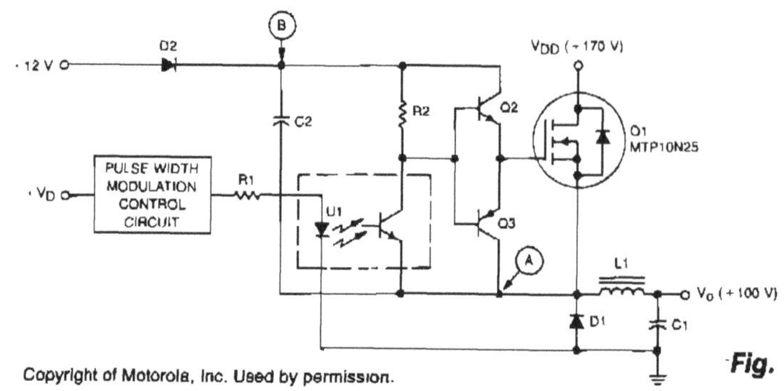

Copyright of Motorola, Inc. Used by permission.

Fig. 74-3

This circuit is basically the classic bucking regulator, except it uses a TMOS N-channel power FET for the chopper and creates its own supply for the gate control.

The unique aspect of this circuit is how it generates a separate supply for the gate circuit, which must be greater than V_{DD}. When power is applied, C2 charges, through D2, to +12 V. At this time, Q1 is off and the voltage at point A is just below zero. When the pulse-modulated signal is applied, the optoisolator transistors, Q2 and Q3, supply a signal to Q1 that turns it on. The voltage at point A then goes to V_{DD}, C2 back-biases D2, and the voltage at point B becomes 12 V above V_{DD}.

After Q1 is turned on, current starts to flow through L1 into C1, increasing until Q1 turns off. The current still wants to flow through L1, so the voltage at point A moves toward negative infinity, but is clamped by D1 to just below zero. Current flows less and less into C1, until Q1 turns on again. Q2 and Q3 drive Q1's gate between the voltages at point A and B, which is always a 12 V swing, so V_{GS} max. is never exceeded. For proper operation, the 12-V supply has to be established before the pulse-width modulator signal is applied.

HIGH-VOLTAGE DC GENERATOR

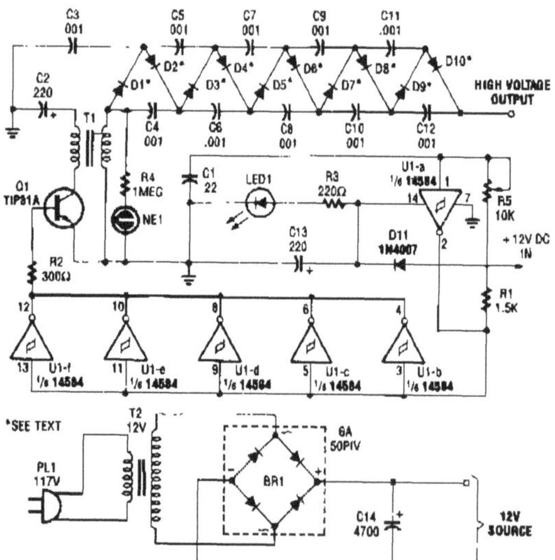

POPULAR ELECTRONICS

Fig. 74-4

This circuit is fed from a 12-Vdc power supply. The input to the circuit is then amplified to provide a 10,000-Vdc output. The output of the up-converter is then fed into a 10 stage, high-voltage multiplier to produce an output of 10,000 Vdc.

Marx Generators

A Marx Generator is a clever way of charging a number of capacitors in parallel, then discharging them in series. Originally described by E. Marx in 1924, Marx generators are probably the most common way of generating high voltage impulses for testing when the voltage level required is higher than available charging supply voltages. Furthermore, above about 200 kV, the discharge capacitor becomes very expensive and bulky. The Fitch circuit is becoming popular where very good control over impulse voltage is required.

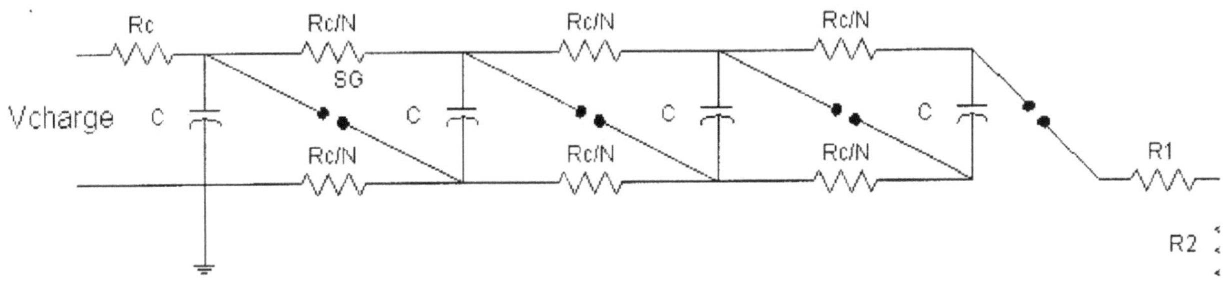

How it works

The charging voltage is applied to the system. The stage capacitors charge through the charging resistors (Rc). When fully charged, either the lowest gap is allowed to breakdown from overvoltage or it is triggered by an external source (if the gap spacing is set greater than the charging voltage breakdown spacing). This effectively puts the bottom two capacitors in series, overvoltaging the next gap up, which then puts the bottom three capacitors in series, which overvoltages the next gap, and so forth. This process is referred to as "erecting". A common specification is the erected capacitance of the bank, equal to the stage capacitance divided by the number of stages.

The charging resistors are chosen to provide a typical charging time constant of several seconds. A typical charging current would be in the 50-100 mA range. The charging resistors also provide a current path to keep the arc in the spark gaps alive, and so, should be chosen to provide a current of 5-10 amps through the gap. The resistors are sometimes called "feed forward" resistors for this reason. The discharge through the charging resistors sets an upper bound on the impulse fall time, although usually, the impulse fall time is set by external resistors in parallel with the load (or integrated into the generator, as described below).

For example, with a stage voltage of 100 kV, a desired output voltage of 1 MV (i.e. 10 stages), the charging resistors should be about 20-40 kOhms (corresponding to an arc current of 5 to 10 Amps). If the capacitors were 1 uF, then the discharge time constant would be 20 milliseconds, much, much longer than the 50 microsecond time constant of the standard test impulse. This example generator would have a stored energy of 5 kJ/stage or 50 kJ for the total system. At a charging current of 50 mA, it would take at least 20 seconds to charge the entire stack.

If a constant voltage charging source is used, significant energy is dissipated in the charging resistors, equal to the stored energy in the capacitors.

Design enhancements and considerations

Charging with a constant current source

If the Marx generator is charged from a constant voltage source, the energy dissipated in the charging resistors will be equal to that stored in the Marx capacitors. If the bank is charged with a constant current source, this energy loss can be substantially reduced.

Integrating the waveshaping resistors into the generator

In the classic capacitor discharge impulse generator, the shape of the pulse is controlled by external impedances (usually resistors) at the "output" of the pulse generator. As voltages get higher, it gets harder to build practical resistors with low parasitic inductance that will also withstand the full impulse voltage. The usual remedy for this is to include the wave-shaping resistors in the Marx bank itself, as illustrated in the following figure.

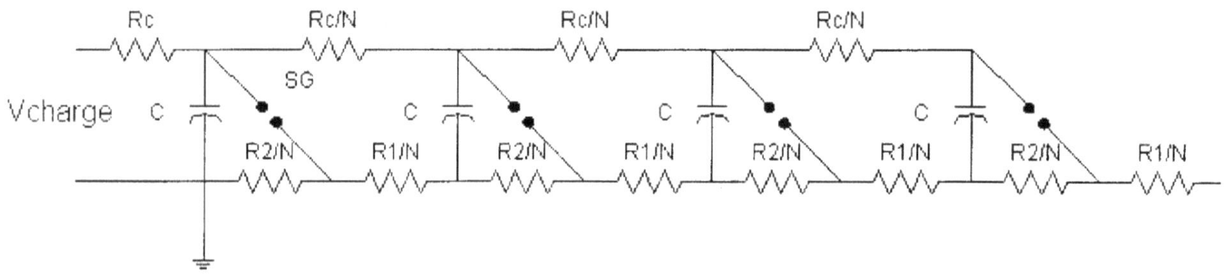

Reducing the jitter

If the gaps in the Marx generator don't all fire at exactly the same time, the leading edge of the impulse will have steps and glitches as the gaps fire. These delays also result in an overall longer rise time for the impulse. If the jitter in the gaps is reduced, the overall performance is improved.

The traditional Marx generator operating in air has all the gaps in a line with the electrodes operating horizontally opposed. This allows the UV from bottom gap to irradiate the upper gaps, reducing their jitter. Tests reported in Craggs and Meek showed that obstructing the UV led to greatly increased jitter in the bank output, which they attribute to the lack of UV irradiation on the upper gaps.

For a Marx generator which is immersed in oil, or using enclosed spark gaps, resistor or capacitor networks can be used to propagate the trigger pulse to all the gaps, rather than relying on the overvoltaging of the upper gaps to fire them.. A design from Maxwell labs uses a series of resistors to apply the trigger impulse to all the gaps.

Laser irradiation or triggering of the gaps could also be used.

Craggs and Meek also report the use of radioactive sources included within the gap electrodes to reduce the jitter.

Other switching devices

Fitch Impulse Generators

The Fitch circuit is used when better control of the impulse voltage is required than can be provided by the Marx circuit.

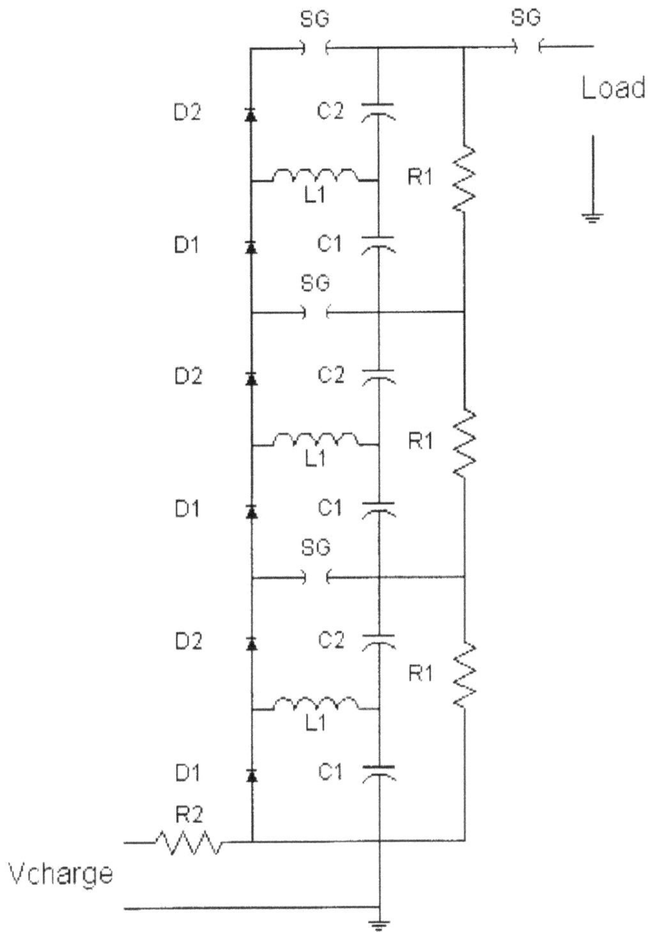

There's an error in this drawing.. there should not be a short from R2 to ground.

How it works

Let's just consider a couple of stages.

Charging

During charging, the capacitors charge through D1 and L1. Current passes to the next stage through D2. The polarity of the voltages on C1 and C2 are reversed, so there is no net voltage on the two.

Ia is the charging current throug

D1, L, and C1. Ib is the charging current through D1, L, C2, and R2. Ic is the charging current through D1, D2, D1 (in the second stage), L (second stage), and C1(second stage). And so forth.

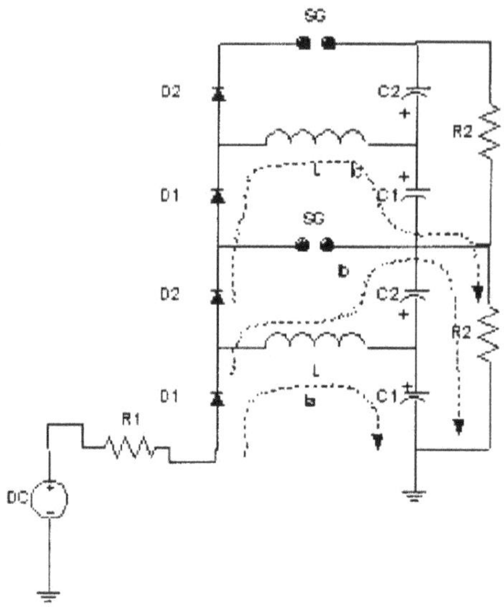

Now we are charged and ready to fire.

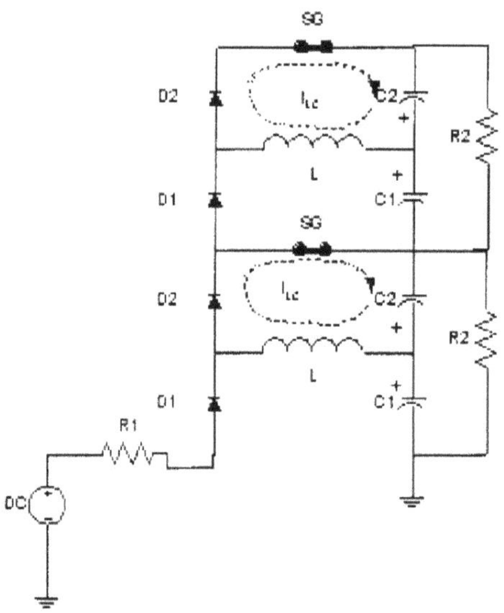

Erection, phase 1

Now, we trigger the gaps, so current starts to flow in the resonant circuit C2, L1, D2, SG, shown by loop ILC. D1 is reverse biased during this time: C1, L, D1, SG.

Erection, phase 2

A half period later, the voltage on C2 is entirely reversed, but it cannot discharge, because D2 would be reverse biased. The spark gap extinguishes.

C1 and C2 are now in series with the same polarity, so the stage voltage is twice the original charge voltage. The fall time is set by the resistance of R2

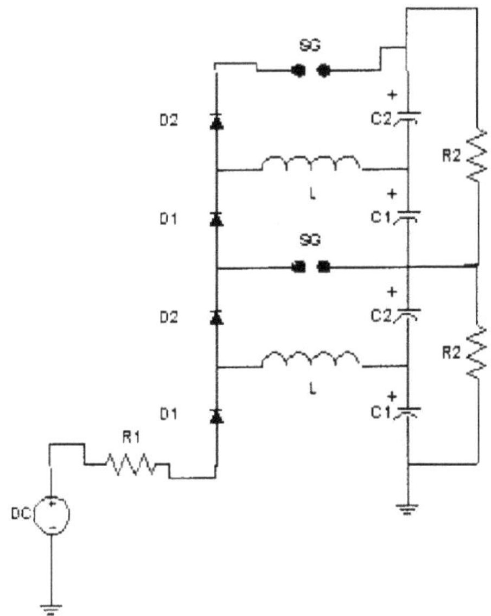

Design ideas

L must be able to store all the energy in C2 in this circuit. One would also want to keep the loop current high enough that the spark gap stays "ignited" (on the order of 4-5 Amps), but low enough that the diode D2 doesn't disappear in a puff of smoke. Most diodes, though, have a Ipk substantially higher (10-20 times) than their continuous rating (1N400x series of 1Amp diodes have an I_{FSM} of 30A, 1N540x (3Amp) series has I_{FSM} of 200A).

Say we have a stage voltage of 50 kV, and 1 uF capacitors (i.e. a stored energy of 1.25 kJ). We want to keep the peak current below 150A (we're using some form of 3Amp diodes..), and since E=0.5 L I^2, we can calculate the inductance as L = 2500/(150*150) or 0.11 Henry, which seems quite high...

Perhaps, if one used much smaller capacitors... 10 nF, for example. Then, the inductance would be 1.1 mH, which, although high, would start to be more practical. In this case, the half period (2*pi*sqrt (L*C2)) would be about 20 microseconds.

One would probably want erection times substantially less than a microsecond, though, so a more reasonable inductance would be even smaller, although, that will raise the peak diode current.

References

Bazelyan and Raizer, Spark Discharge, pp 97-98

ELECTRICAL PRINCIPLES, TERMINOLOGY & SAFETY

The use of electricity is so common place that most people assume that it will always be available on demand. To fully realize the dependence upon electricity, survey the ways electricity is being used each day in the home and on the farm and ranch. Electricity is doing more to increase work efficiency and promote enjoyable living than any other single factor. The use of electricity has grown to the extent that an increasing portion of the home or business budget is used in paying for this source of energy.

1. **Definition of Electricity**

   Electricity can be defined in several ways. The layman defines electricity as a source of energy that can be converted to light, heat, or power. Electrical engineers define electricity as a movement of electrons caused by electrical pressure or voltage. The amount of energy produced depends on the number of electrons in motion.

2. **The Manufacture and Distribution of Electricity**

   Electricity is produced from generators that are run by water, steam, or internal combustion engines. If water is used as a source of power to turn generators, it is referred to as hydroelectric generation. There are a number of this type located in areas where huge dams have been built across large streams.

   Steam is used as a source of power for generating much of today's electricity. Water is heated to a high temperature, and the steam pressure is used to turn turbines which generate electricity. These are referred to as thermal-powered generators. Fuels used to heat the water are coal, natural gas, and/or fuel oil.

   Generators at the power plant generate from 13,800 to 22,000 volts of electricity. From the power plant, electricity is carried to a step-up substation which, through the use of transformers, increases the voltage from 69,000 to 750,000 volts. This increase in voltage is necessary for the efficient transmission of electricity over long distances. From the step-up substation, the electricity is carried on transmission lines to a step-down substation which reduces the voltage to 7,200 to 14,000 volts for distribution to rural and city areas.

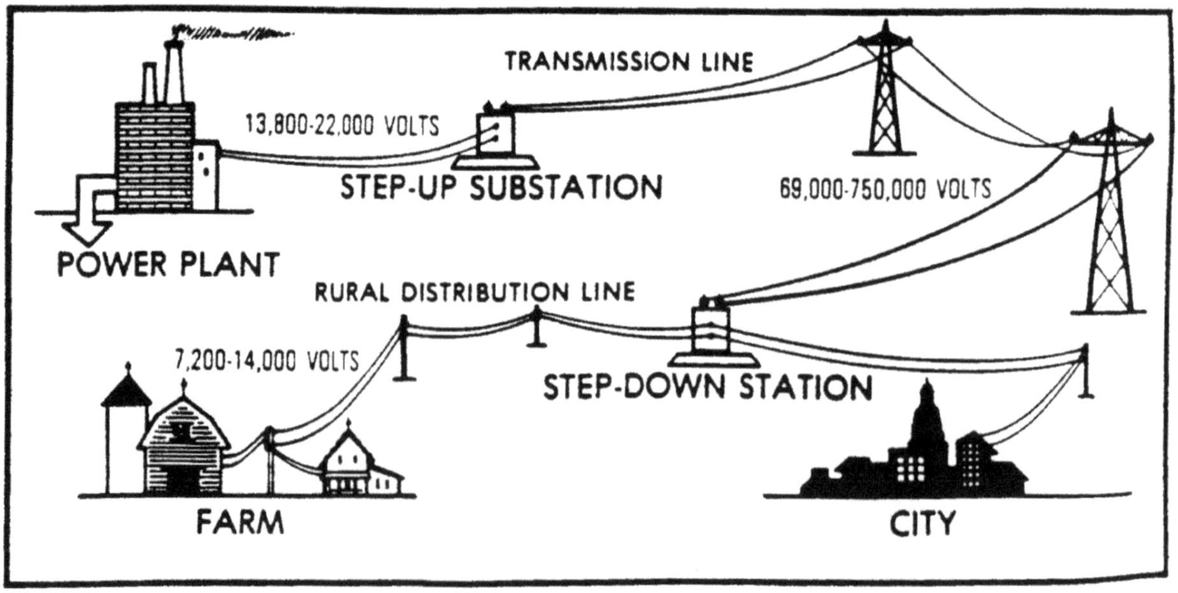

Transformers at the business or residence reduce the voltage to 120 or 240 volts to the meter of the customer.

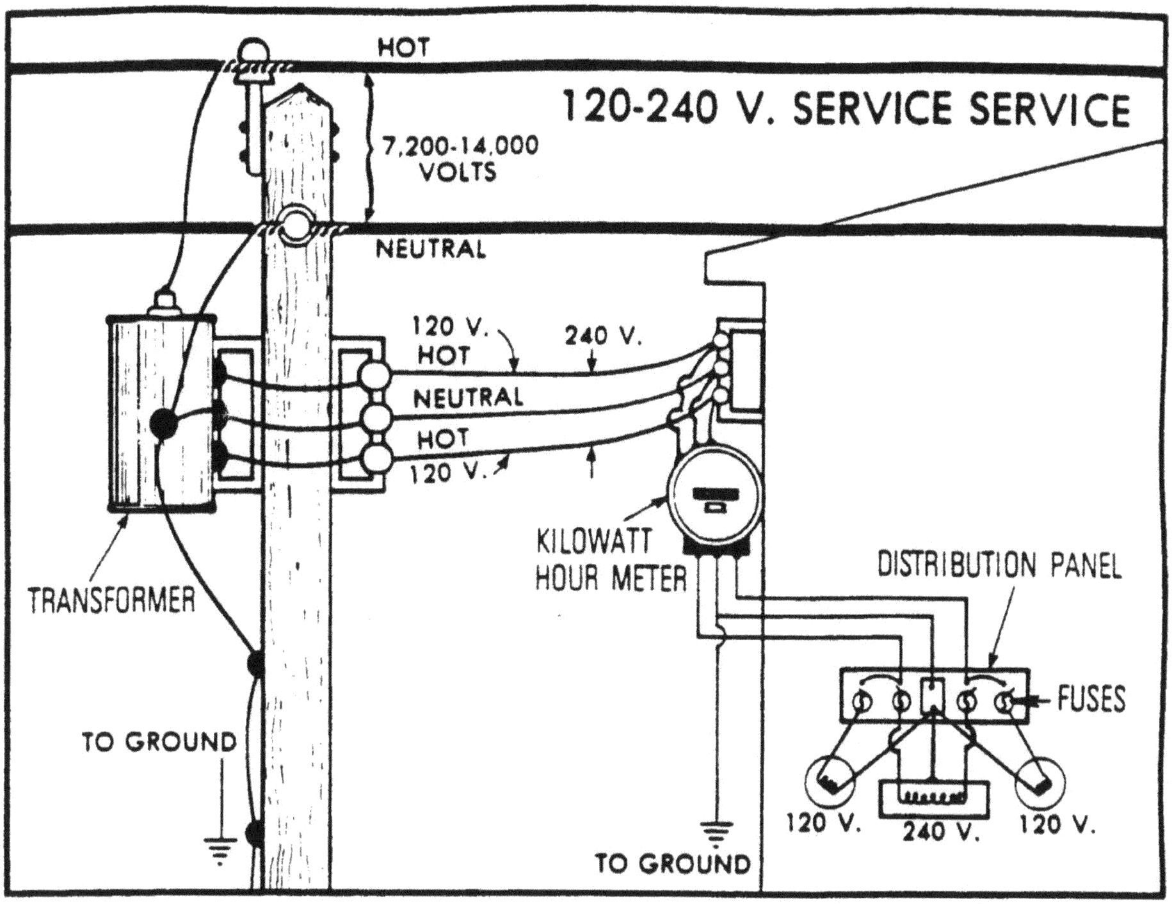

3. **Common Electrical Terms**

In order to work safely and efficiently with electricity and have the ability to converse on the subject, the following terms should be understood:

Ampere (Amp) - A measurement in units of the rate of flow of electrical current. This may be compared with the rate of flow of water in gallons per minute.

Example: A 60-watt incandescent lamp on a 120V circuit would pull 1/2 ampere of electricity (60 divided by 120 = .5 or 1/2). (Formula: Amperes = Watts divided by Volts

Volt (V) - A unit of measure of electrical pressure. A given electrical pressure (V) causes a given amount of electrical current (Amps) to flow through a load of given resistance. Voltage may be compared with water pressure in pounds per square inch in a water system. Common service voltages are 120 volts for lighting and small appliance circuits and 240 volts for heating, air conditioning, and large equipment circuits.

Watt (W) - A unit of measure of electrical power. When applied to electrical equipment, it is the rate that electrical energy is transformed into some other form of energy such as light. Watts may be compared to the work done by water in washing a car. (Formula: Volts x Amps = Watts)

Kilowatt (KW) - A unit of measurement used in computing electrical energy used. Kilowatts are determined by dividing the number of watts by 1000 (1 KW = 1000 W).

Kilowatt Hour (KWH) - A measure of electricity in terms of power in kilowatts and time in hours. A KWH is 1000 watts used for one hour.

Alternating Current (A.C.) - Electrical current that alternates or changes direction several times per second. The direction current moves depends on the direction the voltage forces it.

Cycle - The flow of electricity in one direction, the reverse flow of electricity in the other direction, and the start of the flow back in the other direction.

The cycles per second are regulated by the power supplier and are usually 60. Most electric clocks are built to operate on 60 cycles. More or less cycles would cause the clocks to gain or lose time. The present practice is to use the term Hertz (Hz) rather than cycles per second.

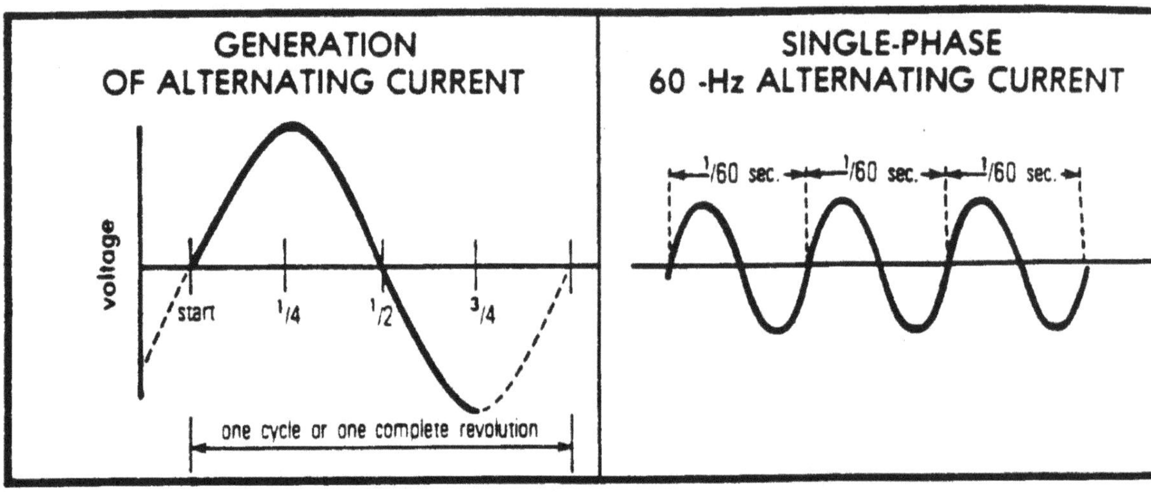

Direct Current (D.C.) - Electrical current flowing in one direction. **Example:** electrical circuit in automobiles and tractors.

Transformer - A device used to increase or decrease voltage.

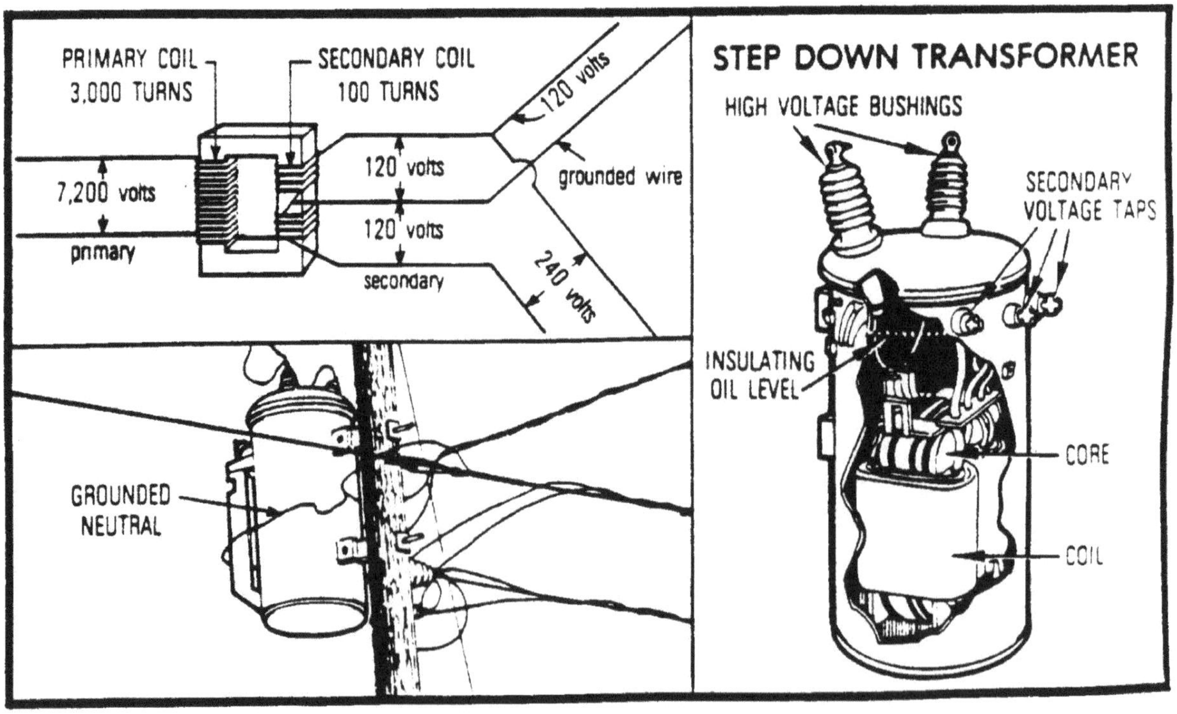

Single Phase - The most common type of electrical service or power available to consumers. One transformer is used between the distribution line and the meter. Usually three wires, two "hot" and one neutral, are installed to provide 120V and 240V single-phase service. Single-phase service may also be supplied with three-phase service.

Three-Phase - This type of service is designed especially for large electrical loads. It is a more expensive installation due to three wires and three transformers. The important advantage of three-phase power is that the total electrical load is divided among the three phases; consequently, the wire and transformers can be smaller. Other advantages exist in the design of three-phase motors.

Short Circuit - A direct connection (before current flows through an appliance) between two "hot" wires, between a "hot" and neutral wire, or between a "hot" wire and ground.

Voltage Drop - A reduction of current between the power supply and the load. Due to resistance, there will be a loss of voltage any time electricity flows through a conductor (wire). Factors that influence voltage drop are size of wire, length of wire, and the number of amps flowing. A drop in voltage may cause a loss of heat, light, or power output of a motor. It could cause motor burn-out unless the motor is properly protected (time-delay fuse).

Fuse - A device used to protect circuits from an overload of current.

Circuit Breaker - A device used to protect circuits from an overload of current. May be manually reset.

Time-Delay Fuse - A fuse with the ability to carry an overload of current for a short duration without disengaging the contacts or melting the fuse link.

Horsepower (hp) - A unit of mechanical power equal to 746 watts of electrical power (assuming 74.6% electric motor efficiency). One hp and above motors are rated at 1000 watts per hp; motors below one hp are rated at 1200 watts per hp.

Conductor - The wire used to carry electricity (copper or aluminum). Copper and aluminum should not be spliced together due to their incompatibility resulting in deterioration and oxidation.

Insulator - A material which will not conduct electricity and is usually made of glass, bakelite, porcelain, rubber, or thermo-plastic.

"Hot" Wire - A current-carrying conductor under electrical pressure and connected to a fuse or circuit breaker at the distribution panel. (Color Code: usually black or red)

Neutral Wire - A current-carrying conductor not under electrical pressure and connected to the neutral bar at the distribution panel. (Color Code: usually white)

Grounding - The connection of the neutral part of the electrical system to the earth to reduce the possibility of damage from lightning and the connection of electrical equipment housings to the earth to minimize the danger from electrical shock. (Color Code: can be green or bare wire)

Underwriters' Laboratory (U.L.) - A national organization which tests all types of wiring materials and electrical devices to insure that they meet minimum standards for safety and quality.

National Electric Code (N.E.C.) - Regulations approved by the National Board of Fire Underwriters primarily for safety in electrical wiring installations. All wiring should meet the requirements of the national as well as the local code.

4. Computing Electrical Energy Use and Cost

If an estimate of cost for electricity used is desired, the name plate data on appliances and equipment and an estimate of operating time may be used. The following formulas should be used for determining watts, amps, volts, watt-hours, kilowatt-hours, and cost.

$$\text{WATTS} = \text{VOLTS} \times \text{AMPERES}$$

$$\text{AMPERES} = \frac{\text{WATTS}}{\text{VOLTS}}$$

$$\text{VOLTS} = \frac{\text{WATTS}}{\text{AMPERES}}$$

$$\text{WATTS} \times \text{HOURS OF OPERATION} = \text{WATT HOURS}$$

$$\text{KILOWATT-HOURS} = \frac{\text{WATT-HOURS}}{1000}$$

$$\text{COST} = \text{KWH} \times \text{LOCAL RATE PER KWH}$$

EXAMPLE COST PROBLEM:

LOCAL RATE PER KWH USED - 8 CENTS
NAME PLATE DATA - 120 VOLTS, 5 AMPS
MONTHLY HOURS OF OPERATION - 10

(1) $(W = V \times A)$ $W = 120 \times 5$ $W = 600$

(2) $(\text{WATT-HOURS} = W \times \text{HOURS})$ $\text{WATT-HOURS} = 600 \times 10$ $\text{WATT-HOURS} = 6000$

(3) $(\text{KWH} = \frac{\text{WATT-HOURS}}{1000})$ $\text{KWH} = \frac{6000}{1000}$ $\text{KWH} = 6$

(4) $(\text{COST} = \text{KWH} \times \text{RATE})$ $\text{COST} = 6 \times 8$ $\text{COST} = 48$ CENTS

5. Electrical Circuits

An electrical circuit is a completed path through which electricity flows. Insulated conductors (wires) provide the path for the flow of electricity. A water system and an electrical circuit are similar in many respects. Water flows through pipes and is measured in gallons per minute, and electricity flows through conductors and is measured in amperes. A simple circuit is diagrammed here:

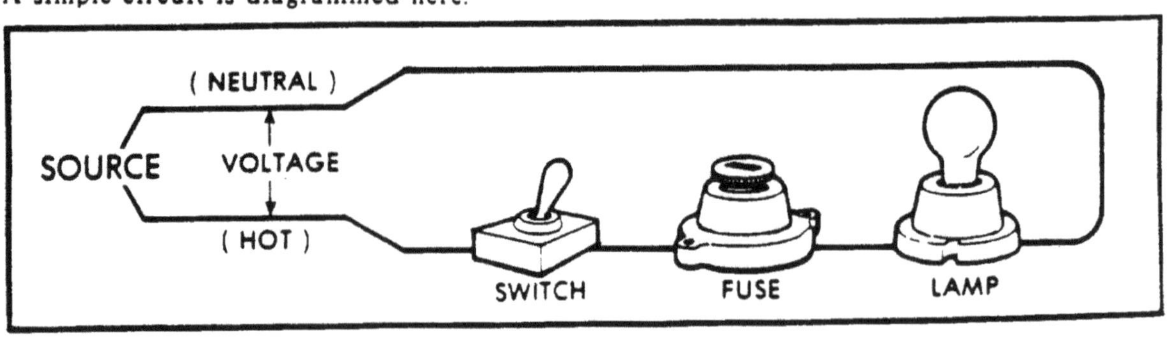

A circuit includes a "hot" wire (red or black) carrying current from the source through a switch, circuit protector (fuse or circuit breaker), and an appliance. The neutral wire (white) conducts the current from the appliance to the source (ground).

There are two methods for connecting devices in a circuit--series and parallel. In a series circuit all the current must flow through each device in the circuit. Removing or opening any one of the devices in the series circuit will stop the flow of current. In parallel circuits the load (lights or appliances) are connected between the two wires of the circuit providing an independent path for the flow of current, and removing a lamp has no effect on the other lamps in the circuit.

Switches, fuses, and circuit breakers are always connected in series. In most cases, except for some Christmas tree lights, appliances and lights are connected in parallel.

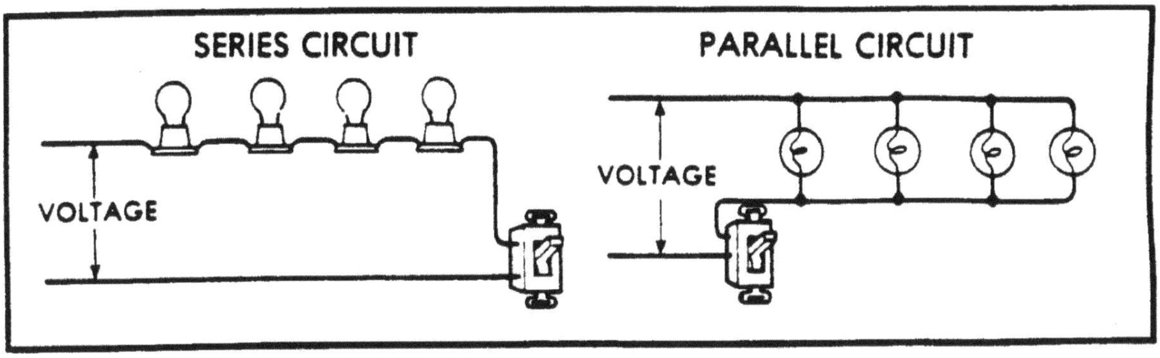

6. **120 Volt and 240 Volt Circuits**

The 120V circuit has one "hot" and one neutral wire with the switch and circuit protector in the hot line. The neutral wire from the appliance is connected to the neutral bar in the fuse or breaker box. For safety, the neutral wire should never be broken or interrupted with a switch or fuse.

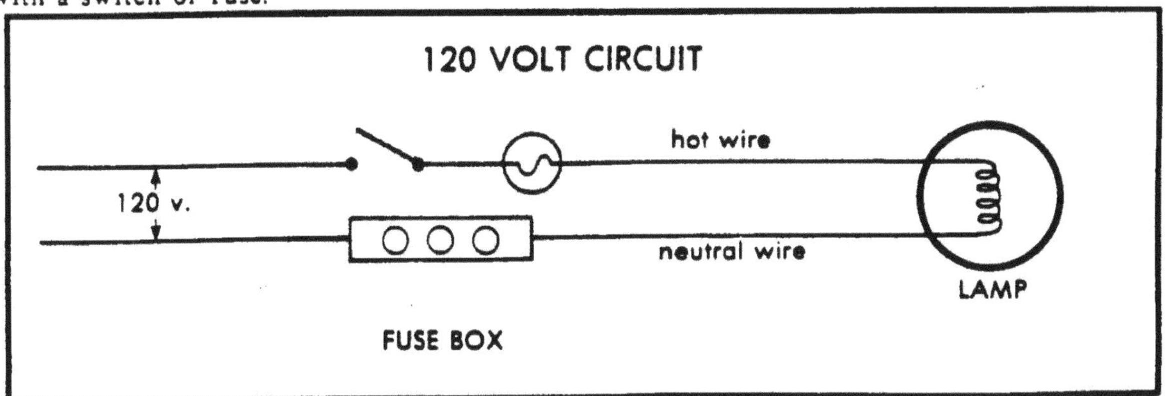

The voltage in a 120V circuit is measured with a voltmeter with one lead on the hot terminal and the other lead on the neutral bar. The number of amperes flowing may be measured with a clamp-on ammeter by encircling the hot or neutral wire with the jaws of the ammeter.

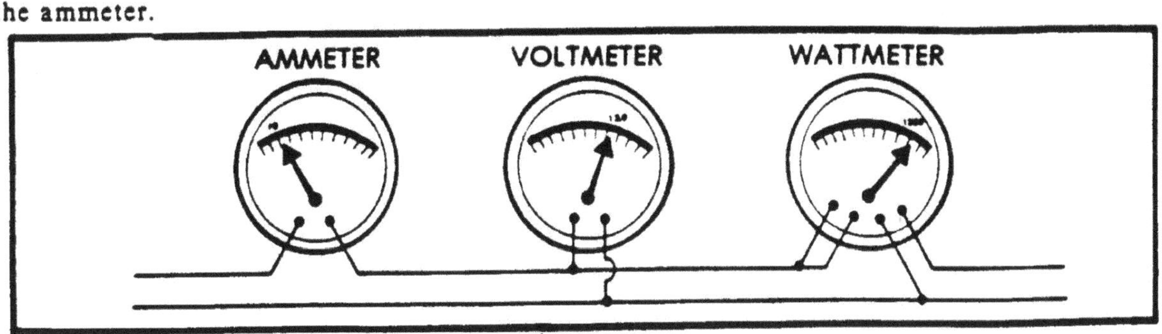

The 240V circuit has two hot wires and one safety-ground wire. Switches and fuses are installed in the hot lines. The two hot wires are necessary for the operation of 240V welders and motors. The safety-ground wire, connected to the metal frame of the equipment or motor and to the neutral bar, does not carry current unless a "short" develops in the motor or welder. If a short should occur, one of the circuit protectors will burn-out or open, thus opening the circuit.

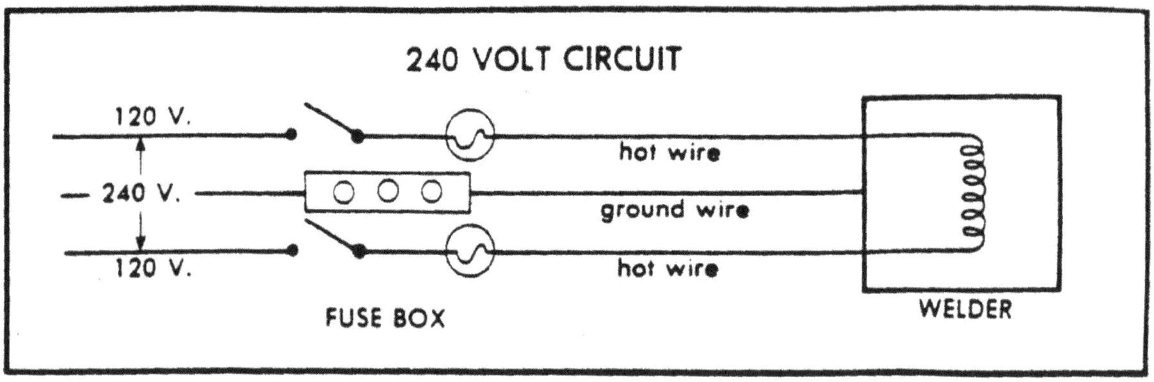

The voltage on a 240V circuit is measured by fastening a lead on the voltmeter to each of the hot wires. Voltage between either hot terminal and the neutral bar will be one-half of the voltage between the two hot wires. The number of amperes flowing can be measured by clamping an ammeter around either of the hot wires.

7. **Safety Grounding Electrical Equipment**

Refer back to the 240V circuit and note the ground wire from the metal frame to the neutral bar. The following illustration shows proper safety grounding when operating a drill in a 120V circuit. The safety-ground wire may be bare, but a three-wire romex is recommended. Safety-ground wire in three-wire romex is usually green in color. A current-carrying neutral wire should never be used for a safety-ground. Likewise, a safety-ground wire should never be used as a current-carrying hot or neutral wire.

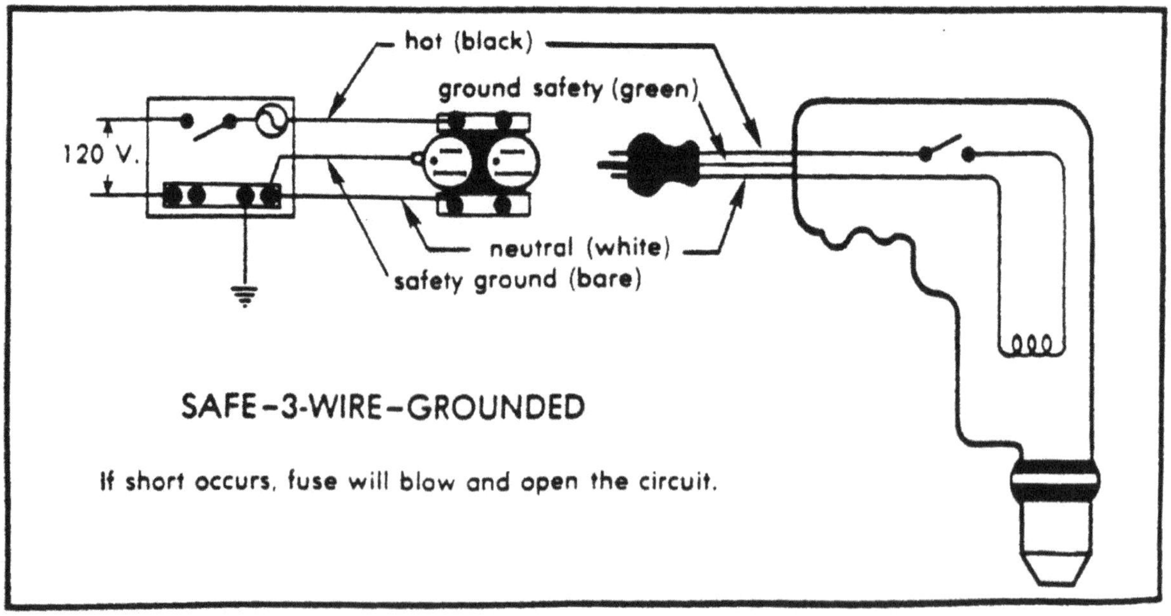

Using grounded receptacles and a safety-ground on all circuits will allow the safety-grounding of appliances when they are plugged into the outlet. An adapter must be used to properly ground appliances connected to receptacles not safety-grounded. If an adapter is used, the green pigtail wire must be connected to a known ground to give protection from electrical shock should a short occur.

GROUND PLUG AND ADAPTER

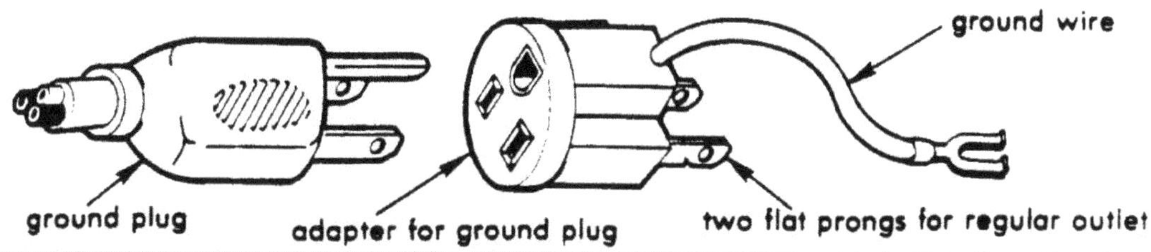

A test lamp can be used to check a circuit completed between a "hot" wire and a neutral wire. Use the test lamp to check appliances for shorts. With the appliance plugged into an outlet, touch the appliance frame with one lead of the test lamp while the other lead of the test lamp is grounded to a water or gas line. If the test light does not burn, reverse the appliance plug and check with the test lamp again. If the light burns, a short exists. (Hot wire is touching the frame of the appliance.) Unplug the appliance and repair or discard it.

8. **Electrical Circuit Protection**

Electrical circuits should be protected from an overload of amperes. Too many amperes flowing through an unprotected circuit will generate heat, which will deteriorate or melt the insulation and possibly cause a fire. The number of amperes that a given conductor can carry safely depends upon the kind and size of wire, type of insulation, length of run in feet, and type of installation. Charts are available in reference texts giving allowable current-carrying capabilities of various conductors.

The four types of circuit protection are common fuses, fusetrons (time-delay), fustats (two-part time-delay), and circuit breakers. Fuses are of two basic types, plug and cartridge.

Common fuses contain a link made from a low melting alloy which is designed to carry current up to the rating of the fuse. Current higher than the amperage rating causes the link to heat above its melting point. When the fuse "blows", the link melts and opens the circuit.

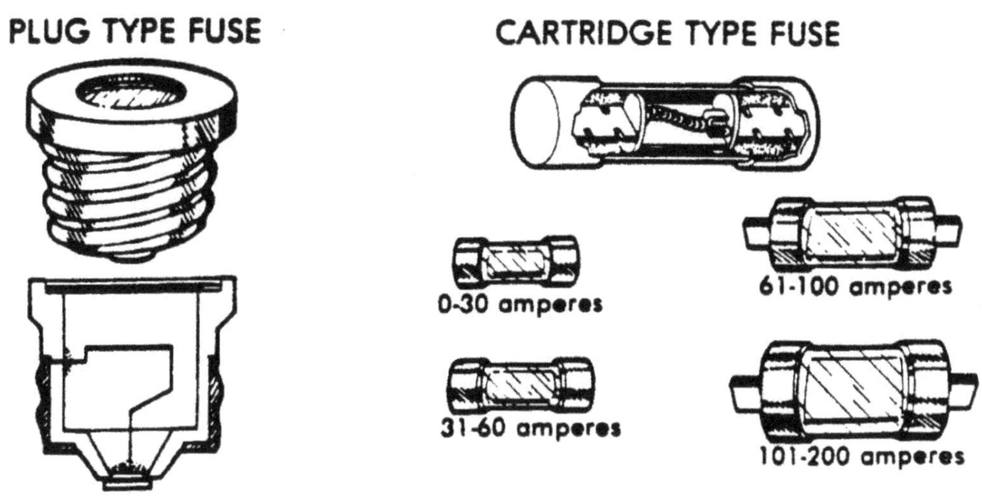

Fusetrons (time-delay fuses) are made to carry a temporary overload, such as the overload caused by the starting of an electric motor. The fuse, however, still provides protection for the circuit, and a short circuit will melt the fuse link. If a common fuse is used, the fuse link will melt every time an electric motor starts. The use of a larger ampere common fuse will prevent the "blow" resulting from the temporary overload, but will not provide protection for the motor or the circuit.

OPERATING PRINCIPLE OF DELAYED ACTION FUSE

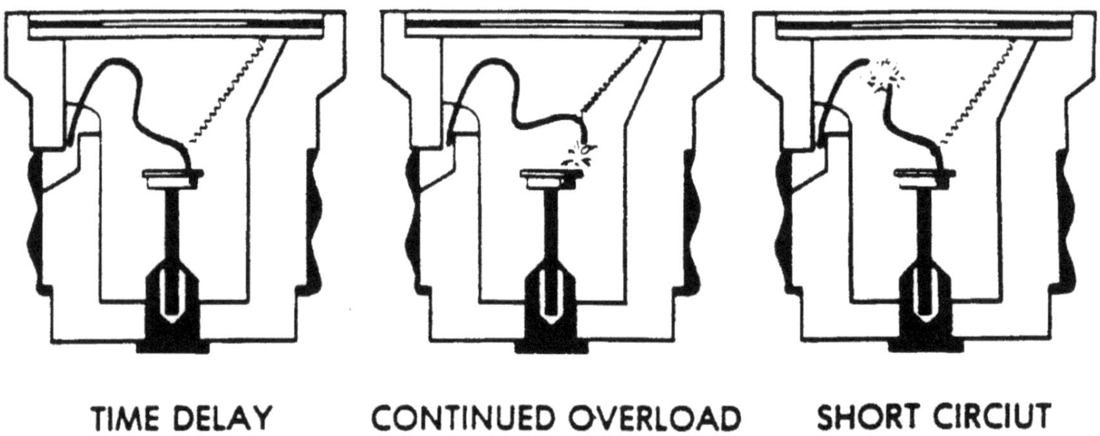

Fustats, nontamperable fuses of the time-delay type, have a different size base and require a special adapter that is screwed into the standard fuse socket. After the adapter is installed, it cannot be removed. For example, the installation of a 15-ampere adapter allows only the use of 15-ampere or smaller fuses.

FUSTATS

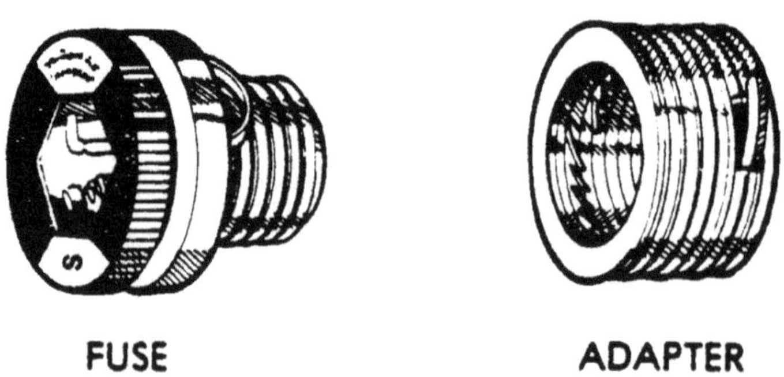

Circuit breakers eliminate the replacement of fuses and are commonly used even though a circuit breaker box costs more than a fuse box. Circuit breakers are of two types, thermal and magnetic. The thermal breaker has two contacts held together by a bi-metal latch. An overload of current causes the bi-metal strip to become heated, the latch releases, and the points spring open. After the bi-metal strip cools; the switch is reset; and service is restored.

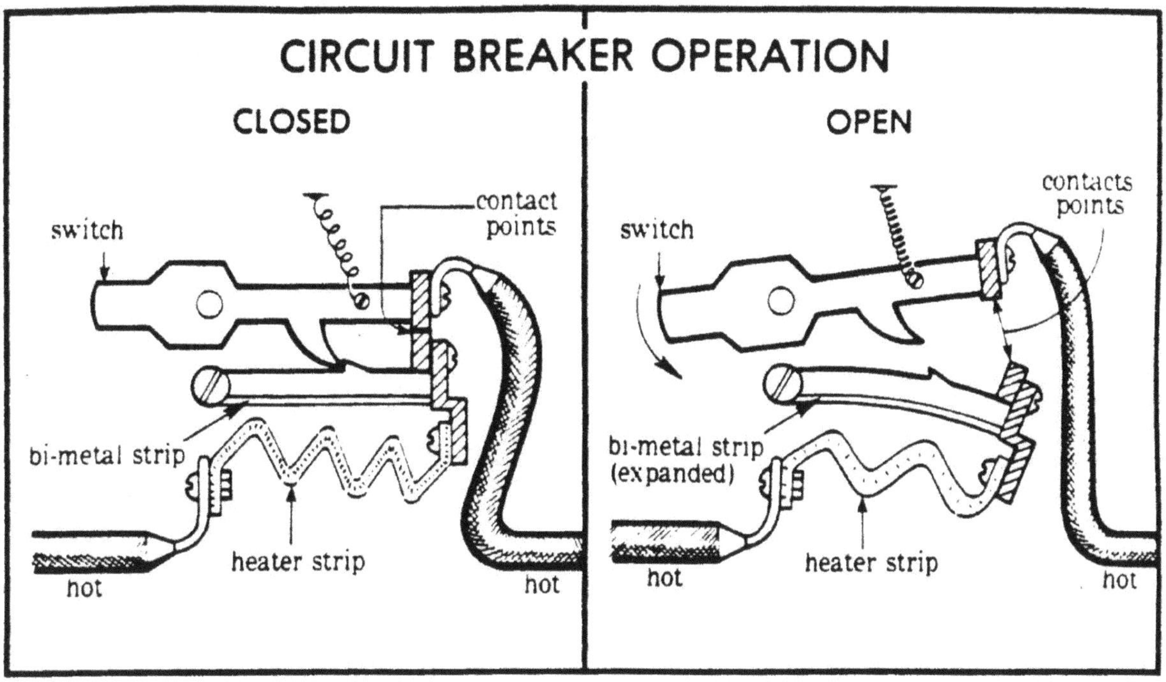

The magnetic breaker has contacts that are held together by a latch which is released by the action of an electromagnet. The amount of current flowing through the circuit will determine the size of the electromagnet. This type of breaker is reset by moving the toggle switch to the "on" position.

The following diagram shows the parts of a circuit breaker.

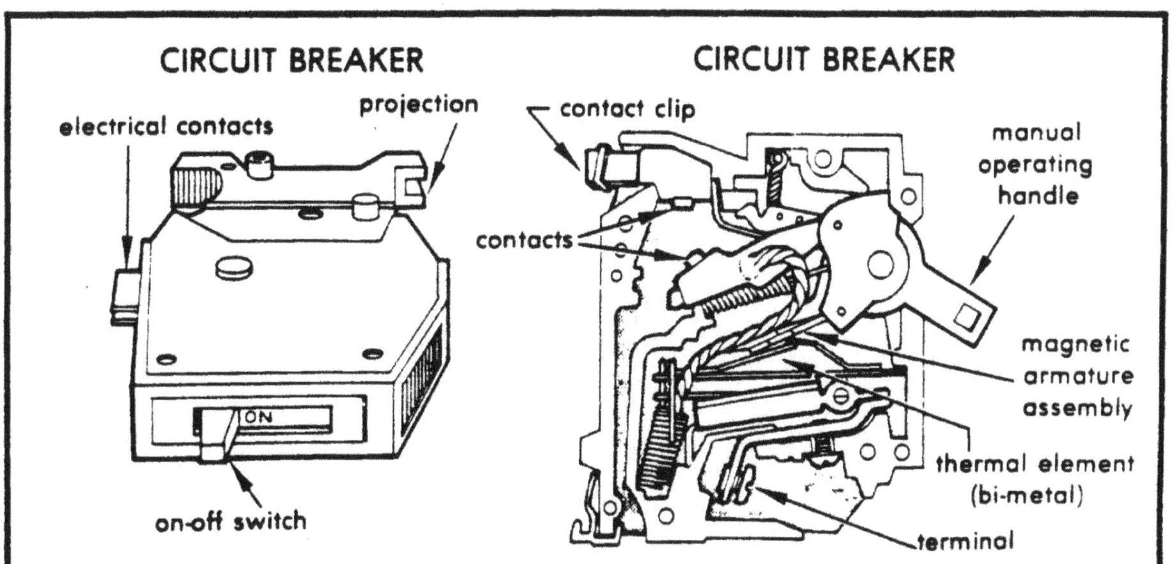

9. No Fault Grounding

Fuses and circuit breakers are safety devices which limit current (amperage) in a circuit. Their main function is to protect equipment and wiring from overload. Ground fault circuit interrupters (GFI) are designed to protect humans, equipment, and/or electrical systems from injury or damage if electricity flows in an unintended path (a short).

A GFI is a very sensitive device that functions by comparing the current moving in the "hot" wire with that in the neutral wire. If these two currents are not equal, a fault exists, and current is "leaking" out of the circuit. If the difference in current between the two wires is 5/1000 of an ampere or greater, the GFI will open the circuit, shutting off the power and eliminating any shock hazard.

The National Electrical Code requires GFI's for all 120V, single phase, 15 and 20 amp receptacles installed outdoors, in bathrooms, and in garages for residential buildings. A GFI is required at construction sites and some other applications. After correcting the circuit fault, the GFI may be reset for further use.

A variety of GFI equipment is made for 120 and 240 volt circuits.

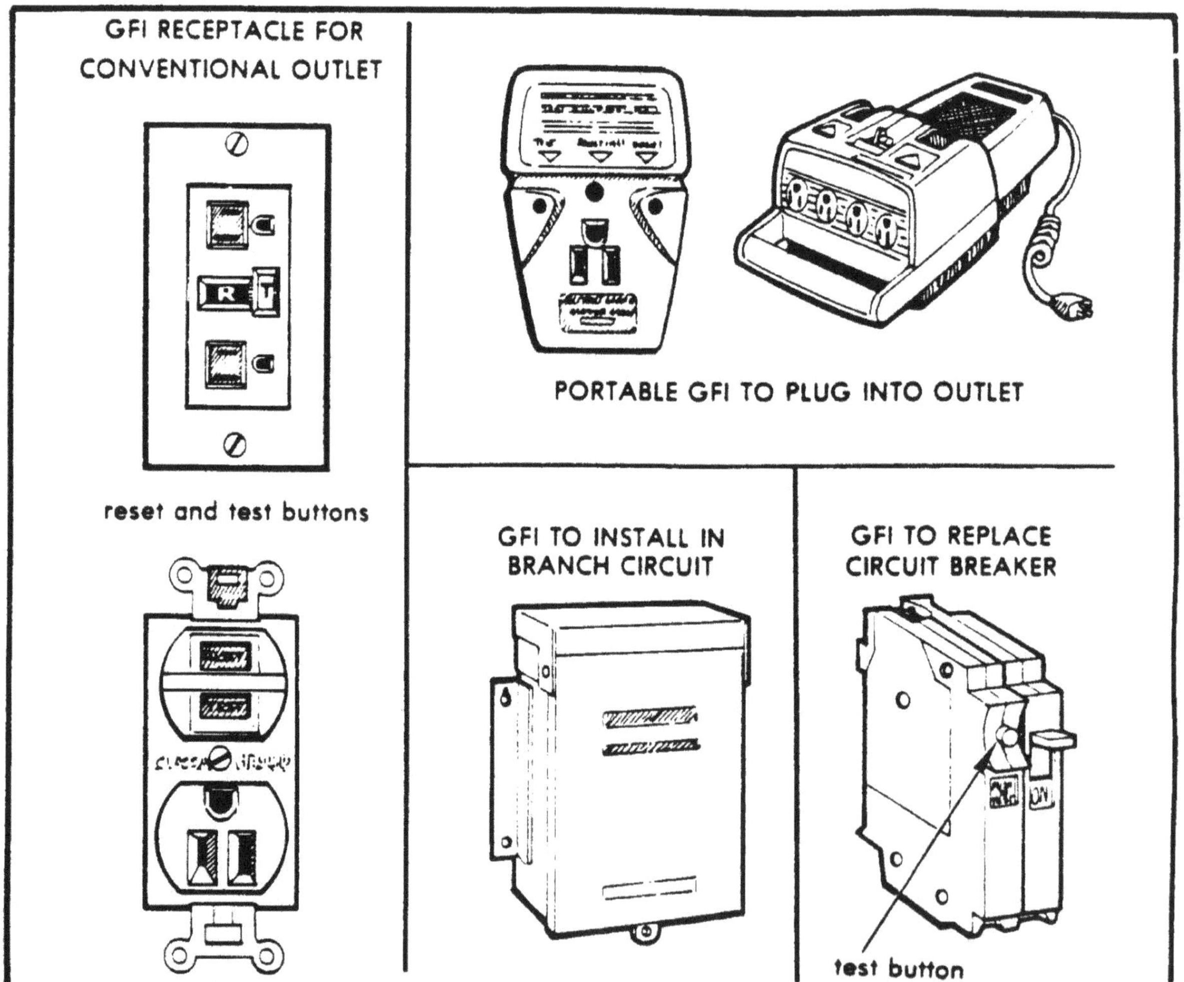

References:

Cooper, Elmer L., *Agricultural Mechanics: Fundamentals and Applications*. Delmar Publishers, Inc., Albany, New York.

Electrical Wiring - Residential, Utility Buildings, Service Areas, AAVIM, Athens, Georgia.

Don Smith TransWorld Website in 2001

Don is a University degreed professional in science & engineering. Served in three wars; WWII, Korea and Vietnam. He has enjoyed a long and successful profession as an executive in the petroleum industry. His life long outside interests include electronics and engineering. A popular walk on substitute for University Professors in physics, chemistry, biology and computer assisted drafting. Teaching style similar to Richard Feynman. He has celebrity status in Japan and Saudi Arabia. His website http://altenergy-pro.com is worldwide and is on more than 2000 separate servers. He also serves on several multibillion dollar corporation amongst their esteemed Board of Directors.

Don has designed and built fully functional energy related devices with very little outside assistance. It is quite the honor to have him as a presenter for our Inventor's Weekend.

Abstract

Don states that useful electricity is obtained from disturbed electrons, which radiate magnetic fields and waves. He says that a generator disturbs and collects this emitted electron energy. These electrons remain fully in tact and undiminished until the end of time. This Energy is free and the cost of accessing it depends on the ignorance of the collector. This none ionic electron cloud permeates the cosmos and with an appropriate collector system provides an unending source of environmentally benign, inexpensive, available energy everywhere. The Earth's ambient background is an enormous unending source of electrical energy. More details are made available on Don's website. Devices and methods not related to the Rube Goldberg energy systems are currently in use by special interest will be examined and tested.

INTRODUCTION: The Ultimate Energy Source should be never ending, fully self renewable and not compromise the environment or the Worlds oxygen supply. Electrons fluxed by magnetic wave resonance is the only option that meets these requirements. This electromagnetic system is available universally. Electron spin phenomenon produces magnetic and electrical waves, when intercepted naturally or with man made devices, provides useful energy. All present and future electrical energy is sourced from this unending reservoir. Conventional energy is captured when magnets and coils are moved relative to each other. Faraday's action at a distance is provided by magnetic waves which replicate when encountering a receptor. Capture of these waves at resonance does not diminish the source. For example, multiple capture devices with exact copies of the magnetic resonance source provides greater output than input. Self sustaining electrical devices then become a reality. Electrons have been fluxed since the beginning of time and remain undiminished. Prescription physics, special interest version is intentionally blind to the natural behavior of the electron.

Electrical generators, dynamos and such, do not make electricity, they simply collect it from magnetic waves which spin the adjacent electrons. Electrical waves always diminish when returning to ambient. Magnetic waves however freely replicate when resonate, providing multiple exact copies of the originating source, without diminishing or degrading. Action at a distance by this magnetic resonance multiplies the energy present through electron spin. Devices constructed based on this knowledge provides self sustainable electrical motors and generators. Once the mental block put there by special interest is recognized and dispensed with, a whole new world of electrical devices exist.

The Presenter is an open minded Science and Engineering Professional with more than 60 years of highly successful background. During the past ten years many futuristic energy related devices based on the knowledge herein have materialized as practical. Those blind to reality will have difficulty allowing that their concept of electrical energy is very wrong. The purpose of this Web Site is awaken and share knowledge in a beneficial and magnanimous way, such as to benefit Humanity.

ELECTROMAGNETIC SPECTRUM ENERGY: This Energy has four physical expressions, being radiated, reflected, deflected and absorbed. This Energy becomes useful when displaced from ambient background. Being other than ambient, a pressure or potential develops, which needs to return to it's background. This Leg between displacement and equilibrum is then useful Energy. Physics of useful energy then requires disturbing the ambient background. In Electrical Systems this occures when none ionic electrons, present everywhere, are disturbed, spining left and right by alternating the electromagnetic wave environment present. Pulsating magnetic waves, when deflected become electrical waves and vice versa. This is observed throughout nature and in man made devices. The greater the spin rate (frequency) the greater the displacement from ambient.

The electrical side of the equation is fully described in Conventional Physics. The magnetic side is intentionaly clouded to prevent access to clearly obvious benefits present. Some of these benefits are:

1. Multiple copies of the magnetic wave energy present can be made without depleting the source. Recptors present in the resonance field duplicate the field without dimenishing it.

2. In conventional systems, the equal magnetic side of the equation is radiated and waisted.

3. The magnetic resonance is not dimenished when copied and feed back into the system. Multiple copy's of same source, when feed back into the system provides Self Sustainable Devices.

RESONANCE ENERGY CONCEPT: A human is a speck of dust on Earth, the Solar System is a speck of dust in the Galaxy and in turn it is a speck of dust in the Universe (Cosmos). All of these respectively represent vast ambient energy reservoirs. Awareness of the Sun opens doors into other energy sources. Electromagnetic Energy present everywhere throughout the Universe is accessed by catalytic activity, directly as in Solar Cells or indirectly as by mechanical means. Resonate Magnetic Waves (Faraday's action at a distance) allows energy activation transfer to remote points of usage. Capturing and use of this energy (Free) is optional as to method and therefore it's cost is a function of human stupidity.

Direct access is more desirable and technology transfer from Solar Cell type Devices provides the Catalyst. Enormously high ambient energy levels are not detected by instruments that use the ambient background as a reference plane.

A spoon full of water lifted from the Ocean does not define the Ocean. Incoming magnetic waves spin electrons sideways producing useful electrical energy. Absorbed wave energy produces heat, therefore a hot interior for the Earth. Physical movement in electrical systems is in the direction of current flow, frictional drag from inflow current defines gravity. Accretion masses (Energy Sinks) provides all solid entities with their respective gravitational fields.

Increasing the tolerance level for intellectual awakening opens doors of reality. These doors blink into and out of existence and upon recognition benefit mankind. Opening some of these doors, presently seen through a deep fog is our purpose. Exploring unrecognized energy sources, being a part of the ambient background is another goal. Vast energy sources that totally surround are available through technology transfer. They are inexpensive (Free), fully self renewable and are environmentally benign.

Incoming magnetic wave energy with Faraday's action at a distance will be looked at closely. Particle physics will be left to Astrophysics. Excited electrons at point "A" the Sun (also includes the Galaxy and Cosmos) do not travel to point "B" the Earth, however a corresponding action occurs at point "B". Electrons being disturbed at the Central Power Plant in the same manner excite the electrons at Your House, upon switching into an Earth grounding (known as flipping the switch).

Correspondingly there are four major power sources providing enormous amounts of ambient background resonant magnetic wave energy. They are Cosmic, Galactic, Solar and the Earth's ambient electromagnetic backgrounds.

The Earth's background comes from reflection, deflection and absorption as a result of action at a distance from the above.

Prescription Physics mandates that the Earth's background is of little interest. When We have considered the evidence herein, it will become obvious that Special Interest's effort at keeping the People ignorant has until now largely succeeded. World wide information regarding the Earth's magnetic flux field is available (US Geological Survey, Colorado, USA, Office). When examined and properly understood, these Maps yield important information regarding reflection, defraction and absorption of incoming magnetic waves, plus action at a distance. When properly understood these Maps yield an ambient electromagnetic flux background o 7.2 Billion Volts at point of measure. This is the part of the Earth's Energy System that relates to the Bird on the High Voltage Line.

When deflected magnetic flux from electrons changes to electrical flux, providing the Motor System that spins/rotates the Earth. Physical movement by electrical systems is from inflow current movement. What level of electrical energy is required to spin the Earth. The Earth's mass in Kg is 5.9 X 10 to the 24 Th., Page 288, "Physics for Scientist and Engineers", 2nd. Ed., Edited by Ramond A. Serway, Pub. Saunders, USA. From this information the Watts of Electricity required is calculated! Absorbed flux energy heats from the inside out, therefore a hot interior for the Earth. Water is strongly diamagnetic and ocean waves present on windless days provides visible proof of the overhead incoming magnetic flux. The Earth's weight and rate of spin allows the calculation of the incoming ambient background energy required. As You can see it's not inconsequential as Prescription Physics mandates.

Astrophysicists are concerned with charged particles that whiz by once every one hundred years, rather than Wave Phenomenon associated with action at a distance. This highly active Wave Energy translates into Electrical Energy at point "B". The Galaxy is alive with Magnetic Wave Energy, being Billions of times greater than the Sun. Visible Light is a very small part of the Electromagnetic Energy Spectrum! Frequencies (Magnetic) present in the Galaxy and Cosmos allows Radio Telescope Photographs of their existence and magnitude. One such Photograph being at 408 MHz (Radio Sky, Haslam 408 MHz .)

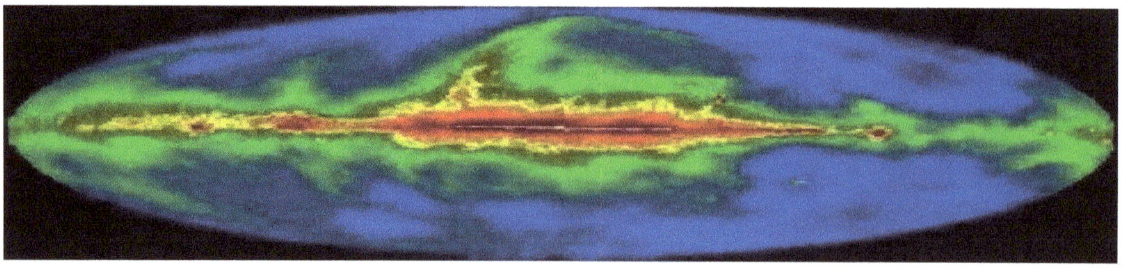

(Data) http://www.tuc.nrao.edu/ (Our Galaxy Disk, end to end measures 150 Light Years , Showing as red and yellow, with outer radiated energy in green).

The edge view of our Galaxy seen at 408 MHz, shows the Earth as a tiny speck of dust in this Enormous Ocean of Energy, near the left end of the Central High Energy Area. This Energy extends in all directions. Accretion and formation of Galaxies, Suns and Planets are the results of energy sinks and variable sized black holes. Mass retains heat and is cooked from the inside out by the microwave back- ground provided by the Universe. Flux movement into energy sinks provides the frictional force known as gravity. Spinning mass in the presence of incoming magnetic flux amplifies the gravitational effect. At present only Solar Energy is recognized. It is inconsistent, flaky and a very small part of the magnetic Wave Energy present. Technology transfer from Solar Power type Devices provides uncomplicated and inexpensive direct access to other Greater Energy Sources.

All electromagnetic energy harvesting methods include a catalyst, a collector and a pump. Catalyst include sensitization through doping with certain elements, air and earth grounding. Collectors include temporary storage as in capacitors, coils and transformers. The pump system includes induced movement onward to point of use. Conventional rotating coils and magnet systems activate electrons present, such that action at a distance can occur, therefore being an energy activation pump. Direct access systems, such as Solar Cells have the same effect without mechanical action. Magnetic Waves impacting and deflecting spin local electrons sideways producing useful electrical energy. Indirect acquisition of electrical energy by mechanical means is wasteful, troublesome and degrades the environment. The dynamo is a combination collector and pump of energy which is collected from the Earth's Ambient Energy Background, then transferred as in Faraday's action at a distance.

Energy Conservation Laws related to the Systems herein, relate to gray areas and when understood "are excluded because of the existence of External Forces, Open and Nonlinear Systems" a la Einstein. The knowledge base here viewed provides a Direct Understanding of the requirements for harvesting unending, fully renewable, environmentally benign Sources of Electrical Energy.

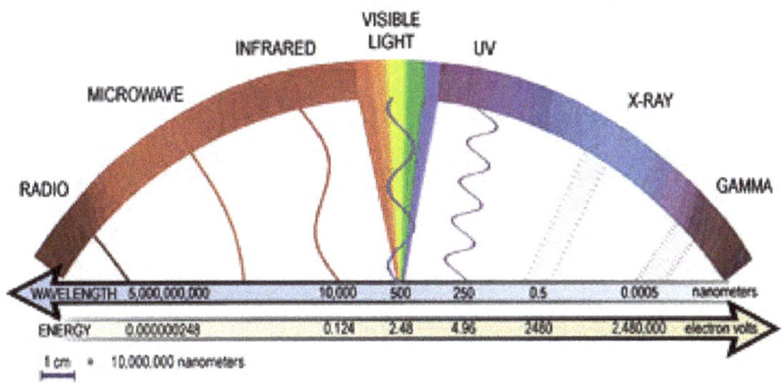

Electromagnetic Spectrum Energy . . .

. . . has four physical expressions: Radiated, Reflected, Deflected, Absorbed; and it becomes useful when displaced from its ambient background. Being other than ambient, a pressure, or potential, develops which needs to return to its background ambience. The Leg between displacement and equilibrium is then Useful Energy. The Physics of Useful Energy requires disturbing the ambient background. In electrical systems this occurs when non ionic electrons, present everywhere, are disturbed, spinning left and right by alternating the electromagnetic wave environment. Pulsating magnetic waves, when deflected, become electrical waves, and vice versa. The greater the spin rate (frequency) the greater the displacement from ambient.

The eletrical side of the equation is fully described via convential physics. The magnetic side, and its obvious benefits, are only now coming to light. Benefits include:

Multiple copies of the existing magnetic wave energy can be produced without depleting the source. Receptors present in the resonance field duplicate the field without diminishing it. In convential systems the equal magnetic side of the equation is radiated and wasted.

The magnetic resonance is not diminished when it is copied and fed back into the system. Multiple copies of the source, when fed back into the system, enables Self Sustainable Devices.

Magnetic Resonance Energy - Spectrum . . .

. . . Being four physical expressions: Radiated, Reflected, Deflected and Absorbed; and it becomes useful only when displaced from it's ambient background. Being other than ambient, a pressure, or potential, develops which needs to return to its background. The Leg between displacement and equilibrium is then Useful Energy. The Physics of Useful Energy requires disturbing the ambient background. In Magnetic Systems this occurs when non ionic electrons, present every where, are disturbed by alternating the magnetic wave environment causing left or right spin in the reverse of the acting force. Pulsating magnetic flux, when deflected, become electrical flux, and vice versa. The greater the spin rate (frequency) the greater the displacement from ambient, the greater the magnetic flux produced by this disturbance, when captured becomes useful energy.

The electrical side of the equation is fully described via conventional physics. It's obvious heat death curtsey of Ohm's Law is wrongly accepted as proof that over unity is impossible. The magnetic side and its obvious benefits, are only now coming to light. Benefits include:

Multiple copies of the existing magnetic flux energy can be produced without depleting the source. Receptors present in the magnetic resonance flux field duplicate the field without diminishing it. In conventional systems the magnetic side of the equation is radiated, ignored and wasted.

Magnetic resonance is not diminished when it is amplified, copied and fed back into the system. Multiple copies of this source, when fed back into the system, enables Self Sustainable Energy Devices.

Sample Website Prototypes

Device 1 . . . PROOF OF PRINCIPAL . . . Tesla Flat Coil Transformer System Demo, Energy Amplification is the Ratio is the in / out Diameter, times the power input .

From video 5, 2001: "The ratio of the diameter of the inside versus the outside turn on it is the improvement, or the advantage that you gain by doing it that way. In the one that's in the picture in the book there the ratio is one to five. So not only do you get the L1 and L2 gain, you get five times that. That's the advantage of the pancake coil. The the L1 one wire should be an even division of whatever the length of the L2 is, that way they are able to talk to each other. The frequency, if you worry about the frequency, don't worry about that. Just build the thing so it looks nice and make an even division. Even multiple or even division of the lengths of the wires if you want to talk to each other, and they'll be talking to each other. [Question. What's the ratio?] The ratio in the case of the one that's in the book, is one to five. So it's five times, plus the fact that it's a normal Tesla coil, so normal gain times five. [Question. Do you ever do quarter of the wave length, like Tesla did?] They will be able to talk to each other. You can calculate the wavelength of those, you take the speed of light and divide that by a certain factor and that will tell you how long the wire will be for a certain frequency. What I did on that particular one, I used about two hundred megahertz which meant the wire is going to be about eighteen to twenty feet long. I think that's eighteen something. Once I had the main frequency set up, then I made the other coil. As long as it's a even division or multiple of you know, which ever way you go, they're gonna talk to each other and that's all you need." Page 283.

Device 1 . . .

PROOF OF PRINCIPAL . . . Tesla Flat Coil Transformer System Demo, Energy Amplification is the Ratio is the in / out Diameter, times the power input .

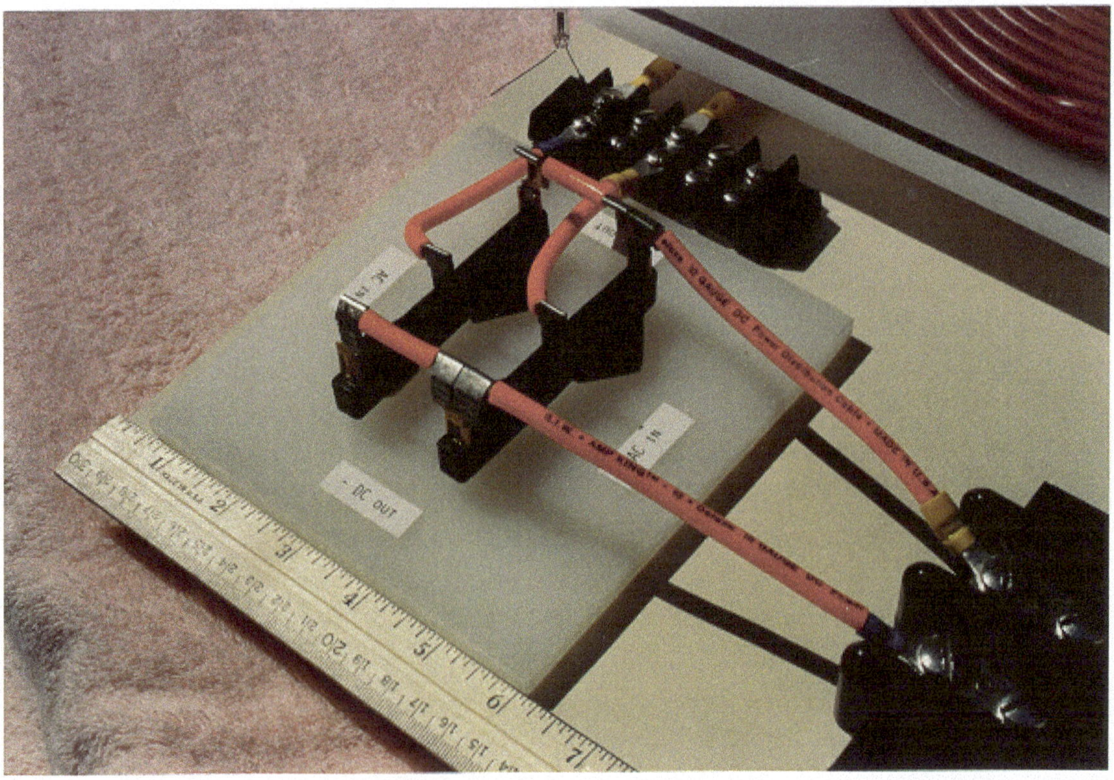

Device 2 . . . PROOF OF PRINCIPAL . . . Energy Duplication by Magnetic Flux Resonance, Circa 1983, as many full Duplicates as You may wish with out diminishing the source. One Transmitter, three Receivers .

"It was the first device that bill because I was told that you couldn't do it."

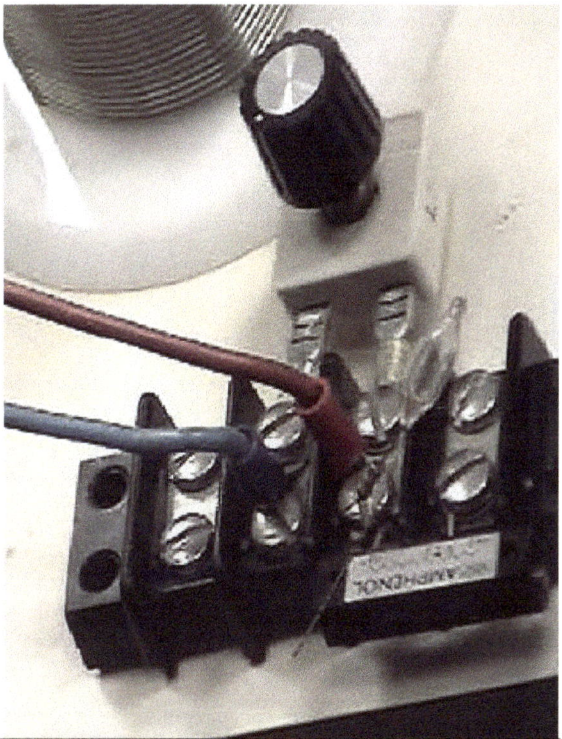

Device 3-1 . . . PROOF OF PRINCIPAL . . . Resonant Energy Device: Magnetic Flux Amplified to 8000 Volts at 20 Amps from 12 Volts at One Ampere input .

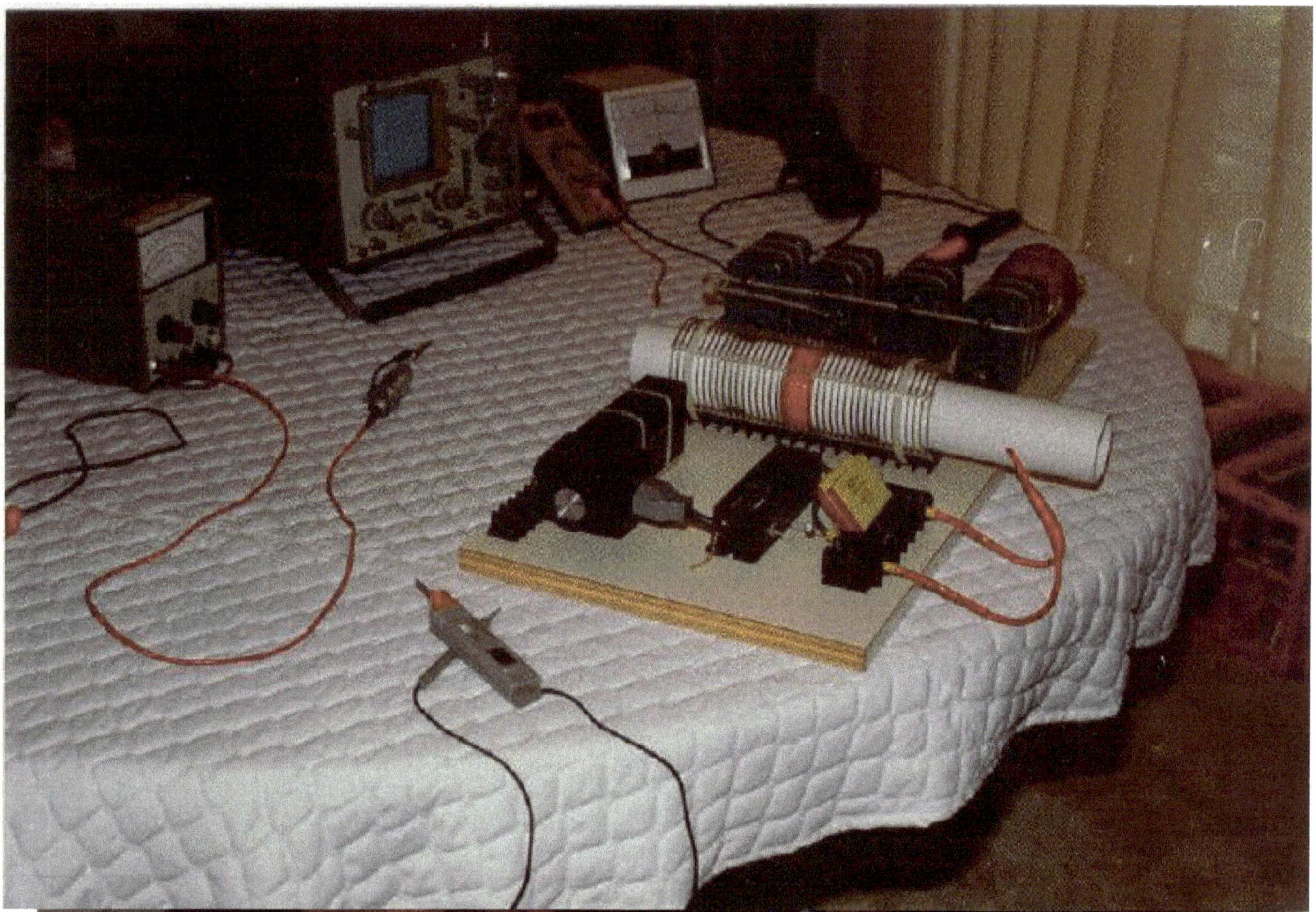

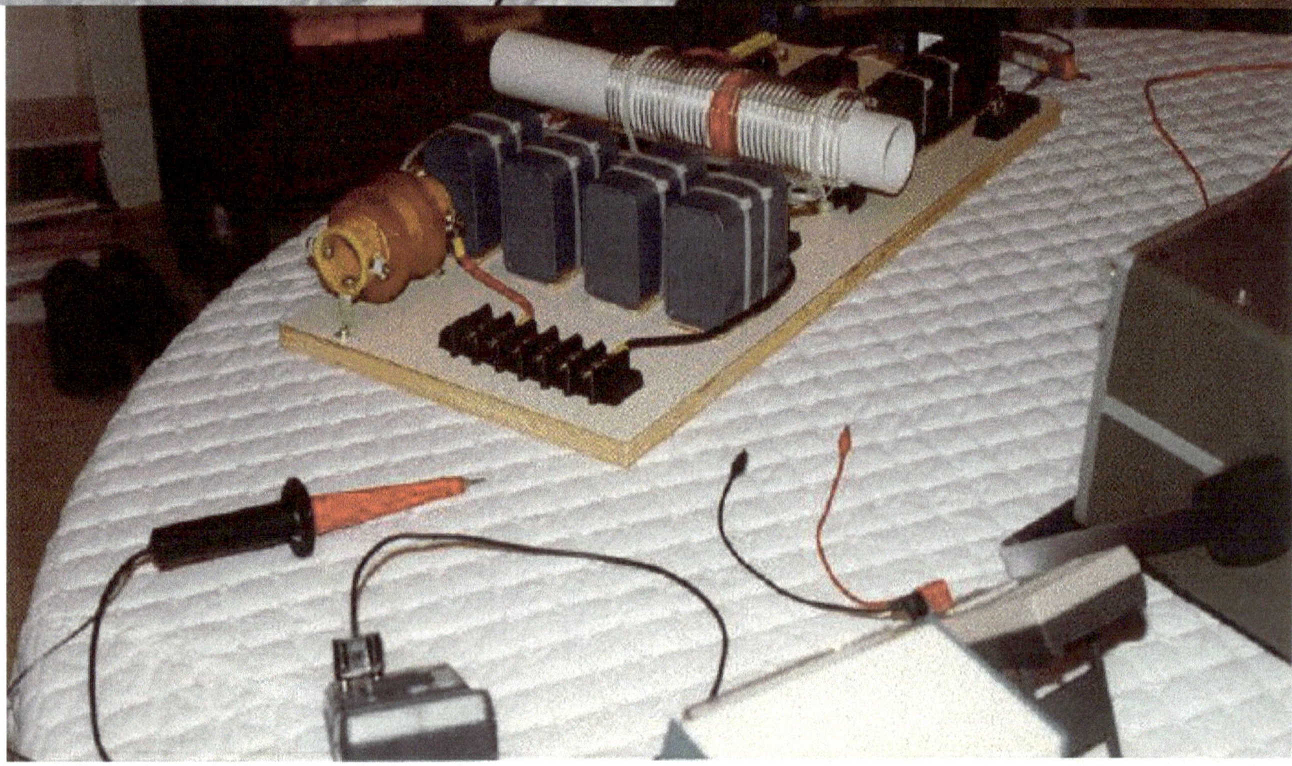

Device 4 . . . A High Gauss Source . . . Magnetic Flux from The Plasma, Generates Useful Energy. One Gauss equals the Magnetic Flux equivalent equal too 100,000. Volts of Electrical Energy. When Transformed back into Electrical Flux it provides a very Large increase, becoming a Self Powered Energy Source .

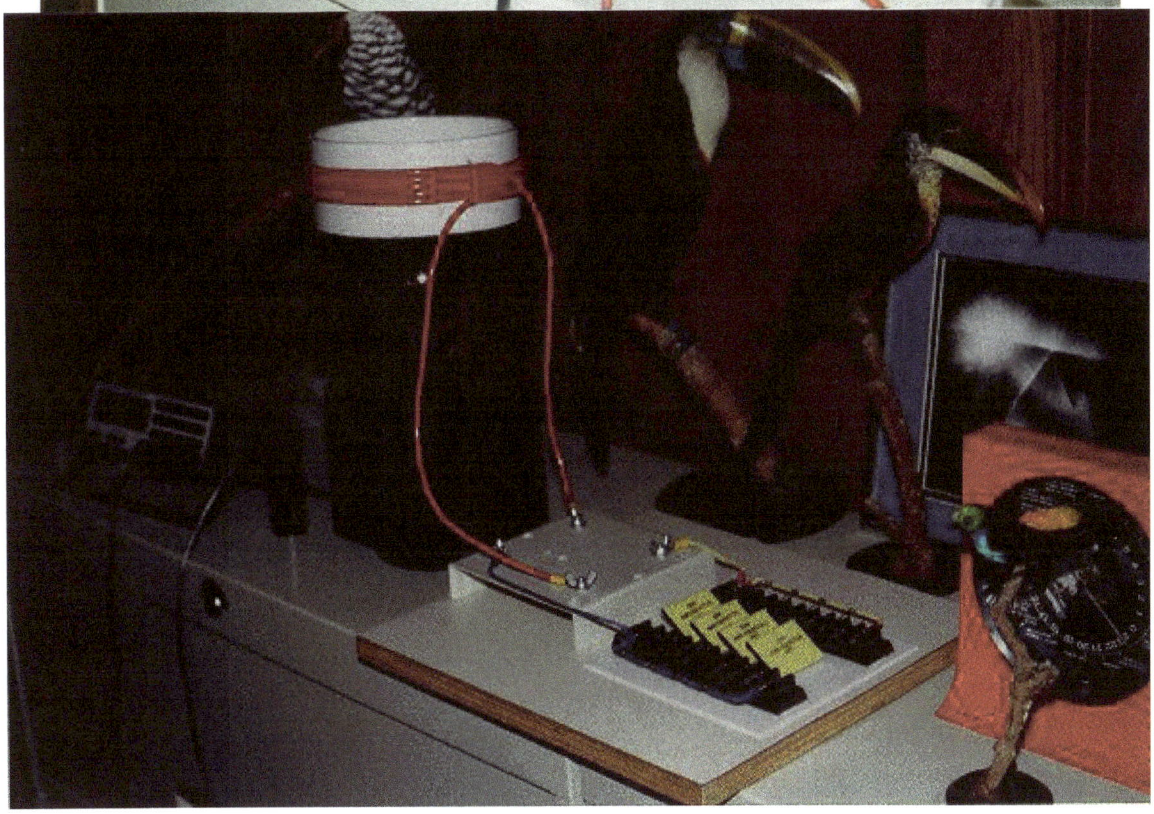

Device 5-1 . . . PROOF OF PRINCEPAL . . . Prototype Self Powered System for Vending Machines in Japan

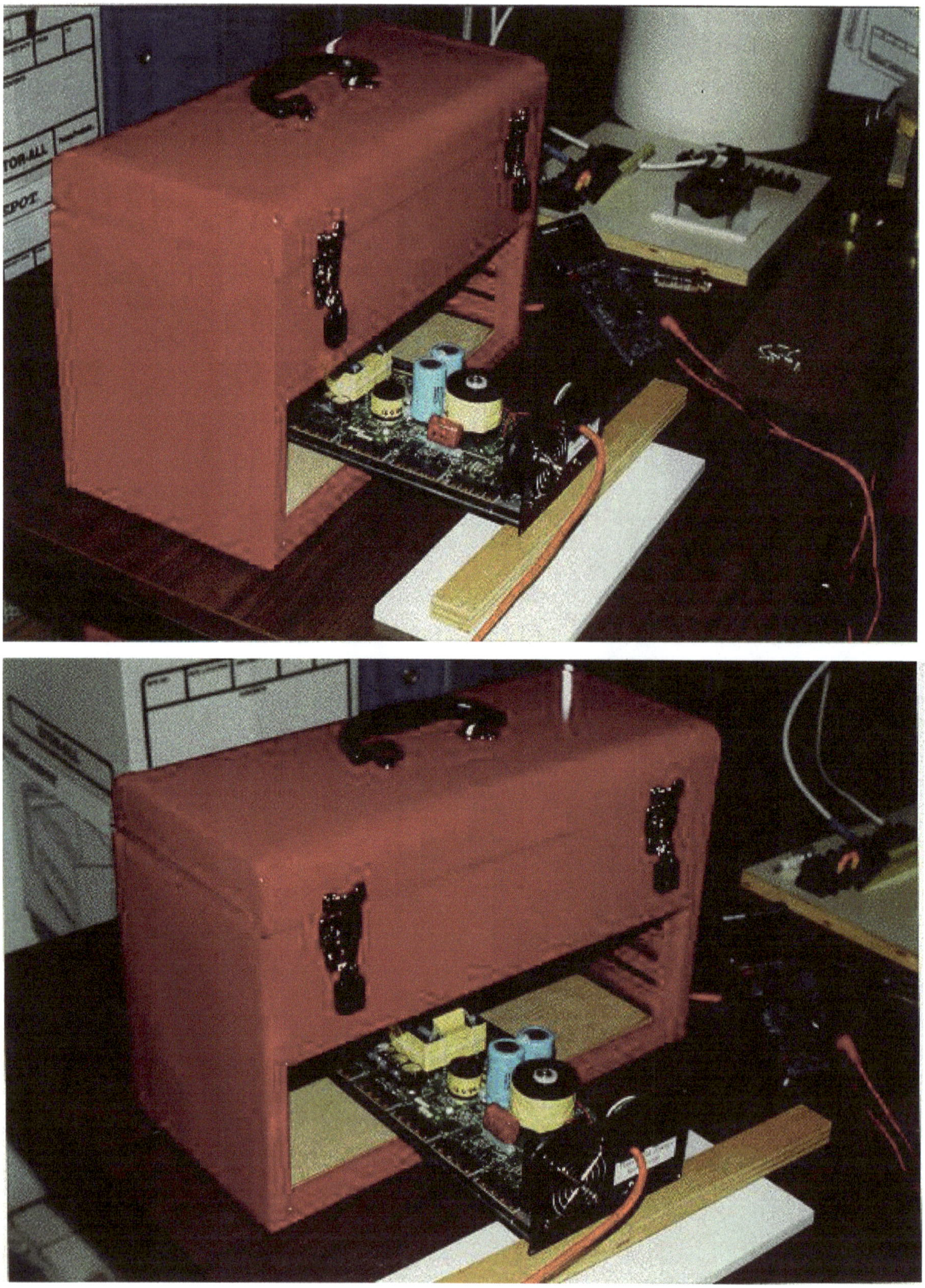

Device 6-1 . . . PROOF OF PRINCIPALE . . . Prototype Power System for Brazil: note Tesla Coil L-1 and L-2 wound on large diameter Cobalt Nickel Core. The Magnetic Flux Energy is Amplified then Transformed into Electrical Flux becoming useful Power.

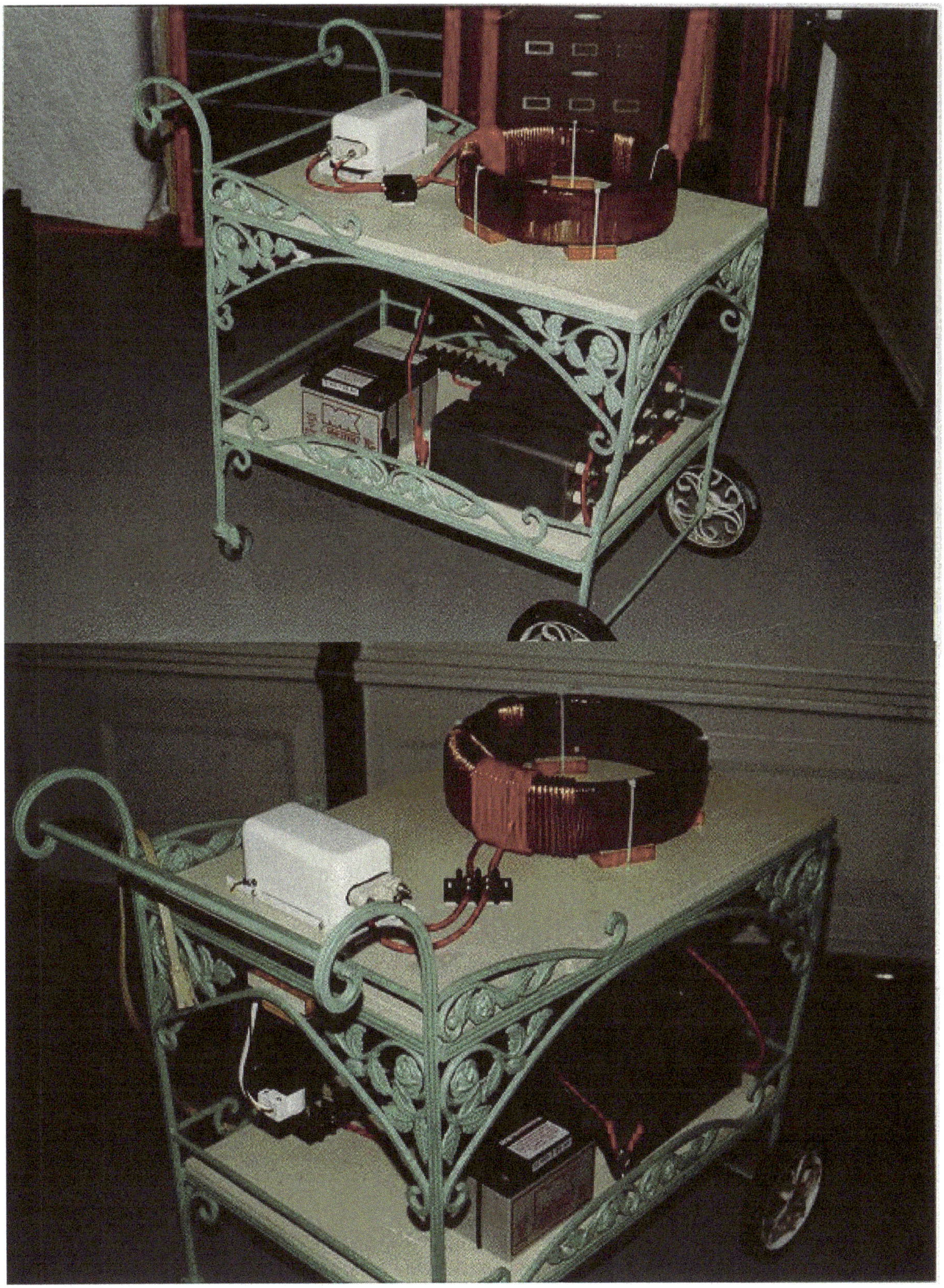

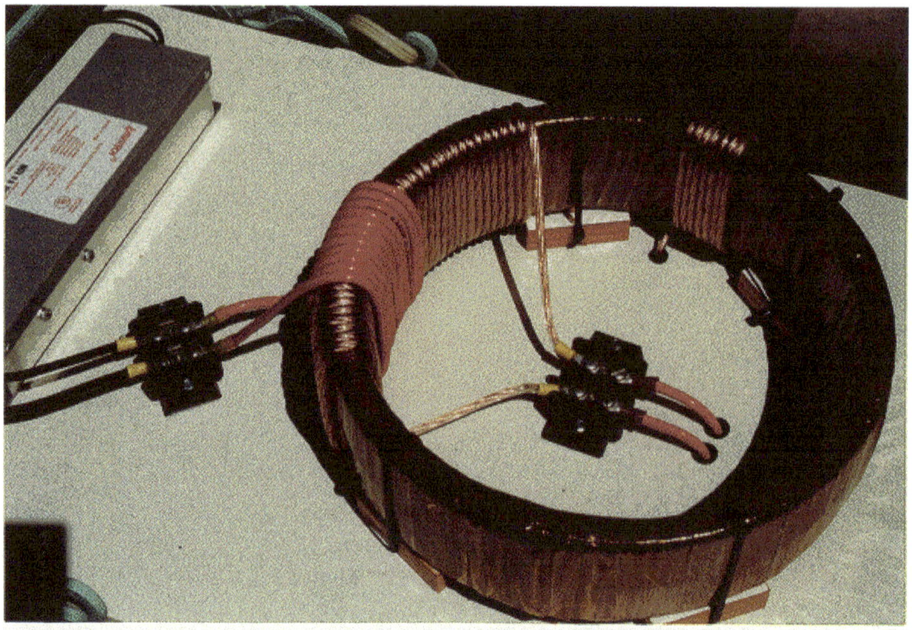

TransWorld Energy
Electrical Power Systems

Device 8-1 . . . . . . 14" Diameter Tesla Coil System

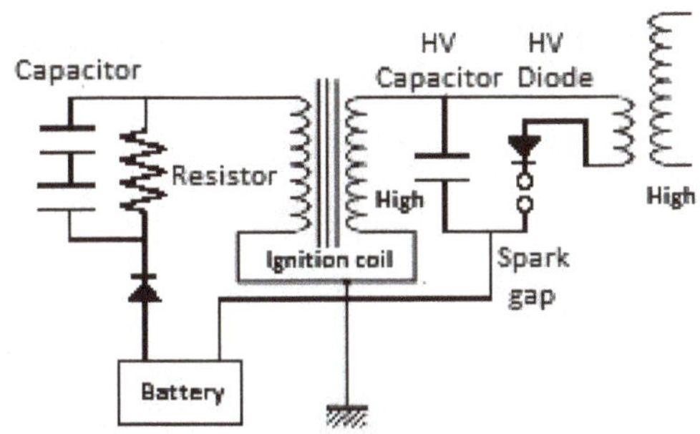

Device 9-1 PROOF OF PRINCIPLE Prototype Energy System for Brazil and Saudi Arabian Markets: Note the coils in pairs left and right. Each pair produces 1000 Volts at 50 Amps; with output in to separate Bus Bars in parallel. Dimensions are 14.5" High/16" Long/10" Wide. A milliwatt motor turns the magnetic shield which separates the magnet pairs at their Bloch wall separation providing magnetic fluxing and useful electrical energy.

Question: "Hi Don, Is it possible that you vary the speed of the motor to find/make the two coils and the two Caps. all in series (series LC circuit) resonant thereby increasing their efficiency/output?" You are correct. The variable speed switch is from an automobile dimmer. The drawing is the simplest possible and is therefore incomplete. The basic idea is clearly shown. Don Smith

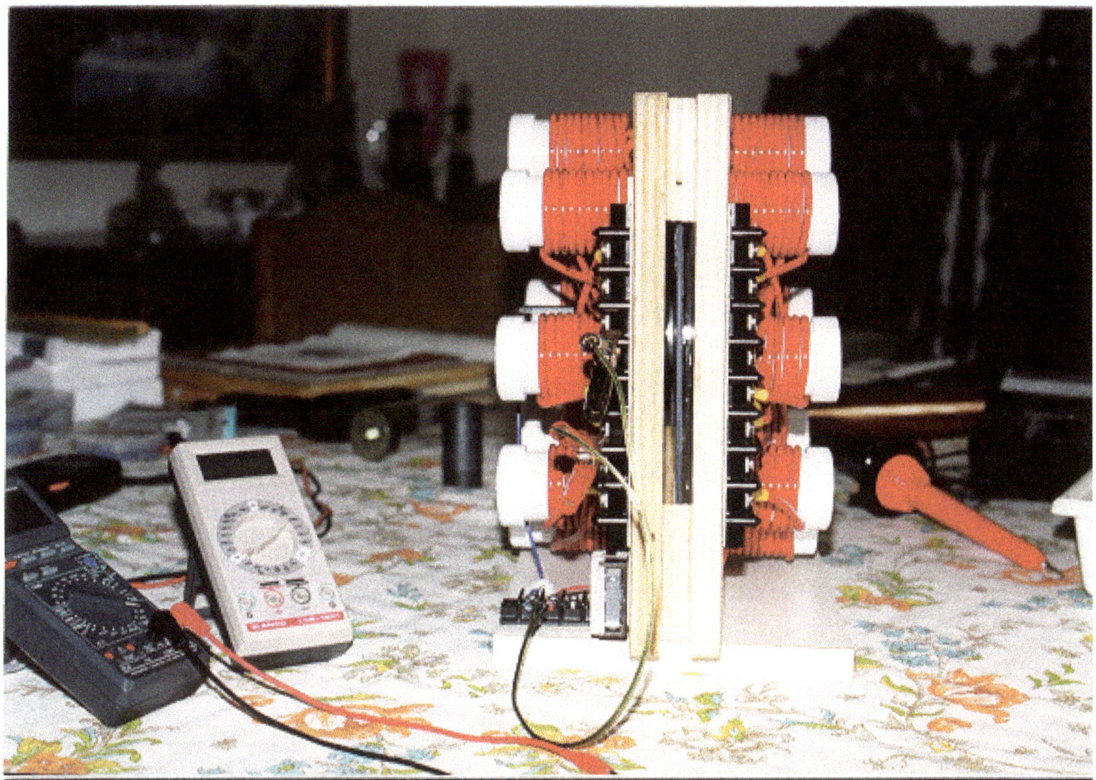

The small motor turns on 28 mW at 12 volts and just rotates the phonograph record. It creates a disrupter, interrupter. Terfenol-D is the most advantageous material, although bismuth or powdered neodymium will work. Spray adhesive on an old phonograph record and sprinkle it with the neodymium dust. It's enough to disrupt it. If you have the holes equally spaced there's little or no load on the phonograph record while it rotates disrupting the field.

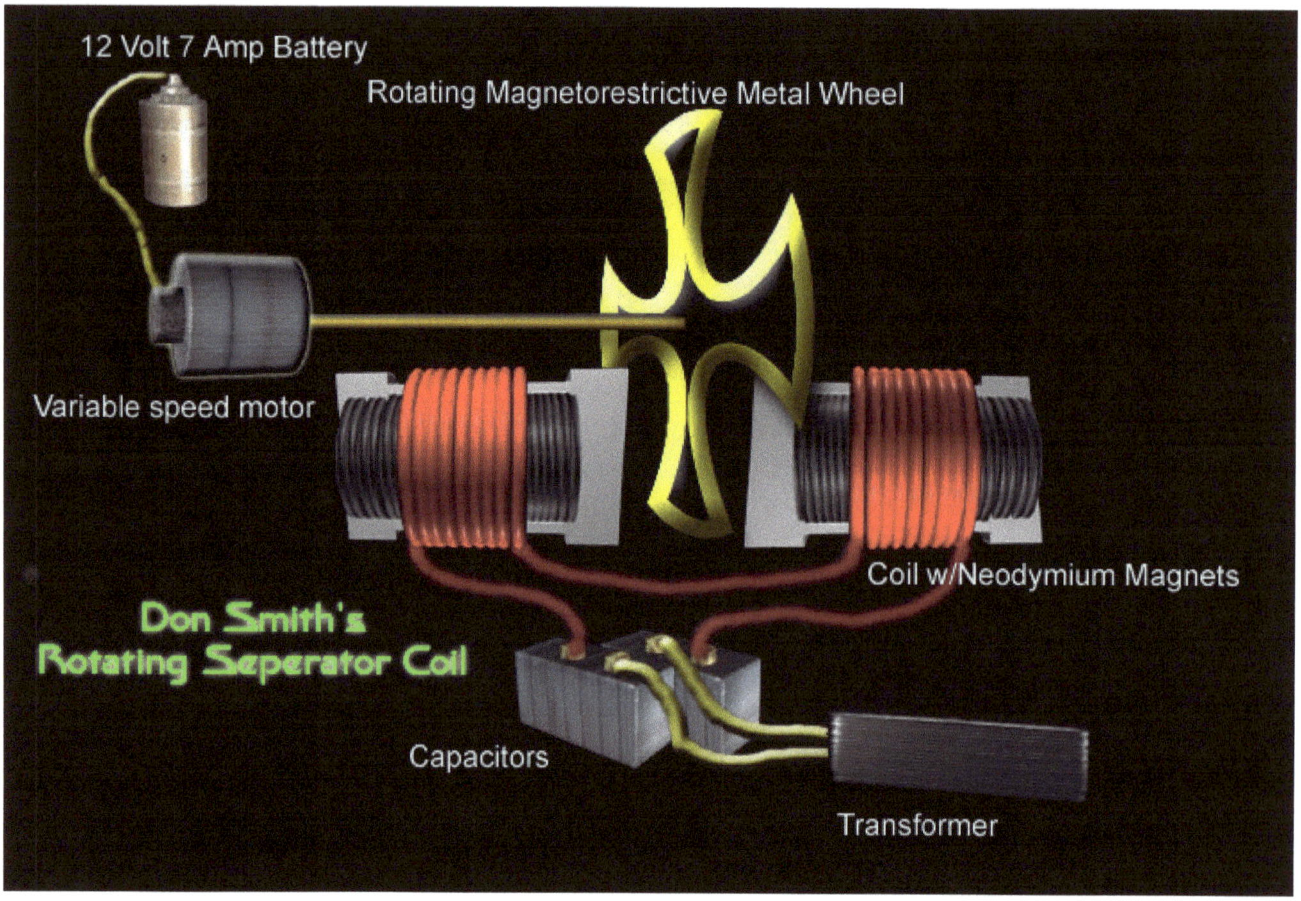

"The disc in the middle, all it has is just voids with a material like bismuth, neodymium or there's actually another material that even works better, that are sprayed on something as simple as a Plexiglas disc or a phonograph record." Video 7

"Try several thin layers of magnetostictive material alternating with thin layers of electrically conductive material. For example stainless steel and copper and or aluminum. This is not yet an exact science and requires a skill level. You may get some weird results at first. It requires mixing electrically conductive with none conductive which scatters-disorganizes the magnetic waves."

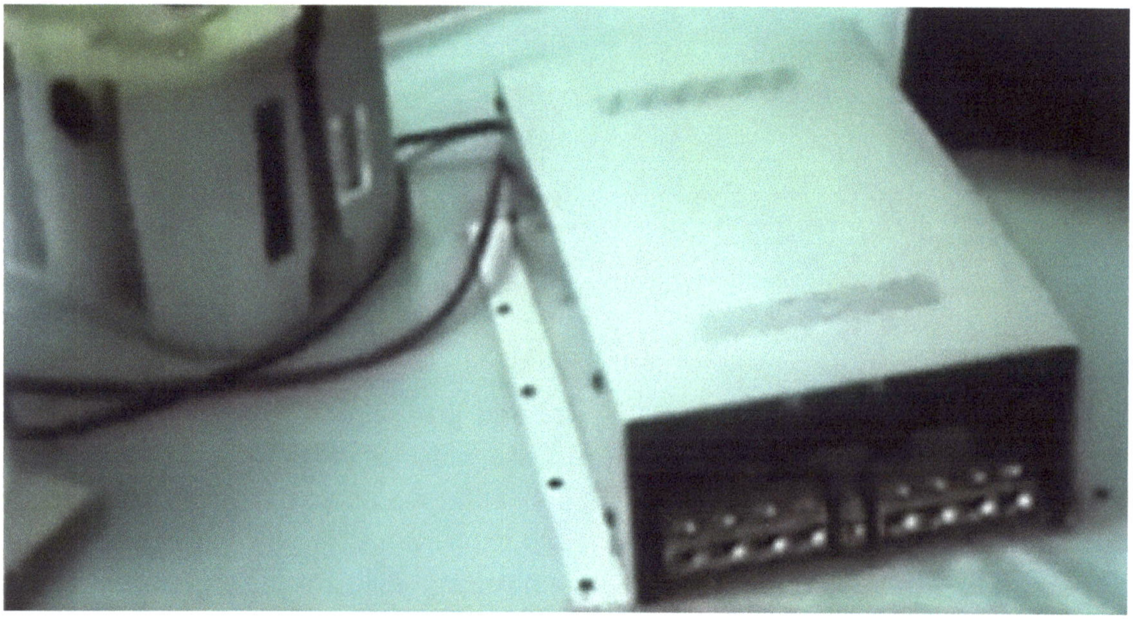

50F Capacitor bank off the Dipole Device.

Additional Don Smith Prototype Pictures

Barium Titinate Transformer

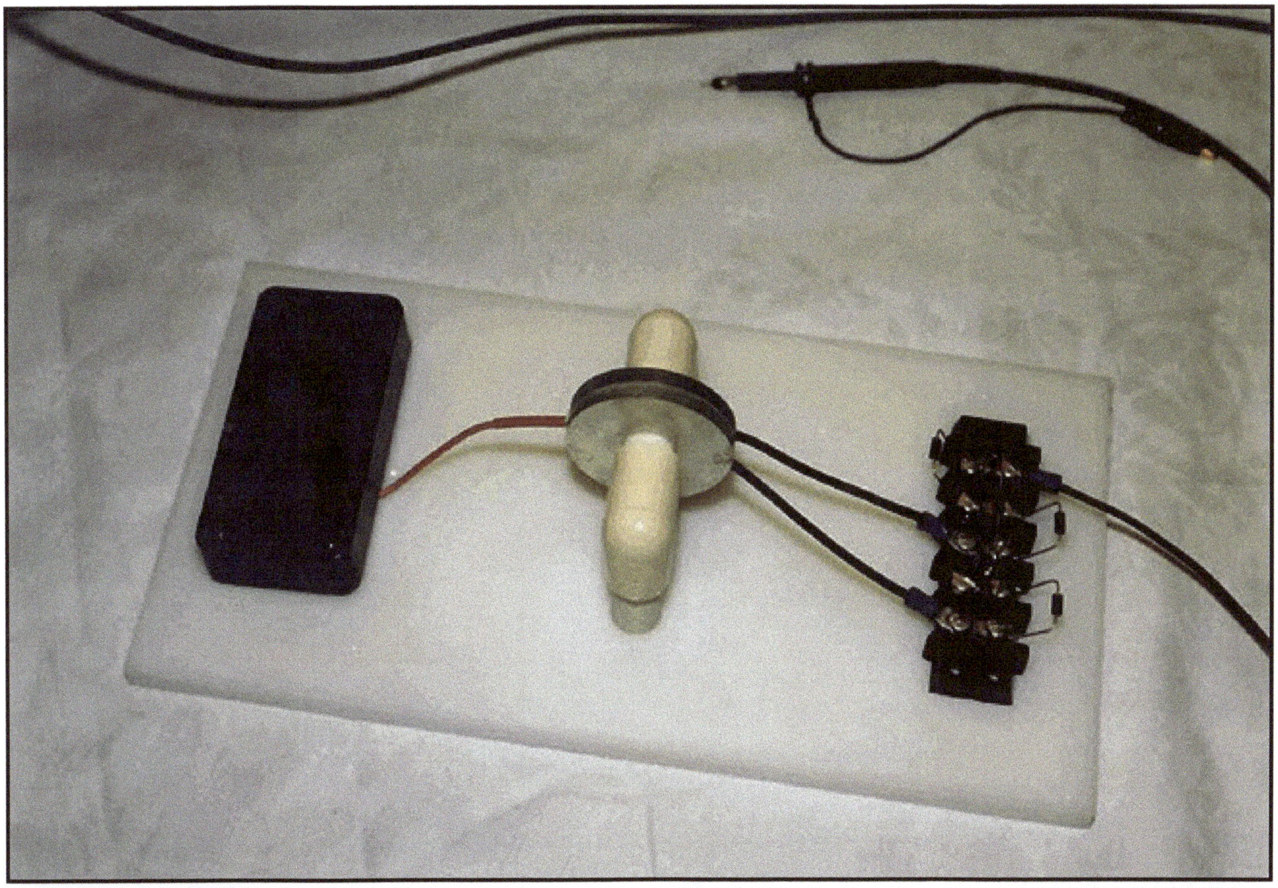

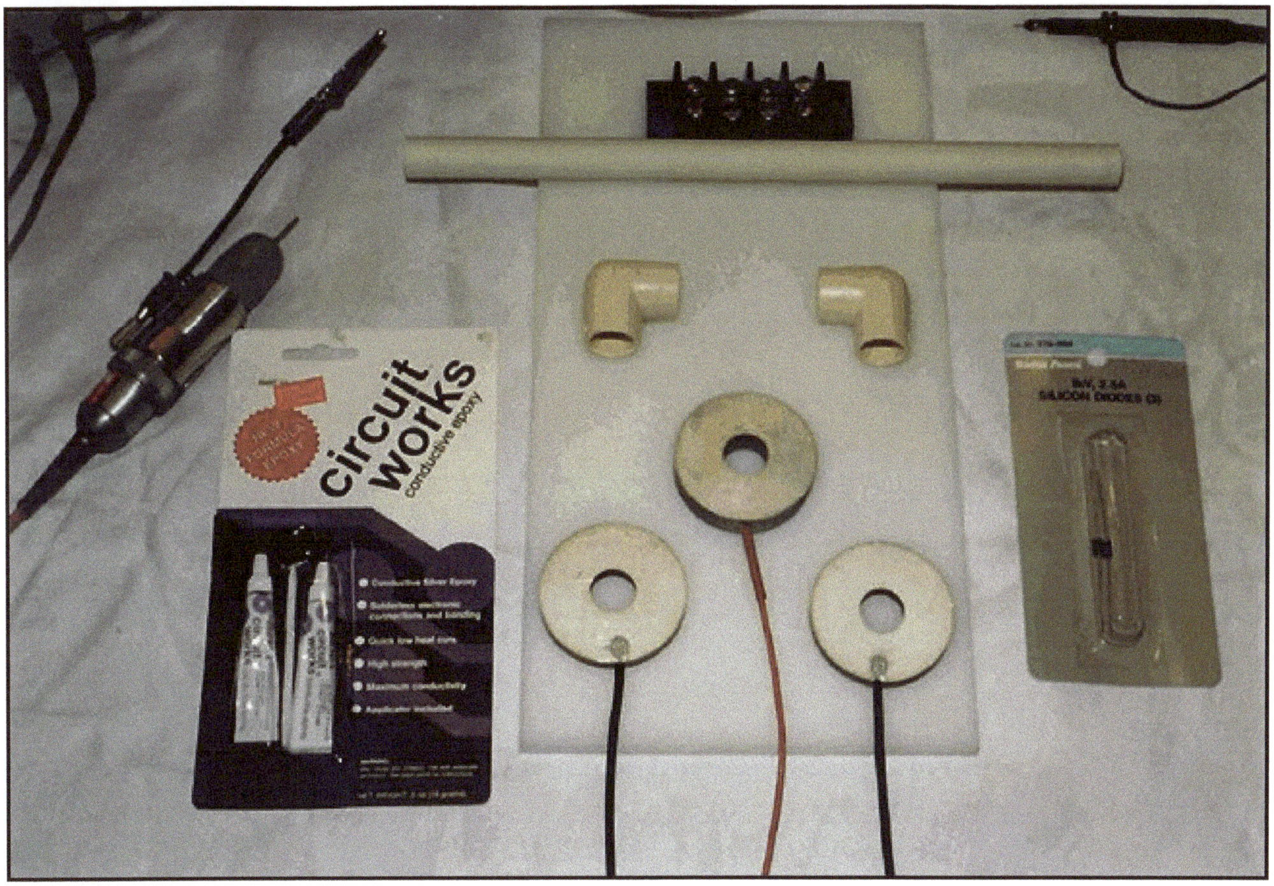

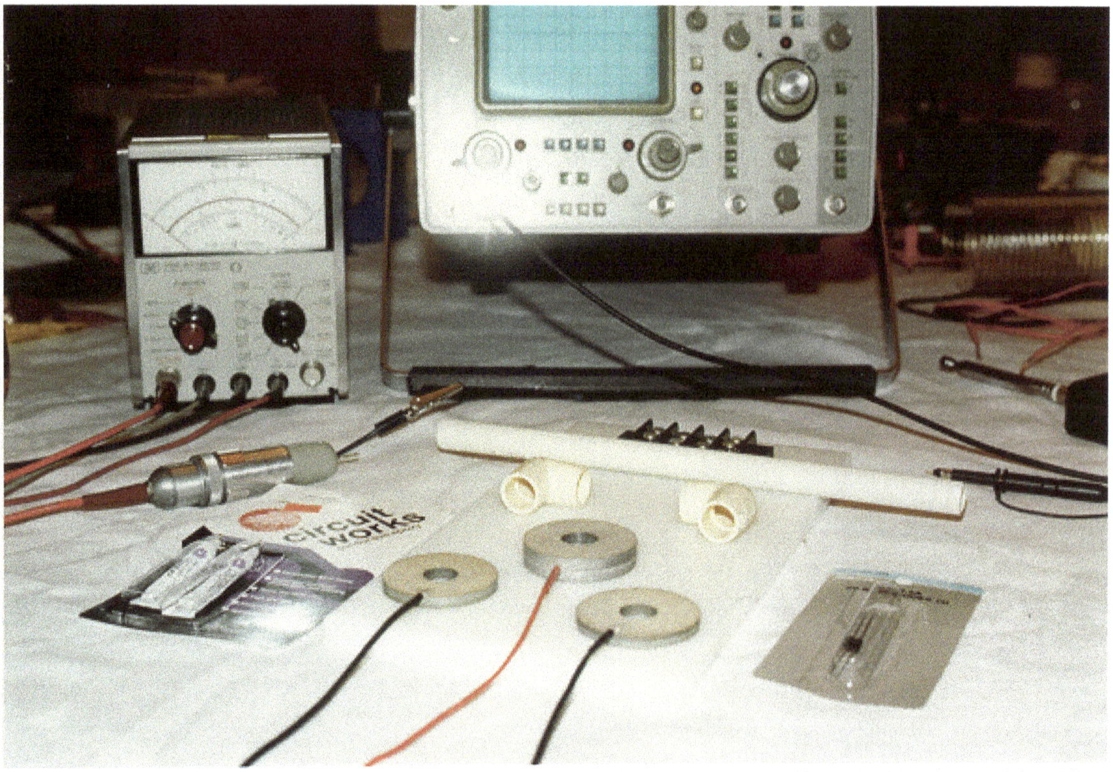

Cockcroft-Walton Device

Plasma Column Dipole and Model 29 from 1995 Video

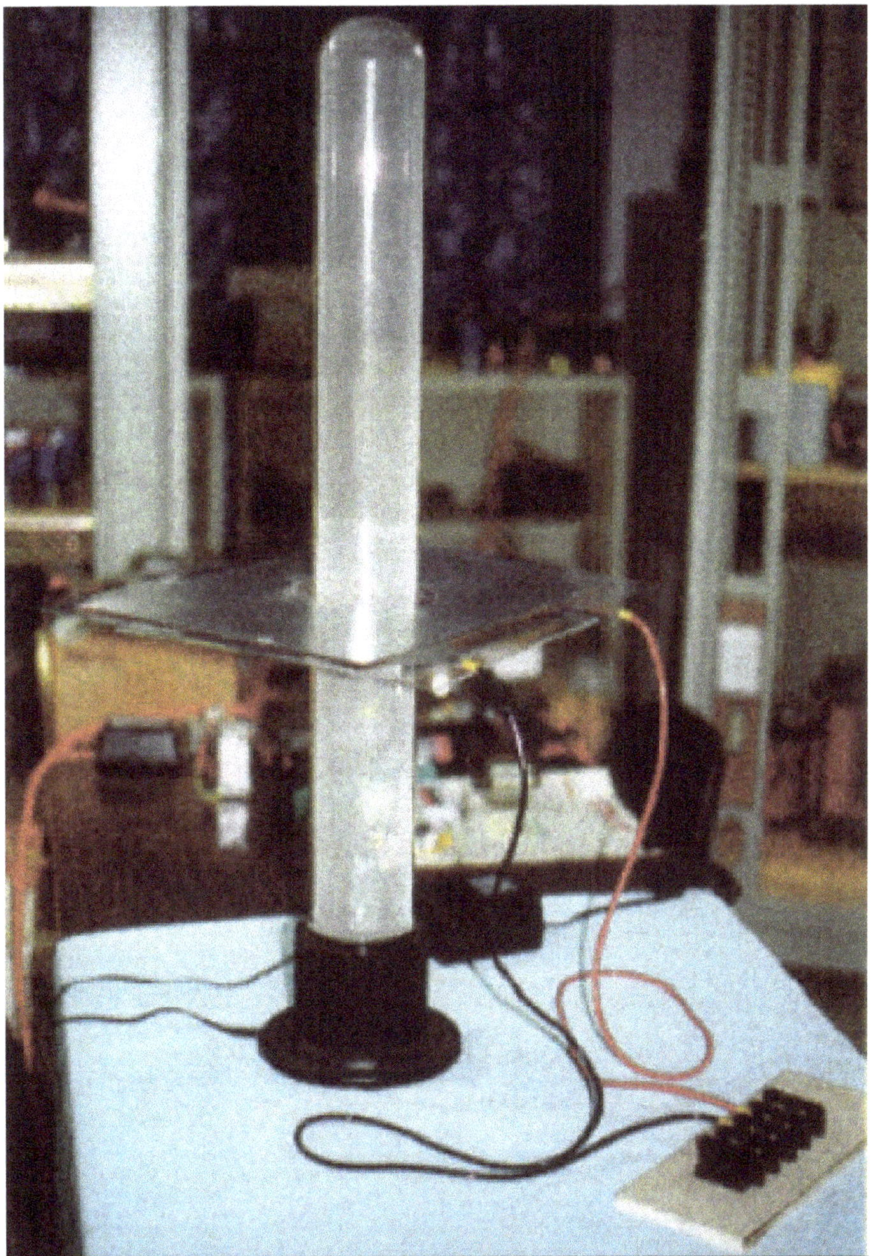

45KW Commercial Prototype

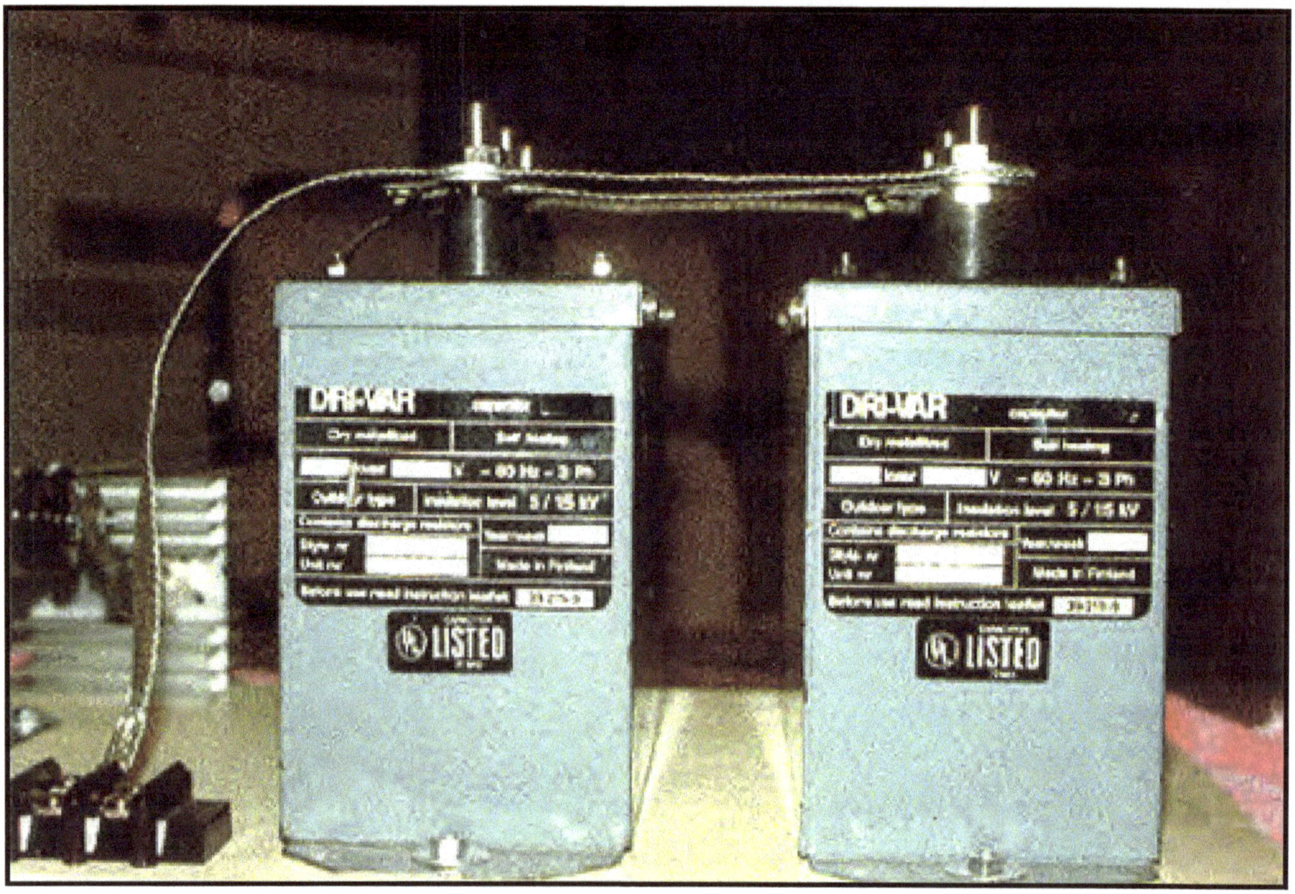

For Japan

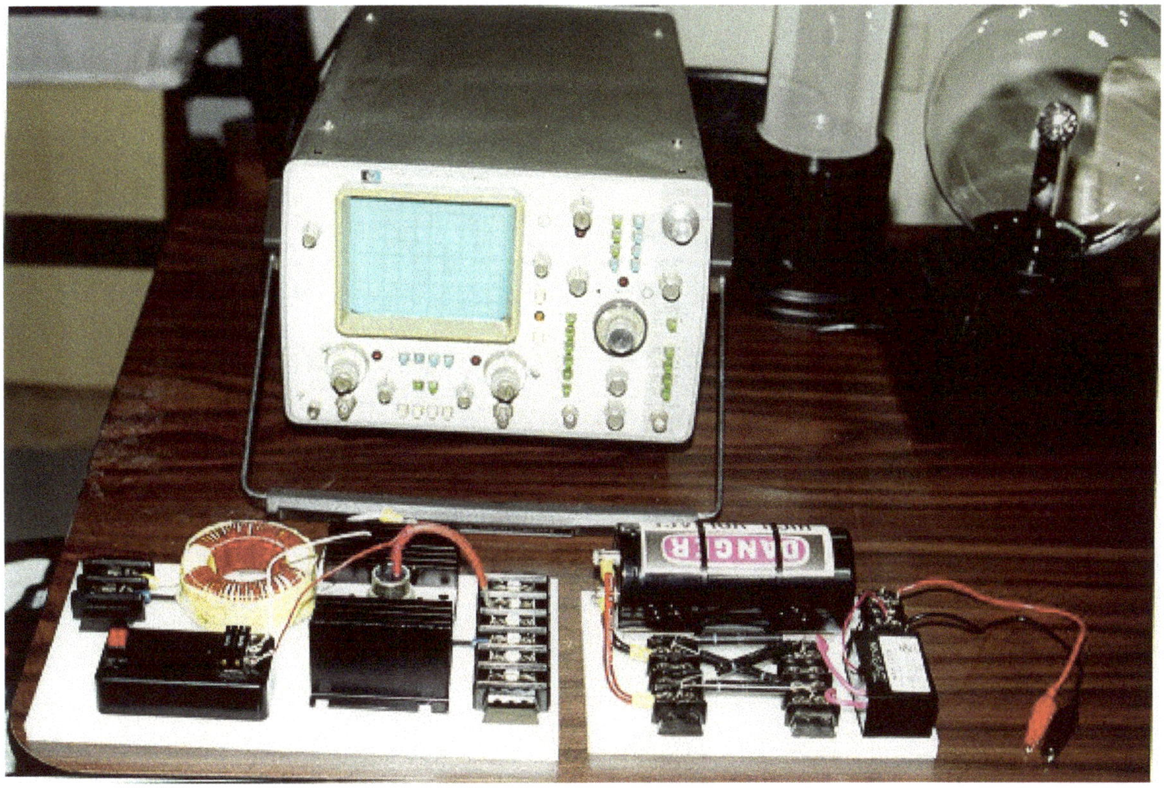

Resonate Transformer

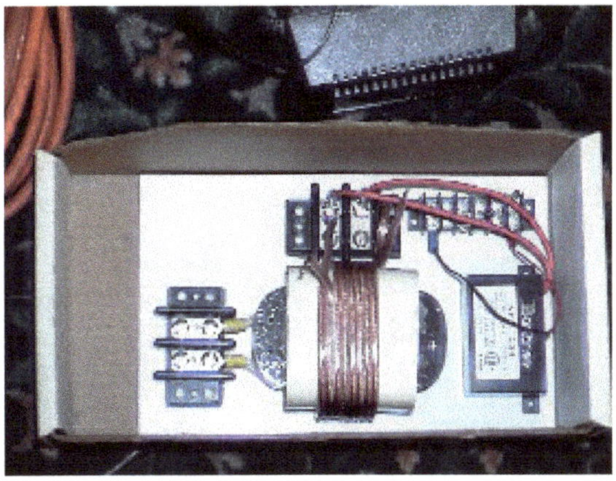

"The coupling on the left [below] is Bearden's [MEG] and the one on the right [above] is Don Smith's. The Smith Device precedes Bearden by more than 8 years."

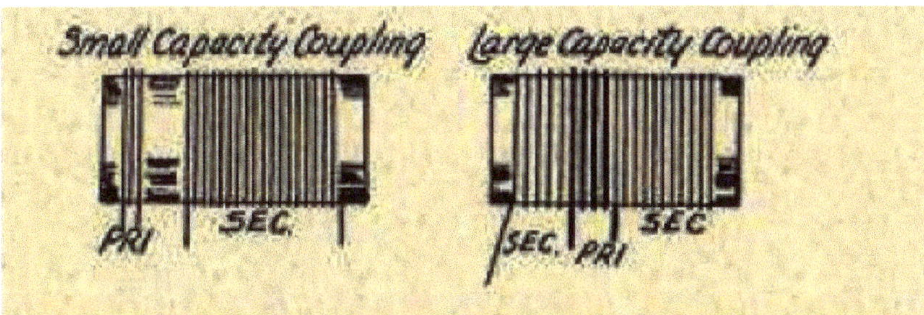

FIG. 4.—Effect of Primary Position on Capacity Coupling in Radio Amplifying Transformer.

Output is larger with coil position of L1 in above right.
From page 53 of *Radio Cyclopedia* 1928
https://www.americanradiohistory.com/Archive-Early-Radio/Call-Book-1926-Amateur.pdf

4KW Generator.

"This is one of my early devices with the shielding material on it which keeps the flux inside the box."

Below is *Transformer End Module* for schematic from Video 1, 1994. Minute 18 Video 6.

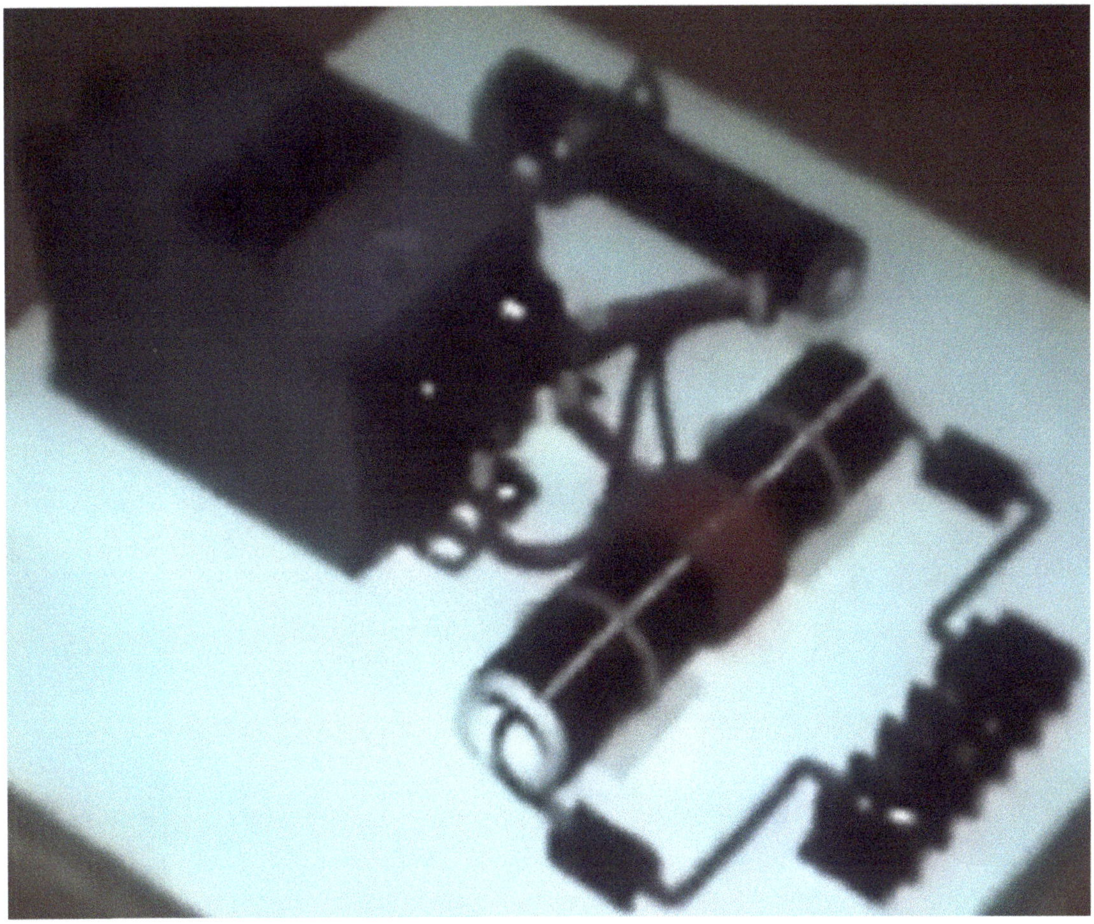

Dr. Smith.

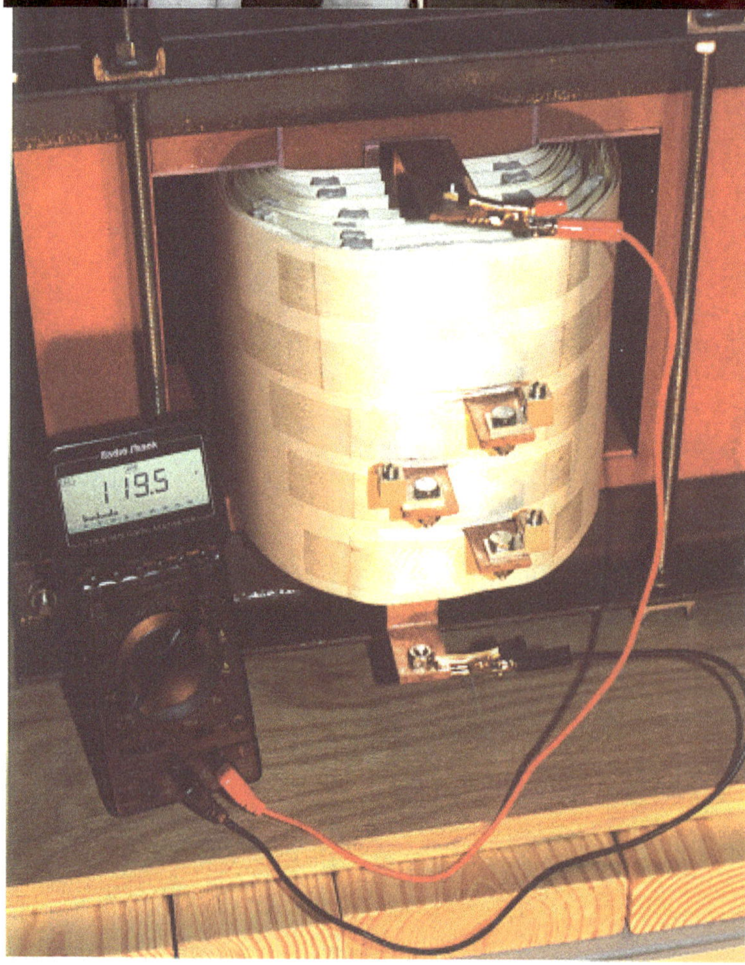

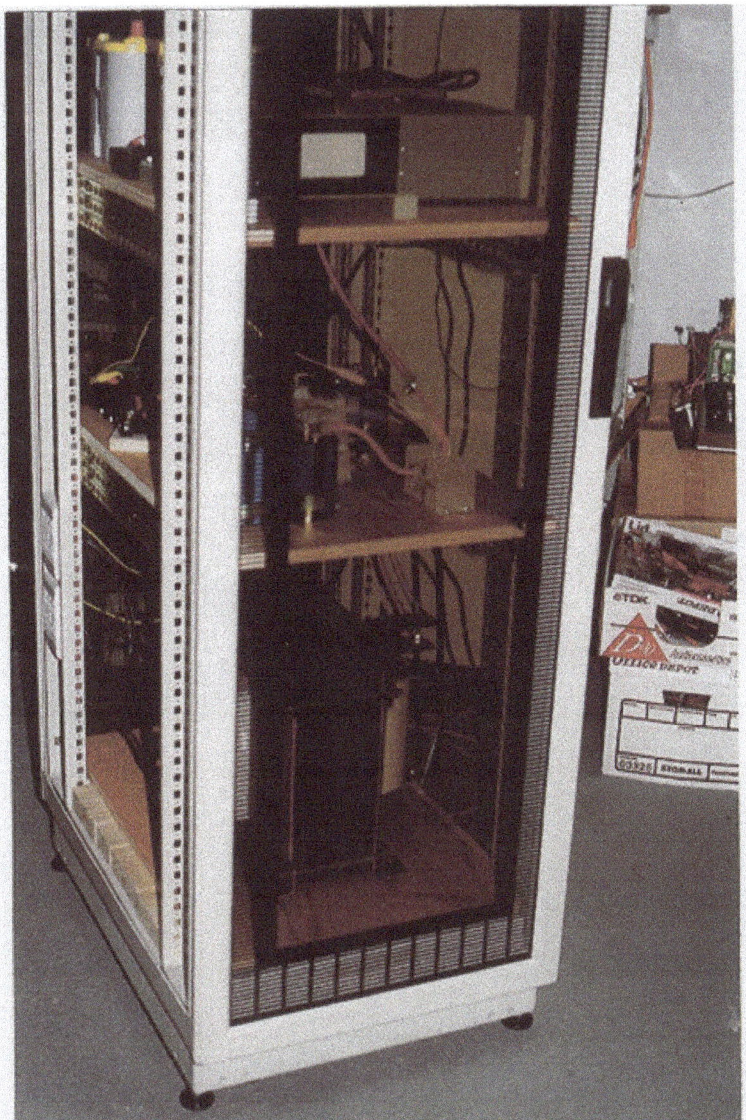

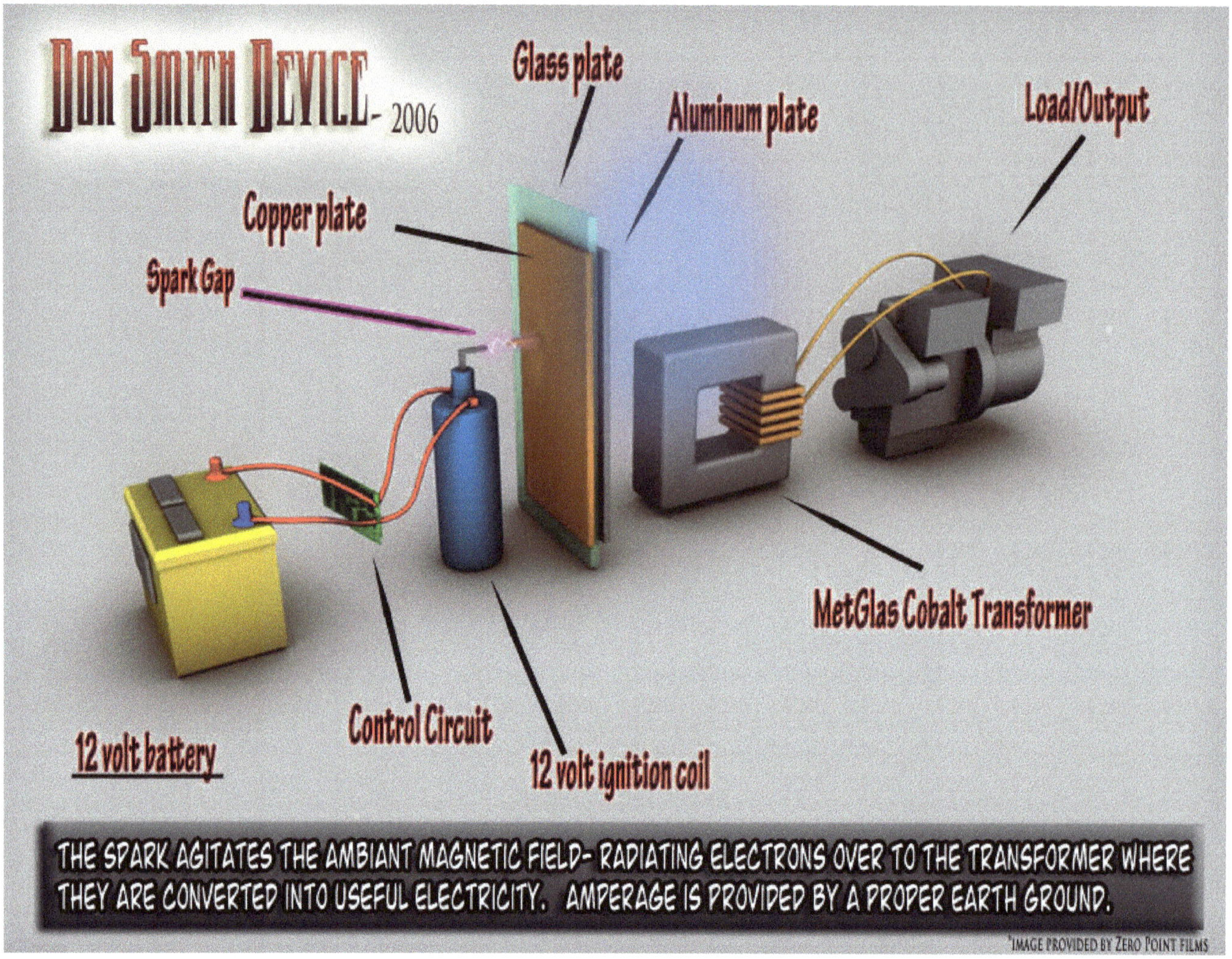

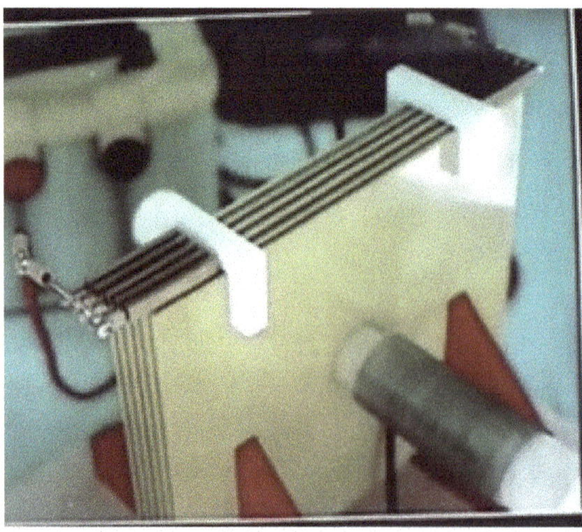

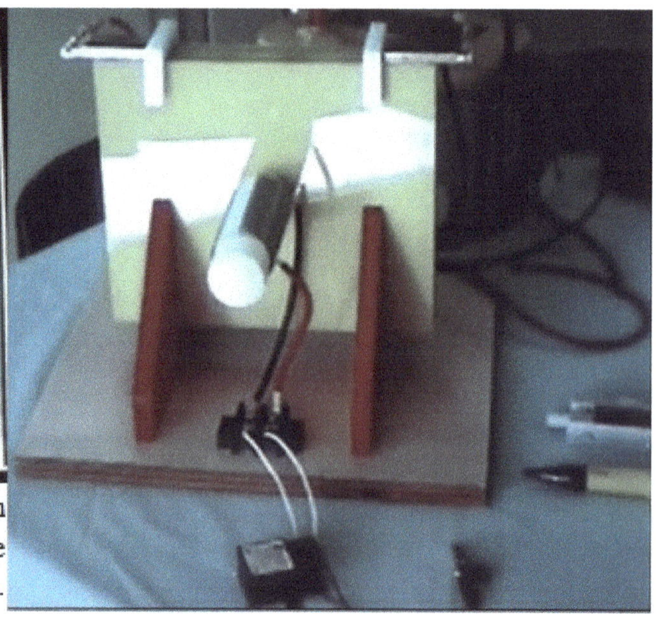

Terfanol-D rod Dipole device with collection capacitors at right angles capturing Heaviside current. 50kW system out, 20-30ma, 12V in.

Video 1, 1994

Okay excellent. These guys are really fast on the videos. So anyway, Don Smith virtually needs no introduction after this afternoon's presentation. It was excellent. He was our surprise speaker when somebody didn't show up and didn't tell us he wasn't showing up. So Don came through in fine style and he's going to tell us a few finer points that he didn't have a chance to cover this afternoon. That's gonna be part of what he does and then he's also going to show us some hands-on demos tonight. Don has a background in geology. He's done some incredible work. And I think this is really neat because the insights into free energy can come from any discipline literally. I've gotten insights from music he's got them from geology of course he's got insights. But anybody can help in this stuff right now and Don has really done a fine job helping out so let's welcome Don Smith for our workshop.

Minute 1

Thank you we only got to page three on my speech this afternoon so do you want to maybe finish up a little bit on that and then go into the other things? If someone hand me this little blue thing here we'll see what we can do about it. Thank you very much. When you're writing a speech you never really know how long it's going to take, and this one apparently took longer than I thought. Let see, here we go, we didn't actually get to look at any of the devices as such on... Is that working? [Picture as shown on page 51.]

2

You want to look at it for a couple of minutes here and then we'll start talking about it? There's a number of unnecessary parts there that were actually put in there because people expect to see them. Actually on this you can take everything back this way and take it off of there and it will work just fine. [Points to the transistor and three variable devices.] You can also take the spark gap off of it and it will still work just fine. And the reason for that is that in the coils they have their own specific resonance due to the fact that they have both capacitance and inductance. And when you cross that on a chart you find that that particular length of wire will have a specific frequency.

3

And it will load and unload by itself. And you don't need spark gaps or you don't need anything else. Your Tesla coils can be built the very same way. This is a Tesla coil device. The Tesla coil in this case is in number seven. It's potted inside that laser module. And that laser module is about three inches long and about an inch and a half wide there, and about an inch deep. So that'll give you some idea of the size of them. I'm going to show you after a while when we get into the actual core part of this thing some little modules that are about the size of a calling card, and they put out thousand volts. And they don't look at all like what you think a tesla coil might look like but there still, that's exactly what they are.

4

Is it okay to go ahead and move it? The one you're looking at there can have a very wide range of voltage. It can be up to five million depending on the parameters that you build into it. The one that you're actually seeing there is the one that was on the board this afternoon which I showed you the larger one. That's the device right there, except that it's got fewer windings on this L2 thing, and it's also wound in a spiral way. We'll just sort of take this as it comes here and see what happens.

5

We were talking about in resonant circuits that your impedance or your resistivity, or whatever you want to call it, which takes the place of ohms becomes zero and therefore you're at superconductor conditions at that point. The last statement over on this end where it says resonant energy, explains it all and we'll go into it in some detail in a little bit. Basically what happens when the volt and amperes becoming equal your into what's coiled a volt, VAR System. Volts Amperes Reactive and you can look in as many radio books as you want to and you're not going to find and hardly any of them that will give you an explanation of that.

6

And if they do it's misleading in almost all cases. But basically what it means is that your system resistivity becomes zero at resonance. Which means that your at superconductor conditions at that point at room temperature. Excuse me. Should we do something else? In over-unity devices, the statement at the top of this particular slide. [Shown as page 37.] A real fine example of overunity would be the fact that in induction the rate of induction is directly related to the cycles per second.

7

So if you want to establish at a reference point there at 60 cycles a second, square that. And then if you want to see the relative position of say 220 megacycles, square 220 megacycles, and compare the ratio between the two squares and you've got the ratio of improvement with the higher frequency. So if you look at that from a standpoint of overunity production you'll see that on this curve here that you've gone from a certain level of under unity type situation down here, which is what it's normal to most things, and then all of a sudden whenever you get into above 20,000 cycles per second and start climbing up between that and 200,000 cycles per second the curve does this.

8

It's just above 200,000 cycles per second and it stays that way all the rest of the way. What we're talking about, like if you made a chart and you showed the efficiency of the device, and the efficiency of the device as far as induction is concerned at sixty cycles per second is here, and the efficiency at two hundred million cycles per second is up there somewhere. The ratio between those two is telling you your efficiency. That's one of the reasons why I built this device to get out of way from the ranges which the Tesla coils normally operate in which is down below two hundred thousand cycles per second.

9

The Tesla coils are disruptive and they have other habits which are not enduring, and if you get up into the radio frequency you got a device which is not harmful to humans and you can you can handle it with millions of volts going through it and you want to have anything happen except maybe the hair on your hand stand up a little bit. And you can also hold light bulbs and light them and that sort of thing. The reason is at radio frequency the electrons are running around on the outside and they do not penetrate to any depth. That's one of the reasons for the high efficiency rating. In the lower frequencies the electrons are running through the wire, and they're bumping into things as they go, and they're not functioning very well they're under unity. Once they get spinning and outside the wire then they're kind of free to do their own thing.

10

They don't have anything diminishing their effectiveness. Did we talk about dominoes this afternoon? We did. We talked about Ohm's law, and we'll also talk about how to recognize over unity. I'll put this back on and let you look at it a little bit more if you want. [Page 98 shown.] Let me step out of the way here. You can see where the envelope is set up and where the thing has passed outside the envelope there, and it continues even greater distance outside the envelope. As it goes along and it gets out to a certain point here and it collapses and then it comes back and it starts the whole thing over again.

11

This is a particular resonant circuit that you find depending on how you work your shunt on the resonant circuit. If the first one, there were three of them if you remember. [Shows page 99.] This the one at the top is the one in which you normally expect to see. That's where you're dampening effect happens. You're seeing these things on an oscilloscope screen and it's showing the analog, or whatever you want to call, it's showing this cycling. It starts out with certain amount of energy and that gradually diminishes. The second one there is another setting on that same resonant circuit with a slightly different adjustment. It shows an envelope of energy very similar to FM carrier.

12

In FM you have a compression of these cycles which takes the place of in the analog system where you have a various modulations on it. But in the third one is a particular arrangement of a grounding shunt, which means that its ability to communicate with the electron source, which would be your grounding. The resistance there is like a dam on a large body of water and it lets the electrons through and they get caught behind this dam. This whole device is like a pump. All it's doing is like the dominoes we were talking about, its tripping the electrons here and that electrons trip in the next one, and this whole bunch of electrons in all directions are being tripped. Each time one of those would be tripped it gives off a magnetic impulse first.

13

Then, when it comes back on that infinity loop [page 43], it gives off an electrical impulse. This just goes on perpetually, the electron is never diminished in anyway. It's giving you life and a lot of other things. That is also an over unity device if there ever was one. So if the physics books have a problem with that, so be it. Question from audience: "Have you written a book, which incorparates this information? And is it in the bookstore?" Yes, yes, well show you what it looks like. I've got one here. They are getting some printed up just now and they may have them before we're finished here. They may not be quite as nice as this one but these I did myself.

14

You remember they have nice colored photographs in them and that sort of thing. That was the actual device you were seeing this afternoon. One of them. Let's see what else we have here. I'm going to take the speech which I didn't get to make. I'm not going to make a speech, I'm just going to check and see if some of these things I didn't didn't talk about. Whenever we get into how we get these magnetic flux measurements we go back to years 1777 through 1855 and about the same time for another fella. We're talking about C. F. Gauss and H.C. Oersted.

15
They were both trying to establish the electrical background field for the earth with the external influences removed. As it turns out one of them did his measurements at night and came up with up with 50,000 volts. The other one did his measurements during the day and he came up with a hundred thousand volts. And that's per square meter of surface area. This is above the ground, with the ground influence removed, and it's a normal natural electrical charge that's in the air at that point. Now this varies depending on the time of day the influence of the Sun and sunspots and a lot of other things. Sometimes instead of being those units or I just gave you it goes up into two and a half to maybe five million volts per square meter of plate surface.
16
This does not take into consideration the other measurements which I was telling you about from the colored maps. The influence there is strictly from the graph character of the minerals and the ground and that sort of thing. There is a huge amount of energy out there and you need a device which can pump it or move it from one spot to the other and make these voltage differences so that you have a potential. If you have the same voltage here and the same voltage here you don't have any potential difference so nothing's going to happen. Yes sir? [Audience:] "In the schematic that you showed of the coil, approximately of the device you had up there, it showed a capacitor connected to ground?" That is a blocking device it could either be a resistor or a capacitor, or variable of either one of them.
17
[Audience:] "Is that grounding device the source of negative impedance?" Well see you've got two coil systems there that are working so you have to tune one of them to make, you have to have one of the coils with a tunable device on it so the grounding you can stick it in the grounding end of it or you can put it somewhere else. But it's convenient to put it into the ground. [Audience:] "You mentioned that the source of electrons?" Would be coming from the earth. [Audience:] "Right, is that the source of your effect that you get in your system?" Yes, if the electrons are not coming out of the ground they're coming from either an air grounding or an earth grounding. [Audience:] "But there is no other device that you require on your coil?" No it's very simple. As a matter of fact when I started building these things I put all kinds of things in there that everyone thought should be there and which the books the on coil-building and Tesla coils and everything said well we need a spark gap.
18
We need this and so on and so on. When I realized that wire has its own induction and capacitance and that sort of thing, well that told me immediately that I did not need a spark gap because it's going to load and unload based on the frequency that's attributable to that piece of wire. The spark gap it's not a necessity. [Audience:] "Do you need to put it in a particular geological area?" Well it started out that way because of these high concentrations of negative electricity in certain areas the problem in the very beginning was: You've got these two areas of different potential so how do you turn that into something useful? It became obvious that you are not going to be able to put a cable between the two because the cable of any size regardless of how large was

going to vaporize like a flash bulb whenever you grounded it.

19

With that in mind and it became obvious that there's more than abundant electrons everywhere so you really don't have to be concerned about the fact that you're in an area of higher concentration except that your devices will work a lot better. For example that where I showed you the 16 trillion deal. If you put a radio station in an area like that and ground it they will hear you throughout the universe and several other universes. That may be why NORAD is over here in Cheyenne Mountain. It's most likely reason that they can have a command thing there that they can talk to anywhere anywhere in this world and any other world you might want to think about with the potential that they've got in that rock mass over there.

20

We're talking about Carl Gauss and Oersted, and units of magnetic flux bear their names. We have to relate those to other units of magnetic flux in order to see how this thing all ties together. These maps that you were looking at were made in gammas. They sometimes are in nano Teslas and sometimes other units. But the most common is say gamma. Nano Tesla and a gamma or equal in range. One gamma would be equivalent to ten active volts of electricity, or ten volt-amperes of electricity if you want put it that way. When you get into the other units, Gauss of course is 100,000 volts.

21

It's a unit of measure of flux density and Oersted is 50,000 volts. Now 1 Tesla is equal to 10,000 Gauss. Would somebody real quick-like like to tell me how many volts 1 Tesla is? It's a bunch of them isn't it? A big bunch. Is it 10 million or 10 what? A Tesla is how many volts equivalent? I think it's probably ten million. [It's 1 billion] It's a rather large unit and in practical terms I'm not quite sure what you can do with the unit that large. Of course the entire Earth has been surveyed due to the desire for atomic minerals and cold war related things.

22

These background maps on the magnetic flux were used for reconnaissance survey for zeroing in own areas which had promised as far as certain types of minerals are concerned. Where I come in they are also used for locating oil deposits. It's a real fine tool for locating very large oil fields without drilling for them. There's certain characteristics you look for. I see. [Audience:] "Can you purchase those maps?" Those maps? The United States Geological Survey in Denver has catalogs and you can get maps on any square inch of the earth that you want to put your finger on. There are several other groups that do them. Also the Russians do them and a lot of other world governments for various reasons.

23

Originally they were mostly done for military reasons or for some kind of strategic mineral that was required. Whenever we get into energy and mass, and that sort of thing, it becomes obvious that we have basically about three different situations that we can describe. We have a steady state energy which would be static or pre-energy and it's where mass attracts mass, and where gravity dominates. In the unsteady state we have two different types of energy derived from mass. We have two kinds of kinetic energy. We have expanding magnetic energy and it's dominate for that.

24

It gives you your electrons moving apart, pressure decreasing. It gives a cooling effect. That explains why it's two hundred and seventy-six degrees minus out in an open space away from the earth. When the electrons are moving apart or expanding it's cooling, and sometimes very cooling. There's less scattering of energy in an expanding situation. In addition to that you have negative resistance as dominant. If we go into the contraction stage of energy or mass we get electrons moving together. Pressure is increasing, heating effect dominates and you have a scattering of energy, and positive resistance results. You're under unity people work primarily and in that particular set of circumstances.

25

Most of the things that the under unity organization involves themselves in relates in contraction of electrical our dissipation of it or that sort of thing. Functions of active electrons. Electrons have provide us with a lot of things other than electricity. Among those we get magnetic effects which are useful and causing motors to turn. The the magnetic thrust there is against gravity which gives the motor the kick to start it turning and keep it going. That's the gravitational thrust as an electric motors. Electrons are the source of visible light. When two electrons get too close together they immediately try to do something else, and this results in light being emitted.

26

This in no way diminishes the electrons. You get your photons, or whatever, which in a lot of books are subscribed, they say are a particle or mass in itself. which I don't really know for sure and I'm a little bit skeptical about it. I don't see how you could if you have a particle given off that you'd be diminishing the electron. But that doesn't occur. The charge of course is negative. Any time you have an electron involved you don't really have a choice, it's it's going to be negative. We discussed about the fact that the movement it's not a circular movement in the manner which most of the books describe it but it's sort of a closed lazy eight, sort of thing, that's compressed in on itself and this also happens to be the symbol for infinity. The infinity symbol was put in effect probably in biblical times.

27

I wonder what the connection between the infinity symbol and electron which goes essentially forever would be. Something that's producing energy electrical magnetic energy, and it does this continuously over billions and billions of years, and is in no way diminished. How come it does the infinity symbol. If somebody has an idea I'd like to know. I have a different concept of how the forces of nature to operate. I believe that we mentioned that that's composed essentially of a weak force and a strong force. In addition each of these forms of energy occur as fields and waves depending on what the resonant frequency is at a particular level.

28

Both of these have their own separate set of circumstances and they are parallel to each other. For the weak force for example you would have your gravity, you would have your electricity, and you in have your magnetic field that would the weak force would have admit. That's of course from your electrons. Now if you get over into the strong force you're into atomic energy at that point because you have to tear the atom up in order to get the center out of it– if you absolutely insist on a positive electrical charge. Which some people call it antimatter because it cancels out the electrons. If they cancel each other out if they get in the same boat at the same time.

29

In the strong force it's the one that controls the universe and it's the clock motor that runs the universe so to speak. And the weak force is the one that you could compete with on a daily basis. It's the one that you really have to be friendly with. It's the one that's going to be doing your work for you. Unless these particles electrons, unless they're discharged, are dislodged from their source and free to move out, which there are huge and immens quantity of these electrons available to you, and the source is the Sun. The radiation from the Sun it throws this things out, and they're part of the solar plasma. When the Earth passes through the solar plasma the magnetic field of the earth pulls these electrons in in immense amounts.

30

It's equivalent to, I'll give you some numbers in a little bit on the barrels of oil equivalent and other things, but it's astronomical. This has been going on since the since the beginning of the earth. And all of these electrons have been accumulating with nowhere to do that nothing thing to do and nowhere to go and they're just sitting here looking at you. Some of the people, there's a book called, I think, it's *Sea of Energy,* Moray. There's a number of other books where they acknowledge the idea that we have a huge envelope of electrons or coronal-like effective electrons which surround the earth. That go up hundreds of miles. You can tell how far they go up because where the light begins to form the impact of the earth passing through the solar plasma, the electrons of the earth are banging into the electrons which are already out in the solar plasma.

31

And when those electrons get too close to each other they form visible light and that's where your day light comes from. Its electrons that have gotten too close to each other and they're trying to get away and when they do they give off this visible light. When you get out into the universe proper there's no reflected light out there at all. It's just as black or dark as things can be, and the only things you see are other bodies that are electronicoily active out in the distance where the electrical forces are reacting, and they give off a point of light that you can identify.

32

We talked about the fact that the electron itself is an excellent example of over unity. Just as with a few moments ago we were saying that it goes through these cycles billions and billions of billions of times over long expenses of time and is never diminished. If that is the case then it is producing energy which is greater than what appears to be there and therefore it is a excellent example of an over unity device. If someone wants to tell you that they don't exist, well the electron is probably the most obvious one of all. We talked about the fact that the induction process itself is an excellent example of overunity due to the squaring of the cycles per second.

33

We also talked about the fact that I believe that mechanical advantage is not considered unusual when it comes to pulleys or levers or gears or things of that nature. But if you use a small amount of energy to extract a larger amount of work, regardless of what the books say, haven't you've got something back more than what you put into it? The physics books don't agree with what I just told you. Think about it, the physics books are wrong. The idea was set up back in 1670 or somewhere back like that.

34

A lot of the actual mechanical age hadn't occurred at that point. The machine age was still somewhere out in the distance, and so was the electronic age. It's not very likely that somebody who wrote a set of physics rules, say in 1670 something, would be able to compete with something that comes along later which they had no way of ever knowing about. So think about that for a while. Another way of looking at this would be that if you take a roomful of ping-pong balls and you get them all going in many directions, and they're bouncing around, there's a certain amount of energy in that room. But if you line these ping-pong balls up, and you can do it magnetically, and then all of a sudden you've got a tremendous amount of energy that was not present before.

35

This is what resonance is. It's what super conductivity is, and that sort of thing. That's overunity. The books won't agree with me on that I can tell you. We're just about to the end of this. Basically electrical energy is achieved when an electron density at point A becomes greater than electron density at point B. This is your potential difference between two points. If that doesn't exist you don't have any energy, regardless of how how many electrons are there. It's just like the high pressure and low pressure areas of the weather system. Things are going to move away from the high pressure area into the low pressure area. The electrons density acts the same way.

36

Some of the benefits of this particular system are some of the points which might be made are: the electrons which cycle through this system after being used are returned intact to their former state for future usage. This aught to delight the environmental people. Electron spin causes current and magnetic lines of force. The effect of current results from unequal distribution of negativity. That's a restatement of something I just said a minute ago, the electron density. The system is an extension of present technology. It is a spin-off or extension picking up and going forward with Tesla type devices.

37

The system can use magnetometer studies for locating areas of greater electron density, or the fact that one has a greater density than another one and event that you want to do your electricity between the two points. For example if you put a contact on top of the Cheyenne Mountain out here, and you put one down at the bottom, and you ran a wire between the two, and you left a place there where you could plug your device into it, you would have over five million volts of usable electricity there that would essentially have the same as five million amperes because your resistivity would not come in, or your amps would not come in there after the fact, until you put a load on it.

38

At that point you would know how many amps or whatever you could push and they would be at the beginning essentially the same as the amount of voltage. That's the reason why you can't run a cable between the two points and do it exactly what we were talking about there because it would vaporize. But there's other ways of getting around it. The system I call my system Earth Electrical System 2, the "EES 2" - and this was the name that was given by the Department of Energy. I got into fisticuffs with them over it and the elections came up and at the very point where they were supposed to sign their name on the check, well they went sideways on me. Basically all I got out of them was a name for it.

39

That was the name that they assigned it for their files: "Earth Electrical System II." The system utilizes a non-polluting energy source. If you want look at the pollution effects you've eliminated internal combustion engines. You've eliminated coal for generating electricity. You've eliminated just a whole slew of things that are poisoning the environment. I think the state California might like that one. The system utilizes a universally available energy source. There's no place you you can go that you won't be able to work it.

40

You can anticipate that areas that have pollution problems, like Mexico City, or Tokyo, or Houston Texas, or wherever, would like to get in line maybe to certify this thing as being there a preferred way of doing their energy. It has some very interesting possibilities. We have one other part to this and then we're ready to go into the coils. I think we've already covered this. That completes the speech part of it. I think at this point we can start thinking about coils. We had some coil guides done and I don't know who has them or what.

41

Do we have the printouts for the coils that we were going to hand out? Is it still open be open? They will be open till 10:00? If someone would step in there and ask them for those things we'll look we'll pick up on the coils. They're selling them? Are you talking about the books or the coil builders guide? It turns out I don't have a copy of it here so I'm gonna have to wing it if I don't have one to look at. I'll get it back to you just a moment.

42

Do we have some markers where we can mark on these boards a little bit here? Right here? And these are the ones where I could wipe them off? I'm gonna pull this back a little bit because as long as this day's been I'm liable to step off the edge there and never know it. I'm going to draw you a diagram of an actual working unit that was presented on the device this afternoon that you saw. I'll do it with a block diagram first so it would be real simple. Then i'll do the individual components. What you need basically is a power source. That will be a battery. It can be anything from say one and a half volt battery up to say 12 volts or anything you want it to be, just about.

43

You can control your output somewhat by the amount of voltage that you're putting into it. The next item which shows up in this would be. In our case we have simplified it. We go into the Tesla coil part. But I'm using a laser module which shows up in the Xerox machines. It takes care of the step-up in the voltage for you. Basically going into that, and then from that we're ready to go into L1 out here. This is your tickler coil, or the kicker coil, or it's the one that's going to set things up for you so the other coil will know what to do. Say that you have over here 12 volts.

44

It comes through this device here. Say that it's coming out 8,000 volts. Immediately someone is going to say, 'well that's 8,000 volts and it's at say 30 micro amps so there's nothing there.' As we found out from Ohm's law when you get into this type of device that dissipation your ohms and amps are dissipation and there after the fact. So if they're putting micro amps in here is simply telling you the amount that's being thrown away, it's not telling you the amount that's there, or was there, or was left. This is the 12, let's see, 30 microamps is amount of dissipation that occurred when it went through that step up coil.

45

That primarily occurred through the resistance of the wire. The wire in this type of device is about the size of your hair, and there are many many turns to it. [Question.] Maam? This one right here? This is a laser module. They show up in the Xerox machines, they show up in laser printers. They're little high voltage module. They show up in all kinds of interesting electronic devices. Any type of a device that needs a high voltage source. Something very similar to this may show up in your bug zapper, or your electric fence, or all other kinds of devices that work on voltage step up. [Question.] What frequency?

46

The output frequency is going to be determined by the length of this wire right here [L1 coil with wire ends, see page 196]. The highest frequency that you have represented in any of the components is going to be the dominant frequency. That's going to be set by the length of the wire. [Question.] Well actually in the laser module you're correct. Actually you have one output and then you have the the ground deal here. Some of them this ground deal that comes back. Here's your input over here which would be, excuse the expression, plus or minus. One is greater negativity than the other. So this comes back instead of being the earth grounded here will come back and ground into the so-called the plus deal.

47

[Question.] Any step up coil. It doesn't have to be a laser module or anything else. I'm going to show you, just in a little bit, examples from automobile transformers and a number of other sources. I have them that put out 30,000 volts, and then you triplet from there, and then you're just then getting into the L1 coil. Then you take it from there. This is going to be our L1 coil here. [Question.] It's not critical. In any event you're going to have more voltage than you want. There's almost no way to do this and and have the voltage that you really want to end up with unless you're wanting to make big Sparks. That's not the object of what I'm doing.

48

I'm trying to tame it, and bring it down, to where you can do something with it which will be useful. [Question: How many turns on your L1?] You get into a division or multiplication situation. Once you know the length of wire you know what frequency you want to run it at, well you cut that wire so that it matches the wavelength of that particular frequency, and then your inductance and capacitance will take care of the off-and-on deal which you would expect the spark gap and some other things to take care of. Say that you have coming into it, for example, say eight thousand volts, we just let that for convenience purposes. We put four turns, say that we put four turns over here, one, two, three, four. You divide eight thousand by four.

49

Each turn there will have two thousand volts

on it. The next coil that we want to do, we can put one turn, or we can put hundred turns. What we want to do basically on that second coil. Say that this piece of wire here is at four point something feet, which corresponds to 220 megacycles. Whenever you get into wave type arrangements you have half waves, quarter waves, eighths waves, and so on, and so on. Basically what you want to do on any turns, or whatever you put on your second coil, they should be a relationship such that it will be either a multiple or a division of the same frequency. The second coil is going to run at a different frequency because the wire length is different. Coils are actually antennas that have been coiled for convenience. They work just fine that is a piece of wire.

50

But when you start winding them you're doing that more for convenience than anything else, and it's going to change your inductive characteristics and a number of other things, but it's not going to be detrimental to what you're doing. Once you've got this thing divided down, and you've got two thousand volts on each turn here, then you're ready to go to your L2 coil which will be the second one. In many cases that is going to be a great many more turns than the L1 coil. But it may be one turn. If you want two thousand volts instead of a great deal more, then you make just one turn, and make it so that it's a division or multiple of from the length of the other coil so that they will react from a frequency standpoint properly.

51

Now you can get computer programs where you can figure these things all out on the computer and you'll end up with the same answers. If you doing the way I'm doing it here you can do it off the top of your head and there's really very little mathematics involved. So it's not that complicated. I've mentioned earlier, this is one of the things which I teach the Cub Scouts to. So if they can do that you to do it. Which you got them and you give them, you've got a group of these kids, and you give them the parts, and tell them what to do, and they do it, and take it home and dad here's my over energy device. We've gone to the next step here. Say that we've got one turn and on this deal it makes one complete circle here.

52

It has to be grounded in order to pick up the maximum effect of electrons. In order to ground this thing–see we've got two different coil sets there. Now anytime you add a capacitor, or any other device, or a resistor, or anything else into this circuit here it's going to lower the frequency. So the highest it can be will be based on the length of the wire. Then if you add something else to it well then it's going to lower the frequency. We've got two coils and they may be very slightly out of tune, not precisely tuned, so you need a variable device here, which can be a variable blocking device, which can be a variable capacitor. Or it could be a variable resistor or a number of other things.

53

You have to do this by experiment and don't put any real heavy resistors in there. I mean any high resistant deals. Put something say like down below 500 Ohms, maybe around a 100 Ohms or something like that. This is going to be that shunt that's in the resonant circuit that we were talking about a little bit earlier. It's going to be the dam that's going to hold the water behind it so to speak. At that point we've got a thousand volts in radio frequency setup which is absolutely useless unless we do something else to it.

54

At this point we're going to pick up where we have the one turn here, and we're going to take it on to the next steps. We've got one turn here, and we've got it grounded with a variable device on it so that we can fine tune it. At that point, from this deal, we've got an output out here which is in an alternating current type situation at radio frequency. [Question. Is that ground symbol, could that be the negative?] Yes that is the shunt in a resonant radio frequency configuration that produces the dam that keeps the energy. Once it gets to oscillating inside the system, it can come in but it can't get out unless it is hooked into another system.

55

[Question.] Yeah well it's simply a blocking device which keeps your resonant coil system. Which normally you have a capacitor and a coil, and you've got them like this. Well in our case we've put both the capacitor and the coil in one wire and we eliminated some components that are subject to fail. When we do that capacitors are subject to fail so you've actually made a device which is less likely to fail. It has greater longevity. We're going to be throwing pieces away as we go through here and see how we can still end up in business. We've got a thousand volts here.

56

It's a radio frequency, and we need to get it into a condition so that we can run something which is 120 volts at 60 cycles. Oh yes sir? [Question: L1 ground and L2 ground?] Well they're grounded in different ways. L1 grounding can be grounded back into the laser module itself, or it can be grounded into the earth, or into the same common ground that L2 is actually grounded to. You'll notice in the Tesla coil books that frequently they both those coils are have the same common ground. We've got a thousand volts and we're going to go to a lower voltage which we can do something with at a waveform which we can do something with.

57

At this point we're going to have to go through a diode-bridge. We're going in with alternating current, and we're coming out to DC. At this point we've got something which we can take through one or more steps and we're where we want to be. We're very close to where we want to be. We take off from the DC here and we come out like so. We're going to inject a coil in here which is actually two Tesla coils "bottom to bottom." This is the middle here. This would be a base of a small Tesla coil, and this would be a base of a small Tesla coil. [Question.] Yes sir, I can hardly hear you. Yeah we've got DC coming out here so we've got to pulse that in order for this coil to work.

58

What we do we put a capacitor in parallel there, which gives a pulsing effect, or an off-and-on effect, which will cause the magnetic fields to be, or induction to pick up in these two other coils. Do what? [Question about DC bridge drawn wrong.] Well actually I'm trying to get around that. Let me come across here and do that sort of thing. Will that be better? The other one I got to do the same thing. Well I'm gonna have to erase it and start over. I'm gonna do it on a larger scale on a larger scale where I can get my lines in there properly.

59

We've got our diode bridge here and we've got AC, and we've got AC, and we've got DC and DC, are you okay so far? We're going to turn these diodes so that the cathode end of it is pointing towards DC right here. We're going to turn this one so that the cathode is pointing towards DC. The two coils here, the cathode is pointing towards the AC and here it's pointing toward the AC. Does anybody know why we do that? What we're going to end up with is basically a spike. A DC of course, if you've got your reference point here. You've got alternating current which is doing that sort of thing. In DC you're gonna be on just one side of that reference point which would be the top side, or the bottom side, and you're going to be on just one side.

1:00

At this point the waveform coming out of here is going to be a DC spike and we're going to change that when we go through this thing right here that's going to change. Say that this is 1000 volts up here. We've got a DC spike with 1000 volts on it. When we go through this system here we're gonna have a DC spike above and below. Only it's not going to be DC anymore, it's going to be alternating current. It's going to be double the voltage that you had here. So we've gone through a voltage doubling maneuver here. We're coming to the waveform that we're really looking for now. We're getting very close to the waveform that we're going to need in order to run our 60 cycles with the device.

1:01

This double Tesla coil thing here has a variable blocking device on it, and it's grounded. Two bases of the Tesla coil are grounded. The top of each Tesla-coil would be these ends right here. At this point I'm going to have to erase this in the center because I'm going to superimpose something. You just have to keep in mind that you've got a double Tesla-coil there. It's oscillating, and it's forming a double spike deal, which is very close to what we're going to go for. I'm going to turn this around a little more. Is it safe to say that everybody can see it?

1:02

Yes sir. There are 50 turns on each end. The diameter of the PVC that I wounded on is 1 1/4 inches. There's 50 turns on each end and add an extra two on one of the ends. The reason for that is that you want one end to be delayed just slightly more than the other so that you're going to have an offset on those spikes when you see them on the oscilloscope. On that coil we've got two spikes. One above and one below now. We're getting you very close to the waveform that we're going to be wanting.

1:03

At that point we assume that the grounding and other things and the bases are still there. At this point we've got 2,000 volts on this arrangement right here. Now say that we want a hundred volts, instead of–I'm going to round it off–instead of saying 120 which is going to be created a mathematical problem shooting from the hip so to speak. I'm going to say that we want a hundred volts. In order to do that we need 20–divide 100 into 2000–and we need 20 turns there. What we're going to do is come in from this side with a piece of wire and we're going to–it's going to be loosely looped around. This is going to be very informal and it's going to come off like this.

1:04

Once we're off out here then we have another wave-shaping manoeuvre that's going to occur. At this point what we want to do is round those spikes off so that we get what we're looking for. Out here is a normal sine wave thing. We need to delay the time a little bit there so that when it's up at the top there that it's going to spread out in this top area. When we do that we end up with this type of thing. Now we're at the wave form where we can do something. We've got a hundred volts and we've got the proper wave form. We're at the point now where we're going to set the frequency. There's going to be another maneuver at this point.

1:05

Basically this is where we go back into the overunity thing, so overunity people will be happy about that maybe. We're going into a resonant iron core transformer system, and this is where we're going to end up with all the amps and other things that the electrical people would like to see maybe. Say that we go into an iron core transformer here [see page 196], and we're working off of it. This device is going to–it's resonant transformer. It's the self resonant isolated transformer which means that the other side of it which I'm going to put up in a moment is not connected in any wires. The only connection is through the induction. So we're running it through this thing that's going to set the 60 cycles at this point.

1:06

Then we're going to take the one which we're going to come off of, and here's where a load is going to be added. This is where the wall plug in the house is. So we have a magnetic flux field on this one, which is driving this one. And the self resonance of that particular transformer is 60 cycles per second. I've got one over here on the floor I can show you. It works at 4,000 volts. It's a real heavy looking kind of block iron looking deal. Yeah that's it. This one is good for four thousand volts. You see there's two different sets of coils here. One sitting on top of the other. This is an isolation transformer.

1:07

[Question: Is the ratio 1:1?] No this one is–the top coil is set for 4000 volts. The bottom coil is set for 2,000 volts. In our case over here both coils are set for the same. You can either step it up or step it down at this point. Depending on what kind of output you want, if you want 440, or 6 whatever, or 220, or that sort of thing. If the two units are equal you will end up with equal amount of say a 100V. Sir? [Question.] How did you guess. Well the price is right, that's what it is.

1:08

I had a neighbor who was an electrical engineer who knew what I was doing, and he and his son I switched out their microwave oven, and they were curious as to what was inside. After they tore it apart and found this thing they came bouncing over to my house and said, 'well we think you need this.' This was about six months ago and I didn't know that I needed it. But whenever I got to the point to where I knew that I needed to do something like that well I realized I had this thing out in the garage, and rather than go out and spend $400 buying one, I already had one. So this can drive a very substantial output of electricity. [Question.] Because the resonance is built into the transformer. You don't see it here but there's a capacitor that goes with this.

1:09

You see this has a certain amount of induction, and there's a capacitor that you put with it that forms your resonant circuit. [Question: In series or in parallel?] Well in this case it's in parallel. Almost all the things that go on here will be essentially in parallel because when you put them in series you get a degradation. [Question: Input side or load side?] If we're talking about the one that

that we have here we have four sets of wires. You notice that with these little snap-on deals here. Well those are the output deals, and these other wires these two heavy ones are the input. This is a grounding. It's going to be in parallel on the on the input.

1:10

At this point we've got our 100 volts. We've got 60 cycles. We've got the correct waveform and we're in business. Yes sir? There's basically there's 50 turns, and on one side add two additional which gives a delaying effect on the second spike there so that you've got an offset on it. 52 on one end and 50 on the other. It doesn't matter which end. One is wound clockwise and one is wound counterclockwise. That as part of the slowing down of the electron so that you end up with a waveform that you want.

1:11

We will see in just a little bit. [Question: You had the primaries tied together on the back-to-back Tesla coil, and that coil that is going into iscolation?] Well remember now what we are working with is a DC spike going into those things and what we're trying to do is get rid of that DC spike. [Question: Is that coil going into that isolation coil is an addition to that? That's the third coil around that. Is that right?] Tell me when I get there. This one right here? The 20 turns in the center is lowering the 2,000 volts. You've got a transformer there.

1:12

By keeping this sort of informal it will make it a lot easier on me. This is the wrong time of day for me. [Question: On the tesla coil is that a closed wound? On the 50-52?] Because you get a lowering and you get your Q factor messing you up a little bit, but the induction on this thing is so strong that you're not really having a problem. Yes sir? [Question: What's the amperage of the load?] It will be the same as the voltage until you put a load on it. Once you put a load on it that set your amperage. That doesn't make sense to you? Earlier, you probably weren't here whenever we talked about it, but ohms and amperage are after the fact, and it's not until you put the load on there, or the resistance, or the dissipation factor, that you have ohms and amps.

1:13

So you have 8,000 volts or 2,000 volts, or whatever, that's how much you've got. You have potentially the same amount of amps as you have volts if you work it correctly. [Question: You can run anything?] You could run just about anything you want to. One of the units I had up this afternoon will run a diesel electric train. You can disconnect the diesel part and put as many coaches on it as you want. That unit the induction we were talking about, the induction rate being so much higher than 60 cycles, well something millions of times smaller can produce the same amount of electricity as a very large 60 cycle generator. That small unit would produce the amount of electricity that a conventional generator the size of this whole room would produce. [Question: Can we see it?]

1:14

You want a demonstration? You bring me the train and we'll do it. An airplane? Yeah we can fly a DC-10 if you want to. It looks like some of the units are missing. There's two devices which are in the in the office in there, the bookstore. Yeah the two that I need right now. Well somebody in there knows this is the stuff we brought in from the automobile. While that's going on let me get something here.

1:15

I'm going to show you how this induction stuff works. [Question: I'm wondering if you could quickly comment on the size of wire that's that's been used in this so far?] When you're working with radio frequency you'd expect at very high voltages to have very large wires. That is very true if you're down in the lower frequencies, but when you get up into the radio frequencies the diameter of the wire can be very tiny. It has to do with the size of electrons, and how they're spinning around, and that sort of thing. You can have a wire that's about the size of one of your hairs in your head, and you can have basically a million volts on it, because the electrons are not going through the wire and there's no dissipation factor, and they're spinning on the outside of the wire so you can have this corona effect.

1:16

See I told you they were somewhere else. Just put them on the table here and whenever I get to them. I've got some magnets here which I want to demonstrate something for you. See, how much voltage would you think it would take to penetrate a quarter of an inch of PVC? Take a lot of it wouldn't it? Maybe even a million volts. What I'm going to do here, I'm going to show you how the magnetic effect varies from. Your induction is from magnetic effect, so you could take these wires and they can have a covering on them. They can be encapsulated. You can put them in solid concrete, or plastic, or anything else inside completely insulated from everything and these wires, when the induction effect starts taking place, which is magnetic, it doesn't see any of it.

1:17

That's why your induction works so well. That's going to lead us right into what you were asking about. Everybody see that? There's magnets inside and on the outside. That's the magnetic flux thing that we take a reading on for these maps that we're looking at with a Gauss meter or other kinds of magnetic flux devices. By the reading we can directly convert into the active electrical energy that's present.

1:18

At that point before we get off on this device. Now there's a little lunch pail like thing over there somewhere, and it may be back in the office also. It's got the battery for the device. It's a little lunch pail thing sort of interesting colors. You can probably tell who it was designed to sell to. That wasn't a very nice comment. I'm going to show the battery. This is a calcium lead or lead calcium battery. It's 12 volts at 7 amp hours. That's good.

1: 19

In my experience with this device, when I hook it up to it, it turns out that the L1 coil on this device has a four point something feet length of wire. It turns out that the wire which hooks onto the battery here has that same length. We're dealing with a radio device so we've got an L1 there which is oscillating at the same frequency that this wire here that goes to the battery would oscillate at. So guess what? When I hook it on to this battery here it's going to be generating electricity, and feeding back into the device, and it's not going to be using the electricity out of this. How I know that? These devices have been ran for as much as a month at a time with no lowering of the capacity of the battery.

1:20

So if you're looking for an overunity device you've got one right there. and there's other ways of making it do the same thing. There's also a thing called a capacitor battery which is a new invention, which is on the market just this year I think. It shows up in your some of the computers and other things. Basically what it does it stores electricity so that if the source misses a beat it picks it up and keeps things going. By putting a capassive battery in parallel on these input wires here you could have an additional kick which would do the same thing as the wires were doing on their own. If the wires were not the proper length so that they would pick up the radio frequency energy then you would have your overunity type device in one case, and in the case if the wires were two different lengths you wouldn't be getting that feedback.

1:21

Remember they have to be a multiple of our division of frequency which is there. They don't have to be the exact length. If they had been two point something instead of four point something that would still be a division of the frequency so therefore they would still have worked. But with the fourth thing you've got more feedback. That's one advantage of operating up in a higher frequency because if you were run down below two hundred thousand cycles this cable here would have to be maybe four hundred feet long to do the same thing. In the frequencies witch I'm operating it works very well. It had a surprise effect which I only discovered after a certain period of time.

1:22

I knew that something was wrong my batteries were never running down. The books tell you that that can't happen. Here I'm looking at the silly thing and and you're doing something you're not supposed to. What is it? It turned out okay.

That sent me running to the Patent files to find out if there were any patents that were related to that sort of thing. There was a thing I think it's Baldini is magic battery. Back in nineteen and forty six there were a lot of articles appeared in the newspapers throughout the United States about this magic battery that would never run down. You didn't have to charge it but he never ran down. But they didn't bother to tell how it was accomplished.

1:23

There had to be some sort of thing happening similar to what I was doing in order for that battery to function the way it was. It turns out that it was never patented. There's no patent of record and whatever became of it I don't know. Somebody apparently bought it out in order to keep it off the market who paid the guy you know a few hundred thousand dollars or something and was able to stick it in a file somewhere and keep it off the market for some business advantage whatever it might be.

With that in mind I'm going to hook this thing up and we're gonna see if it works. It's been bouncing up and down the road. It was in the back of the car from Houston so we will see whether I still got a device or not here in a second.

1:24

They're pretty durable. There's not much here that can go wrong. There's nothing unusual here that you can't find down at Ace Hardware or at a local level Radio Shack. Radio Shack, I should own some of their stock. This wire here? This, if you recognize it, it's what's in your chandelier. I think they pulled it apart. I thought with a transparent gold tint on it it looked nice. That's what determined what I use. [Question: 18 gauge?]

1:25
That is probably correct but it's a normal lamp cord and with this in mind we have to be very careful when we hook this thing up because the laser device there, if I hook it up wrong I've blown it. You have to get your plus and minus in there correctly or you you've lost your device. If you go to the store to buy these laser modules they can get very expensive. Or you can find of it surplus places, and sometimes you can get them at very nice price you know if you're real lucky. If you just happen to luck into it you might get them at very reasonable price, but if you get caught in the market, and you pay the full price, what you're looking at there is about three hundred dollars. You can wipe that out so easy you can't believe it. If you reverse these two things. That's why the colors are on here.
1:26
You'll notice down here where they go into the into the block they're also colored. And you'll notice on the battery they're also colored. You get into this situation where you try to idiot-proof what you're doing so if somebody is going to be duplicating. [Question. Why don't you just solder your bridge diode on there for protection, that way no matter how you hook...] There is a diode right here and this is the kickback protection right here [see pages 100, 101, 151, and 152 for pictures of this model]. It's a little tiny diode that comes from Radio Shack. You can get about 50 of them for 50 cents. If this thing is working properly here–where you see these little the bean looking deals that are on the end here? These are 60 watt neon bulbs, and they've got a resistor on the deal here.
1:27
We're working with radio frequency remember now here so we're crossing over and we've got a radio frequency device. We've got these little bulbs that are built in various parts of the circuit so it'll tell you that it's working up to that point. With that in mind let's see what we've got here. Something's not functioning. [On page 100 the schematic shows the laser runs at 24V so Don forgot that that module does not run on 12V like most of the others did.] These things should light up. No the battery's not. It's been sitting out in the hot though would that to mess it up? I've never had one goes dead over yet, and it was fully charged when I came. Basically what at this point we'll learn how to fix one I guess.
1:28
There's only a couple of things that can go wrong here and we'll find out real quick like. Let's see, let me make sure that I get these on correctly. What we're going to do is check and see if we're getting a voltage up to this first first point here. Which we are, we've got 12.94 volts there. What that's telling me immediately is that this little diode here is out and if you want to suffer with me we'll change it. Over in those little plastic boxes you'll find a whole bunch of them. No those are vacuum arresters these are little tiny things, they look like. Well we just determined it was okay.
1:29
They're little tiny things. It looked like a little hair or piece of. Yes that's them. We're not that bad off we're going to have, it'll be operating here in about two seconds. This is not a terribly unusual thing to happen.
1:30
Actually we can put a different kind of diode in here and it probably won't happen very often but the fact that it does happen has saved us our laser module.

We'll go to that in just a little bit let's do this one and and see what happens and then I'll show you how we get through those different stages.

1:31

I've got the parts of here. We should be in business. If these lights light up you know that we got there. If they're not lighting up the on a radio device the energy would be hitting me and running off. Now the the only other component that can fail, and this thing would be the spark gap here.

1:32

So somehow another something has touched this thing and got the spark gap out. It's bad also? I thought I saw a flash for a moment. I saw a spark there. When it flash they probably burned, so you need another diode.

1:33

Well those were in it when I brought it so someone may have been playing with it before I got it. It's in there to prevent kick-back and destroying the laser module. Can you use two of them together?

1:34

I was going to get some 4000's and I didn't get there. If you move that wire over there you're bypassing the diode, but we can do that.

1:35

It turns out that the diodes that we're trying to put in there are no better than the one we took out, which was bad. The device itself doesn't really have anything wrong with it. It's a safety diode in there to protect the laser module was burned.

1:36

This this is what I was using, which is obviously too small with those. It should be 4000 series. This puts out enough energy to drive our least diesel electric automobile. I haven't yet because it has been my show-and-tell thing.

1:37

In any sort of radio device you'll have things that will fail on you and they're relatively few things in this device so there's only two three things that could go wrong.

1:38

There's a spark plug in this system which it doesn't need and that is one of the things in favor of getting rid of some of the components that fail on you.

1:39

No, ordinary spark plugs are made to run for over 40,000 hours or something, so that is plenty long. No it's just an ordinary spark-plug.

We're going to go to another device, this one I'm having problems with it. I've got another one over here which we're going to have some fun with. We're not lost yet. It's one it's got three coils on it and I don't see it.

1:40

It's real obvious, when you find it you have three coils that are about five inches tall. Yeah this in the office there was what I had in here for another deal and it was not in the trunk of the car so we've go everything. The one I'm going to bring in now for the office has five 1000 volt lightning arresters in the circuit in place of the spark plug, and they're gonna light up like a Christmas tree. I've eliminated a spark plug and I've stuck lightning arresters in the circuit in place of it.

1:41

I don't need them at all except to show you that the voltage is there and that things are going on. But for someone who's not sure these devices work it's a cure for it. [Question.] Let me show you what's gone wrong there. This one right here is L2 this is L3 and L4. And out here this transformer would be L5 and L6.

1:42

We should say that is L5 then, and this is L6 and L7. That better? Remember now this is the wrong time of day for me and I'm not functioning 100% here. I'm running about 20%, so if I do something like that you want to tell me about it. While they're working on that one we're gonna bring the other one up and I think we're going to have a very very interesting difference in what's going on.

1:43

Someone may want to, once we get this thing going, to lower the lights a little bit so you can see these things flashing. Lower the lights until everyone can see those little lightning arresters working. They were working a second ago. I saw them when I first hooked it up.

1:44

You can put it back up. I wonder if we have a diode this bad on this one. It's the diode. Because the moment I touched it I saw the flash, so I know that. Let me hook it back up here.

1:45

You notice that there are four coils on this. The center one here is actually the one that's doing the work, and it's a miniature radio broadcast station and it's putting out induction. There's a small number of turns down at the base of it here, and then there's a certain number of other turns up above. Basically what we're doing we're we're raising it to about–this is eight thousand volts going in from this module here. We're running it through six turns. They're on an L1, so divide 6 into 8,000 and that's gonna.

1:46

It's okay now? Each one of those lightning takes a thousand volts of actual real electricity, not diminished, in any way to trigger each one of those. So at that point we know that our main coil system is working there and we know that the outside coils are tuned to the same frequency, so if we have 40,000 volts on that inside coil we have 40,000 volts on each of those outside coils. This is built to show what an overunity device and several aspects does.

1:47

Now see the little 60 watt neon deal here? See, that's lighting up because of the radio frequency electricity that's running along my skin. That's actually 60 watts right there and this battery is not putting out that type of deal. You can put a load up to whatever the voltage is. [Question. Can you plug the fan to that?] In order to do some wave shaping–actually we've got a situation here where we're going to be running the fan, but we're going to be doing it a little bit different than you think. Let me show you a part of the components over here on what you're looking at: the actual device. This thing, see the box is the diode bridge, and here's the system. But it's wound a little bit differently.

1:48

As a matter of fact I've turned this around backwards. I've got the windings on there a little bit different. But this is actually the mechanism that does the wave shaping that gets your double spike that gets you set up for doing your wave form that you're really looking for. So this is one of the wave shaping modules and the windings are–this thing is on there backwards. I've got it hooked up backwards. This is the resistor component, adjustable resistor component. It works on the resonant circuit. Yes sir? [Question: Could you run a house?] Oh yes, what we're looking at there is a great deal more than you would need for a house.

1:49

What you're looking at running here will run a large office building. It's due to the fact that your induction is working at a radio frequency, which gives you your higher efficiency of induction that makes the small device possible. Because it's doing millions of times more efficiency and output in comparison to what's going in than your 60 cycle device would be doing. So that means that it could be millions of times smaller. You have to go through the wave shaping and set things up so that it matches the devices that you want to run on it. Then you're you're home free after that.

1:50

[Question: I don't see and earth ground on here?] It's actually grounded in a way that you wouldn't suspect. The ground in this case it's inside the module here and it's coming back through one of these hookup wires. [Question: But how is it getting to earth ground?] You can ground these things into the air, or into the earth. You have a great deal of more efficiency if you ground them into the earth. If somebody will produce a metal pipe for me we'll ground it into the earth. Well we need one that goes down inside the earth, and that's what this is all about. Yeah I know what you're talking about. I don't want to use the microphone here, does it have three prongs?

1:51

It looks like somebody got to it. Somebody's been helping me! I wonder who. But the idea is that you have a much more efficient system if you ground it directly into something that's into the earth. You'll notice that on your fuse box at your house for the electrical system which you have.

1:52

You wanted to see a fan run and some other things I understand. Before we can do that there's a little square box over the here that's kind of dirty looking, kind of small. No it's about twice that size. That's it. This is an inverter. An inverter is basically what the circuit we're looking at over there. But this is one that came off the shelf. My gold cords!

1:53

See if we can figure out how to do this. It's going to be rotating. Maybe I can move this over a little bit. We're not going to be getting exactly the circuit that you asked for, but I'm going to show you how you can run a 300 watt device off of a 12 volt battery. Basically what it amounts to. Which would be equivalent of tapping in on this thing at this point. So I'm not going to be able to run it with this device at this point because you have to be hooked into to this device and there's another part that's missing off this. But I'm going to show you how you run a 300 watt device off a 12 volt battery through an inverter which is what that circuit is there.

1:54

Well if you want to come up here and hook it up you're welcome! The inverter technology is what enables you to do what we were talking about. How's that? If we can do that with 12 volts, what can we do with the thousand volts with an inverter circuit? Now this inverter circuit I'm going to draw you what's inside this little box here. There's a resonant iron core transformer which we were talking about. This unit right here is the main component of that inverter there, is this unit here.

1:55

There are two transistors which are back-to-back, and I'm going to draw you that circuit. When you put those things into this circuit here, when you couple them into it, you've got the same thing going. You can get rid of the battery there. But they have to be adjusted to the voltage which you're working with, and that device there is only made to run with 12 volts. [Question. Your machine is not functioning.]

1:56

You missed the point, you missed the point. Just back up a second and let me tell you what's going on here. You weren't listening. We have inverter technology which interfaces between this device and any device that we're going to run. So that wave shaping deal that we were showing you over there, this device here with some other things added onto it, is the equivalent of what's in that inverter there, except it's designed without the transistors. Basically what you have here our two transistors. One on each side. And you have a resonant iron transformer in there which gives you your 60 cycles. You have a couple of resistors and that's the whole device. [See pages 109-111.]

1:57

What you do, you take that circuit and you convert it so that you can handle the output of a much larger device. So there's your inverter technology being adapted to work with the output of the Tesla... One problem at a time. What I was showing you there is the inverter technology required to make the other device work. Now you ask why haven't I done that? Because simply because there aren't enough hours in the day is the reason I hadn't done it.

1:58

[Question. You can't prove here tonight.] If you understand the inverter technology you have no problem understanding what's going on. I'll see you after school! [Question. Mr. Smith, y'all need to be careful with this. Give this guy a chance. This was to be set up for tomorrow and the test we're all going to be done, the machine was supposed to be working tomorrow. He did this impromptu. He drove all the way from Houston. This stuff was banging around in his car. He did not have time to test it out. Please don't be doing this. This is not professional. I understand we're all upset cuz it's not working, but there's a lot of things going on here. That you see the wires are cut, the diodes aren't working, the laser power supply's blown. Give the man a chance.] It's unlikely but there's something else there if the spark gap is screwed up, or if any of these screws have gotten loose, and bouncing up and down the road, we've lost contact.

1:59

But we know that the basic idea works because this is the same device right here, except for those lightning . Your laser module, your L1 and L2 coils. Yes sir? [Question: Is there any way you hook the machine up and do a test off it?] For output. If you have the proper instruments to do it with. I have some radio frequency devices which I can tell you what frequency these things are running at. Which I can tell you they will correspond to the length of the wire exactly the way I was telling you, but what you asked about there I don't have with me. You need oscilloscopes, and you need a lot of other things in order to properly prove these. I have engineering friends at home. I'm not an engineer myself.

2:00

I don't have all of the devices that you need to properly test these things, but I have neighbors who are professional electronics engineers, and other kinds, and they do have these things in their home workshops, and I do take them to their house, and I do check them out, and I do know what they're doing.

So even though I may not be able to demonstrate it to you right now, they have been checked, out and I know exactly what they're doing, and I can tell you. Yes sir. [Question.] That is an that is an inverter deal, a commercial version that you buy off-the-shelf. This actually has the sales slip inside the little book here and I could tell you exactly. It was bought in 1985, and in 1985 it cost 83 dollars and 95 cents.

2:01

Every time you go through a change of coils there you're either going to have a step up, step down, or you're going to have an isolation deal where the same amount passes through. [Question.] You're stepping down. What we're trying to do is get it down to a level where it's compatible with the devices. So if I got them turned around there I didn't intend to. [Question.] It was intended to be stepped down, so if I made an error and lets credit it to the fact that the hour is wrong and a number of other things. But we'll try to get through this.

2:02

We're getting to the point where we're going to have to wrap this up. Depending on how serious you are about it, and what you need to know about it, you can locate me. I'm easy to be found, and I'd be more than happy to sit down and work with you, and what to do next, and that sort of thing. [Question: Mr. Smith one of the things that really interested me is because I like working with youth and you were saying that you work with Cub Scouts on these issues and show them how this works.] I give them pieces and they sit down and do the soldering and everything else an put them together. And when they hook it onto the battery their working. [Question: Do you have like a step-by-step video that you've made so that we could bring it and show it to kids that we know?] I tell you what. I'll get a group of Cub Scouts and we'll make you a video.

2:03

Thank you very much Don. We really appreciate you coming in.

Video 1 Schematic Drawn on Board and Explained

By Rick Friedrich, 7/25/18

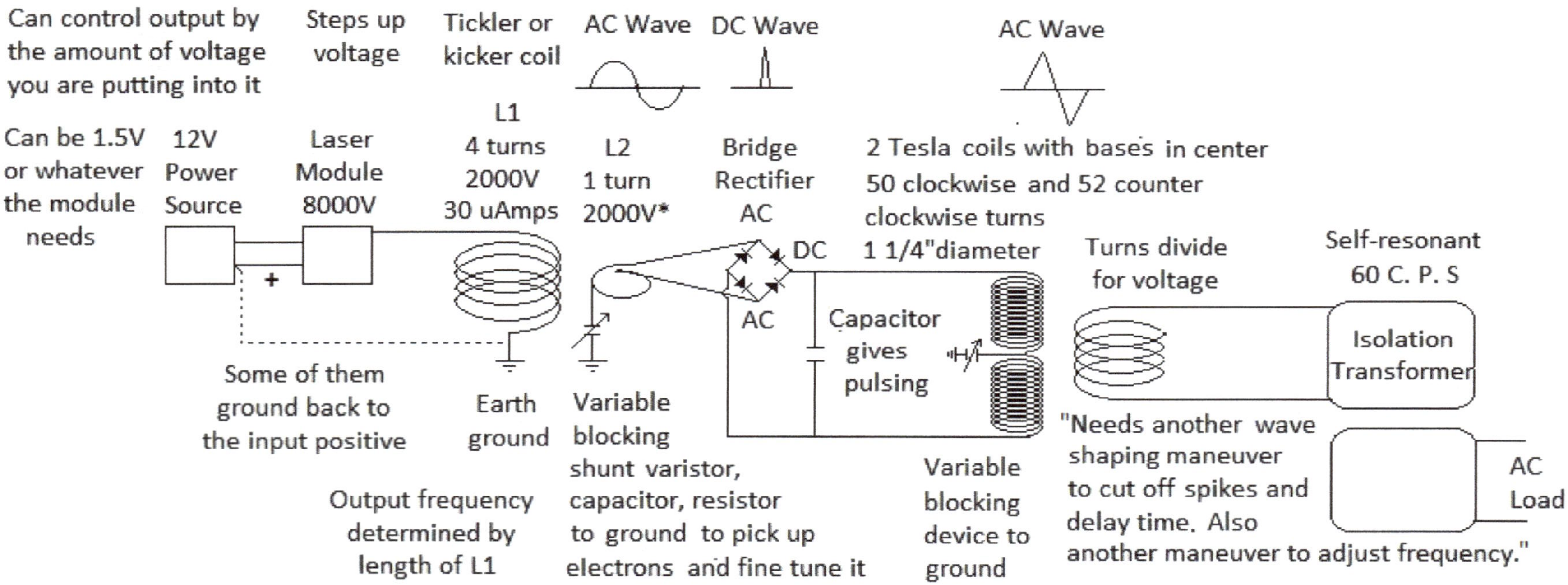

* Don made a mistake in putting 1000V when he went to the next board.

Same board, coils and resistor with bridge below and transformer to the right. Below would be step up and other step down volts.

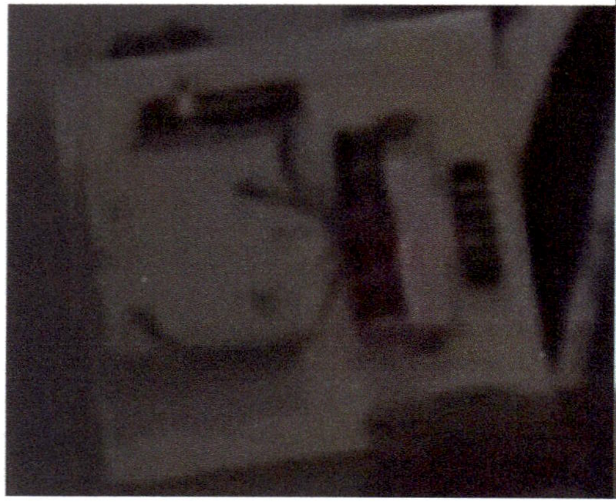

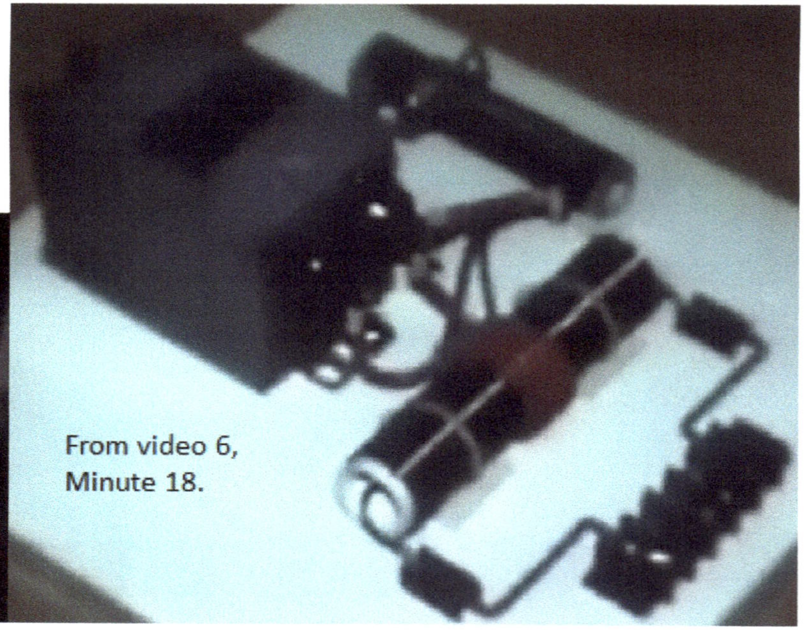

From video 6, Minute 18.

From video 1: "This thing, see the box is the diode bridge, and here's the system but it's wound a little bit differently. As a matter of fact I've turned this around backwards. I've got the windings on there a little bit different. But this is actually the mechanism that does the wave shaping that gets your double spike that gets you set up for doing your wave form that you're really looking for. So this is one of the wave shaping modules and the windings are--this thing is on there backwards. I've got it hooked up backwards. This is the resistor component, adjustable resistor component."

Video 2, 1995

Our first speaker here on Saturday afternoon is Don Smith. Don hails from Texas, and he came up to the Tesla symposium last year for his first time and was kind of an unscheduled unexpected kind of person there, and showed us some very interesting ideas in free energy type motors. He's been working on this for several years and he's now at a position where he's ready to present some of this to the public. So let's welcome Don Smith.

Minute 1

2

3

There's going to be an awful lot of questions about some of the things that we're going to be talking about, and most of them are contained in this booklet which they have out in the bookstore. There's some schematics and a lot of other things in there, formulas and things which are useful in building this type of device. So with that in mind we'll see what we have next. Basically energy, all energy, electrical energy, comes from electrons. In the case of electrons it's all negative. It's either more negative or less negative, in the case of electricity. Now the in case of chemically derived electricity it's a different deal. But in the case of electrical power generation it's all going to be electrons, and the electrons are all going to be negative.

4

Basically what we're doing, if we have ambient, which is background, which where you're sitting is perhaps two hundred thousand volts per square meter. Anything above or below that is going to be useful energy because the electron is just like the snowball that you put on top of the hill, and you roll it down the hill. It's trying to get back to the gravitational level which it was before you put it up on top of the hill. If you raise the electrical potential above the background it's always going to try to go back to the background level. It's the in-between part there that's useful to you and it can either be below ambient or above ambient.

5

Anytime you have electrical energy, in most cases that it's going to be above ambient, which means it's above Unity. If you're talking about overunity, all useful electricity is over unity because it has to get back to ambient which is background. Your over unity people won't be real happy with that thought I'm sure. We'll give you a clue as to one of the very early things which I saw, which got me involved in this type of thing. I'm a geologist by background and work with electromagnetic Maps and locating all deposits, mineral deposits, and other sorts of things. Of course that's a direct reflection of the electrical characteristic of the earth and our portions of the earth.

6

So that got me involved in electricity initially, and I've had a life-long interest in it. About seven or eight years ago when the petroleum industry went sideways, I had a lot of extra time and I was asked by the local school system to help out in their science courses. Of course when you look at the physics books fifty years after you saw them the first time, or you had them in college, you see things which you probably didn't see the first go around. You've had some life experiences and you begin to see how these things all fit together. So it became quite obvious to me that there were some real big holes in the way that electricity was being presented.

7

One of the early on things which I found was this [*Resonance Circuits Demo*]. It's from a science supply. The initial device that I ran into was from a science supply house and it turns out that this company is still around and still makes these things and I'll show you the actual device. What happens here, the larger one up there would be like this. The smaller one would be like this.

8

Basically, what you do you put a 1.5 volt battery in this bracket thing here and it causes this cell to react, and in turn this one is tuned, so that it matches this one. Now on this one you put a 70 volt lightbulb, and this one, or a 70 watt lightbulb, and this one with the one and a half volts transmits the effect that causes this coil to react and generate electricity. The electrons from here are not coming here. They're actually talking to the electrons here and cause this device to generate useful electricity. So you're not diminishing the battery here by running the light over here.

9

If any of you are interested in knowing all of the nice formulas that you need to know to understand this sort of thing, this kit comes with a nine page front and back instruction sheet, and it's got all the formulas related to resonance and how that can occur. This thing should tell you immediately that there's something wrong with the way electricity is taught because you've got a 1.5 volt battery here which is causing this coil to resonate and which in turn is causing this one to resonate. None of the electrons from this electrical system are coming over to this electrical system. Yet this electrical system is lighting a 70 watt [volt] light bulb. That should tell you immediately that there's something not quite right. If you start digging from that perspective it turns out that there's a lot more to it than that. It's all good as far as what you might like to see happen.

10

When you get into building the Tesla coils proper, basically what you're doing, your energizing a smaller coil which is your L1 coil, and causing a larger coil, which is your L2 coil, to react. It's the exact same system there, except that it's put into a more compact and convenient system. Normally the L1 would be three times the diameter of the L2. But that's not really necessary. It can be smaller than. One of the problems that people run into in building these things is getting the 2 coils to react with each other.

11

You've got to have them so that they can talk to each other. So if you use like quarter wave system, your shortest wave will be the longest length of wire and your shortest length of wire will be your higher frequency. With that in mind, one of the problems that people run into immediately is tuning those things. I'm going to show you in a little bit, a slide that shows this exact same thing that from 1893. It was one in the early radio devices how they were able to tune them fairly easy. This is your L1 coil, L2 coil. The way we're going to do this. This is PVC, and I cut a collar here so that I could slide it on there and use it for a spacer, and put another one there as a spacer, and the net result is that this thing fits almost perfectly.

12

It slides in and out of this particular coil here. This is one way to tune these things without having to figure all of the other characteristics that have to go into it–which gets pretty complicated sometimes, and especially because resistors. Although you know think resistors are resistors, if you take your LCR meter and you measure the capacitance and the inductance of almost any resistor you're going to run into, you'll find out that the resistor is a lot better capacitor, and a lot better inductor, than it is a resistor. Those things vary considerably from one resistor to another. The only way you can get around that is get non-inductive, non-capacitance resistors. I can tell you if you go to people have been manufacturing resistors for 50 years they won't know what you're talking about.

13

You will have to find the sources pretty much on your own. They do exist. Radio Shack has a few non-inductive and non-capacitant resistors. Very few though. And they're marked. Don't expect to walk into Radio Shack and tell them it that's what you want, because the people that work there won't know what you're talking about. But if you look on the packages you'll see that there there's a little square ceramic box looking deal, that says on the top of the package that it's a non-inductive resistor. If you're trying to tune these things by changing the resistance or the capacitance and you're adding these things in, you get into a situation where every time you get one thing corrected you've got something else out of balance. And the net result is that doesn't work very well.

14

A lot of people have tried to build the circuits which I presented last year, and several hundred people that I know of, that have tried to do it, there's only about six out of all of those that were able to actually get it done. I'm not quite sure how they were able to do it. Some of them I know the details and some I don't. But using this particular system here you can correct for a lot of things that otherwise you're going to need an engineering degree to figure out probably. This will simply go to your high voltage source. If you want to build a nice coil, Barker Williams makes a large variety of these coils ready-made. They're copper, and they've been dipped in stannous chloride, which tinplates them. And that's how you get the pretty tinplate.

15

They've used a heat process to embed them in slits of lexan, and that's what the ribs are made out of. Now if you want to make a real pretty Tesla coil, this happens to be the cap off of a sewer pipe. They come in all sizes from about an inch and a half up to over 25 inches. If you look at it this way and you see this square thing on the bottom, the square thing on the bottom, you can screw onto a piece of board or something very nicely. You can wind your L1 coil very nicely. See the grooves on it? It will stay exactly where you put it. You can use some little clear ties and run a hole, or two through it, where you can run your ties through it. And all of a sudden you've got you a very nice L1 coil, see.

16

This being a PVC you can further take a piece of PVC pipe that's been cut to the proper length. You can set that thing on it just like that. You can use PVC glue if you want to, and glue these two pieces together. Then you can take your center coil here minus the orange one in this particular case, and you've got a very nice Tesla coil almost ready-made. If you want to get into amps and other sorts of things then we have to start using the larger wires. It's the flux lines that are reacting with these particular coils. The magnetic flux lines that are transferring the energy from one energy triangle to another.

17

If you get into a heavy amperage, and that sort of thing, this is welding cable. It comes in all sizes. This happens to be number eight welding cable. It will make a very nice high amperage type coil, for your L2. Then you can use even larger or copper tubing for your L1 coil. Since we know electrons are our source of electrical energy. Electrons do a lot of things. We get a lot of different types of energy from electrons.

18

It turns out that electrons provide us with a lot of things besides electricity. This is sort of a brief breakdown on it, and there's probably other types of things that they provide us with also. But in the systems of storage there, for example, your capacitors would store the electricity and your coils or your inductors would store the magnetic field. You've got equal amount of each. Anytime you have electricity, it's a two-sided coin. You have equal amounts of each. There's no way around it. Because when the electron moves forward, or is disturbed, it's going to give off magnetic energy.

19

When it pops back to where it was it's going to give off electrical impulse. So you're going to see both of those. I have a number of other slides so we'll need to look at them. You're familiar with the power triangle and how that works [see page 66]. For example, here, the power triangle system 1 there is your L1 coil, and your L2 coil would be the power system 2. Now the electrons are not moving from number 1 to number 2. As you noticed while ago, the L1 coil is covered over with plastic. That's just fine because we don't want any of the electricity to be jumping around, or sparking, or moving.

20

It's the magnetic flux lines that we're interested in. So in [the] case of conventional energy we've got atoms A and B there. A would be your volts times amps. That's what's inside the wall before you turn the switch on. When you ground it, then you've got a time which comes into it, which would be your volts x amps x time, which you've got watts or useful energy at that point. That's the conventional way of doing electricity. In C there we have resonant energy which is not actually going anywhere, or doing anything, except that it's resonating. In the process of resonating it's creating lines of magnetic flux. Your electrical people will tell you that number C there is wattless energy or useless energy.

21
What they're not allowing for is the fact that at resonance you have a tremendous intensity of magnetic flux lines which will transfer beautifully over to another electrical system. That's what's happening between your L1 and your L2 coils. Once you get your L2 coil going it's going to be moving from the base of the coil towards the top of the coil. As the magnetic flux starts building on each turn there it's transferring over to the next turn, and it's carrying with it what it had, and it's multiplying by the flux lines that are being active there. So you're getting an amplification of the magnetic flux lines. These magnetic flux lines are building in intensity as they go towards the top of the coil and they reach their maximum at or near the top of the coil.
22
At that point you've got high voltage with not a lot of amps. But the amp business is not going to be a problem for us, because we're not going to be using the system that we have at the top there. We're only going to be using the lines of magnetic flux. In a system which I'm going to show you in a little bit, there's a Tesla third coil, which comes into play, and it acts like the little receiver that did the light bulb thing that I showed you earlier. It is situated in the highest density of magnetic flux, which means that it's reacting a like a power system 3 there. It's got the benefit of the resonant energy that's been amplified at a very high level. So at that point what's transferred over into it becomes an entire separate electrical system. None of the electrons have ever actually gone from the L1 to the L2 to the L3. It's the magnetic flux which is reacting with the different coil systems which is powering the next stage on, or the next layer of turns as you go towards the top of the Tesla coil.
23
At this point you're free of the power system that was present at the beginning of your experiment. You've got a whole new power system with a whole new a bunch of lines of magnetic flux, which did not exist previously. Therefore you have gained in energy potential. Your potential, at that point, you have a lot more energy than you ever had when you first started out. If we look at the law of squares in the Fibonacci series and other things in nature. For example, if we take the beginning of a human being we have one cell, and then we have two cells. Then the two turn to four, and four turns to 16.
24
After that has happened a few number of times there you've got a million things going for what you had initially just one of. The same thing is true in the case of flux lines. The flux lines are actually what will give you your energy back because they translate back into the energy system as electricity. We will get you some more overhead things in just a moment here and see how things go from here. [Question: How close do the coils have to be?] They don't necessarily have to be very close at all. For example, if you have a resonant system, of course the flux lines are determined by the square the distance from the source.
25
They're diminishing in value as you move away from the device so it's probably better if they're within a relatively short space if you want to benefit 100% from the flux lines

that are there. However, with a resonant system the flux lines will give you your energy that's totally separate. See none of the electrons from the original system are actually present in the second or third system. If someone tells you that what you get out depends on what you put into it– once you've generated those flux lines well you're into a different language system which will translate back into electrical energy. Since you have a lot more of the flux lines, you have a lot more potential than you had to begin with.

26

This law of square business you see that anything that's nonlinear, for example, when Galileo had his first two laws of motion–and of course he got rewarded for his effort because you don't go around saying that the earth goes around the Sun and things like that. So they had a barbecue in his honor and you know the rest of the story on that. But anyhow, we go along to Cambridge University about a hundred or so years later and we have a farm youth of the name of Isaac Newton who's studying mathematics, who's not a very good student. He gets in a fight one day with one of his friends and something about his whole chemistry changed after that and he became very competitive, and a very good student.

27

Of course while he was studying mathematics at Cambridge University he came up with the third law of motion. Then they renamed the whole system to the kinetic energy system. The rest is history. Some time later a fellow by the name Albert, a clerk in the Patent Office, decided to paraphrase it and came up with $E=MC^2$. Of course he got the Nobel Prize for his paraphrasing of the of the kinetic energy system. With that in mind let's see what else we have. We have all kinds of things to talk about. Here's a drawing showing what happens with magnetic lines, whenever you start getting additional ones.

28

It goes to millions in just a very short period of time. When that thing flips over a very few times it's into the millions of flux lines. At that point you can disturb an awful lot of electrons. That's what it's all about. Your magnetic flux lines will translate into electrical flux lines and that sort of thing. Then if you've got your coil system where your Henry's and capacitance, and that sort of thing, are all working for you, and you know how to hook into it. Well then you've got a lot more energy than you started out with. There's no doubt that it works. There's all kinds of evidence for it. Getting into the formula for finding what type of useful energy is available to you through this system.

29

This will give you a little bit of an idea [see page 71]. Joule was a English brewmaster who had all kinds of machinery that ran on steam, and some electrical, and this and that. He had to come up with a system which would make translating one system of energy to another more universal. So that's how he comes into this. A Joule is 1 volt times 1 amp times 1 second which is a watt second. That also happens to be the definition of a Coulomb and a whole slew of other things. These things all translate in and out of each other. Joule was the one that came up with the system and that's where we got the name Joule.

30

Let's see what that can do for us in the case of a Tesla coil. Say that we have a Tesla coil that is running at 100,000 volts. Well you might say, well we're not going to get a megawatt out of something like that. But what happens is that it's the cycles per second that come in on that. See you remember a watt was 1 Volt times 1 Amp times 1 second. So if this thing repeats itself many times per second then you have automatically got a lot more than one watt there. This explains how 100,000 volts can become a lot more energy than say 100,000 volts.

31

What you have to do in order to collect this energy is that you have to have capacitor banks to collect it for you, and store it temporarily, in order to get onto the next stage. Because if it's not in a useful form it doesn't really mean anything. You notice on this that we're going from a hundred thousand volts into the hundreds of mega watts. Of course this system has to be built to contain or to operate safely at whatever level it's operating at. But if you've got proper construction for a hundred thousand volts, you can pump forward many megawatts of energy. I might say that Hoover Dam is approximately five hundred megawatts, and I'm showing here that you can approximate Hoover Dam with a unit which will sit on your coffee table.

32

This hasn't been actually done yet. I've done it at a certain level and I know that it will probably work at that level. I'm in the process of building larger units with welding cable that goes up to about the size two or even a zero. The way you do that, you use some of the technology if you were going to build yourself your own welding machine. Some of the mechanics that are involved in building a welding machine are also involved in building one of these things that will have very high amperage. I'm going to show you some devices in a little bit that are related to this that have thousands, many thousands of amps, plus, say a million volts, at three or four thousand amps or something like that. So how many megawatts is that?

33

I'll let you look just momentarily at some of the benefits of this particular system. While you're looking at that I'll get some more things for you. Anyhow, this is in one of the books. Thought you might like to see what some of these systems look like. The one that you see there on your right. That's 12 inch PVC sewer pipe [see page 159]. The top piece is about 16 inches long and the bottom piece of exact duplicate of the top piece.

34

The center device there is a Rubbermaid oil barrel dolly with the rollers taken off of it. It turned out to be 25 inches in diameter, and it worked out just fine for the L1 coil. Below you see a power system. I'm going to Center it now so you can see that power system a little bit closer. What you're looking at here is an automobile ignition coil. What you're looking at there where the 12-volt electricity comes in, on that connector block, basically I've used a timing circuit which is very common. You take a capacitor with a resistor shunt across the legs of the electrolytic capacitor and you run electricity through it.

35
You're going to end up with a timing deal. It's going to be cutting it off and on. So it's the resistance times the capacitance of the capacitor that tells you how many times per second this particular ignition coil is coming off and on. Since it's an automobile ignition coil it runs at 12 volts. The timing system is set up to match the timing system which it normally accepts. If you have an eight cylinder automobile, and you have 2,000 rpm on the automobile, that would be 8 x 2000 would give you your number of times per second this thing is going off and on. And also your automobile ignition coil is made to last for about 40,000 hours in some cases.
36
It has longevity. It uses a common energy source. You don't need a complicated device to turn the 12 volts off, and on, and your off running. So I might say that the large coil which we were looking at there, I tried to use the computer programs. There's two of them that are available through the bookstore here. In most cases they work real well. I tried to use that on the larger coil and it didn't quite turn out the way that the program said. But that is neither hear or there, it's a good guide to what you might expect out of your coils.
37
The three coil system that you see is [see page 153], the center one here is the one that's powering it. And you have your L1 coil and your L2 coil. You have a high voltage system. You have three coils that are around it there that are tuned to the same frequency. Remember the little model that I showed you at the beginning there? Well actually this is duplicating that sort of thing. The center coil is putting out about thirty thousand volts. That means that each of those three on the outside they are doing the same thing. You connect those up just like there were batteries or such and you end up with about over a hundred thousand volts real quick like. [Question: are they that far away?] No they're not. That whole board that you see there is about 14 inches square.
38
It happens that the coils that you see there are actually about an inch and a half in diameter, and four or five inches long. That will give you a perspective on the size of it. This was made to demonstrate over unity. Since people are real skeptical about over unity, well this is intended to really mess them up. Most of these things are in the book in the bookstore. The device which you're looking at now [see pages 67-69] is this particular device.
39
Basically what you have, you have your 12 volt battery, you have a diode which serves to protect the laser module from a kickback. This is a high-voltage laser module. You have a capacitor that's in parallel there. Then it goes into your L1 coil and then comes back. There's a spark gap which helps control the energy level that's passing. The spacing on the spark gap there will control the level of the energy that's going into your L1 coil. Then of course number 7 is your L2 coil and number 6A up there is Tesla's third coil. I'm going to show you something from eighteen and ninety-three that relates very closely to what we're doing here [page 1].
40

These are useful circuits which I gathered up out of one of the Tesla publications [*The Inventions Researches and Writings of Nikola Tesla:* https://tinyurl.com/yczkbrh2 pages 170-197]. If you notice number one there is showing that tuning circuit which I showed you early on, where we slide the L1 coil inside the L2 coil. You know when you've got it tuned, because you start getting sparks. We go to drawing number two there. If you've ever wanted to know how Tesla lit incandescent lightbulbs and how he powered transformers to run motors and do other sorts of things, you have the basic circuits for doing that. This particular case, the Tesla coil is built laying on its side and the L1 coil is in the center of it, and it moves in and out. You notice the up-and-down thing?

41

This is a capacitor or a bank of capacitors. Whatever energy level you need that's the temporary storage there that gets the amount of electricity that you need in order to do the particular job that you're intending to later on. Then we go to drawing number three there. The coil on your right, is wound in one direction, the coil on your left is wound in another direction and the L1 coil is run in and out. The layers of the windings, if you could see them, there's layers and layers of windings here on top of each other instead of having an extended coil they're wound in more compact.

42

The L1 coil is moved in and out there so that you get it tuned properly. That center deal, if you put it on a timing circuit and ground it, you can get 60 cycles direct from this thing on the first go-around without going through any other sort of transformer or anything else. You can generate sixty cycle electricity direct with the Tesla coil. Number four is another illustration of how to light incandescent lightbulbs. The trick is knowing what capacitance to use and a number of other things. But this is actually how it's done, and this was 1893. You're familiar with how Tesla was able to take lightbulbs and globes and things and walk around the room and they would light up.

43

This bottom drawing, number five, is how he did it. He put sheets of metal on the opposite ends and hooked those into his high voltage system and was able to get enough intensity of the magnetic field so that he could walk around in a rather large space and these light bulbs would light up. Now if you look at this drawing properly you'll see that we have a generator there that's generating electricity, that's going through a capacitor, which is running on an L1 coil, which is powering an L2 coil. Which takes it to a higher level. That L2 coil is in turn activating those two metal plates which cause the lights to go on.

44

You can take that same circuit there and use it in a number of other ways. How's that for a spark gap? From top to bottom it's nine inches. So they give you some idea. This thing is made to run at in excess of 200,000 volts. This is going to fit into something I'm going to show you next. There's a thing called a Marx generator. All the things which people think are not present in a Tesla coil, suddenly become available. The deals here are that spark gap that we just saw.

45

When you take this spark gap, and you put it with the capacitors, just a few stages and you're into hundreds of thousands of volts, with thousands of amps. The thing's called a Marx generator. They're used in plasma studies by universities and other sorts of organizations. Inside the round thing there is a real Marx generator. The distance across here is a little over three feet, so you can see that the Marx generator is not very large. It's very effective. The one that's in there is multiple stages, and it goes into a couple of million volts real quick-like.

46.

With about maybe two to three thousand amps. That creates a problem because you can't have all of that voltage and amperage at the same time. So you have to put it inside this metal contrivance, and you have to fill it with oil, and you have to pressurize it. Otherwise you can't do it because the thing will just turn into one great big plasma ball. Once you've got it pressurized and inside oil then you can contain it and take it on to your next stage. Now you can use something like this as your high voltage module to power a very large Tesla coil that goes into maybe five to ten million volts real quick-like, with very heavy amperage on it.

47

I'll show you some other kinds of spark gaps. These are commercially available. The ones that you're seeing in the illustration there go to about three hundred and seventy thousand volts. Of course they make them for much higher voltage. If you had a ruler on there, the top one up there is about, from one end to the other is probably about six inches long. You have to pressurize these things. You put a gas inside them in order to keep them from burning up. It's a sulfur compound of some sort. It's a gaseous form of sulfur and the balls are made out of carbon. They're made to last for a long period of time.

48

These things are commercially available, just like the other ones I showed you. In conventional measurement of electricity or high voltage, you have to use the voltage dividers in order to get your instruments to work for you. As the voltage gets higher these voltage dividers get quite large. In this case in order to use conventional measuring instruments this thing is more than nine feet tall. It's less than a million volts. That means that you can measure up to a million volts with your regular measuring devices by using this divider in order to bring it down to where your instruments can work with you. But that's not necessary at all. If you remember that for each line of electrical flux we have a line of magnetic flux.

49

It turns out that your emi measuring instruments are your magnetic instruments which measure magnetic flux, some of them are actually calibrated in volts. You have an equal amount of magnetic as well as electrical, so you can use a magnetic detector in order to read the voltage indirectly. The way this works you have coulombs or Newton's inverse square law. The square of the distance times the remote reading will tell you what the full reading would be. So that if you're say a hundred feet away, or something like that, and you're getting a reading maybe 10–and so you have to square one hundred and then times 10. The first thing you know well you're reading a million volts and you've got this little hand-held gadget and you didn't have to even go near the device that you were measuring the energy from.

50

From a safety standpoint and a cost standpoint for about $35 you can get the device that looks like little handheld computer. It reads it directly in volts. You square the distance times your remote reading and you know what the voltage would be from the device that you're interested in.

Okay well thank you very much. [End of later presentation.]

55

Good morning. Welcome to the 1995 extraordinary science conference, our morning workshop. The workshop schedules, we have a quick note the one that says mark Hendershot, Mr. Hendershot was not able to be here this morning so Joseph Newman took his place.

56

This workshop is labeled John Picard, and he was not able to be here, so we have Don Smith in his place. And interestingly enough Mr. Smith and Mr. Newman are talking about pretty much the same type of thing, just two different approaches, and that's a really neat thing. I first met Mr. Smith last year at the International Tesla Symposium. He was quite a hit last year. Just really bent a lot of people's minds and erased a lot of preconceptions. Now we're going to have a chance to explore Don Smith's work in detail. So please let's welcome Don Smith.

57

I'm going to work largely from overheads and the floor is going to be more or less open to questions as we go along. Then at one point along the way we'll stop and we'll look at some of the overhead devices. How many of you were here last year so I know what not to repeat?

58

It looks like I could probably repeat a number of things that we talked about last year. There's a book on this which is available out of the bookstore, and I have a few copies maybe myself. I'm going to show you what the front cover looks like on it so you know what to look for. In case you don't get all your notes on it, it will have almost everything that we talk about here in it. If it doesn't well it's got my address and phone number in the front of it, and I get lots of phone calls on these things. Last year after I went home, by the time I got there there were several hundred phone calls on the answering machine. I got from 20 to 40, I guess the highest maybe eighty phone calls from people from all over the world, sometimes in the wee hours of the morning.

59

The book that we're talking about, they have some of them in the library, and I have a few copies myself in case you miss out on the ones over there. I noticed they didn't appear to have very many of them so they must have another box somewhere.

That's the front page on the book that we we're talking about [similar to page 25]. It's growing. There's going to be about 400 pages in the book, and it's going to look real nice when it gets finished.

1:00

I noticed that Joe Newman over there has already got his book finished pretty much. It looks real impressive. He's doing some of the things which I'm involved in. I noticed he's using the 200kV capacitor there which is one of the tricks to getting useful energy out of one of these devices. Basically you have to have some way of storing the energy in order for it to become useful. There's several ways of going about making it useful. One would be using a diode bridge. The problem there is that most of the diodes that are available aren't high enough frequency or high enough voltage or amperage. You can get diodes that are above 2000 amps, but generally speaking the frequency rating on them eliminates the use of them.

1:01

There's an interesting way around that however because the equivalent circuit of a diode is basically a capacitor with a resistor shunt on it across the two poles of the capacitor. [Question: Maybe we use the old vacuum diodes?] Yes, you can you can use those, they'll get you into some higher frequencies and maybe even some higher voltage. How to identify a diode that works at high voltage, the length of the diode has to do with the voltage, and the diameter has to do with the amperage that it will stand generally speaking. That's correct. I'm going to start putting these slides on for you. I did a thing here last year and some of the stuff that we're going to be doing will be coming from that.

1:02

This will give you a little bit of a general summary as to what to expect from this thing and what the ongoing situation is. There were numerous people who approached me wanting rights to manufacture the thing commercially after I was here last year, and a lot of them turned out to be promoters. A lot of them turned out that they didn't get along with each other, or the money source, or me either, and I had to run some of them off. For the last six months I've been trying to get rid of one group, and I did get rid of them. I got along real fine with the funding source but I didn't get along well with the guys that had gotten in between.

1:03

The funding source basically is the part of the World Bank and some of the other people, and it involves the rural electrification of India, Pakistan, and China. I'm their consultant on how to do it sort of thing. So then, the funds are there, the contracts have been writ, and everything is in place. Within the next year we'll see some very interesting developments as far as this technology goes.

Most people think that lots of volts and no amps, and that's the thing that they'll try to get you with every time. However, volts are basically the water that's behind the dam, until you poke a hole in the dam, you don't get any flow rate or amperage and that sort of thing. Whatever your voltage is it's whatever your potential is, and how you go about letting that thing through the dam, is what's going to give you your useful energy.

1:04

In this case the useful energy is accumulated in capacitor banks, and a mega Joule range is not unusual at all. I'm going to show you some things after a while that relate to this that are in place and you can actually buy them off-the-shelf so to speak. They work for railguns, plasma studies and a whole slew of things. Most of the universities have these capacitor banks. With that we'll move on to something else.

One thing that we need to take note of is that you're always faced in this type of thing with whether something is unity or over unity, or what happens in that case.

1:05

Basically anytime something is nonlinear it has potential for over unity. If we go back to the time of Galileo and we find that he had the first two laws of motion in place. Along later came fellow the name of Isaac Newton, a farm boy from studying mathematics in Cambridge University. He added the third law and also changed it from the laws of motion to the conservation of energy laws. Of course he got knighted for that, among some other things, and became one of the professors of physics at Cambridge University. But his basic background, his whole family were just trades people and farm people. As a student at Cambridge he was not particularly noteworthy in any respect.

1:06

One day he got in a fight with one of the other students and something about that fight changed his whole chemistry, and after that he became one of the best students in Cambridge. Which, if you go back to the 1600s Cambridge is probably like one of the little agricultural colleges out here in the backwoods somewhere. So it wasn't the Cambridge that we know about today. Sometime later a fellow that was a bookkeeping clerk in the Patent Office by the name of Albert decided to paraphrase the whole thing into energy equals mass times velocity squared and of course he got the Nobel Prize for that. So if you fellows want to do something great you've got some good company there.

1:07

Anyhow, once we get into the laws of motion and into the EMC^2 thing, it's really easy to convert that over to what we want to do. Because basically in the energy business you've got magnetic and you've got electrical energy. They're both the same coin, just different sides of it. You don't have one without the other. If you got electricity you've got magnetic energy, and if you've got magnetic energy you've got electric electricity. So that's going to come back in a little bit and we're going to find out how to measure the large amounts of electricity that we have very safely. What that means basically is, that your electrostatic volt meters and other kinds of things that measure magnetic flux, can be used due to Coulomb's inverse square law.

1:08

It was Newton's before it was Coulomb's inverse square law. Which means basically you take the remote reading, and the square of the distance times the remote reading, and you've got a direct reading on the magnetic flux, which converts directly into electrical flux. So you haven't had to poke your finger into a two million volt device in order to measure it. If you try to measure it with the conventional systems–I'll show you some after a while–you'll have a resistor stacked as high as this ceiling here and then some in order to divide the voltage down to where you can measure it with your instruments. That gets to be very complicated, it's also very dangerous. It's much easier to use the magnetic flux, which is the actual other side of the coin. But it's not dangerous like the electrical flux.

1:09

Electrons provide us with a lot of different kinds of energy. We see that we have a source of electricity. We have a source of magnetic energy. We also have spin. When the electron spins it gives off energy. It also gives off light. When two electrons get too close together they push apart. When that happens you get a flash of light. That's where your daylight comes from. It's the electrons that are coming in from the solar plasma that impact on the side of the earth that's spinning into the solar plasma. When they get too close together you get flashes of light. And that's where your daylight comes from.

1:10

If you get outside the Earth's field you just have dark, it's very black out there. You see the stars off in the distance and they're just pinpoints of light. The Sun itself just looks like another star. So that will give you some idea as to where the energy is coming from and what happens then.

The establishment has a problem with this term unity. Basically, what happens in unity is that ambient or background is what you're really seeing most of the time.

1:11

In the case of sitting in this room here, or anywhere in the general area here, the voltage background voltage is several hundred thousand volts. Which you're not aware of it, because it's the same all around you. So if you want useful energy, basically what you have to do, you have to get above ambient or below ambient. The water has try to go back down the hill so to speak. Or things to try to equalize, otherwise you'd have no useful energy. So the fact that you have all this electrical energy around you, which can be measured fairly easy, doesn't really mean anything because it's not useful to you. Apparently it's not all that harmful either.

1:12

Whenever we start doing our conversion from EMC^2 to our formula which we're going to use, remember now that half the energy is magnetic and half of it is electrical. Half the energy is electrical and half is magnetic. So we're going to have to multiply our formula by 0.5, which is half, in order to get say the electrical energy, which is what we're actually looking for. The energy we're going to let equal joules, which happens to be a universally convertible unit. We're going to say that joules equal mass. In our case mass is going to be coulombs or electrons or whatever you want to call it.

1:13

The way we're going to collect those things is in a capacitor. So our formula is going to read joules equal 0.5 x capacitance times, in our case velocity is going to be potential or voltage. Those things convert over relatively easy. The next thing we have to do in order to ascertain how much energy is there, is to take the number of cycles per second. So we multiply what we have just got by cycles per second. And for capacitance we're using farad's. So we've got something like a point three or four zeros and then we've got a number. Then we're multiplying that times the voltage which say is a hundred thousand volts or more, and that sort of thing. That's going to move the decimal point over some in your favor.

1:14

Then that doesn't really amount to a whole lot of electrical energy at that point. But whenever you start multiplying the number of cycles per second times that then all of a sudden you've got an awful lot of energy. Basically what happens at that point, you pump it into a set of capacitors or a capacitor bank and that capacitor bank takes it on from there. One of the slides I'm going to show you in a little while, shows you that in 1893 that that's exactly what Tesla was doing. It shows these circuits for running motors and electric lights and that sort of thing. He was actually doing the same thing then. It's not anything new. It's how you go about doing it and the type of capacitors and a few other things that you use. Now where you get into trouble on this sort of thing is the fact that you think a resistor is a resistor.

1:15

Well it doesn't really turn out that way, because most of the resistors are better capacitors, or they have more coulombs and more Henry's then the actual things that you want to get those out of. You have to have resistors that are non-inductive and non-capacitive and those are a little bit hard to come by. They exist. If you go into a electronics store and try to buy that sort of thing you're going to get very interesting reaction because they've been in the business of about 50 years and they have never picked up a resistor and measured the LC factor on it. Huh?

1:16

[Question. This problem presents itself very clearly in radio work where you try and dump a lot of power out of a transmitter to load. If you use a large, say 50 ohm wirewound resistors, the inductance of that wirewound resistor will present a different impedance to your...] Yeah it messes up your impedance and your device doesn't work and whole slew of other things. Now since last year there have been hundreds of people that have tried to duplicate what I did. There's about 12, I guess, out of several hundred that were successful. The main reason most of them failed was the impedance problem. Because when they started putting capacitors and resistors and you know various other things in there, which controlled the timing circuit, so that the different coils could talk to each other, but basically they blew themselves out of the water. And none of them, or very few of them, ever bothered to check to see if there was something there other than resistance or capacitance or whatever. Or even in your L1 and your L2 coils. For example, some of the L2 coils I have, have like three or four hundred microfarads. You can change that very quickly. Basically what that tells you is that instead of having four hundred thousand cycles a second you got four hundred cycles per second. Which actually that's kind of interesting because that's down in the range where you can start running electric motors and such. So the potential exists there for going directly to the motors instead of having to go through several stages to get to them. Because a lot of motors run at 400 cycles per second. The ones that are used in airplanes and the controls of airplanes and a number of other places. They have a lot of torque in them. They're small, they have a lot of torque and a lot of horsepower for the size.

1:18

Whenever we talk about ambient and that sort of thing we get into a situation where as the frequency increases the magnetic lines increase. It's those magnetic lines of force that disturb the electrons which cause more electricity become available to you. That's basically what's happening on your L2 coil as you start from the bottom of the L2 coil and go towards the top of it. Each winding there has capacitance and other types of things. Those things start squaring as you go up towards the top end of the L2 coil. The net result is that you have a tremendous magnetic flux field going on at the top of that coil there. That also presents you with some real interesting possibilities. Because that's the end that you really want to get your flux off of in order to go ahead to your next stage.

1:19

A small coil we'll give you a for instance on it, say that L2 coil has a hundred thousand volts potential, and it's cycling's say, well it's normally going to be above 20,000 cycles per second. In order to ascertain how much energy you've got available to you have to multiply it by the number of cycles per second. You can see that with a hundred thousand volts output on a Tesla coil that you've got way into the high mega watts or mega joules. If you got a way of containing that energy so that you can get it on to the next stage properly, which that's your capacitor banks. A real nice capacitor bank for a mega Joule production deal, they are available.

1:20

I want to say off-the-shelf, that's not really true because there's no two of them that are alike. But I'm going to show you something in a little bit, and you'll see what what the size of them would be. Some of them are about the size of that podium there if you double the height on it, and that would be approximately one mega Joule. Its consists of numerous banks of capacitors that are in the say the two hundred thousand volt range and that sort of thing. You might say well how do you get capacitors in that range? Technically it's not as difficult as you might think. Your dielectric or your separation deal that separates the plates there. There are several types of things. Kapton is one of them. It has about seven to eight thousand volts per mill.

1:21

Several mills on that, you're beginning to move up into the voltage, and by the time you get up to say a sixteenth of an inch, or something like that you got real high voltage. Kapton, it looks like mylar and it's kind of amber color, and DOW chemical company makes it. A lot of people sell it. It's not always used for capacitors. It's used for a lot of other things. You could put printed circuits on it and things like that, and you can build a radio device that you can roll up inside something the size of a pencil that's equal to a large color television or something like that. You print your solid-state devices right on to the circuit, and on top of that, you put another sheet of it and then you press it, so that they seal against each other.

1:22

And the net result is the whole circuit is inside this very high voltage dielectric which doesn't really–all it does is just protect the device.

In transfer of energy you have what's called *Volt-ampere reactive*. What I'm showing here [page 66] is the conventional power triangle in energy transfer and number C on there's considered to be wattless energy or useless energy, or whatever, and that's your volt-ampere reactive part of it.

1:23

You're A part there would be volts times amps, which is electricity that's in the wall, that until you ground it by closing the switch is not useful. And at the time you ground it or run it through a device to an earth ground, or whatever, then you've got number B there. So anyhow, in this power triangle you can take a small power unit and you can power much larger power unit. An example that might be an atomic bomb because you use a small bomb to set off a larger bomb which in turn sets off the atomic bomb. In this case here you can use a small power system, and you can pump a much larger power system. And doing that you really don't have to pump it all that much because like I mentioned while ago, at a hundred thousand volts you're into the mega joules range already.

1:24

A hundred thousand volts is relatively easy to obtain using a Tesla coil. Even a small one can do that. The large ones you probably have too much energy and it's self-defeating after you get past a certain point. Because even if you are able to contain it, the containment problems would be so great that you'd be better off to try to do it at a lower level. Because you can use multiples of these units instead of one gigantic one. If you need mega joules, for example, one of these units that has about a hundred thousand volts output with the proper capacitor bank, and that sort of thing, can equal Hoover Dam which is about 500 mega joules.

1:25

So you've got something that will sit on a coffee table that will equal Hoover Dam. Or it's as large as maybe 2 or 3 atomic energy plants. Yet you're actually using a small amount of energy in order to activate a larger system which in turn gives you off your potential and the potential is captured in capacitor banks, and that sort of thing, and at that point well then you can process it further. I'm going to show you the formula that relates to what we're talking about [page 73]. Then we're going to run some case histories and let you see what happens from there.

1:26

If we have any questions as we go along here be sure to let me know. We're going to stop after a while anyhow and take questions. We've got enough material here to go for several days so we're not about to run out of things to talk about. So as you see Joules is a universally usable term for energy. An English brewmaster in his brewery had different kinds of machinery, including electric and steam and other types of things. He came up with a unit which was convertible into other systems.

1:27

Basically a Joule is a watt second. It's a volt times 1 amp times 1 second. That's not a lot, but when you get into millions of them it becomes a lot real fast. We're going to be seeing this same formula over and over again as we go through the documents. We're going to skip over onto something else and then come back to that formula. In the law of squares where you see this a lot is in nature. For example, in the case of a human being you take one cell and it divides, it becomes two cells, and then two become four, and four overcome sixteen. By the time you've squared that thing of several times well you're into a million. This same thing applies out into the universe, in a lot of other areas that we know about.

1:28

In our case the electronic flux that we were talking about, as you move from one unit electronic flux, to a larger unit of electronic flux, these things are squaring. As the flux squares the number of electrons are being disturbed is squared. The net result is that you end up with a lot of energy that you did not have there previously. So the people that have a problem with overunity, basically everything that is useful is above ambient so everything that is useful is overunity. Whenever you start squaring it again, the laws of nature dictate, not only in living creatures, but in the field of electricity, and other areas, we see that it doesn't take very many of these leaps until you have a very large amount of electricity looking at you.

1:29

As to the magnetic force, it's governed by the law of squares again which means the square the distance times the point that you're reading [Question. Inverse square law.] Yeah. If you move your things away from it. Now you get around that to a certain extent. I'm going to show you a device, after a little bit, that works like a radio station. Where, if you have one coil running at a certain level, and you have other coils in the general area if they're capable of responding to that frequency then they will duplicate exactly what's going on in that first coil. So you can have a one coil that's running in a certain energy level and you can have dozens of other coils sitting around it.

1:30

They may be at different distances, but the fact that they're being stimulated by this central source they start reacting just like they were that original source.

We're back to our formula again there and we're actually substituting some things into it [page 71]. See I haven't added the 0.5 yet. You'll see that come in in just a little bit. All this information is in that booklet that I mentioned earlier. There's also some schematics and things which all along I've tried to get people to build their own device.

1:31

Here's where our point 0.5 comes in [page 73]. You're going to see the same formula, but it's going to be with some things added to it. I'm beginning to show you what happens and then I want to take a real big jump just momentarily.

1:32

I had mentioned earlier I'm trying to demonstrate what happens with the 60 cycle system and why the amount of the energy that comes out of it is not as high as it might be otherwise. That number of cycles per second is what you want to watch for. Because it's building or ratcheting up the power level for you. So we don't see a lot of energy on this thing, but we're going to very shortly see what happens. Where the same formula is going to turn up again [page 75].

1:33

You'll notice that what happens whenever you have a hundred thousand volts of electricity which I had mentioned earlier. When you start multiplying it by the cycles per second, and the other factors that enter into there. You can use much larger capacitors than I've used there. I have used a relatively small capacitance and you can switch higher capacitance in there, but your amperage is going to burn up your coil almost immediately. The moment you start raising that capacitance your amperage is going up tremendously and you may end up with four or five thousand amps almost immediately. It's obvious that the equipment that you have, including your capacitors probably couldn't handle it.

1:34

So basically what you would have, you would have hot spots on your coils which would end up melting off the protective coating and maybe even melting your coil. The capacitors would probably fail because the dielectrics and other things would be heated up so much inside that you couldn't build a capacitor which would function with that kind of amperage. So how you get around that, you have banks of capacitors in parallel, and none of them individually are loaded to the extent which would cause failure, but by the same token all added together you end up with a product that's very high. You end up with an awful awful lot of amps and an awful lot of volts. Where you see this in practice, the railguns which the military works with is one case in point.

1:35

All the universities have high energy departments and most of them have marx generators, and other types of things, which produce very high amounts of energy from a relatively small device. I'm going to show you some of those after a while, and you'll see some multi-mega Joule devices that are not very large at all.

1:36

We've got more material to see and look at and, that sort of thing, then can imagine. Most of you have been wondering what these things look like when you actually build your own. I'm going to show you some right now [page 159]. Those did not come out as clear as I would have liked for them to. You're looking at one there that's in the megajoule range.

1:37

Basically what you're seeing is a 12 inch PVC pipe. Top piece is 16 inches or so long. The bottom piece is about 16 inches or so long. The larger round thing that you see there is a rubbermaid drum dolly which I took the rollers off of. It happened to be 25 inches in diameter which was in the acceptable range for the L1 coil. What you're looking at down below, your seeing, this is an automobile ignition coil from a 1968 Pontiac. It's brand new however, the coil is. I've built a timing device which is built onto this particular deal here. It consists basically of an electrolytic capacitor that's, I think it's 4700μF.

1:38

It's not very large. It came from Radio shack. I think it was was maybe two dollars and something. There's a shunt across the two arms of the capacitor, the two arms of capacitor with a resistor on it. It doesn't take a very large resistor because your basic cycles per second there are going to be the multiplication of the resistance times the capacitance that you have. That works backwards so that if you had say ten thousand microfarads, and you had say, a hundred ohms, that's moving the decimal point in such a way that you'd actually multiply 100 times the capacitance there, and you'd see the number of cycles per second.

1:39

Now in this particular case here, since we're using an automobile ignition coil, we know that the RPM on an automobile, when it's running at normal speed is maybe 2,000 or more rpm, and you've got eight cylinders there. So you would think that the times off and on of that particular automobile ignition coil at its normal frequency would be about say two thousand or more times your capacitance. Your cycles per second there is going to end up being like say 14,000 or 16,000 or something like that. Which is not very many cycles per second, but it turns out that that's enough to make a dent in things. If you multiply that times some of your other factors you will end up with a lot of energy out of this particular coil. So this is one that you can build yourself real easy. Yes?

1:40

[Question. How much energy are you getting out of that?] Well this particular one I used–there's two computer programs which are available normally, and then I have a couple of others which aren't normally available, and I used every one of them, and ran the thing, and I came out with very very different answers from each one of them. This one was supposed to have come out 400,000 volts at fairly high cycles per second. Whenever I got it built it turned out that the capacitance by the large diameter, and the capacitance between those wires, it came out with a natural frequency of 400 cycles per second. Which awhile ago I was mentioning that it's possible to build this coils so that they will run down in the range where you can get electric motors that are set for that particular cycles per second.

1:41

Yes? [Question. What kind of wire are you using on there?] What you have to watch out for is the fact the voltage difference between two wires. So if they're 750 volts difference–protection on one wire, between the two wires you would have 1500 volts possibility that you could play with. So that you could have a differential of voltage going as you go up of 1500 volts and you'd probably want to stay at about a thousand which will give you a five hundred volt safety factor. In this case there's a hundred and thirty turns on that thing so it'd be a hundred and thirty times your 1000, say, if that's what you put on each turn of your L1 coil, which would be this one. It has three turns. [Question. Do all of you coils have 3 turns?] Yeah, this one has three turns of number-8 wire. [Question. #8 wire?] Yeah.

1:42

[Question. The L2 coil has how many turns?] The L2 coil has a hundred and thirty turns on it, and it's number fourteen I believe. Its multi-strand wire, because if you use the solid wire it's very difficult to control and make it look nice. The solid wire goes its own way, it's very difficult to make it. Sir? [Question. What type of wire?] What type of wire? Yeah, it's not truly litz type wire because to be litz type wire each individual strand, when there are many strands, has to be individually wrapped with a little piece of cloth or something to be a true litz wire. But it would qualify as a litz wire under the terms that we would use maybe today. [Question. Like grounding cable of ligthning rod?] Yeah.

1:43

Basically what happens is that these multi-strand wires are made for DC generally speaking, which any sort of resistance knocks the DC down real quick-like. So you're using a DC type wire which you're using it with AC. [Question. What are you using there on your secundary?] On the secundary it's a multi-stranded wire. I got that at one of the electronic supply houses. But I noticed that Ace Hardware and some of the lumber supply places have very similar wire. [Question. What gauge is it?] It's number 14. [Question. What voltages are developed?] On the secondary, that depends on how much you put in from the input source, and what kind of resistance you put between that and the other thing. Now there's some diodes and other things that are in circuit there that control what's going in.

1:44

So the output on this thing was, even after you go through all the commotion here, and that sort of thing, is like maybe seventy eighty thousand volts. [Question. What's on the top terminal block there?] Here's the detail on this thing right here. Well the spark plug is goes to the earth grounding. That is you've already passed through all the circuits, and your timing is all set, and that sort of thing. As a control on the load that's going to be built up, your potential that's built up, this thing, it goes to an earth grounding see. If you notice on the output coil here there's only one wire coming out.

1:45

Somehow you have to get back to your negative pole on your battery or a ground plane, or something like that. In this case here there's another cord that hooks onto this, and it goes to an earth grounding which your–there's an earth grounding that comes off of one corner of that high voltage coil there. [Question. On the terminal block, what is the small orange and yellow mechanism?] The small orange and yellow? Actually what I've done here, you can't really see it. But I've got a circuit that's going along here with a capacitor on it, and a resistor. I folded it back instead of having in a series thing looking at you that would go out like this. I've gone down skipping the contacts there and then I've come back to a certain point.

1:46

So I've got a resistor and a capacitor in that deal. It turns out the capacitor is about this long and the resistors about that long. So the net result is I loop back in that orange thing you see there is where I had a blank space that I had to bridge in order to get the circuit closed. That's what that orange thing is. There's nothing magic about it. Yes? [Question. Is that a capacitor?] Yeah, you're looking at ten thousand volt capacitor there. It's a reconstituted mica and it's building up. The blue one here is the negative lead coming back from the L2 coil.

1:47

The one that's white here is your positive lead going out. So between those two, in relation to your other things, you've got this ten thousand volt reconstituted mica capacitor. How that happens, they take mica and they break it up in little bitty pieces, and they mix it into a glass ribbon type deal. That becomes your dielectric in this particular case. They go to very high voltages. I'm going to show you something a little bit that go to over 250,000 volts. It's 0.01 microfarad at 10,000 volts. I think it cost about $200. The mica capacitors are a little bit hard to find and no one likes to admit that they have them.

1:48

Even if you find somebody that has them, they'll be looking right at it, and they'll be looking at you, and you'll be asking them, if you tell them what you want to buy, and they don't have it. In this particular case here this one came up from New Jersey. I had the listing on it, and I knew what they had. I called them and they told me that they had them on the shelf and that they could send me one. I said well fine I can use it as soon as possible. And they weren't doing anything. They were not busy at all. And they said it would take about six weeks. They could reach over and touch it, and it took them six weeks to reach over and touch it and put it in an envelope and UPS it to me. So anyhow we went round and around and I got it down to where it took me about it was a week getting to me after I located it. Yes?

1:49

[Question. Did you have any load on this device at all?] Well before you can put a load on them you have to put it into a capacitor bank and get your thing built up so that you have something to put a load against. If you try to put a load directly onto it it's going to detuning it and you'll end up with nothing. We have some more. That's that particular device.

Here's one [model 29] that I'm going to show you the schematic on in a little bit. The top one there with the yellow deal on it. Actually, this is going to be your L1. This is going to be your L2, and this is going to be an L3. You remember I said that the lines of magnetic flux on the distal end, or the busy end of the Tesla coil are very large in number. Basically what's happening there is that this is the off take that goes into the diode bridge that you see there. That diode bridge is built on the underside of a plastic case. One half of that plastic case, it's a little box that you build electronic projects in. Coming out, what you're interested in here. This is going to be your plus DC and this is going to be your minus DC. These are your two AC things going into the diode bridge. This particular one is set up so that you get 60 cycles at a usable level or you can just plug directly into it and in your off running.

1:51

It's intended to go directly to–without the capacitors and other things. It works fairly well but it would work a lot better if there was a nice capacitor bridge in there somewhere before you get to the area where you're using it. Now from this you go from your plus DC–you've come over to your center deal here where the jumper is. And that enters into this inverter circuit which is the device that you see here–it's what you get down at Radio Shack that you plug in to your 12-volt deal for running things from your automobile. It's 60 cycles at 120 volts. The resistors there, and those transistors you see, those happen to be very high voltage and amperage transistors.

1:52

You can see that they're on a heat shield tap deal which helps control them. If you run it for any period of time of course it would probably overheat. So in order to get a practical unit which you can use over a long period of time there's still some engineering to be done there. But the basic idea works. Yes? [Question. What are the small knob like devices on the lower picture?] This is a system I mentioned earlier that you could have individual coils sitting around that were tuned to a coil that was active. They will copy exactly what the other one is doing. In this particular case here, this is the one that the Tesla coil with your L1 and your L2.

1:53

It's the one that's hooked into the system. It's the one that's active. The other three that you see there are simply tuned to that center or L2 coil there and they do exactly what the center one does. If the center one has like thirty thousand volts on it, well the other three will develop independently thirty thousand volts, and they do not draw down on the power from the center one. It's the electrons are moving around and the natural frequency of those coils, actually they have variable mica capacitors. High-voltage mica capacitors on them. And they have also a variable resistor on it. So you can tune them individually. And how you know when they are tuned or when they're actually functioning: there's a little neon light that you get that's got one leg with a resistor about a forty seven thousand volt [ohm] resistor on it.

1:54

On one leg of the wire that's on this neon bulb, and then there's another leg there that just sort sticks out like an antenna. When this thing is functioning properly well that bulb lights up. So you know immediately. [Question. You said a 47K ohm resistor?] I measured one of them and that's what it came out. They didn't specify on the package. [Question. That's standard I think for a 240V neon indicator] Yeah, anyhow the bulb lights up very brightly. And you have no trouble at all, it takes 70 watts [volts] I think to light the thing, something like that. I'm going to show you a thing in a little bit where you can order–as a matter of fact it's the next slide. There's a company in Waldenboro Maine who manufactures a device [model 30] which actually illustrates everything that we're talking about here, with all the formulas and everything else.

1:55

So that if you're interested in studying it, you can get this kit. It's got the price on it I think there. Give you some ideas. It's not very large. It's a little bakelite tube that's about two inches or so in diameter. That's what you've got here. You've got a tuner deal on one of them. And the other one, it is already tuned. You've got a one and a half volt battery. The output lights this 70 watt [volt] neon bulb. So if you have trouble, that you know. If you're uncomfortable with what's going on there. Well this actually illustrates your overunity.

1:56

As a matter of fact it was one of the first things that I knew about. I'm a geologist actually. In about ten years ago when the petroleum industry went sideways, friends of mine try to get me to assist in the high school science department in this large high school that was close by. I did, and of course I ended up teaching physics and chemistry and a bunch of other things. The students in our particular neighborhood, we live in a upper white collar neighborhood. Most of the parents are all advanced degrees from the larger Eastern Universities. So you've got a bunch of kids, they're pretty sharp. If you don't defend yourself well you're in trouble real quick-like. I was very successful at it. Much more so than the regular teachers. This was one of the devices that I used to teach radio characteristics in physics.

1:57

That was the first time that I was aware of the fact that this overunity type thing was a possibility. That was more than seven years ago. I was very much into it, far advanced before. I knew about Tesla, I knew him, but I didn't really remember much about him. People started telling me well that's the sort of thing that Tesla does. So I started to getting every book that I could find on the subject. We have a neighborhood library that's on a computer system that's hooked into all the libraries across the United States. They didn't take very long to end up with books from the city library of New York, Harvard University, Caltech and just about everywhere else.

1:58

One of the books I got was purchased by Harvard University back in 1937. I was the third person to check it out. So you can see that the subject was not all that popular apparently. I didn't quite know what to think whenever I saw that. You know there had been only two people who checked this book out, and it was a very important book, and it was from a major university. So it seems like in this particular group here we have a hotbed of people that are interested in that sort of thing. But of people in general apparently they're not all that interested in it. Yes? [Question. I have a question about that other slide if this is a good time?] Which one you want to see, the circuit? Yeah, I've got it. It's coming up. It's in the stack here of things which are going to be showing you.

1:59

Yeah, we're coming back to it. As a matter of fact we have so many things to look at, it will kind of boggle your mind. [Question. What frequency range is that device?] That device. If you look closely at this one, this is running down in the AM radio range. You can tell that by the size of the coils and the number of turns. Just looking at it you can tell that it's going to be in the AM radio range. I think I got your phone number on there, and address and everything. I took this out of a science catalog, and Xerox it, and then cut the pieces up and paste them on another piece of paper in order to get this slide for you. It's one of the major scientific supply houses.

2:00

It normally has it, but this particular one is on the front sheet of the instruction thing which is about nine pages long, front and back, that comes with it. This is the bottom part there. The top part is out of the catalog. [Question. What's on the very left end of the board?] That's a 1.5 volt flashlight battery. Well if you notice what that 1.5V is doing you'll notice immediately there's something there that doesn't match up with what you would expect to happen. [Question. Well there has to be some circuit interruptor or something else that goes with it.] Well there's a capacitor in there. Yeah it's a small capacitor. Well actually, you can see there's a butterfly capacitor there that's 200 to 250 micro farad's or something like that.

2:01

Or pico farad's it's probably pico farad's, 250 pico farad's. Which would get you into the AM radio frequency with the particular coil winding. It's a tuned tank circuit. The other one, the second one there, is on a block of wood. You can move it around the room. You know we were talking a while ago about the coils being able to be somewhere else and they work just fine. If you decide to build yourself a Tesla coil.

2:02

One nice way to do that is to get one of these little push carts like two rubbermade makes there. Especially this one would be very nice. I think this even has the prices on it, if you want to see how much they cost. If you want to go first class, well that's one way to do it. Of course the insulation of the plastic push cart there is very helpful to you. You can build a little bit larger platform on the top of it or just use it the way it is. We're going to come back to that schematic in a little bit. But if we don't be sure to let me know. If we get close to our second hour ending here, also let me know and we'll take a ten-minute break. If you want to see what was going on in 1893 here's some circuits here [page 1] which are not at all different than what they would be today except that you have better capacitors and other things.

2:03

You notice there's not very many parts there. Tuning is built into these devices to a large extent. You notice on this particular one, here, this is your L1 right here. This is your L2 here. You're actually tuning this thing by pushing the L1 backwards and forwards in there, till it actually matches up with the tuning deal. Then you've eliminated the possibility of having to put capacitors and a number of other things. Because the capacitance is already built into the wire that's part of the coils.

2:04

So you've simplified and you've taken out some stages there which would create a problem for you unless you just happen to have a lot of nice test instruments and you're able to match up your capacitors, resistors and the characteristics of the particular coils, so that you end up getting what you want. That's your number one drawing. Number two drawing there is showing the capacitors combined with the output thing. The top one there is set up so that you can run a conventional transformer directly. If the frequency of the coil is something other than what you want, on that number two, just slightly above where the number two thing is there, you can put a resistor or a capacitor in parallel across that particular coil there. And you can bring the frequency down to where it match your transformer.

2:05
The bottom number two there is showing you how to light lightbulbs. Conventionally incandescent lightbulbs. Have you ever wondered how to light conventional light bulbs? You can light fluorescent bulbs fairly easy, you don't even have to plug them in. You can just hold them in the general area where this thing is activated and the fluorescent bulbs will light up. But to light incandescent bulbs gets to be a real problem and the thing there that makes that possible, is the capacitor. We're going to substitute that capacitor with a capacitor bank, in a little bit, that has in the megajoule ranges. At that point well we're off running and we can produce any amount of useful energy that you can imagine.
2:06
[Question. What's he doing on figure 2 at the top?] The top figure 2? [Question. Between the left and right coil, there's something in between, what's that?] The tall thing that is kind of in between there? That's a capacitor or a capacitor bank. These are actually Tesla's drawings. You're looking at something that Tesla himself actually drew. [Question. What's the spark gap actually for?] The spark gap is actually controlling the level of energy that's passing through there. You can do it without the spark gap. I'm going to show you some fancy spark gaps in a little bit, they can be used in a real high voltage. You know commercial type spark gaps.
2:07
Come down to number three there, that is a device which I've built a model of and demonstrated last year. What it amounts to is that you've got a transformer. You've got your L1 coil and you got two L2 coils there. The base of both the L2 coils are hooked together, and a ground actually comes off of this thing. You put a timing device on that ground there so that these two coils are activated for 60 cycles. One of them is wound reverse winding. What it does on part of the cycle when you get your conventional cycling, and doing this sort of thing, well one of these coils takes it up like that and then the other takes over and takes it down like that. So you get your conventional wave, sinewave deal at 60 cycles directly from a Tesla coil from base one.
2:08
Where you go to from there, depending on the amps that you want to be bothered with, you use a very heavy wire maybe a size four or something like that. Or maybe even size 2 or 0 or something like that. You make these things quite large and heavy. Then you make your L1 coil out of rather heavy stuff also, and the net result is when you pump the capacitor bank with this, you've got so many amps that you have trouble controlling them.
2:09
[Question. This operates at 60 cycles per second with no iron?] Well remember now I said that on the grounding wire, that center one in the middle there, it has to go to an earth ground, and you have to put a timing on that earth ground which will pulse across the two coils that you see there. Actually you have to do a hundred and twenty cycles because 60 cycles up and 60 down, and the net result is it that adds up to 120 cycles and you come out with the actual full cycle as 60 cycles. [Question. What is that upper half of that, is that a toroid?] Well those are wound

somewhat, you see the layers of wire going across it. It's not a true toroid. Exactly how to describe it, it's layers of windings on top of layers of winding. On a true toroid you got... Yeah, well you know what you've done you slice through the center of it. You're looking at like the inside of a cake or something.

2:10

Yeah, it's round coming out this way and it's round going back that way. And it's got a round L1 that's being pushed in and out of the center there, which is tuning it at the same time. No the winding is in each case, it's winding and it's coming across, and then it's going back and coming across, and it's layers on top of layers. This same wire that you see right here is the same wire that's up here. It just went around the thing.

2:11

We go to number four there. It shows another circuit for lighting lightbulbs. We go to number five and what you see there basically you know that Tesla had this thing rigged up where he could walk around the room with light bulbs and they would light up. You have grounding plates on each side of the room. You have this high voltage generator here, which is running this capacitor deal. Which in turn there's got a step up into your conventional Tesla coil-like arrangement. These coils are actually not the way you see them, they're wound around each other. Of course that's going on and that's causing the flux lines of the room to be so heavy that the light bulbs are lighting up without being plugged in.

2:12

This immediately will tell you something interesting. If a generator can do this, and this is your Tesla coil here. A generator and an electric motor are really basically the same thing. It's just a slight difference in the way they're put together. So if you can generate electricity, and put it into something like this, then you can reverse that situation and you can run electric motor. So this same circuit here if you do the right things with it, will run your electric motor for you. That's basically what's going into the electric golf cart that the California electric Automobile Association sent me to play with. A group of people out in California are ready to start manufacturing electric automobiles and they want to use the device. It's not very complicated.

2:13

We're going to come back to the schematic that you wanted. Just to show you for sure what kinds of capacitors. Now these are just some of the capacitors that are available. If you read across the.. I'm trying to line it up for you. You notice at the top there where you've got kilovolts and kilojoules, and that sort of thing?

2:14

These are capasitors that you got to be real serious to get involved with. I'm just showing it that they do exist. [Question. What's there primary use in conventionel technology?] These? Universities like to use them to build for their plasma experiments. The military likes to use them to build rail guns. [Question. I notice that's there no manufactures name on..] Well it's Maxwell. Yeah, there are a number of people that make them, this just happens to be one of them. I wanted you to see that they really do exist. The dimensions, if you want to see the size go over to the inches column there. When you put these things in parallel in banks you can get a mega Joule real quick-like.

2:15

They're temporary storage that sort of thing. [Question. With these capacitors, you connect them all in parallel?] Well just like you would batteries. If you want amps out of batteries you connect them in probably parallel. [Question. How do you reduce the voltage?] How do you get it down to where you can use it? We haven't got to that part yet. We're coming. That's a good question and it's one that's going to be answered.

2:16

This is what one of those capacitor banks looks like. We're going to see some more of them. This one is one kilojoule I believe. Have you spotted the capacitors yet? They're there. That's all these people do, they build these capacitor banks. They do a booming business worldwide. You'll see some more in a little bit that will probably look more like what what you're thinking that they should look like. What you're seeing here mostly are spark gaps and the bank. Those are a different kind of capacitor. There's a picture of one of those. You have the smaller ones and the larger ones.

2:17

This is an engineering drawing, of those small biscuits looking things, that you're seeing there. Remember the things that look like biscuits that we're kind of hooked together. Those are banks of those things. So that's got the dimensions on it, and the physical appearance and everything else, if you're used to looking at engineering drawings. In this particular group of things, we don't use the Tesla coil, we use a Marx generator. It doesn't take very many stages of this Marx generator and it produces actually in the amps and everything there to begin with.

2:18

You get very exceedingly high amperage. You get like maybe into the thousands and thousands of amps plus the voltage with this particular type of device. What you're seeing here, these are engineering drawings of a spark gap. A very high amperage spark gaps. A Marx generator is composed.. This is a schematic drawing of one of one of these things. So you see basically a Marx generator is a group of spark gaps, very high voltage and high amperage spark gaps, with capacitors. High-voltage capacitors in between and resistors. I'm going to show you an actual photograph of one of these things next.

2:19

Where we can see it better. There's other kinds of spark gaps now. This is one of the type that we're talking about. See the spacing between there and there? It's also got this center deal which prevents an overflow if you reach proportions which are dangerous to destroying your spark gap for you. While that thing partially unloads it does so that it doesn't destroy your spark gap. Because you go to extremely high voltage of amps almost immediately. You don't go through very many stages. So the time you've gone through the the first stages or so, well you're into very high voltage in very high amperage. This is doing the same thing that your Tesla coil does but in a different way. The amps are coming at you.

2:20

I mean to tell you there there there and there you've got oceans of them. You're going to see some more of them. You're actually going to see a complete Marx generator sitting out looking at you and that sort of thing. You can actually build one you know, like that connector block that I showed you that had the capacitor with the resistor for the timing circuit. If you get a large connector block like that and you get to your capacitors and diodes, and other things to go with it. Well you can build your Marx generator on the little connector block directly, and have a small-scale one that still does an awful lot of damage.

2:21

Several of you wanted to see that circuit for the Marx generator. There's a German fella came up with this and back before the war. The basic circuit is in the upper right hand corner, and it consists of one of those spark gaps that we were looking at. It consists of a capacitor and resistors and other capacitors so as to control the level of energy and direct it, so that most of it goes in one direction. One building block there. By adding additional building blocks on it you up the voltage and the amperage and it doesn't take very many of them to get into the very high level of energy.

2:22

This is used largely in plasma studies, university level plasma high-energy studies, and in the railguns and a number of other things. [Question. It also appears that you control the frequency of the spark gap.] Well, you can by changing the spark gap adjusting. I'm going to show you some in a little bit where you can adjust all kinds of things on them. You can control the level of energy that's going forward. They control the amps and the volts and a whole slew of other things. Depending on how sophisticated the spark gap gets. This one is basically a off-the-shelf sort of thing where everything is already built into it for particular uses. You'll see the thing I'll come together just a little bit.

2:23

As to how the things go. I'm going to show you a different kind of spark gaps. I'm trying to get it to where you can see the whole thing. Where you can see the dimensions. Those balls that you see in there are carbon balls. And the reason for that is that they don't evaporate like your metal ones do. They have a much higher capacity for electrical current and other types of things. So you'll see that there's all kinds of adjustments on these. For example, about screwing these and you can change the length on it. On this one here, where we have the three deal, one of these is a ground.

2:24

One of those three would be an overflow ground or one that would control the level of current so that you have a great deal more voltage and current coming through there then you want to go forward. By adjusting those things you can adjust it to the level which you want to go forward. The size on these things is not particularly large. You see in inches down here the description of them. These are actually smaller than the other one. The first one I showed you there was about nine inches long. It was quite a bit larger than these. You can use these with your Tesla coils also. Any questions on these? You see these are made to go up to about

370,000 kV, and that is way beyond what you need to get a megajoule or a megawatt.

2:25

You could almost go into the hundreds of thousands of megawatts with something like this. We're going to see a Marx generator, or what it actually looks like. That thing in the center there, that's round that's got the device sitting inside it. This is a Marx generator. This is a large metal tube that has a cap that goes on the end of it, and you fill it with oil. Very much like the oil that's in your automobile. The reason for that is that a Marx generator has a lot of plasma display going on around it. You have the ionization of the air and other things around you in such a way that it will interfere with a lot of things that will short out the Marx generator.

2:26

You create all kinds of interesting electrical problems. So in order to contain that thing, and control it, you put it inside a tank and you fill the tank with oil. With that you can get up to million or several million volts fairly easy. From this relatively small device. The measurements on this–this is about a foot and half right there. You can see that on the end of it. This one is a very complex one and a very high voltage one. This one probably goes up to maybe four million volts or more. So if we were talking about generating electricity like the number of cycles per second, and that sort of thing, and with the capacitor banks, and other sorts of things. 2:27

With the output of this particular device here–which regardless of your inputs–you have a tremendous amperage going along with your voltage. So you get a huge amount of joules from one of these devices. It's very difficult to control. That's the reason why you put it inside the large metal tanks and you have to pressure it up on top of that. It could be like four hundred pounds pressure. This is sulfur fluoride gas, which you can use either that, or you can use the shell oil that it's mentioning there. Or some other similar oil.

2:28

So the sulfur fluoride six thing is a gas. Yeah, it's a very highly insulating gas. I might mention something else if you get the dielectric constant of various things, your very highest ones that you normally see are like eight or nine or something like that. But it turns out that the dielectric constant for distilled water is about two thousand. It's so high that it's way out of range, which means if you use distilled water and you used gold plates and things in there, so that you wouldn't get a chemical reaction. That you could, by using distilled water, you could get voltage that you wouldn't believe. You'll see the dielectric constant here is probably maybe six or seven or something like that, or maybe eight.

2:29

[Question. Is that deionised distilled water?] Well I never actually done it myself. I just know from the University of Chicago physics thing on experimental physics that it lists is the dielectric constant for distilled water is 2200 I think. And it shows mica which is probably the one of the highest ones otherwise, it's been around eight or nine or something. There are some clays that I know of that go probably ten times higher than the distilled water. So depending on the circuit you're building.

2:30

Here's the back end there where the hinges are was where we were looking into this device just a few moments ago.

2:30

If you want to fix on the size on this thing, these wheels are basically probably about five or six inches in diameter. The Marx generator of course doesn't take that part to any of the inside of this thing. It comes that way because if you want to add additional stages to it that you can go up into the megamega-joules. The output on it? It would be through some of the things here. They have to be very careful on the output outlet on it because the sparking distance on something like a million volts there, it creates some real interesting problems. Of course the frequency it runs at, and a lot of other things change that.

2:31

Some of the test equipment that we're talking about that we normally use on high voltage devices. These things are voltage dividers. It depends on the amount of voltage they're made to handle. These come in units between thirty kilovolts and two hundred and forty kilovolts. They look very similar to the one you see there, except they get start getting longer if you get the higher voltage. They have certain kinds of filaments inside them, and they also pressure it up. You notice the pressure gauge here? When you've got a lot of voltage and current flowing, it also produces some heat.

2:32

So your pressure thing, they change on you while you're working on it. But the general idea on this is to get your division on your electrical output down to where you can measure it with your meters. The type of device that you would end up measuring with would be something like this. And I'm going to show you more details on that in a moment. This, with the proper deal like that, can go up to over a million volts. But the million volt deal is taller than this room here. This device, it's built up in sections. I'll show you one of those in a little bit. These are some of the other devices. You're off and on switches for high voltages.

2:33

Those toroids that you see down in the lower corner come up in sizes up to 52 inches in diameter. What they are they're capacitance end that you put on the busy end of your Tesla coil to keep it from sparking unless you really want it to. It's a capacitance device which you load. It helps you control your Tesla coil and the voltage that's on it. Now this is the same device. I'll have to show you. This is not the voltage divider that goes with the million volts. I'm going to show you one of those in just a little bit. This has the metal balls and the toroids and things like that.

2:34

Of course they come in quite large sizes sometimes. The man there with the switching device, that's a high voltage switch that he's working with. So whenever you move hundreds of thousands of volts from one source to another, you have to have a very special switch in order to do it. Otherwise the thing will vaporize on you. This is the instrument giving a technical description. It's the same same device we've been seeing all along with more technical details on it. Which you see it does go to a million volts

with the proper voltage dividers.
2:35
Here's a 200,000 volt one. You notice the 12-inch scale on there? We have a different slide that we see the 900,000 volt one. But you can tell that this one has a number of stages on it, and it goes to 200 kV. And this is 120 kV. These are switches, different kinds of switches. This is the instrument with a particular voltage divider that gives you a 200 kv peak to work with. Here comes the nine hundred thousand volt one.
2:36
See that it gets out of control real fast like as to the voltage dividers and the instruments that you have to use and that sort of thing. There's actually a much easier way to do this. Since magnetic flux and electrical flux are part of the same coin, it's real easy to measure the magnetic flux with the relatively cheap instruments. You know hand-size calculator-size deal. You can measure the magnetic flux then you can convert that directly into voltage. Some of those meters that read the magnetic flux actually read in voltage. So if you're say, 10 feet away, and you're reading say 200 magnetic flux if say 200 volts at that point, then you have to square of the ten, which would be a hundred and a hundred times your whatever and then you see immediately that that source has a rather large amount of voltage. You didn't get close to it. You didn't get shocked.
2:37
Yet you got an accurate reading on it, close enough for whatever you need. So it would be recommended that these high voltage devices, that we see there, are more of a nuisance than a help. They're terribly expensive. That meter deal that we're looking at there seven thousand dollars. The voltage divider things can run up to over a hundred thousand dollars real quick-like. And you can do the same thing with a $35 magnetic flux meter that's got a chart–you know it's calibrated in volts on the front of it. And by knowing the distance that you are from the source, and squaring that times your reading, you can read the voltage direct without even getting close to the device. Because it's a two-sided coin, if you have magnetic flux you have electrical flux.
2:38
They're both exactly related to each other. You don't have more of one than you do of the other. It's a exact copy now. Here's what somebody was asking me about.
2:39
I'm kind of going to open the floor at this time to questions, and we'll take off from this device [model 29, page 67]. This is the one which I showed you earlier that had the yellow L3 coil on it. That had the real nice-looking metal coil, that's real shiny looking. This is the one that you key on to. To kind of give you an idea as to what some of these things do there of course. Here's your gel cell battery. Here's a diode that you put on your plus side so that you don't get a kickback. If you get a kickback it damages your high-voltage module.
2:40
The high-voltage module is a laser high voltage source like you'd have in your laser printer or whatever. They're available off the shelf. And if you buy them direct, they're a couple hundred dollars or three hundred maybe, or something like that. If you're lucky

enough to find one in a electronic surplus place it might be as low as 50 dollars. Or maybe even lower. So needless to say I've got pretty good at finding the surplus ones. I've burned a lot of them up figuring out how to get things done. This diode here was one thing that pretty well saves me, it prevents kickback which destroys your high-voltage laser module.

2:41

It's on your plus side outlet of the battery. That does something else very interesting also. If the wire here and here is a component of the frequency which your six is there. It loads up like a coil, and the current goes towards the battery. The electrons are trying to go that way, to go that way, and the current is going the opposite direction, which goes to the gel cell. What is it that charges your battery is current. So it turns out that through accident or otherwise. This may not be true of all kinds of batteries, but I know in my case where I had been using the battery for more than a year and it hasn't changed one bit, consistently over a period of year. So that's a pretty good indication that something is going on there which doesn't match up.

2:42

Yeah? [Question. I was just curious, you mentioned earlier that that inverter, the output is ...coil going to that bridge.] yeah? [Question. You said that load reduces the output.] Well, what's happened there is that this 6a is detuned so that even though there's a tremendous amount of magnetic flux going on the busy end of the coil there, this 6a is detuned in such a way that it doesn't permit all the power to come through it. If all the power is coming through it would vaporize everything that's down strained from it. [Question. It brings it down to some voltage so it could drive that inverter?] Yeah, you've got voltage dividers. This is a frequency adjust deal here, if you need it. You may not need it. It depends on the input to the diode bridge.

2:43

The diodes which I use are a high voltage radio frequency diodes which show up in your television sets and things like that. And they start at about 25,000 volts and go up. And a hundred thousand it's not uncommon. So you've got say, four of them they're at 25,000 volts in the system. They're radio frequency. They'll operate very well say it 2 megahertz or 4 megahertz or whatever. The particular ones that I have used are 200 nanoseconds. Which that would be roughly about 4 or 5 megahertz, or if any one correct me on that? It will be about five megahertz? So that's high enough in your radio frequency so that if you multiply five million times your output there you can see real quickly that if your output was even low like a few volts, and you know partial amps, that all of a sudden you've got something that's much more potent than what would seem to be.

2.44

[Question. What size of resistors do you use for the voltage divider?] You'll have to determine those experimentally. Again, as we were talking earlier with some of the people, resistors are not necessarily resistors. If you use your LCR meter you'll find that almost 100% of your resistors have capacitance and inductance. A lot of times they're much better inductor or capacitor than they are resistor. In this particular case the deals which are used here are 220 ohms. They're 1% accuracy and that sort of thing. It turns out at 220 ohms they've got about 30 microfarad each.

2:45

And they've got some other characteristics which change it. But it turns out that–and also these transistors here have capacitance and inductance too. So if you're trying to balance up your impedance on something like this. You can be balancing all day long, and a lot longer than that, and still not get anything. You've shot yourself out of the water, unless you just happen to lock into it. It gets to be very skillful. Now the way you get around that, is that there are capacitors and inductors and resistors that are actually just that and nothing else. And if you go to your electronic supply and try to buy one, you get very interesting response. I can tell you, because most of them even though they've been in business for 50 years, won't know that. [Question. Is there some place to start with the values of resistors, like number elevens?]

2:46

On the dividers, these voltage dividers here? What those are doing are correcting the voltage, so that when you get downstream you've got what you need for downstream. You notice there's one on the plus side, one on the minus side. There's also a capacitor in here to take the ripple out. Because when you come through that diode bridge, you've got rippling DC. [Question. How much voltage are you dealing with down there?] When it comes out, well down here, at this point, you're at 60 cycles and 120 volts. [Question. Before capacitor?] Going backwards? Well you have to allow for some degrading, but when it reaches this deal here–the basic deal here is the inverter that you buy down at RadioShack. Which you plug in your automobile cigarette lighter and you get from 12 volts DC, you get whatever.

2:47

What I've basically done is tapped out of it and tapped on to this system and put those dividers in there so that everything is adjusted to what you want to have coming out at you. [Question. With a small battery input, how much output can you get out of that thing?] This particular one here I think it's about 12 amps and about 120 volts. You could get a much larger amount. You can get probably 30,000 watts or more out of it. [Question. How much input?] The input that's the 12-volt gel cell, so it's doing things, which unless you know what's going on there, can't be done.

2:48

There's a lot of things happening there and you would have to see a speech I gave last year, or the book which has got the details in it, which explained that. [Question. Is this something that everybody can build?] Well anybody should be able to build it. But it doesn't always work out that way. There's not very many pieces there. The source of those pieces is not very complicated. The only thing on there that would create much of a problem would be that number three. The rest of it you can get down at Ace Hardware, or your local lumber company, or Radio Shack. You would think that would be very easy but I can tell you that after I presented this thing last year that there were hundreds of people that tried to build it.

2:49

And there were an awful lot of frustrated people. Out of that several hundred people I guess, that I'm aware of that tried to build it, there were six to maybe nine that were successful. [Question. This is described in your book?] Yeah, and I can tell you that the PhD's were the ones that flunked out the fastest. The ham radio operators and people like that were the ones that were successful. Yes? [Question. Can you put the output of transformer to 12V converter and replace that battery?] Yeah, we can start all over again. You can put these things in stages and do all kinds of things. But you can actually have a much higher output than this. You can have up to say 30 kV from this particular device. [What kind of wire did you use for the L1, 2 and 3 on this one?]

2:50

The number 7 there is from Barker Williamson's in Pennsylvania. They make professional coil makers. They were the ones that made the real pretty silver coil that was in one of those earlier ones. They make them up to six inches in diameter and normally about ten inches long. But you can get them in greater lengths. The larger ones run into like $250 or more each, plus $50 setup charge. The smaller ones like the one I was using was about I think maybe forty-fifty dollars plus the setup charge. You don't have to use those, you can build your own though. [Question. What kind of wire for L1?] Well the other windings that you saw in there was basically a 750 volt multi-stranded wire like you would find down at the hardware store.

2:51

With a 750 volt rating. So between the two of them that would be 1500 volts. So as long as you keep the division there, so that only a thousand volts is on each term, even though at the top you may have a hundred thousand volts, but the difference between the two of them is say a thousand volts or less. The one that's next to it, and that one is say a hundred thousand. This one would be ninety nine thousand and this would be a hundred thousand whether you're okay. Some more questions? [Question. The spark gap on #5, you mentioned a mill thickness of paper.] I think it's 0.0025 or something like that. I think it's written on the number at number five there.

2:52

[Question. What's the duty cycle? Can this device work continuously?] Oh yeah, there's no moving parts there to wear out. [Question. Is there power dissipation any where in substanial quantity? The only thing I'd be concerned would be with 11 and 14.] Those voltage dividers, that's going to be tricky because I could have a saying a while ago the the resistors aren't necessarily just resistors. [Question. Given the assumption that you used pure resistance in those locations dealing with the wattages that you described..] Well the particular instances going on here we have to look out for about–let's say that we got–did I say 20 amps or 8 amps or what? [Question. You said about 15A]

2:53

Well 15 times 100 would be about 1500 wouldn't it, watts? So that means that on the resistors there between the two of them, you have to have some way of containing that. You have to work out the technical problem on that. For the short period of time that I've ran the thing, I haven't, you know, ran it over an extended period of time, just long enough to get a reading on it, well it works fairly well. I'm using a very fancy resistors. They're the aluminium ones they're gold plated with fins on them, and they're 1% accuracy and that sort of thing. Typically, you can get them up to about 500 watts or more. And they do cost a bundle. [Question. Well you kept 14 low, I noticed you got 2 of them marked 11.]

2:54

Well there's two number 11s and two number 14's [Question. But that wouldn't necessarily mean the two 11's are the identical in value?] That's not necessarily so. What you do, you position them so that by taking off in between them, you have divided the voltage. Part of that goes on, and only the part that you really want to keep goes off. [Question. Yeah well at this point the emitor-collector breakdown voltage of your 2 transistors become a factor.] Well these particular transistors are good to 2500V, there TO3's. [Question. You're using flyback type transistors?] They're unusual, quite unusual. As a matter of fact you'll spend quite some time trying to figure out where to find them. [Question. Numbers?] Well they are Motorola, and I do have some numbers, but I don't know whether I've got them with me or not. There's some other things I need to to show you.

2:55

There is one book which has your timing constants and a lot of other things, and it's the Sam's book. Radio Shack used to sell it, book stop has copies of it, and it has your basic circuits and other types of things, which you need in order to proceed. [Question. Could you just look at the ISBN# and read it for us?] Sure. Let's see the ISBN number on this particular book should be right here. I've got it here. The international standard book number is a 0-672-22469-0

2:56

With that in mind you should be able to get the identical exact book. There's some other books which are quite interesting, and it turns out you would think that there would be all kinds of books on resistors. There's really only two that are in print, and I've got about a half a dozen or more at home that are out of print. But this is one of the ones that's still in print. It's $27. And you'll have a terrible time explaining to the book people what it is you want because they'll be looking right at the name and the number, and everything–and I'll give you a nice ISBN number on that also. It is somewhat,.. as far as high-voltage capacitors, it's not particularly helpful. But it has a lot of other things in there that are.

2:57

The ISBN number on that is 0-442-01558-5 This is an inter-library loan. If you have trouble with getting books that you need–these things are terribly expensive some of them. You can spend 80 or 90 dollars before you even blink an eye. And this is one of the university textbooks on the subject which was commonly used and it has a lot of nice things in it. The way I work it my neighborhood library is fairly small but they have a computer system for inter-library loans. They're hooked into just about everywhere. Most of the books I get, nobody wants to really read. So there are very few copies of them out there, and they find the real quick-like with the computer. Here comes from City Library of New York, from Harvard University, from Cal Tech, Topeka and Southern or whatever, that sort of thing.

2:58

Some more questions? [Question. Were you going to go into those capacitor banks?] We still got some bunch more things that we haven't looked at. We were beginning to run out of time. I tell you what I'll try to switch those into the speech when I do it here in the next day or so. Try to remember to put those in it. We're getting pretty close to close down time. So let's have some more questions and a couple more [Question. The sources of the energy, where it's coming from?]

2:59

Basically this goes back to the speech that I gave last year here. Everywhere, all about you in huge numbers are electrons. These electrons basically came from the disintegration of the Sun, spewing out into the solar area, and became solar plasma. As the Earth spins through the solar plasma, the gravitational field and otherwise, pull these free electrons–they're not chemically bound electrons–into the Earth's atmosphere. The way you know where they are is by the level where your visible light starts forming. It's that maybe 50 miles or a couple hundred miles up in the air. [Question. In other words, the solar plasma which consists both of electrons and alpha particles..] It's got all kinds of particles there.

3:00

Most generally the electrons out-number everything else. Well basically, what happens, you end up with your shear zones or your Van Allen radiation belts. And various other radiation belts. Those are kind of shear zones where the electrons are being disturbed. So anytime you disturb an electron, when you push it forward, it gives off magnetic impulses. When it pops back to where it was it gives off electrical impulse. This can go on forever and ever. That electron is not diminished in any way. When you disturb the electron you're causing electricity. In a conventional system you've got your coils and magnets whizzing by each other, and that's disturbing the electrons.

3:01

That's where your electricity is coming from. [Question. Is there a particular frequency that's seems to best for coupling?] Not really. The frequencies which conventional generation has set at, are horribly wasteful, and very very inefficient. I don't know whether that was intentional or otherwise, but if they intended to do it wrong they succeeded. Basically you're disturbing the electrons and the magnetic flux lines are greatest up around 6a there. None of the electricity from the battery has ever passed through any of those coils. It's just magnetic flux has transferred and caused the other coils to react in proportion to the amount of magnetic flux that's there.

3:02

[Question. What kind of frequencies might be the best?] Well actually once you get the magnetic flux going, you can have coils that are not tuned to it, that are in that magnetic flux, and they will oscillate at the resonant frequency that they are meant to resonate at. So you can go from one frequency to another and still have your electricity there because actually the electrons are not going through your system, they're just being disturbed by the magnetic flux lines of another electrical system. And that goes back to those power triangles we were talking about, where you can use a smaller power triangle to excite a larger one. In your L1, you're number six there, say the small power triangle, then you're number seven is your L2, which would be a larger power triangle. That's just like in the atomic bomb we were talking about using a small explosion to make a larger explosion, which makes a larger explosion.

3:03

So that is if you want to call it is overunity. Because really, basically anything that's above ambient, is overunity. [Question. Duplication factor there?] It's squaring according to the law of squares, and you see that Fibonacci series. You see it in all kinds of interesting things. Like in human development from one little cell that splits to two, and then two becomes four, and four becomes 16. By the time you've done this several times your up to a million of whatever you start out with. You're flux units will become so many of them you can't believe it. Once those flux units are going then you can transfer this over to another power source and it won't diminish the one that's running. The one that's running will just maintain itself. But it will transfer–cause the other power deal to activate.

3:04

[Question. But we are under the assumption that we are tapping into an energy source here?] Yeah, well that's the whole story. It's the fact that the electrons when you bother them they give you a useful energy. And it's a beyond ambient and therefore the snowball, you've carried the snow up to the top of the hill, the little snowball. You've turned loose of it and it's rolling down to the bottom of the hill and it's getting bigger and rolling faster as it goes. [Question. Do you have ... to the mechanics of the ..?] Well it's best to think about that after you've got one up and running. The trick is to get one that's up and running and is behaving according to what a Tesla coil is supposed to do. Forget about all these other things or where you're headed. Just get the thing going.

3:05

Once you get that part done, well then you can start worrying about other things that you might need to do in order to make it useful. Otherwise you never get there. You get hung up on some problem along the way. Chances are you have a fair chance of getting there the first time if you, you know, if you're lucky. [Question. How does this relate to standing waves on the earth?] Standing waves? The earth has its own cycling and has its own resonant frequencies. As a matter of fact one of the offshoot inventions on this thing–you know how your MRI works, where you can take pictures of inside people without cutting them up? Well you have a very large magnet that's coming off and on. Then it's disturbing the hydrogen molecules. Then they spin out of position and when they spin back they give off an impulse.

3:06

You're photographing that impulse. It turns out that the earth is a very large magnet, and it's pulsing off and on. With a simple coil or wire you can do the same thing with the earth that you do with MRI and you can see the rock formations. You can see the tops and bottoms of them. You can tell whether there's fluid in it. And what kind of fluids in it. [Question. Telsa calculated the earth resonant frequencies very closely, he found out that they where slightly different.] Well actually there quite a bit more than what Tesla did. I'll tell you how you know this, and why. It turns out that the Van Allen radiation belts are one side of a big capacitor. The surface of the earth is the other side of that big capacitor. So the Van Allen radiation belts are–the oscillation there is say between two thousand and eight thousand cycles per second, it varies depending on the time of day, and a number of other things.

3:07

So the surface of the earth starts out at whatever frequency the radiation belts are running at. And it goes down from that as it goes inside the earth. The capacitance picks up, and that slows the frequency down. So by that, by tuning in on these different frequencies as you go down, you listen to the noise level and that tells you what depth you're at. [Question. Have they ever done any calculations on the possible interplanetary charge density differences?] Well it's very obvious that all your galaxies and planets and everything else all have an immense electrical charges on them. That's what's holding them pretty much in place. [Question. If you take a sphere as the earth and you get a large charge on it and you take another sphere and put different charge on it, you end up with a large capacitor.]

3:08

Well, actually whenever you get hit by a meteor or comet or something like that, they're coming in, they have a different electrical charge on them. As a matter of fact, part of that is the reason why I say the space capsules coming in they heat up. The friction, the electrical difference there, causes the friction. You know the electrons are banging against each other at such a rate that they're generating heat and that's where your heat comes from. We're going to have to close it down. Maybe one more question here and we'll do that. Did I leave anyone out? [Question. Are you speaking later today?] Not today, but if you look on your schedule for Mark Hendershot there, I'll be giving a speech in his place and I wouldn't be surprised if they don't ask me to repeat this since we didn't have a real large turnout.

3:09

Last year we had this room and that room all together filled up on this deal when I gave it. [Question. Do you know Mark is going to be here this week?] I really don't know why. I asked why he wasn't here, and I think someone said he wasn't feeling well or something, I don't know the details. I wouldn't know the details. Yes? [Question. You said quite a few people built these and they had more failures than successes.] Yeah, and the people got real upset with me. They call me on the phone and just talk for hours of time, and were terribly upset. You don't know what you're talking about, you're naive, bla bla this and that. I said well let's see if we can get it right and then we will talk about all these other things. [Question. If a person has done it.. it should work shouldn't it?]

3:10
Well if you're lucky enough to get all the things in the right spot, it'll work the first time. Well thank you and I'm gonna put this up and this is the end of the speech and we'll talk about it a little bit. Thank you.

Video 3, 1996

Don is from Spring Texas, which is a suburb of Houston. He escaped Houston to escape the crime. The crime followed him anyway. Which is what it seems to do in those parts. He's an engineer by background. Works as a petroleum geologist for many years. About eight years ago he got out of that field. That gave him time to start working with physics, and free energy, and chemistry, and things he really wanted to do. Sometimes he works with the local school systems down in Texas. He teaches physics and chemistry, and other subjects for the kids there. He's doing quite a job.

1

He's been investigating various aspects of free energy, which he's going to give you a big presentation about today. So please welcome Don Smith.

Thank you very much. You're not going to be disappointed today. Since we met last year I realized that there are a lot of people that had trouble understanding things unless you could go turn the wall switch on and that was the extent of it. So this particular speech is built around that particular aspect of it. Plus the fact that I'm going to give you a lot of backup information which you would probably never in a lifetime find, if you went looking for it yourself.

2

You just have to kind of fall into some of this stuff with dumb luck. And it seems like every time I've gone around a corner there was something waiting there which was what I was looking for. Now those of you that have seen the article, and the recent or the present issue of the Tesla magazine, will have a lot of the theory that's here. I will sort of bounce through that. The speech is not written so we'll have to work off of that particular aspect of it. What I'll do is go through some photographs here, and if we can dim the lights down, let's see what we can do with them.

3

I'm going to walk you through the different parts on this [page 66]. This is actually a commercial model. It's 30kVA, and it's not very large. It's a little bit heavy, because of the copper in the transformer. But the transformer is driven by an inverter, which is here. Which runs of 12 volts and kicks it up to 112 volts.

4

Then in turn comes through a variable deal that is very similar to what you have on the light dimmer in your house. Except that it won't work in this case. It's a special dimmer that's made for a neon tube transformers. You can vary it from one volt to whatever you have, like a 112 volts. So that means basically that the forward going part of it, going into the–you have a couple of 200 amp diodes here, one on each side. There's one here and then there's one there. In this particular case the diodes are not critical, they're silicon. But they're good for several thousand volts and 200 amps.

5

It goes from these diodes–it becomes pulsating DC at this point. One diode will give you a sawtooth pattern. The second one there kicks in and cause the other one to kick out at certain points. So you have a continuous pulsating DC on a straight line instead of a sawtooth type arrangement. That goes to these 30,000 volt capacitors that are there. They're set up to take hundreds of amps. They're lifetime capacitors, they're self-healing. I'll show you in a little bit the reading on one. This is the transformer, and it may be dark, but you can possibly see what's going on.

6

The closest to you here is your inverter. In this case it's about 240 watts inverter. It goes into fifteen thousand volts neon tube transformer. The neon tube transformer, on each end here, is kicking out the same amount, where those knobs are. There's one on each end. Each one goes to one of the diodes. The earth ground on this, is here. So this fulfills something, which we are going to talk about after a little while. There are many ways to wind coils. You can wind them where you have voltage only, or where you can have very little amperage or where you have 100% amperage and very low voltage. The way the coils are wound will determine what you get.

7

It has nothing at all to do with the fact, that what you're putting in, is say strict voltage to begin with. It's going to come out, after it goes through a certain type of coil system. With this center ground thing here, you're guaranteed that you're going to get amperage. We're going to walk around the commercial model and look at it from different angles.

8

You see the 200 amp diodes sticking through these cooling fins. You see the take off from those two, going up to the capacitor. So it's going to be a lot simpler than the one you're looking at on the floor here. Here's the same thing from a slightly different view. Below here, you see a radio frequency high voltage probe, that goes into a meter that measures high voltage at high frequencies. There's about twenty of these sheets and then we'll go to some other things.

9

See those? We're gonna take a look at one end of them now. If you got your glasses on, you can gather a good bit of information from this. You notice the point of origin there is Finland.

10

They give some idea from where those capacitors came from. They're self-healing if you notice, and they also have some other characteristics. It's got the kVA written on there if you can read it. But the combined kVA of those two capacitors is about 30,000. Your input is about 12 volts at maybe 30mA or something like that. That goes into the inverter. The inverter is tuned back so only a tiny bit of that is going forward. So whenever you come to the output you've basically sent forward maybe 2 or 3 volts and you're getting back say 30,000 kVA.

11

So the computer program that I like to use with this type of situation, won't work with this one. But it will work with that one down there. We'll talk about it in a little bit. But I sit down at it for days. Run computer programs on these things, before I even start building them. Then I correct as I go along. I start taking readings as I go along. Then I start building corrections into the device. We'll see that in some detail a little bit.

We're going to the one that's on the floor, down in front of us [page 154]. Let's see what happened here.

12

The parts of this particular device here, up to this point right here, is what you're already familiar with. There's 2-3 things, that are built into this which are new and different from what you might be thinking about. Here's the inverter again. A smaller sized one. Here's the variable device on the end. It's coming into a neon tube transformer which is here. That neon tube transformer has two high voltage outputs. Those two things are brought over to here, and underneath here, and they're joined together by a couple of diodes. So the high voltage output is going through this one right here, and coming in at the base of the coil that's inside the spring there.

13

It's coming out at the top end, and coming back this way, and going to here. There's a lightning volt protector here, that's serving as the case of the spark gap. So there's not a spark gap as anything that you would recognize. Once you get over to this point here, two things have happened. One–I calculated the L1 coil which is on this sliding 2-inch PVC tube inside the 3-inch coil. It was calculated to be one-fourth the length of the wire that's involved in this L2 coil, or the outer coil. The reason for that is that coils have both capacitance inductance and resistance. All those things play part in the tuning of the thing. But since you started out with about one quarter length of the longer coil, you're pretty much tuned. You're almost tuned immediately.

14

As it turned out–of course, you have to put a capacitor on one of the two coils. The easier one to work with is the L1 coil over here. You can put a capacitor in there that will correct for any slight difference that might be. Those things are usually into like 1/10 of a micro farad or something along that line. As it turned out, when I put that thing in and then I took a reading on the circuit, well it turned out that I had a 0.5 6000 volt capacitor that I could put over here. And I had an exact tune on this thing. Even though sliding this coil in and out is meant as a tuning device I could have tuned it by doing that. As it turned out the thing was exactly tuned without even moving the thing by using the particular combination of capacitors that I used.

15

With that in mind the next thing is that in order to have useful energy, you can't go directly from this coil system and use it. You have to have some way of storing it. The storing options are basically either battery or capacitors. And it turns out capacitors are–we'll come back to that later. Capacitors are probably the best option, because they can be built to where they're self healing. They can be built to where they have over 40 years of useful life that's known. They've tested some of them and they know that they will last 40 years without any sort of damage, if used in the proper manner. So a battery has a useful life of about a thousand cycles. So if you're going to use a battery system, you're going to be changing your batteries pretty often, if you're going to be using a lot of electricity.

16

Because you're going to be cycling them quite often. The capacitor however does not wear out. It cycles at millions and millions of cycles a second. It goes well beyond 40 years. So that's the correct option, and I'll show you a chart later on from one of the battery manufacturers, that backs me up on that. We're going to another photograph here. We're walking around this one. We're going to walk a sort of a circle around it, and see what's there. We were talking about the variable neon transformer deal for making the tubes brighter or dimmer.

17

It's made different from the dimmer switch that's in your house. You can switch the two, but you've probably come to grief if you use the house dimmer. These things are maybe thirty bucks a piece, depending on where you're at. Of course here's the neon tube transformer and it's built with an earth ground coming out on the other end there. So it's center-tapped, which means it has a capability of making amperage for you. Without that center tap, you wouldn't have the amperage available to you. As a matter of fact, as you lower the voltage, the amperage goes up on this thing. So if you run it at less than 110 volts on this inverter here, and then you back it off again with the variable deal, you're going to end up in some cases with more amperage than the device can handle.

18

You can melt it. You can see the outside coil here. The center part of it is missing some turns. I took the turns out of the center part of it. Basically the electrons are moving this way, and then the tail back here, that hooks the two together, it goes to an earth ground, is where your amperage comes in. So whenever you take your amps and volts apart, and you start bringing them back together, you get volts times amps equal watts. So that's your useful energy, and you do this as you go into the capacitor bank. So coulombs and watt seconds are all basically one Coulomb. And one volt second is basically one volt at one amp at one second. So there are all the same thing.

19

[Question. Neon transformer, is that solid state?] This one looks like it's solid state but it performs as though it's not. Being solid state, I'm not quite sure what it would do to it, but I've used them, and for some reason they work. I don't know what the reason is, but I do know they still work. This one looks like a solid state because that's a 9000 volt transformer, that small one, that's there. See there. 9000 volts right there. If you look at the label on this thing you're in for a surprise. Because they tell you there's so many volts in and at so many amps, and so many volts out, at so many amps. It turns out that this device right here is an over energy device, very noticeably.

20

Because if you look at the information on the one that's down here, you're looking at the one that's on the floor. So if you take the numbers off of the thing, the volts in and the amps in, and the volts out and the amps out, that little device has produced noticeably more energy than it had going into it. So it's already beginning to behave as an over energy device. They actually stamped it on the label. They don't say that it does that, but the guys that are curious can find out for themselves if they want to. We're going to move up closer on some of the photographs now and take a little bit closer look at some of the components that are on the device, that we were just looking at.

21

You see the L1 coil here, it's three-inch Baker Williamson coil.

22

It had the four turns taken out of the center here. They're made into a pigtail, which forms the negative ground out. Or the amperage end of the deal. On the ends here, there coming off, and going out and coming through–this is a series of very high-voltage capacitors that are normally found in flyback transformer circuits in televisions. So we're into the very high voltage on those diodes. Basically what they're doing, is what the two diodes were doing in the other device. They're taking a sawtooth pattern and changing it into a flat pulsating DC at very high voltage. It's coming out of there and there's two outs here.

23

One is flipping around and coming across here, to here. Then it's going to the output over here. Which, there's a coil system. You see a double coil system there. It's going through this first coil system. At that point it's smoothing it out, is what it's done. Then it's going on to the output block which is back there. At that point you're ready to go the transformer. This wire here has very interesting characteristics. It's extremely high voltage wire. It has this thick layer of silicon over it, and then it's got a another envelope over that on the outside that looks like clear plastic. Some of these wires that I use have over eighty thousand volts breakdown on them. In the bottom one down here is the same type of wiring, same manufacturer, but it's a different color.

24

What it does, is basically pick up–it doesn't come over to the device here. It really stops on this end of the device, and goes this way and then it becomes the output on the negative side there. So you've got your very high potential one, and then you've got your other–you've got two of them. In the sense of plus and minus, which you normally think of, you really don't want to think of it here. You want to say that one area is higher pressure electricity and the other area is lower pressure. Since the electrons are in an awkward position, they will try to get back to ambient. In the process of getting back to ambient, they're going to be doing the things that you want them to do.

25

It's just like water that rains on the side of a mountain. It's going to find its way back to the ocean. Once you get the energy balance working for you then you're on your way. We're looking at it from another angle. You can very easily make the complete system out of a three-inch piece of PVC. The one that's on the inside here, is a two inch, out OD PVC. It's got the windings for the L1 coil on it, as you can see, through it.

26

The reason for putting it on the slide deal is the fact that, that becomes a tuning device if your device is not properly tuned. You can build without taking any measurements or doing anything. Simply by calculating the length of wire that's in the L1 [L2 was what he meant] coil and then taking one-fourth of that and measuring a piece of wire and then using that for your L1 coil. You will be very close to the tune, almost going into it without any capacitors or anything. But you're going to probably want to put at least a capacitor either on your L1 or your L2. It's much simpler to put it on the L1. But if you put it on both of them, of course you're kicking the voltage up in both cases. It takes high-voltage capacitors which are a little bit hard to come by. The ones that I have here were custom-made for me by Cornell Dubilier.

27

You won't be able to find them in catalog anymore. They're custom run. The diodes are also not available that I used, except on the large ones. The 200 amps ones, those are commercial variety. But the diodes on the machine that we're looking at here, there's no way on earth that you will probably end up getting those diodes. Unless you get them for me, I got quit a few of them. You have to order a large number of them in order to get a run on them. They stopped making this particular diode back in 1993.

28

You see something here which people say can't happen. If you notice the double coil system. This is the active one going through here, and going out to the transformers. This one is not hooked into the system, except magnetically. It has the same activity as this one, even though it's not hooked into them, due to the magnetic flux working on it. So what that tells you is that this separate coil here is not using any energy. But it's generating energy which you can charge your batteries with. So that your batteries never run down.

29

But this is one of about half a dozen different ways of doing that, recharging your battery. The way which I normally prefer is, I calculate the length of the wire from the battery to the input into the device so it's one quarter wavelength of the L1 coil, and for that reason it picks up the radio frequency. With the kick back diode there, it forces the high frequency electricity back into the battery. The high frequency electricity is charging all the DC that's coming out. Actually it's doing a better job of charging than the outcome. So the batteries–some of the ones that I've used I've had for more than two years, and used them continuously with almost everything imaginable. They measure the 12.97 exactly–like they did two years earlier–today.

30

I may not completely understand why that happens, but I know that it does. [Question. Are your batteries lead acid?] Their calcium lead. They're sealed batteries. They're a very common sort that you find. They're not one which spills out the material. The idea that they can only be recharged 100 times apparently does not necessarily apply. We'll come back to that a little bit later. But the battery manufacturer says that that's what will happen, but I don't find it happening.

31

We have another one here. You can see that one, it's a little bit better shot of the piece of PVC. Of course I very carefully sanded these things, so there's no markings on them. It doesn't look like PVC at all if you notice whenever you see the real device. It looks like it was meant to be what it is. Again, we're looking at the neon tube transformer with the two high voltage outputs. Then we're looking at the block here which has two high-voltage diodes here, and has these capacitors here. The capacitors are between the–where these two come together and they pass through the diodes.

32

It's between that point and the negative or the earth ground which is on the end. There's a little screw on the end of this block which is the center tap or the earth ground deal. That's the thing that ensures your amperage, whenever you get to that. We'll go into the various kinds of coil windings in a little bit and see what happens there. You can see the block and I don't believe that you can see the two diodes, but they're there. High voltage diodes and there's one on each of these outlets to the high voltage side of the transformer.

33

Once it becomes DC well then I tie the capacitors across that DC. And where the earth ground comes back through here and where the lightning arrestor is, where it returns–this is your earth ground right here. Where it returns to the transformer. This is another angle of it. This is the variable deal. The plug socket, the inverter, the lightning arrestor in the earth ground. Which has to be tied to an external earth ground. Here's the point where it goes forward into the L1 on the bottom side.

34

Here's where it comes back from the L1 on the top side, into the earth ground. You can see where I cut holes in the device here. The bottom one goes underneath and comes up to here, and then comes through and then goes around this thing, and then goes back inside and then comes back through this one to the earth ground.

36

We'll zip right on through these and we've got so many other things to talk about that you can't believe it. One of the things which we're going to be seeing is this device. It's a 1/4 size version of the one that we've just been looking at. It's built in this suitcase. It actually totally takes up the top half of it. The suitcase is mostly empty. That will give you some idea of the size. It was just a little bit too long to put into an ordinary briefcase and that's why I went through the suitcase.

37

But the suitcase has wheels on it and rolls and has a handle that pops out. That makes it real easy for me to carry with me when I go overseas and other places with this. You can see the handle. I guess, well, maybe you can. We're going on to something else. We have so many things to look at that you can't believe. There's it standing up in the den at home. I have four of these fans that I normally use, but due to space limitations, I only brought one.

38

These are some of the tools sitting beside the large capacitors that we were talking about a while ago. They came from Finland. Yes sir? [Question. What does that suitcase do?] This particular one, how much output? We're going to take the suitcase and play with it a little bit. It has a complete unit in it, that's fully functional and does everything that anyone would imagine.

39

Basically what I've done is reduced it to an off-and-on switch, which is the level which most people seem to be interested in. So I'm going to accommodate those fellows in a little bit. We're going to run some tests on it to see for sure that it's doing what I'm saying it's capable of. Here's that radio frequency high voltage meter, that the probe that we saw earlier was hooked on to. Those are hewlett-packard instruments that we're seeing. There's more of them. This is just for you fellows that are thinking that somebody fell out of a tree sort of thing.

40

I might mention that you see the high voltage probe with it for high frequency–high voltage probe. Then here's some of the other instruments that are common. That center one goes to 600 amps plus or minus so there's a 1200 amp spread on it. Yes? [Question. Do you have a schematic?] Schematic of this? At this point I'm going to be talking about this up to a certain point and then there's going to be a wall put up, and that's about as far as I'll go. Because this thing is highly commercial, it's in more than half of the world right now. It's being manufactured and the monetary interest which I have in it is substantial. It's not to my advantage to be overly talkative. I'll try to be direct, straight, so that you know that it's really happening.

41

If you're real sharp, you can probably figure the parts that I'm not telling you, out. So that's to eliminate the competition. Which basically I don't need to worry about that because the backing which I have is of such a nature, that there's nobody in the world that can compete with me anyhow. So it would be foolish for anyone, regardless of how big they think they are to even try. Most of you probably got your Tesla magazine this year. The current issue, and starting on page 33, there's what was 26 pages printed by laser printer has been reduced down to about five pages.

42

It has a lot of update information which will tell you why these things will work. It basically blows the conventional approach to electricity out the window completely and forever, and in a way which it could never get back in. Because if you read the article, you will never be able to go back to where you were before. So if that makes you uncomfortable, you don't want to read it. We're going into some of the other background that led into this thing. There are a number of publications in Europe which–in Europe they don't take such a dim view of what I'm trying to do.

43

Those of you that can read French I'll move this up a little bit. We're going to skip through a book that was published quite some time ago, and we're going to show you some devices which have predominantly amperage and not necessarily voltage. They look a lot like the Tesla coils which you are accustomed to see. There's a reason for that, and we'll go into a little while. This particular fellow D'Arsonval was the one that gave us the thing for looking at heartbeats and you're cardiograms, and gave a lot of thermo devices which killed germs. They were using the thermal effect of the high-voltage electricity to kill germs back in the late 1800s. They didn't have disinfectants and things like that in that time. But they were very successful with it.

44

Some of the illustrations from this particular book I'm going to show you. Then we'll skip on to get back to English. The particular book I'm talking about there's three copies of in the United States. That may look somewhat familiar–but if you know the year that it was built–it was built before Tesla was actually building coils. It has a huge amount of amperage. It has the center tap, like I'm telling you about on my coil system which makes it into a voltage and amperage type of device. They had a great deal of difficulty with the amperage, because they were all the time getting problems from it.

45

There's another version of this particular coil which was easier to manage. See that's your L2 coil is on the inside of this glass deal, with oil in it. Your L1 coil is on the outside. It's not necessary to have all of those turns that the outside coil had. It could have only maybe a one turn or three or four turns. Actually by having fewer turns that kicks the voltage up on the inside one. Instead of say having a 30.000 volts coming out, if you reduce the number of outside turns to about maybe three, you probably would have maybe a 100,000 volts, or more, coming out of that inside thing.

46

In order to get useful energy, you have to go from the high frequency, high voltage, into some sort of storage situation. In this particular case the light bulb is being lit with what we would like to say is a Tesla coil, which is actually the D'Arsonval coil. You can see that the Leyden bottle there is the capacitor. You see the smaller winding and the larger winding, and then you see the air antenna on one end. Then you have the light bulb sticking out. When you bring your hand close to it the light bulb lights up. Yes? [Question. What's the name of the book that you're ...?]

47

That was the first sheet that I showed you and we'll go back to it in a little bit. It's a book on D'arsonval, and if you read French you'll like it. There's only three copies in the United States. One's in the Library of Congress, and I know where the other two are. But it's out of print, been out of print since 1930's something. One may turn up in your attic somewhere but that's the only place to look for it, I guess. There's a number of other books written in different languages, and European languages. They go in great detail on this sort of thing. This is all before Tesla came a long.

48

When Tesla came to Paris on his beginning, to come to the United States sort of deal, he ran into D'Arsonval. There's a whole group of people, the Donkin's, the Curie's, the Ruhmkorff's, and it just goes on and on. There's a long list of people that were working with these high voltage high frequency coils. Tesla saw him for the first time when he came to Paris. Of course he later made one which has high voltage but not much amperage. But we're going to see in a little bit that he knew the difference. He actually made them both ways. I'm going to show you the picture of one that's in a patent filing which has all of those things built into it. This is D'arsonval's, the first electrocardiograph that were made in the world.

49

This is a set of instruments that are used with the high-voltage, high-frequency coils. When these sets were manufactured and sold to other doctors and practitioners, and they got a set of these particular kinds of probes which are basically needles and little flat surfaces, which they touched the different spots. We'll see in just a little bit how that worked. This is an alternating current electric motor from 1840, which was in common usage in Europe.

50

This was already out and being in common usage when Tesla was still a little boy. There's a particular person that gets credit for it his name is [Moritz Hermann] Jacobi. I can't think the last name just now. But he is actually the source of it. The device when it gets to the doctor's office looks like this. It's inside that cabinet. It's got controls for the voltage and the amperage, and other things on the top there. It has separate amperage and separate voltage controls.

51

This will give you an idea as to the sophistication of what was inside that cabinet that you're looking at a little bit ago. This is taking you back into a period. You can see that there are spark gaps. You can see a number of different chokes and the variable thing for controlling the output and other things. This is a coil inside a coil deal.

52

I promised you to show you what some of these things look like in use. Here's a fella getting treated for rheumatism I guess. If nothing else happens I imagine that the sensations that he was feeling, he would forget about his rheumatism. Probably that was the least of his worries, so at least in one sense, he felt like he had been cured. If you feel like you're being left out on questions, we really aren't, because I'm coming back to all of that very shortly. Here's another version of what the doctor's office would have in it, with the controls for the voltage and the amperage on it, and other kinds of controls.

53

Here's one of them. This is actually a long row of treatment places. These things were very popular, and they had large following. This was the doctor D'Arsonval–one of his offices. This is one of the areas in a hotel where D'Arsonval and his helpers performed services for a fee. He was professor of experimental medicine for the college of France. He spent most of his time working on just about anything that was not standard.

54

One of the things he developed of course was these thermal uses of this thing. Now Dr. D'Arsonval also was a real medical doctor and professor of experimental medicine. This is one of the laboratories. The inside of one of the laboratories which he had. You can see those high voltage devices there. See a very large Leyden bottle down there with the water in it. The bottom thing would be a very large capacitor, see. Just like if you use your swimming pool for a capacitor, which you can. You see the very high voltage dividers here.

55

This is in itself is a device that builds high voltage. About 5 million volts I believe. That's what the output on this particular device was. As one might gather D'Arsonval had a lot of very interesting friends. These are some of them. If you can read their names, they probably don't mean much today, but at that time they were heads of state and others. Here's one with his class of students at the college. He's in the center, the fellow in the center.

56

This is 1933. At that time he was well into his 80's. He was still in very good health and involved in a lot of things. But those are some of his helpers at the hospital where he worked, one of the hospitals. This is the corner view of some of the buildings at the college of France where he was a professor of experimental medicine.

57

This is the evidence that Tesla knew that the coils could be wound for amperage [page 1 number 6 shows one of the diagrams]. There's three different illustrations here. But you see the center tab? The device in the middle, or the one with the H on it, is your armature, or your spinning deal. Like you have in a starter motor, or in a charger, say an electric charger in an automobile. The outlets there, if you notice, one of them, the center tap, is going to be your negative. The other two are going to be the high voltage output. What they're basically doing is storing electricity into the system in such a way that these high-voltage arc lamps, which over on D there, D prime the little round things. Those are all arc lamps.

58

You can put any number of them there, and it will automatically adjust to them. What is happening in my case, I'm removing that armature there. I'm keeping the center tab, which is a negative, and the two output things. Basically that becomes my L1 coil. I've taken my L2 coil. The L1 coil I put in place of the charging device, which is in the center. By the magnetic radio frequency pulsing of it I'm running that coil system in the same manner which is being ran with the generator, a conventional generator. Except that what I'm doing runs at a much better rate, and it runs at over energy rates and things like that. Which this thing–I don't know you could calculate it, except that it does have output and it does run devices and do useful work.

59

The center one here, is a slightly different arrangement. But it has the center tab which is important to getting amperage. One part of the coil there is wound in one direction, one is wound in another direction. That means that the one that's going the opposite direction is made to produce amperage instead of voltage. The top part produces voltage, the bottom part produces amperage. Then there's a second or third way. Then there's about a dozen other ways. But the bottom one will produce huge amounts of amperage. It's got a double back on the thing. Yes? [Question. What's the patent number on that page?]

1:00

[Answer. 336961] This is from a book that's available in the store here, with complete patents of Nikola Tesla, it's page number 8. If you want to actually go and see that particular book. It definitely is not a Tesla coil in the normal sense at all. It does produce amperage, and lots of it. I told you that I would show you a bit of information on batteries, as to why batteries were not desirable as a way to store energy with these particular devices. On the very bottom down there, you have the very best type of batteries that are available.

1:01

They're the sealed nickel cadmium batteries, and the sealed lead-acid batteries. Those have, if you look out to right there you see the recycle. They can be recycled between 100 and 300 times for the sealed lead-acid. And from 300 times to over a thousand times for the nickel cadmium. Now even at a thousand recycles, this type of device would use that thousand recycles up so fast you can't believe it. You could pretty well forget about conventional batteries, except for the short term experimental type of things. Now that is the reason why I have picked the capacitors instead of batteries, for storing the energy in order to get it to do useful work.

1:02

Last year I had a particular illustration which I used to explain why over unity is possible [page 99]. These are three different conditions under which you get resonant circuits. There are other conditions. The bottom one there is the one what you're really looking for. You could see it on your oscilloscope, and you can measure it. I'm going to show you a more detailed breakdown on it next. You can see that it's building energy. It builds it up to a certain level. Then it cuts off, and then it cycles back.

1:03

The resistor shunt in the middle there, is basically what helps cause this cycling effect. It's part of the frequency in the cycling deal along with the capacitor and the inductor, which would be your coil. You look at that closer you can see what each part of it is [page 98]. What is causing this to happen. What's pulling it out each time. This is your overunity looking at you straight in the face, and there's no way to deny it.

1:04

This is one illustration that I used last year [page 1]. If you understand what's going on on this, you can build any kind of useful energy device that you want to. This has the whole story on it. As a matter of fact you even see the adjustable deal on the top there, where the coil is inserted in it. Which is what you see on this thing down here. You also see the capacitor bank. You see on the number 2 there, you see an actual transformer out there, which is changing it to whatever level you want it to be, to go on, and to useful work. The one just below there, the capacitor or the storage thing, is that long narrow thing immediatly after you come out of the coil.

1:05

That's helping to feed those lightbulbs there. So a light system with any amount of lightbulbs out there would work with that type of set up. The bottom one there, to the left, is basically the way to go. The center tap there actually becomes your negative. The two output things on those coils become your high-voltage thing, just like the earlier coil winding that we were talking about. This type of device is actually the thing that's got two coils on it on that board right there. That's that very same device. By putting the L1 winding in between those two coils there, and hooking up the ground part of this, and the two outlet deals, you've got immediately a working Tesla coil in a very good range with amperage and voltage.

1:06

Here's Tesla's deal on lighting up fluorescent light tubes. He used two metal plates that were behind the draperies, that you didn't see, and used the basic system that we saw on the top right a while ago, to put charge on those two opposing plates. By walking around over the stage with fluorescent lightbulbs and things would light up. He also happened to be the inventor of the fluorescent tube. He had hundreds and hundreds of different versions of it.

1:07

I'm going to show you a large scale Tesla coil, which I did [page 159]. What you're looking at there are two 18 inch pieces of 12 inch PVC. There's one on the top and one underneath there that it's sitting on. The thing in the center is the tub that goes under whiskey barrel planter. It has rollers on it. I took the rollers off and it happened to be the right diameter to build the L1 coil on. Of course it was black so I changed the color on it.

1:08

This particular one has an output of 400,000 volts. It has quite a bit of amperage also. The instrumentation that you see below down there, you see the little 12 volt battery sitting beside it. It's very obvious that if you take what's out and what's going in, then there's a very large difference between the two. I'm going to show you one which was built to demonstrate over unity [page 153].

1:09

If you have separate coils which are no way connected except through the magnetic flux, that are tuned to the same frequency of the output. When in the center here you have the output coil, and you see the L1 and the ring around it, and see the L2 in the center. It's tuned. The tuning parts are right here, see the variable capacitor–it's a mica–variable mica capacitor. Then there's a variable resistor here, which is a part of the circuit. That goes back to the circuit I showed you over the overunity thing awhile ago, where you have a resistivity in there. But when this thing becomes resonant, of course the resistivity all becomes negative. Therefore you've got energy which is greater than what you have to begin with. This thing behaves exactly like a radio broadcast station. A very simple radio broadcast station, about 1811 [1911] or somewhere along through there.

1:10

These other coils are tuned to the same frequency. So being tuned to the same frequency this thing puts out a 100,000 volts at a certain amperage. Regardless of how many of these you put around there, if they're tuned to this one, they're not going to take away from the voltage that's there, or the energy that's there. They're going to duplicate it. The reason is that the electrons that are active here are not the same electrons that are active here. Through resonant frequency, those coils are tuned so that they react just like if something or someone tapes on your knee and your foot goes up, sort of deal. People think that the electrons go from point A to point B, that is not true at all.

1:11

Electrons are activated at point A, and electrons move tremendously slow speeds. So the electrons which are at the generating station in your village are not the electrons which are at your house. When you turn your house switch on, your grounding the system that is connected, and the electrons at your house start acting up. That's where you get your energy at your house. It does not come from the power company.

1:12

Those of you there asking what the basic circuitry on these things boil down to. This one [page 268] is probably easier to understand then some of the ones that we've been looking at.

1:13

But the basic idea is present in this schematic. You see your number five is your L1. The 3 there, or the first block, the large block, that's parallel with the number five to the left of it would be your high voltage transformer. Whether it's a laser module or whatever. This spark gap, number four there, would be the Lightning arrestor, which I showed on the other device. You have a voltage limiting device, which would be– number eight is the voltage limiting device on that particular thing, to keep it down at a level which you can manage.

1:14

There's only one diode showing in this case which is sending forward pulsating DC, a kind of a sawtooth type deal. By hooking another diode down on the bottom corner there and whipping it back up in front of where the number 9 is now going towards the transformer, you would have smoothed out pulsating DC which is what you really want to go forward to your transformer. Number 10 is matching the transformer to the frequency which you want it to run at. Once you've taken measurements on number 11 there on the input side, you can calculate what number 10 will be. Which will give you your 120 cycles per second, which is 60 up and 60 down.

1:15

Which is your 60 cycle electricity which you're normally accustomed to. Number 12 is a voltage control to keep the thing from running away from you, and burning the transformer up and those things. There are many, many different ways to go on this. Another way would be to go with the toroidal transformers and that's what I'm going to show you next [page 268]. The same situation set up with a bank of toroidal transformers. There's almost no end to the types of things that you can come up with. They all seem to work fairly nicely.

1:16

Take a look at some of the measuring devices which are involved. You can see that it says that that device will measure to 1 million volts. It goes to 10 megahertz. With the proper voltage divider–there's a 400 volt divider that's sitting beside it there. There are dividers which go up to about 5 million volts.

1:17

That's the conventional way to measure it. My preferred way to measure it is with a magnetometer and electroscope. Various sorts of devices that measure magnetic fields. The magnetic field is directly related to the amount of amperage that's present. If your amp turns in your coil represent a magnetic field, and it's very measurable. It's a very safe way to go. Here's one form of a voltage divider.

1:18

See all the resistors that are hooked in series and parallel, and everything else. Then you have the the thing on the top there that helps dissipate the energy so that you don't burn everything up. LCR meters can get pretty sophisticated. This is one of the more sophisticated ones. That's from Stanford research systems in California.

1:19

Here's for people that just want to find out if this sort of thing can really happen without spending very much money [showing the *Resonance Induction Coupler Kit* we are selling now]. Here's what fifty bucks will get you. We were talking about the separate coil systems little bit ago, where one was running and the other would copy the amount of power on it without depleting the energy from the first one. That's what this is made to demonstrate and it's used in physics classes in high school and college. What happens is that second block there, which is in no way connected, except by the radio frequency, will light a conventional 60 watt [volt] electric light bulb.

1:20

The input on this is one and a half volts on the left-hand side there. One and a half volts input on the left-hand side powers that coil system that's on the block on the left side. You can move this other coil some distance away, and you can hook an electric light bulb on to it. It comes with a 60 volt neon tube which lights up whenever you tune the frequency of the second coil to match the first coil. There's the hookup information showing how the two blocks work. So that's your basic wireless transmission of energy device.

1:21

This [page 51] is one of the early ones showing the transistor that's being oscillated by the variable capacitor and variable resistor on the back of it. The number 7 thing actually happens to be a high voltage transformer. That could be the ones from the television set. The flyback transformer from the television. It can be a transformer from an automobile, one of the early ones.

1:22

The number 6 down there is basically an off-and-on switch, down to the very bottom in the middle. It causes coil number 9 there to react. Which in turn number 11 over there, can be tuned by the grounding capacitor. So that it will match the number 9 deal, and it will become active, so your energy will be coming out on number 12 there at the top, and the bottom one down below. You can center tap this thing and you'll have your amperage. Hook it up like we were talking about. This is the postage card size that will run an automobile from say maybe a six volt battery–flashlight battery, not automobile battery.

1:23

It's one that can also recharge the battery so that the battery basically never runs down. We were talking about voltage dividers. This one I'm going to have to put it on in more than one position because you can see the man down below. I'll turn him sideways. You can see the thing is over 12 feet tall. The man is 6 feet tall. That's using your mega voltages. You don't have to use that at all. If you use my system of magnetic fields and electrostatic and electromagnetic field measurements, you don't have to use that at all.

1:24

This is something that comes–we can think Romania for this [page 63]. This is from a translation from a book in Romanian, which I have or had, that tells what happens in magnets and a Tesla coil. Or the center tap on the Tesla coil behaves just exactly like this magnet. The Bloch Wall system there becomes the negative or earth ground. That's the place where the electrons are getting organized. On one side they spin right, on the other side they spin left. You can take your hand and cusp it. The current, direction of the electron movement would be like in this plane. The spinning would be around that particular plane.

1:25

Then on the other end of it you have to reverse it. This would be your north end, and this will be your south end on this particular thing. Now coils behave exactly like magnets, they have a more-or-less a north end a south end. As far as I know this has never been published in the English language. This was actually a Russian publication in Romania, from back in the 30's. We're going to see how that ties together with some of the other things.

1:26

What I've done there [page 65], I've taken the Tesla coil with the center tap on it, and I've shown what actually happens when electrons move in one direction the amperage part goes the other way. As it turns out, there's basically, there's two types of electrons. There's one that's a little bit smaller and weaker than the other. When it spins in its direction it becomes the amperage part. The stronger of the two electrons spins and makes the voltage part.

1:27

In nature they are coupled together and that's the reason why they're inactive and floating about in space without anything happening. It's whenever you disturb them or knock them apart that they try to get back together and that's where you get your useful energy. They're relatively easy to disturb. You can do it either by electromagnetic methods, or you can do it with the conventional coils and magnet system. Which is also electromagnetic. It's much easier much more efficient. The rate of disturbance determines the amount of energy that you end up with. Which means that generating at 60 cycles, you're wasting your time. The amount of energy output is directly related to the square of the cycles per second. So it's 60 cycles square of that, then square 218 million times a second, and you will see that the efficiency on my device has gone up unbelievable.

1:28

As a matter of fact it's gone off the chart, it's gone straight out of the sky. These electrons when you disturb them, they are not in any way diminished. They simply will go back to their normal habits, the first chance they get. They will be there and you could use them over and over again. You're not disturbing the environment or doing anything which is harmful to human beings. Depending on the frequency you're running at, if you're down in the lower frequencies, say down below the radio frequency, which is below 20,000, the lower you go, the more harmful the environmental effects are from electricity. The frequency that was chosen to be the standard was probably the worst possible frequency imaginable. The better version of it is the 400 cycles per second or actually it's 400 up and 400 down which is 800.

1:29

Which would be considerably more efficient and better than the 60 cycle way of going. We see it in the electrical vehicles and the wiring systems, and the airplanes, and the servo mechanisms for controlling the airplanes. The wheels and all parts of it, including a lot of the service vehicles that they use at the airport. They're running at 400 cycles a second which means they're running at a higher voltage, they're running like 440 at 400 cycles. The lower the cycles, the more the heat and the wasting of the energy. So the higher the cycles you run at the less harm to human beings, the environment, and other things, and the greater the efficiency.

1:30

So as you go into the higher frequencies, it's very easy to trip over into the overunity systems which we're seeing here today. This illustration is in the current issue of the Tesla magazine. Which they have some copies in the bookshop, in case you want to look closer at it. The electron is a very useful thing. It produces a number of things which we normally don't think about. As you can see there [page 54], we have two basic types of related energy. We have the resistance or the impact and the heat deal. Then we have the electromagnetic reaction, which is up at the top.

1:31

There's methods of storing this energy without disturbing or destroying the electrons. The whole idea behind how this thing works hinges around the fact that you can take the energy without diminishing the electrons. The basic storage formulas are these [page 71]. You'll notice that part of the formula gets squared.

1:32

This takes you directly back to $E = MC^2$, or the things that led up to where Einstein later put into a statement which he gets credited for. The idea had been around quite a long time. As a matter of fact, if you go back to Maxwell and the people that were working early on. Not real early but sort of early on. Their ideas about electricity are almost identical to what I have. Reading the early accounts, Maxwell would have no trouble at all agreeing with everything that I've been saying. He doesn't agree with what is taught in the physics books here. They quote Maxwell all the time, but they only quote him out of text. They quote him in such a way that he's saying something, which he would not have said if he was personally physically present.

1:33

How do I know that, I've been reading the documentation done by the people that actually worked with Maxwell and some of his actual writings. The book that you see that's accredited to Maxwell, was not written by Maxwell. He probably would have had some difficulty with it. Now that we know that those formulas exist, let's see what they really can do. A Joule, a Watt-second, a Coulomb, a Farrad, all those things are 1V at 1A at basically one second. They're all the same thing.

1:34

They're used in different statements and ways that appear that they can be different, but really they basically equal 1V at 1 second or 1 watt. So that's a Joule, a watt second, a Coulomb and a Farrad. So with that in mind half the energy is in magnetic, and half of it is in electrical. So the 0.5 there [page 75] is taking care of the fact that we're using the electro end of it. It's half the energy that's there, the other half is in magnetic. So with that in mind we have a capacitor which has a certain portion of a volt.

1:35

If it was 1V it would be one with the decimal point, but the capacitor showing three sets of zeros basically there with the one interspersed, which makes it 10 microfarads. Then the hundred thousand would be the voltage. It could be any voltage. It could be a 100V, 1V or 25V, or whatever. Then if you square that, and then you have to further go to times the cycles per second. After you've done all of that, that tiny charge that was there, each time it's cycled, it adds up. It becomes accumulative. That's why your capacitors–even though they don't appear to have a lot of electricity in them, based on the fact that say they're 8 μf or whatever–if you repeat that enough times in a second it becomes accumulative, and all of a sudden you have hundreds of thousands of volts that are available to you.

1:36

They're available outside of the system from which you were to begin with. The amperage at that point depends on the degree at which you ground that capacitor to the earth. If you should touch it, and touch the earth ground at the same time, it will cook you on the spot. You don't really want it to unload at a very fast rate. So what you do, you choke it back with dividers, so that only a limited amount will pass through the load which you're actually wanting to do the work for you. At that point you have useful energy. Here's some of the benefits [same page].

1:37

Give you something to think about. In the world of politics and social mixing and that sort of thing, as far as governments go and such, there are really only three basic things which are necessary in order for a large group of people to function. One they have to have water, they have to have food, and they have to have energy. The first two are largely solved when the third one there, energy, comes into position. They have-not countries, as far as food and water are concerned, if they have energy they can usually fix the food and the water problem. Most of the international problems as to countries which have and those that have not, all tie into those things indirectly.

1:38

This particular system here has the ability to heal those problems, in such a way that there are no longer third world countries. It basically makes the whole world energy efficient, energy sufficient. With that in mind the food can be produced, and the water would be available for drinking. But you can make it from salt water, or you can drill for water wells, or you can build with your electrical equipment. You can build reservoirs and all sorts of things. So that the three major sources outside of politics and religion, that enter into world politics, would be water food and energy. With that in mind, what we can do if we have some lights now, we'll go to these devices, and I'll demonstrate them.

1:39

Let you see what useful energy can do for you. I'm sure there's going to be a lot of interesting questions. If there aren't I'm going to be disappointed. The device which you see here, in the small suitcase, requires an earth grounding, so I've got a one wire deal, that goes into a three hole plug back there. It's using the center part, or the earth grounding. If anyone has any doubts about it, they're welcome to come up and see that there's no energy coming from there.

1:40

Anybody wanna [shows a drill running]? You have to put a certain amount of pressure against it in order to make it work, and I can't do that. I would estimate that's probably useful energy. [Question. How big is that battery that's running that?] It's a twelve volt, a minimum size the smallest of the 12 volts. [Question. 4 Ampere hours?] It's more like six. I use more than six amps right there.

1:41

I think there's somebody here that brought a meter, and they're going to check me out here, and tell us how many amps this thing has, and how many volts it has. They will come up, I'll stand aside and let them tell you. If anyone else would like to bring their voltmeter and ampere meter go ahead. If anyone else would like to find out what's here. Get what? [Question. I have to get inside it to reach.] No, well right here's where you're going. Well let's see, I tell you what we will do. We'll hook it on like this maybe. Yeah, well we're more interested in amps here than anything else, because that's what... Let me hook this one off first, and then...

1:42

You're on a good scale are you? [Question. Yeah] I think you're okay. If I can touch those things there, we have maybe 30 amps. I think there's something that we're not making a contact with here. We may not be able to measure this because I should have brought along a wire that sticks out, where we can... Let me see if this other one sticks out farther. That one down there, and that one up there. If I can make contact here we'll be able to read the amperage.

1:43

We're not getting contact on these but I can tell you that when it's properly measured there's more than 20A. We'll work on it maybe after next break, we'll have a way to do it. [Question. Are you recharging the battery?] It's recharging itself as it goes. [Question. Can you run that continuously day after day?] Yeah. We're not getting contact, I don't believe. Yeah, it's not making contact.

1:44

What you have to is to program this thing up here. You've already programmed it? [Yes] So it's programmed correctly. I think what the problem is here, we've got too sophisticated a meter. Does somebody have a more simple one? This thing is half computer and half meter, and we might need a more simple one if someone has one. [Question. Are you measuring in series or parallel?] The way this thing is set up it should.. You're measuring ohms aren't you?

1:45

I guess we should have rehearsed this part of it. [We're at 120 Hertz] 120Hz which is correct for 60 cycles at 60 up and 60 down. Let me try something. I tell you what, does somebody have pair of snippers here?

1:46

We'll cut one of the noses off on these things. Well let's do something else here. It's very obvious that we have power, and that it's useful. [Question. Is the output a sine wave or square?] It's exactly the way you would want to see it. It looks, if you didn't know where it was coming from, you wouldn't.. [Question. You have three, can they run at the same time?] Yeah, I can put a lot more, I can do a lot more then that.

1:47

[Question. You can plug in the drill and not dim the lights?] Yeah, I can do all kinds of things. I can light this whole room up with it. [Question. Can you leave everything plugged on for the rest of the days of the conference?] If you want to sit around and look? If you volunteer, we'll do it. Would you come up closer please? This has a minimum of 20 amps. [Question. Minimum of 20 amps?] Yeah [Question. Reactive load?] No, as a matter of fact it will start kicking in on that and it'll go up to more than 30 amps.

1:48

What happens is when you put a load on it it will blink momentarily, and then it will adjust to the next level. [Question. What do you see as the first commercial aplication of your technology, and what is the time frame that will be available on the market?] It's already there and available in a large way. The commercial applications, some of them are electric train system for Brazil. The bullet trains in Japan. There are so many different applications that you can imagine, just almost anything. [Question. How about an automobile?] Automobile? One of the major manufacturers, not American, will maybe before the years over have the automobile out.

1:49

[Question. Are these suitcases going to be for sale?] No, this is the one that I take with me when I travel overseas to visit the President, or the Emperor, or whoever happens to be. [Question. Could you make it for sale?] I wouldn't because there's some proprietary information in there. That suitcase there, if you offer me five million dollars cash right now, I wouldn't take it. It's commercially viable the way it is. You could roll it into your house and disconnect from the system and you would be home running free for the rest of your life. [Question. I don't understand why...] Because I'm not interested in peanuts. The groups which I'm connected with are interested in doing this in large manner which involves entire governments, entire countries.

1:50

Instead of taking the amount of time to talk to one person or deal with somebody on a one-to-one type basis, the same amount of time will cover a lot more area than you can cover with the others. These bulbs burn out pretty fast. Yes, sir? [Question. I've got a simple meter, can I hook it up?] Yeah, you're welcome if you want! [Question. Will that be available in the United States the technology?] Well I encourage everybody to build their own. [Question. Do you have the plans?] They're published in a book that's in in the office there. Where do you want to plug in at? Down there?

1:51

[Question. Yeah I'll just plug in down there. I'm going to measure the voltage across the bank. Are these all in parallel here?] Yeah [Question. And they're 120V?] Yeah, I could put bigger ones in there but... You tell me when you're ready? [Question. Can you buy this parts each individually?] Yes, if you know where to find them, if you know where to locate them you can. [Question. Can we buy this through you?] No, because I get too many inquiries. I get maybe, in one week, 80 people asking me that same question. [Go ahead Don] You're ready? [Question. Is this actual sine waves or square wave?] It's a modified square wave wave. [Question. Your meter is calibrated for sine wave so this could be off.]

1:52

[Question. This is a fluke 79 meter it regular calibrated for sine wave, and if you have a square wave into it, which actually I have an inverter here. This is one of the 12V Radio Shack inverters and it puts out a modified square wave. You don't read the same voltage on square wave than you do sine wave. You typically measure about a hundred volts, and I'm getting a hundred volt right on the money. So it's actually running the light bulbs the way it's supposed to.] [Question. You don't have a current measurement at the moment on your light bulbs.] We tried to do it up here. [Question. You got a whole bunch of stuff on it.] Yeah [Question. What is the current right now?] It's not very much right now, but what happens is when you add the load to it, it adjust to it.

1:53

[Question. 60 Hertz right on the money.] [Question. Now of course I have not tapped, I had to split the wire to put he ampere meter in there. No it's rock steady.] [Question. You can calculate the amperage if they are all equal wattage, which they seem to be.] [Question. Yeah, so you're looking at how many? You got 40 times what?] Well I can put, I've got some 100-watt ones here. I switch them out real quick-like. Actually those are putting out more then 40 watts. Because you see they're too bright. You see they're actually destroying themselves right now [Question. It's coming from the unit in the suitcase here, right?] Yes, it is the top part of the suitcase. The suitcase is mostly empty. [Question. Can you tap into the 12 volt battery that's being recharged there?] By tapping into it, what do you mean? [Question. Can you hook this inverter, which is a regular inverter?] I'm sure you probably could except for the fact that you're running your battery down.

1:54

[Question. Is there a certain amount just coming back to the battery just to keep it in operation?] The battery here is recharging itself. Once this thing starts, the battery isn't being used. [Question. How many editions of your book are out now?] There's about five or six. [Question. What is the last edition in the bookstore?] I'd be afraid to say.

[Question. Five or Six?] It's probably one. Somebody left me with their fingernail clippers. [Question. Can you give us, where me might be able to find this parts?] Well the coils there, the pretty coils, come from B&W [Barker & Williamson].

1:55

I don't have the address with me. The PVC comes from the local hardware store. The pretty orange colored multi-stranded cable there, which is very helpful in this type of device comes from Gateway electronics in Cincinnati Ohio I think, and Denver Colorado and Los Angeles. [Question. How can we get the latest edition of your book?] They have copies of it if you lean on them. I have in some cases sent people copies of it for free. So what you have to do is just keep on top of it. [Question. What type of batteries did you say you prefer to use?] I'm using the led calcium here but they have. [Question. That's what in here?] Yeah. It's a lead calcium battery. It's about the size of one of those capacitors.

1:56

[Question. How long will that operate?] It will go essentially forever. [Question. Forever?] Yeah, the battery is not being used except to start it. [Question. How to obtain one of the commercial devices, you said they where commercially build right now.] It depends on what part of the world your in. [Question. What part of the world do I have to go to?] Japan. I would have to give that to you later. [Question. Is this basically the design that you have in your suitcase?] It's a scale down 1/4 version of it. [Question. It's a 1/4 version of what's in your suitcase?] Yes. [Question. Ok, can we see what's in your suitcase?] No you cannot, there are some other proprietary things there. [Question. Would it be possible to energize this unit here?] This thing has hundreds of billions of dollars in money behind it. There are people who have their money up and put on the line. For that reason, there are some proprietary things which I'm not discussing which really make it.

1:57

[Question. Do you think Tesla was more interested in money or doing the right thing?] I tell you what, if we have a problem with that, we must not discuss it because there's one way that the ball is bouncing and that's not the way which you like. [Question. In order to share with the rest of us, could we have the questions repeated by you?] Let's go to a question-and-answer session. I have a little bit of trouble hearing so if you don't speak loudly you should come up closer.

1:58

One of the questions which was asked me, which you didn't hear, was which do you think Tesla would think more of money or more what the device can do for someone? This device has the potential of ending world conflict in its totality. So the manner in which it's being applied at the present time is to build enough capital so that that goal can be pursued in such a way that nobody on earth has enough money to push it down. So that has just about happened. So at the point where it becomes impossible for anyone to knock it out or turn it down or get rid of it, at that point, well then, of course it's available to anyone and everyone and all aspects.

1:59

[Question. My question is, now that we have

the internet why do we not give full disclosure on the internet?] If you look correctly on the internet you will find it. You don't know where to look. There's a disclosure enough that you can do what I'm doing here. I might tell you that what you're seeing here is obsolete. I'm a couple of light-years farther down the line than what you're seeing here. But I'm doing what you're seeing because most people think that you have to have something where you just turn a wall switch on and that's it.

2:00

That's to accommodate the wall switch crowd that I'm doing. The thing is actually presently a solid-state device which is very small and does all the things that these things do but it's a solid-state device. You'll hear about that more in the future. Which you won't hear much more here, because I won't talk about it. [Question. Can you address patenting and protection which I think is a big issue?] Well the whole patenting process is a big bad joke. But you do it because it's customary to do that sort of thing. But as far as patenting something, if you think the United States Patent Office is going to be helpful, you're wasting your time for sure. But the way I work it, I'm keeping the ball in the air. Keeping it in such a way that, as long as I keep the ball at–once the patent is issued, well then I'm open and subject to all kinds of conflicts.

2:01

So I don't want the patent to be issued for some time yet. So basically what I do, I keep the Patent Office in a state of confusion until I get to the point where I want them to know what I'm actually doing. Then I unload on them. I get my patent at that point. But again, money talks and if I do what most people would like to see happen, this thing will be dead and in the in the trashcan just like it was before. I'm determined that this time that that's not going to happen. I'm very far down the line in that procedure. Not only that, these devices are–you're looking at 1890 type stuff here.

2:02

The solid-state versions of them, there's really nothing there to see. If you saw it you wouldn't know what you were looking at. That's basically what I can tell you. [Question. I'm from Canada and I imagine every farmer in Canada and in the United States would love to have one of these on their farm. But like you, I'm a little bit skeptical about what General Electric and Westinghouse and the big power corporations would say about it. And of course we do have people around us that don't want peace and don't want the people to be prosperous either. They're trying to keep us into poverty. They make more profits out of wars, so how do we get around this? Like Westinghouse is pretty big and there's a lot of power companies out there they're pretty big.] The power companies and the status quo have huge long-term investments which they're amortizing over long periods of time.

2:03

Those things are very much in jeopardy at this point. As a matter of fact they're basically worthless, they're useless. The automobile industry, basically you could just today, knowing what you know from here, you can write the total American automobile industry off. Because the people that come with the electric cars based on this system will totally wipe out the others. It's not an American company that's going to do it. Because the American companies–the attitude of the people that they have to talk to with the people like General Motors and others–they look down their nose at you to begin with. Secondly, they have a huge staffs of people who are highly competent and they depend on them to supply them with what they need. Someone from the outside is not about to upset that balance.

2:04

So there's no way on earth that I could talk to directly with an American automobile company. But I can tell you at this point they are dead and they don't know it. [Question. If you would given the opportunity to sell to an American company would you?] It would not be my first choice, and at this point I can be very choosy. [Question. Why?] Basically because I don't like the attitude. If their attitude was constructive I would be all for it. But the attitude is destructive. [Question. Could we have you come up the microphone and have each of questions on so when you buy the tape we'll have the questions on as well?]

2:05

[Question. Don you mentioned that the battery itself was a 6 amp hour battery?] Yes [Question. After doing a calculation with the 40 watt light bulbs they would have to burn a little over 10 minutes in order to fully exhaust the capacity of a 6 amp battery?] No [Question. That's according to the 6 amp hours, okay, 400 watts, but it's recharging.] [Editor: It is not possible to run a 6AH 12V Lead-calcium battery at 400+ watts without the voltage drop going so low as to not power anything. That would be over 33A.] I have ran these things for several days at a time and they didn't dim. [Question. Yes, my question is just to show the audience that in fact it isn't working the way you're supposed to.] At this point I'm not going to get caught up in any debate or argument. What you see is what you see. If you're uncomfortable with it well I'm going to leave you that way. [Question. No that's all right, is it possible to change the bulbs to the hundred watt bulbs?] I have some higher power bulbs here, and if you look at these you notice that the filaments were burning out very fast. [Yes] As a matter of fact those bulbs would be gone probably before we ran another 10 minutes.

2:06

[Question. Right, the 100W light bulbs only have to go over 4 minutes to exhaust the battery. And if it ran over four minutes there's no question.] We'll do that another day, okay? [Thank you sir.] Other questions? [Question. Do you plan to energize the unit that's laying on the floor?] I did not bring the proper cables to turn it on. I've had it running and it works just fine. When people come to visit me they get a much better show. Whenever you start bringing things out, in the back of an automobile, and you have very limited space, and you have problem bringing things for people to see. I have all the amp meters and everything else to prove my point, but I was thinking that I would be able to use this thing but apparently not.

2:07

[Question. Would this be a one-time

purchase type deal, or would people be charged a monthly service for having the device? Or how would that work?] Basically I leave that up to the people that are dealing. The basic structure on a deal on this type of thing would be a joint venture. There would be the capital to operate it, would be 50%, and the technology is 50%. The policies, the joint venture, I'm the chief executive of it, and also some of the other positions. Then I have people from the other side, the money side, who are in the command structure. It's kind of like a roundtable. They decide in what manner and method which it will be marketed. In most cases there have been no problems at this point.

2:08

[Question. What do you see as a purchase dealer, do you see it as being sold out right, or do you see it being leased, or put a meter on it?] Purchasing it outright? [Question. Yeah, exactly.] What the best deal there is for you to build your own. You do it with my blessings completely and if you use some of the schematics which I have, you can do it. There are a number of people that have done it. [Question. This device here on the floor is the same as in the suitcase, but is just downsized?] This 1/4 scale of this. [Question. How much would that put out?] It will put out more than the components there can handle. The weak spot would be the capacitors. The capacitors, the way they're set up, will handle up to about 8,000 volts at about 10A.

2:09

[Question. Your normal household is a 200A service.] Well you got a 200 amp circuit, but the 200 amp circuit is your main. In most houses. In our house it's 500A service, but in most houses 200A service would be for the whole house. That's divided up between 20 and 30 and 40 and 50A. Maybe 50 amps is about the biggest fuse that's in there. [Question. So a device like that would have to be built bigger to run everything at once?] It depends on how you hook it in. The fact that I'm even letting you see these things at all is violating my own personal relationship with people who have large investments in this.

2:10

So the fact that you're able to see the things as they are, consider that my gift to you. [We sure appreciate that!] It should be enough to get you started on what you want to do. That's the idea behind it. There are other people in the audience out here who have actually tried to build these things, and there are some that have succeeded. But out of several hundred people I know, that have tried it, there's only about six that have actually completely succeeded. So why that happens, I don't know. Why it happens, there's really nothing that complicated. [Question. Do you have an incorporated company of any kind?]

2:11

Yes, I have calling cards and contacts with the international banking group that I work with. Early on in this thing, I had made a lot of presentations. There were, all kinds of people running at me and tell me what all kinds of great things they were going to do for me. The first few tries, it turned out when you got to the table and the money part, the money part wasn't there. Those were American people. The same time that was going on, I had a lot of governments, government officials from Japan and Indonesia, and from Philippines, and from Holland. [Question. Have you issued stock in that company?] No, there's no stock. The way it's standard operating procedure now. It's non-negotiable. It's 50% for technology, 50% for the money source. I'm the CEO of the whole and the chairman of the whole group.

2:12

Then the structure comes in underneath that. Eight of the eight board members, four of which I name, and it was a joint venture company. The funding is in negotiable funds in the bank. That's one of the requirements for the joint venture that exists, it has to be negotiable funds in the bank. [Question. I want to know where I can get the information that you're you mentioned earlier, your schematic?] Where you can find a schematic? [Yes.] Some of the ones I was showing you there, are in a book which the bookstore has. We need some more questions. [Question. Don you mentioned the book is 25 pages but you mentioned that your original notes are many many many pages. Can we get a copy of those original?]

2:13

Inadvertently the book which they have in the bookstore right now has some of those original notes in there. What happened was I left my briefcase in the bookstore and they headed out copying everything in the briefcase. So the book that they have in there, now has maybe 10 or 15 pages or more of actual original notes and things in my own handwriting, in the back of it. [Question. The notes that you writ up in the magazine, the science magazine that we just got out now, those notes are just a fraction of the other notes is that it?] Part of what came before has pulled over into those notes. But if you read the article that's in the magazine, it doesn't have the schematics, but it discusses what's going on, in such a way that if you're familiar with coils and capacitors, you can build it from...

2:14

[Question. Very easily I would say.] Yeah, anyone who is sort of conversant would be able to do that. We need another question. [Question. You had talked about joint ventures with industries, for example if you were to decide to do a joint venture with the elevator industry, getting these devices installed for use on American elevators. What would that take, and what kind of capital would you be looking for?] If there are people who are interested from the joint venture aspect, what I'm going to do is give you a calling card of a man who's involved in international banking and he is one of the group of people that took care of financing this thing and getting it going.

2:15

He is real sharp, and you want to be real serious if you're talking to him. You don't want to be talking about peanuts because the ticket to the ball game starts at about five hundred million dollars, and it goes up to several thousand times that, depending on who you are and and what you want to do. [Question. Is your unit available for the market yet?] It's actually in commerce in some parts of the world. [Question. But what about in the United States here?] United States you have to build your own I'm afraid at this point. [Question. So is any thing you can say in the near future will be available in the market, or what?] If you wait about six months to a year you may start seeing these things coming in from Japan.

2:16

I can tell you the Japanese are quick whenever it comes to seeing how to turn something into money. I need to tell you they can spot it fast. [Question. You mentioned that your latest device wouldn't be recognizable from that one.] The new ones? [Right?] Yes, you wouldn't if you saw it, you wouldn't think it does anything. [Question. There must be something that is the same from that board, looking at that board right there?] No. [The coils perhaps?] No, there's no coils and... [The capacitors?] There's no capacitors. I told you, you wouldn't recognize it. But it becomes obvious, so obvious, so quick once you saw it, and I explain it to you.

2:17

Right now it's like trying to explain color television to American Indian back in 1850's. But it's not that complex, but the same rules that apply to what's happening here told me that you could do this, do it the other way, and so I built some of them, and they work just like the first time, absolutely perfect. I can tell you this, you can have hundreds of thousands of volts out of a tiny little matchbox and at high amperage. [Question. We're going to hold you down and open the suitcase?] No, the suitcase is locked and if you came even close to it, or attempt to pick it up and carry it off, you might not leave this room, ever. [Question. Could you give us an indication of what it cost per kilowatt or per horsepower to construct the equipment? Rule of thumb?]

2:18

Well that depends on how much it costs to build the device. You can tell from looking at it to probably what it costs to build. I can tell you that it's nothing like a conventional. If you have a conventional generator and you want a megawatt, your going to spend about a million dollars. So that's about $1 per kilowatt, or something like that. So these things are like 30 kVA. So that will be, if you priced it according to what the conventional source cost, you'd be charging $30,000 for this device right here.

2:19

But it wouldn't be necessary to do that. Not only that, your conventional device would wear out. You have upkeep, maintenance, and you have personnel to look after it. You'd have fuel cost, and the costs go on forever on a conventional deal. So that million dollars you spent to get the first device, probably the generating equipment itself, might have useful life of five to ten years, maybe. You'd have to spend your money all over again plus all the personnel, and the fuel, and other things. Any more questions? [Question. What you're really saying is, we don't need those atomic power any longer?] No, atomic power–let me explain atomic power this way. If you have a particle which yields energy, and you have to spend more getting the energy out than the energy that you get out, that's atomic power.

2:20

In the case of the electron you have to just simply disturb it. It's relatively easy to do that. But with the atomic power you basically have to tear the atoms apart, and in the process of doing that, the equipment to do it, and to generate electricity which you end up doing with steam, it's sort of a Rube Goldberg way of getting energy. [Question. What is the wattage of the unit laying on the floor?] Wattage of the unit, well you have to calculate that. I gave a challenge to some of you a while ago on that neon tube transformer thing there. If you look on it, it says how much goes into it, and how much comes out. If someone would care to come up and calculate it, you'll find that that's an over unity device itself. It's putting out more wattage than you're putting into it.

2:21

It's written on the label as to what the input is and what the output is. We need some more questions. [Question. The unit you're having manufactured now, are they the electronic ones, the solid state?] What's going to happen is they're all going to do it this way for a while. Then after maybe two, three to five years I might bring the other thing out. But indirectly, I'll bring the other thing out under different operating groups. [Question. The unit that's being manufactured now it's like the size of the suitcase one, correct?] Well the commercial versions, I actually showed you one when we first started here.

2:22

I showed you a commercial version. If you read the name on those capacitors where they came from you know where they're being manufactured. [Question. I notice on the unit on the floor, it appears as though the secondary is all wound the same direction and that's different than what you were talking about.] If you saw the drawing a while ago that I put on the coil, you can wind it in a reverse direction on the bottom half or a forward either. If you want to emphasize the amp's you wind it in a reverse manner. If you want to make a transformer which produces amperage basically only, and not voltage, take a–well let's see if we can find something here. I'll show you how to make one real quick-like.

2:23

One that will do just amperage and nothing else. You can drive this amperage thing, with your L1 coil. See, take this thing right here, and loop it back, and then you wrap it around wrap it around the coil form. So that both, that your in and out or down on the bottom part. At that point where these two are running parallel to each other, there it cancels out the voltage so that only amperage is there and it magnifies [Editor. Wire folded together and is wound together]. So you can make your L2 coil there, if you wind it correctly, you can get only amperage from it, and no voltage to speak of. Just like you get to say, thirty thousand volts, well you get thirty thousand amps. But you don't want that, because the device would melt on you to begin with. Some more questions?

2:24

[Question. If the ground cable is removed from your suitcase will the unit still operate?] Yes. You notice the screws on the top here, that's where the thing is bolted in. It actually takes up just the top half of the suitcase. [Question. But if you pull the actual ground cable off that, that small block cable, will it still operate?] The ground cable is actually necessary for it to operate correctly, if you take it off, it starts building up heat and other things. [Question. What else could be used for a ground if this were to operate a car?] It is grounded into the air as it is. But if you use it for any period of time and it's ungrounded it starts building up heat. There was no heat from what we where doing.

2:25

[Question. What's your web address?] Web address, you want to find it on the computer? Actually I don't have that thing with me. I didn't put it on there. Some people in Houston did, and some people in Dallas did, it's on there. It's in Amsterdam I know because I've had people call me from Amsterdam. I know that it's in England because I've had a man from Croydon England called me the other day, and was asking questions about it. So I know that it's on the Internet in Europe. I know that it's also on the internet in Japan. But I've got phone calls from there.

2:26

The Japanese television did a one-hour documentary on me and showed in February throughout Asia, and I got a lot of phone calls from that. This PBS type deal. There's a professor at the Technical University of Tokyo that's a good friend of mine who's recommending me for the Nobel Prize for this thing. There's another person like that in Amsterdam and one in England that's doing it. Then there's also several other people are involved. This has Nobel impact from the explanation of what the electrons are doing. This business of having electrons coupling and being disturbed and pulled apart. And they have to fight to go back together, and when they come back together, you have volts times amps equal watts. That is an extension of JJ Thompson's description of the electron originally.

2:27

It's expanding onto it, and there's some other things there. That is Nobel prize material. The fact that we were talking about the peace effect of this between nations, making energy available commonly and very economically to all people in the world, also means that they can produce food, and they can take care of their water problems. So it solves three major world problems. So that is Nobel prize material right there. It goes on and on. [Question. What do you call your system, what's the name?] The system, I just call it an energy system, very simply. What type, what category? It's a *Resonant Induction Energy Transfer.* Resonant induction energy transfer.

2:28

Your magnetic changes to electricity, or electricity in turn changes to magnetic. That's how they talk to each other. Depending on what the electron is doing will determine whether you've got electric or magnetic. You want to see it on the video, he can zoom in on it. What I'll do is take it, and it's a little bit heavy. But we'll take it here. Which way do you want me to turn it? Well he can he can zoom in on it and come in at any detail he wants to. I'm sure. I'll turn it around in a little bit and turn in some different angles. I'll hold it for just a little bit in say this position. Then I'll turn it any different angles, so that you can see it.

2:29

The printed information on the capacitors won't do you any good because they're custom run from Cornell Dubilier for me. So you'll have to find your own but they won't be–a custom run [Question. How about the transformer?] The transformers, you can find that at any supply for neon tubes. I'm still waiting for that person to come up and calculate what's reading on the label of the high voltage transformer. Because it's actual printed and it tells you that it's an over unity device right there. It doesn't say that in words, but if you calculate what they put on there, they have printed their test results from that high voltage transformer and it comes out considerably more than what went in.

2:30

I'm going to put it down, and we can continue with questions if you like. [Question. Since approximately 1850, 1875, the days of Maxwell, Henry, and Faraday, the electromotive formula has been written I = E/R. Your device clearly suspends the balance of that equation. My question to you is how would you rewrite the equation to explain the excess energy that's being produced by your device?]

2:31

If you read the early printings from Cavendish laboratory in England–the people there were connected with Maxwell. If you read the books and the printed material that was occurring at the time that Thomson did his work on the electron, and that Hertz did his thing, and the other people, you'll find that Maxwell actually didn't think the way that he's presented at this time. The thinking of Maxwell is almost exactly what I've been telling you, and my written material. What they put in shorthand form to illustrate what was happening, was very much in doubt at the time that it was done. There were serious doubts, and I can show you in print for a number of people who were headmasters in Cambridge who wrote the books and things at the time that Maxwell was coming along, expressed serious doubts about the shorthand methods of which things were presented.

2:32

We need some more questions. I'm sure there must be a lot more. It's going to be difficult for me to respond to you once we get away from here because of the number of people that. I see we have some new light bulbs, that's perfectly fine. [Question. Is that okay if we hook them up?] Yeah! why not? Why not? I'll be glad to burn your light bulbs out. I had some back of the thing here, you didn't have to go get them. [Question. These are 100 watt light bulbs.]

2:33

I tell you why they are–what they are–because what you're fixin to do is burn this man his camera out over here. That's exactly what you fixin to do. Because unless he's got a shield on there to protect it when this thing comes on your camera won't be much use after that. Do you have some kind of filter that you can work with?

2:34

Because this thing is going to knock your camera out. It's going to knock your camera out when I turn these lights on. [We can turn it down.] [Question. I'll hook my meter up to it again.] Okay, turn it around or what? He wants you to move back so that he can take a picture picture of it.

2:35

Now what I'm gonna do is kind of tilt this sideways a little bit so that he's going to be able to see. Are you hooked up? What's your read? [Question. A little drop in voltage. It's about 64 volts, 65 volts, which translates to something near about a 100 on the AC. It's taxing it a little bit.] It'll correct itself. [Question. Yeah, it's coming up, 68.8.] Are you impressed? [Yes!] What's it reading now? [Question. 68, still holding steady.] It'll come up. It'll eventually will come up to about 112 maybe. [Question. That's calibrated for sine wave.] But you have to correct this reading which he's got because the is reading a different method.

2:36

Yes sir, I can be answering questions while this is going on. [Question. You mentioned that there's a battery in the suitcase unit that's being recharged by the power from the system.] Yes. [Could that battery be replaced by capacitors?] Yes, in part it really is, because they... What's it doing? [Frequency is still steady at 60 Hertz.] [The voltage is dropping, I don't know if I want to keep on going, I don't want to burn up the suitcase.] That's not going to happen, the circuit in here will start heating if it's overloaded. [The voltage is dropping, it's now at 58, 61, 62.] I may have to do some adjustments on it.

2:37

Feel more comfortable now? [Question. It turns out that with the 100W light bulbs.] We've got basically a thousand watts there going on. [Question. With a thousand watts of light bulbs there that 12 volt battery inside the suitcase would have to put out about 83 amps of current. I don't know that anybody taking a battery about that big and draw in 83 amps out of it, you can only do it for a very short period of time. We were running over three minutes with that and there's no way that that battery could do what we just saw. But it did it! Thank you Don.] You're welcome. [Question. My question about the capacitors by replacing the battery with capacitors?] Well actually, the two things do somewhat the same thing. A capacitor and a battery serve somewhat the same purpose.

2:38

A capacitor actually does it better than a battery. [Question. So it would be possible then to just replace the battery with capacitor?] In power correction systems, it's a capacitor bank, and if the power goes off the energy is there and it continues on, even though the powers off. I think we need some more questions. You don't take your bulbs back? [Question. No that's fine, you can have them.] We just ran out of time I've been told. Thank you very much.

Video 3 Schematics

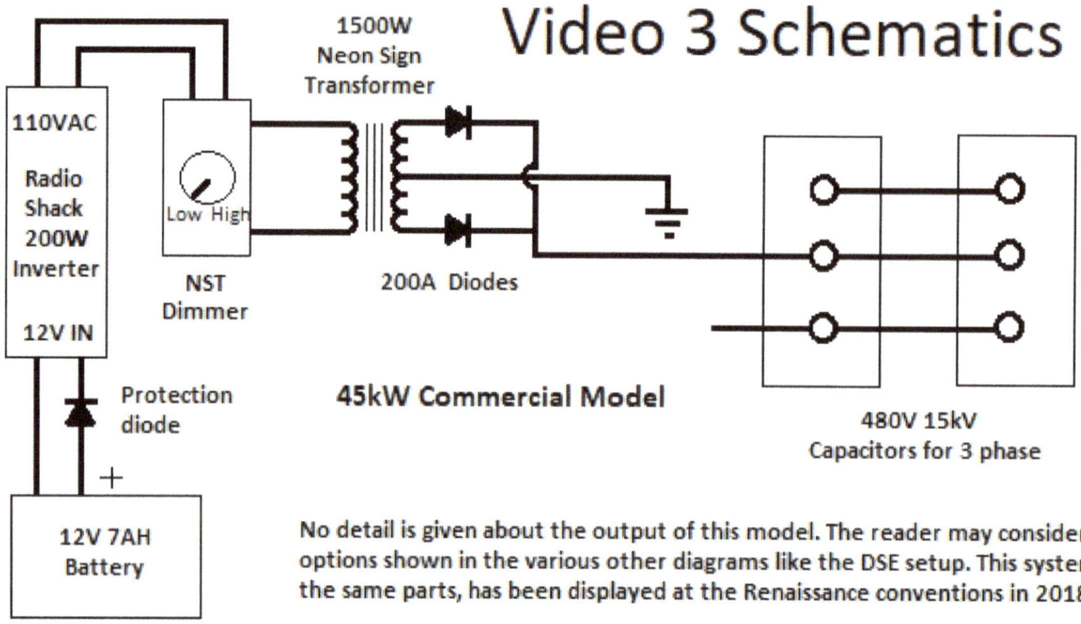

45kW Commercial Model

No detail is given about the output of this model. The reader may consider the options shown in the various other diagrams like the DSE setup. This system, with the same parts, has been displayed at the Renaissance conventions in 2018.

Device 3

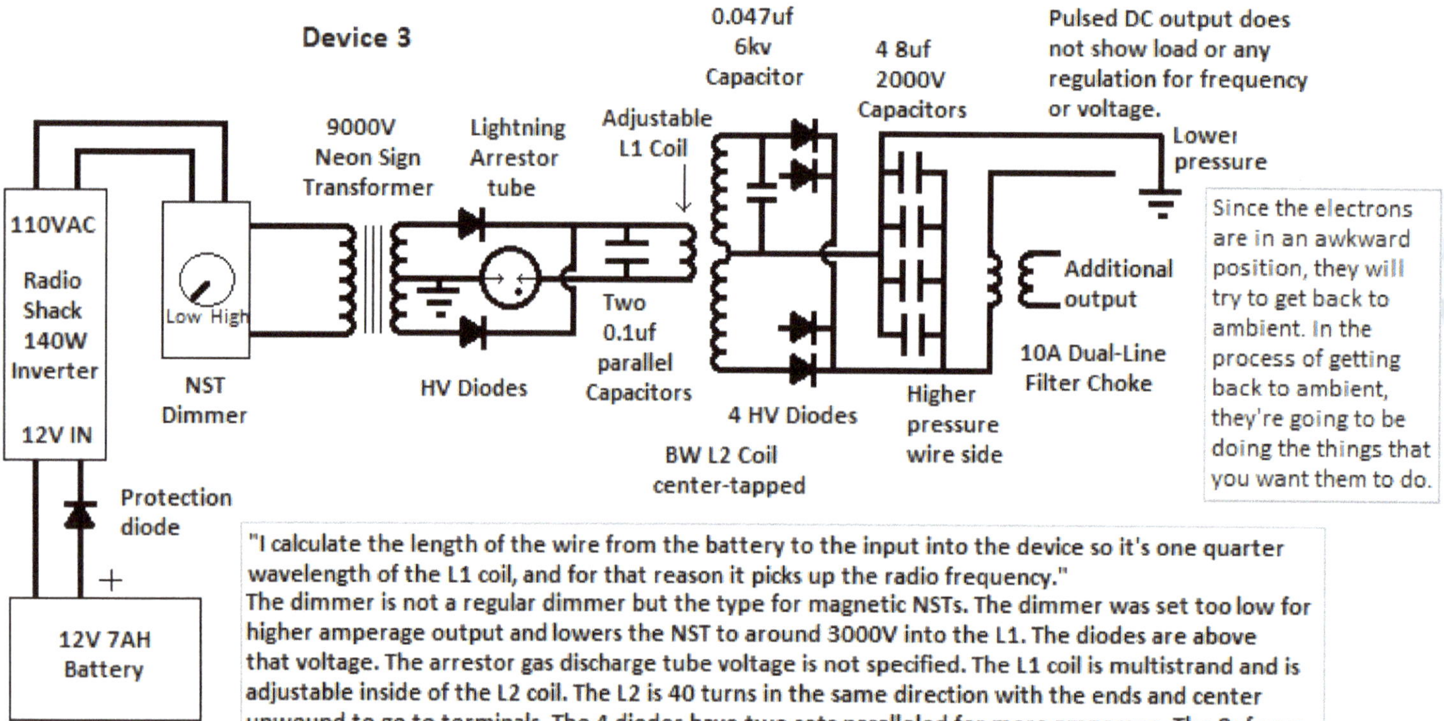

"I calculate the length of the wire from the battery to the input into the device so it's one quarter wavelength of the L1 coil, and for that reason it picks up the radio frequency."
The dimmer is not a regular dimmer but the type for magnetic NSTs. The dimmer was set too low for higher amperage output and lowers the NST to around 3000V into the L1. The diodes are above that voltage. The arrestor gas discharge tube voltage is not specified. The L1 coil is multistrand and is adjustable inside of the L2 coil. The L2 is 40 turns in the same direction with the ends and center unwound to go to terminals. The 4 diodes have two sets paralleled for more amperage. The 8uf caps Don sometimes mistakenly called 8000V. The choke can lower the frequency, and the second side can be used to power additional loads. The output will depend on the load used and grounding setup. The key is to get the L1 and L2 to "talk to each other" in resonance. It is a mistake to try and copy this exactly instead of understanding how it works and using parts available to you.

Resonance Energy Generation System. Patent Pending

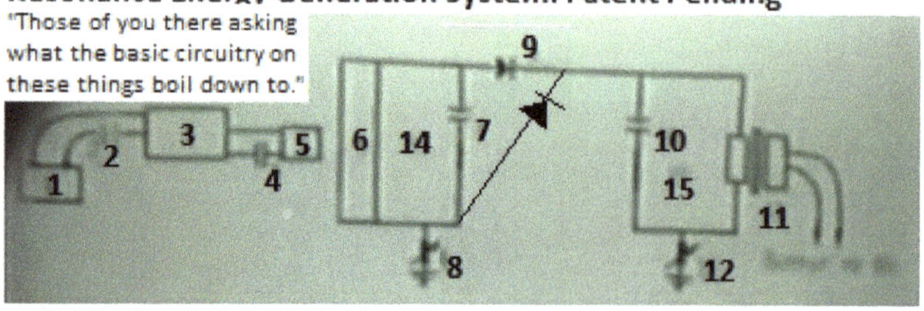

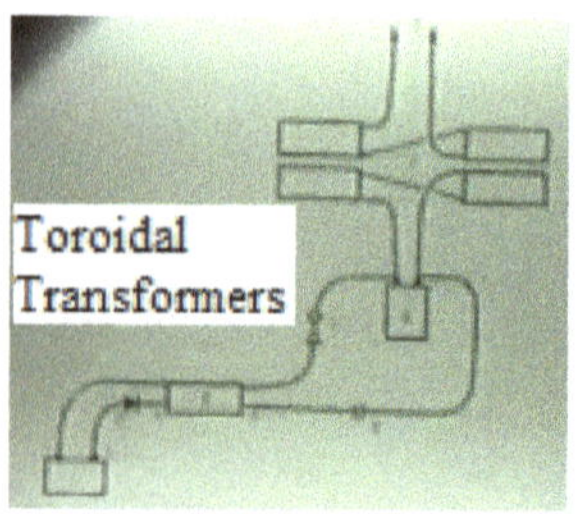

1. 12V battery.
2. Diode
3. HV module
4. Spark gap lightning arrester
5. L1 coil
6. L2 coil
7. Capacitor for resonance
8. Varistor
9. HV diode, adding second diode to eliminate sawtooth wav
10. Cap for resonance with transformer 11
11. Transformer
12. Varistor to keep it from getting out of hand.

Video 4, 1998

[Question. What have you done?] I've been enjoying myself that's what I've been doing. [Question. What have you got going here?] We have an energy device that takes electrons that are available just about everywhere, and converts them into useful energy. [Question. How do you do that?] Once you move the ambient, would be the background, and once you move energy away from our energy source, away from the background, like moving water up a hill, it will always come down. In case of electrons if you excite them, or the level of excitement is above the ambient, or background, you have useful energy. [Question. How did you come up with this conclusion?] When I realized that it was the movement of the electrons, or the acceleration, that was the principal source of the energy.

1

Then I had to develop a way to capture it or transfer it into a more useful source. [Question. What was your intention by doing this?] By doing it, what was my intention? Basically I had heard that it couldn't be done. All the experts told me that and that didn't ring very true to me. So I decided I would see if what they were saying was actually true. When I realized that the components in an electrical field has a alternate component of magnetic field, and that they're equal and opposite.

2

That's your Lorentz or your equal and opposite EMF, opposed sort of thing. They do not appear separate. Being in the field of geology, there were electromagnetic maps which were available, which told me that the Earth's surface had a tremendous amount of electricity in it that was useful. In trying to resolve that in regards to the activity of electrons, I realized that energy is everywhere at all times in great amounts, and that it's dormant until you disturb it or cycle it. The cycling would be the resonant activity seen in radio devices. So you're moving electrons and you're creating electro and magnetic fields.

3

By being able to capture these fields at their higher levels of excitement, you have a gradient that's greater, getting back to ambient. So therefore you have a greater amount of energy. [Question. What's your background? How did you come up with this?] My background? I'm a professional in science. My work history was primarily connected with petroleum industry. I have engineering background. I'm fully familiar with all of the establishments ideas in regards to electricity and electro mechanics and physics and all the other things that relate to it. I have always had a very doubtful eye towards anything that the establishment might consider appropriate.

4

So therefore when people told me that this thing would not happen, to forget about it, that was the match that lit the haystack. I think it's the end of the road as far as energy goes. I think that if there are any other developments, this one is benign as far as the environment is concerned. It's very cost effective. It's something which can be all over the world. It's not something which belongs to any one individual any one country. The one side effect of this, is it generates ozone. Which it actually puts oxygen back into the atmosphere, and none of the energy methods available today do that. There's almost nothing, there's really nothing out there that you can think of that does not use this system.

5

[Question. Is this similar to what Nikola Tesla was trying to achieve with free energy?] It is a spin-off. Basically what I have done is pick up where Nikola Tesla had reached, and I've put myself inside his mind. I've expanded and extended what Nikola Tesla was doing. [Question. Do you think you actually figured out what he was doing?] Yeah, I can read his mind almost perfect. Actually, I did not know about Tesla until I was into this thing more than five years, I guess. [Really?] Yes, and I knew the name, but I had no idea exactly the significance of it. People started saying, well you're doing what Tesla did. Building the same kinds of coils, and doing all the same types of things.

6

At that point, well it really got interesting because I realized that there were other people out there that were thinking along the same line. [Question. Now, you had mentioned something about James Clerk Maxwell, what you know about that?] In case of James Clerk Maxwell, he was professor at Cambridge University. All of the things which Maxwell did had been done previously. He'd put them in a mathematical shape, so that they could be described and repeated. He's one of the fellows that I mentioned that I have told all the pieces of the puzzle except one. His thinking and mine are amazingly close together, because I have researched and gone back to his personal thinking. The books that are out, which I have some of the original ones, original set, were put together after he died, by the faculty at Camebrige University.

7

They reflect what was politically correct to the faculty at Cambridge University. They do not reflect what Maxwell actually said and thought. [Question. So you're saying, if he didn't have to go the whole peer acceptance process we would have had a different book?] Yes. [Really?] What actually came out of it was not Maxwell at all anyhow. It was the faculty in the physics department at Cambridge University with what they considered politically correct. Oliver Heaviside was your $E=MC^2$ source. It was lifted by Einstein, and I can show you exactly where Einstein got it from. [Question. Really, how much earlier are we talking here?] Between fifty and a hundred years earlier. Most people think that I've kind of fell out of a tree.

8

But when they realize that I've had total access to the Library of Congress, to dozens of university libraries. Including here in the United States, overseas, including Italy, Germany and France and England. I've gone through these libraries like you wouldn't believe. I knew from the first moment that it was going to work and how it was going to work. [Question. Really, were you trying to build something specific at the time?] No, all I was trying to do is disturb ambient. [Question. Really, now tell me something about these radio frequencies. I understand you're using radio frequencies to achieve this.] Well the electron, when you cycle it, or move it, you get a magnetic field, and you get an electrical field. The faster you move the electron, the cumulative effect of that magnetic and electrical field becomes greater.

9

If you're cycling at 60 or 120 cycles per second, you'll get some amount of electrons. Then you cycle say at a million cycles per second, the curve on the increase in the amount of magnetic and electrical field there is doing this. When it does this, you capture at a higher point, you've got a lot more energy from the same amount of electrons, then you have at the lower point. This, to some people, would say that you're getting free energy. [Question. What would you call it?] I would say you're getting what's there. You're getting what's there, the term free energy is bastardized terminology, that's used by sour grapes establishment.

10

They're all basically the same same thing. They're called ether, aether, atmosphere, air or whatever. They're all describing the same thing. It's a hula hoop mentality. It's whoever happens to be writing the book or talking at the time, and they're grasping at something at which they don't really understand. In the 1920s people discovered that in studying the Earth's crust that it had electrical characteristics. Most people consider that the Earth's electrical field is useless or near zero. But when you actually look into it you find that the earth has an enormous electrical charge on it.

11

It's just that it's everywhere about you. It doesn't vary in a short distance. So all of the instruments that are made to measure electricity are made to measure a reference point and another point. So the electrical flux, when you start measuring electrical flux or magnetic flux, which is the other side of the coin, you find that the earth has an enormous electrical field. No, I had the conclusion before I ever started. It was what got me into it to begin with. [Question. Really, so this is this is sort of what keyed you into what was going on?] Yeah, I knew that there was an electromagnetic. You can't have one without the other, and it flips. When it flips, at that flipping point is where you have your useful energy. The more you flip it the more you have useful energy.

12

[Question. Does this have anything to do with the planet running at 8 Hertz or anything like that?] Well the death frequency is seven cycles per second. So the earth is, depending on who you talk to, has a frequency of around seven or eight cycles per second. That's consistent. Now if you built a device that would capture that you would have an energy level, but it would be very low because you're not flipping the electrons that much. Now in the other Earth's electrical systems, there are a number of them running all at one time. You have the ionized layers that are up above. There're one side of a capacitor. The Earth's surface is one side of a capacitor. So you have an enormous electrical potential between the ionized layers and the Earth's surface, and you have an exact duplicate of it in the Earth's surface. Because in a capacitor one plate has to equal the other one, or else you've got discharges between them.

13

It's the turbulence in the atmosphere that disturbs this balance. Where you get your lightning and other electrical disturbances. So the Lightning is over unity in itself. Ambient, or earth ground, is unity and anything that you see other than that is overunity. So that means that lightning is over unity. Everything in nature is of course simple, so all the other things including molecules and everything else, they all attach themselves to each other. So in the case of electrons it's unreal to believe that a single electron will be floating out here without being interacting with something else.

14

So as it turns out there's good evidence that some electrons have more negative charge than others. And being more negative or less negative that that's the same as being plus or minus. So therefore the electron with a weaker charge will attract one with a stronger charge, and they will be orbiting in each other. They're relatively easy to pull apart. So the electrical and magnetic charge of the electron, of course, when you spin them, spin them right you've got a magnetic field. When you spin them left you get an electrical field. So when they spin apart you're getting a benefit of both of those. And then whenever they come spinning back together you're getting the volt's and the magnetic field coupling again, which means your amps, amps times volts equal watts, or you've got useful energy.

15

Well actually this thing remakes the entire world of energy. It's a very simple supposition. When put in force, it creates a world in which there's no pollution. It creates a world in which there are no not-haves. All of the political systems are on equal standing, so you don't have a reason for war. So that saves a lot of lives. It also puts oxygen back into useful form which also will save a lot of lives.

This is the state of Colorado. The colors on the map represent levels of electromagnetic energy. You can see there's quite a diversity.

16

[Question. What are all these colors here?] The dark blues are extremely negative, the reds are the lower ranges of negativity. The contour lines on there, each contour represents about 5,000 volts difference in the electrical field. [Question. What would this be used for normally?] These were originally used by the United States government in looking for radioactive materials during the Cold War. [Question. How did you came up with the idea for looking to these?] Well they have been used for locating oil gas deposits since 1920. They've been used since 1920 for locating oil and gas deposits. [Question. Why do you use them, what your looking for with these maps?]

17

Why would I normally use them? [Question. Yeah, what would you be looking for?] Basically about ten years ago I had reached the level in the petroleum industry where my salary was too much, and I had too much experience. The people that were in the company all had less, and they felt uncomfortable around me. So suddenly one day I found myself without a job.

This has to do with the frequencies which things move at, whenever you get into resonant energy. Resonant just simply means that you're flipping the electrons. This is the electromagnetic chart that was worked up by Westinghouse, and can be purchased through the Edmonds Scientific for $11 in case you want to buy one.

18

Or the Exploratorium out in San Francisco. They're the people that had it put together by Westinghouse. [Question. What does this have to do with your technology?] It shows the cycling, or the frequency, radio frequency. It starts at say zero cycles per second and moves up through the various frequencies into the gigahertz, and then it goes on even beyond that. So Farnsworth, for example was the person who actually invented the television tube that we have today. He did that in the process of working with the exact same type of thing which I'm working with. The television tube itself is an over unity device.

19

[Question. Is It?] Yeah, it's an offshoot of a dynode and a number of other things. A dynode basically is where you have a radio tube, a cathode, which is the input side. And you have an anode which the electrons move from the cathode to the anode. In the case of the dynode you have alternate anodes in here which this thing is bouncing off of. If you have a certain type of material on these anodes, one electron leaving here and bouncing to this one knocks a 50.000 electrons loose. Which bounce to the next stage which knocks 5 million electrons loose. Which goes to the next stage, which knocks 50 million electrons loose. Then the anode here collects them and you have 50 million times the amount of electrons active here that you had here.

20

So you have gained 50 million times what you started out with. [Question. It's some kind of multiplier?] It's a multiplier, and it was renamed by a Farnsworth called a multipactor. Basically he was using like a radio emitting substance here. You have this same thing in a light detector in cameras and other things, which open and close the shutter based on the ambient light. A small amount of light hits the cathode which knocks some electrons loose, and they start bouncing through the dynode system. They end up being enough electricity to operate the shutter in your camera. So you have an enormous amount of a output compared to the input. [Question. Interesting! Is there anything else you like to say about this?]

21

Well the same thing applies in changing frequencies here. The same amount of electricity inserted at a lower level raised to a higher frequency, will disturb more of the electro and the magnetic field. Which at that point you capture, and you have a huge gain over what you put in. [Question. Fantastic, so that does fall right in what your talking about.] So you have not violated any laws of physics but you have an enormous amount of energy that you did not have to begin with. [Question. What do you call it when you capture this wavelength? Do you have a name for it?] When you capture it? You can temporarily store it in capacitors, batteries, coils or various other systems temporarily. Then you can convert it to some other system that is useful to the device which you need to power. [Question. So the energy doesn't tare around wildly?] No, it's just running all over the place.

22

Sometimes it gets loose in the form of sparks, or spikes, or whatever. The devices which I have invented, which there are a number of them, they all actually accelerate electrons. They are electron accelerators. One of the outcomes I might mention of the multipactor thing which Farnsworth had, which is a over-unity device without question, is the magnetron in your microwave oven. So all patents are related to magnetrons, multipacktors, dynodes and all related patents are overunity patents. There's no way to deny it. If the person has even slight degree of intellect they will know immediately that that's the case.

23

[Question. So this is sitting in front of us the whole time?] Yeah, it's been here since 1920s and 30s in grand form.

Before we go to the other devices this is a component which shows up on some of the devices. Can you see the reading on it? I'll bring it even closer if you want, so that you can actually read it [Bertonee Gas Tube Sign Power Supply: NPS-12D10]? The reason I want you to read it is because, calculate the input which the manufacturer specifies. Then calculate the output which the manufacturer puts on there for that same amount of voltage. You'll see immediately that this thing is putting out more energy than was put into it. The manufacturer actually states the input and the output and it's on the label. They have tested it or they would not have put on there. [Question. So it exists?] Here's an over unity device looking right at you.

24

There are a lot of them that are that way if you bother to calculate the input and the output. [Question. What is this?] This is one of many devices which I have built [see page 154]. I would say one of several hundred. This is one of the intermediate ones. The technology present here is largely 1800's. [Question. What is this here?] What is the device? It's a resonant electrical energy generator. [Question. What does it do?] Basically what I do, I take a low voltage 12 volts at about seven amp hours from a battery, and I run it through the device. The first thing I do, the high voltage module here has to run off 120 volts.

25

So since I'm intending to use the 12 volt battery I have to bring the voltage up, so that the high voltage module can run. So this is an inverter here. [Question. This is just a Radio Shack part?] Yeah, everything you see here, except these–most anyone could get–and these. Now, this and this, unless you get them from me, you don't get them. [Question. So this are your homemade ones here?] Well they were special made for me. [Question. What are they?] Well these are 10,000 volt capacitors. This is a 9000 volt high voltage module that is similar to the little one that I showed just a moment ago.

26

This is from the same people. This is about 6,000 volts, and this is nine thousand volts. This is essentially a solid-state device, and so is this. [Question. So how does it go here?] Where does what go? [Question. Where does it start?] It comes in here, and comes from a 9 [12] volt battery. There's a cord that's calculated to a certain length because it's a frequency that works with the quarter wavelength of this coil here [red L1 coil]. The length, physical length of the input from the 12-volt battery. The reason for that is, that it will recharge the battery if it can communicate with the radio frequency magnetic field that's in the area. [Question. So it's constantly recharging?] Yeah, the magnetic field from what's going on here will recharge the battery faster than it's being discharged.

27

How do I know that? If you take a light emitting diode–electricity can only flow in one direction in a light emitting diode. Take two of them, and reverse them. Make a pigtail and put them in the line that comes from the battery to the device. Depending on which ones lighting up you know which way the electricity is flowing. When they're both lit up you're taking out and you're putting in at the same time. The reason you can do that is that they are different frequencies. We've got 12 volts coming in. We have a kickback diode here which prevents a kickback from the inverter to the battery.

28

We also have on the input where the 12 volt comes in, this pigtail set of a light emitting diodes. Which tell you that you're charging your battery at the same time that you're using it. Once we get past that point and get into this inverter, this is 9000 volt high voltage module which would be a version of the Tesla coil except it's a solid-state version. It does the same thing as a Tesla coil except that it's solid-state version of the same action. So in order to control that, since we don't want to use the full 9000 volts. I have a neon tube transformer control here. Which controls the amount of the 110V that comes out of this, and into this. I can use the whole 9000 volts or I can use any part of it.

29

The reason I have to cut it back is because what's over here cannot take what's over here. I'm going to end up with a lot more voltage than I can manage, so I have to I have to reduce it. It's easier to reduce it on the low end. Once it gets to the high end you have trouble, special equipment and a lot of trouble of varying it. So I vary on the input, on the low side. I get to the high voltage module and on neon tube transformers they have two output wires. So I have to take the electricity that's on the two output wires, and put it back together as one source of electricity. Then I have to pump it through the L1 coil, and then back into the system. Then I have the capacitor bank here in order to help regulate that.

30

[Question. I'm sorry you said that is 100,000?] These are 10,000 or maybe 15, I don't recall just at the moment [4000V 0.1uf]. But they're more like 15. But they were done special order by Cornell Dubilier for me. They were free. Because most of the companies like that have a department where they will make up samples for you. If you're doing research and development and you need something which they don't have in the catalog, they will make them up special for you. That's exactly what happened here. This is the broadcast end of it, or the part that sets the electromagnetic field into oscillation. [Question. How do you do that? Is it sending a radio wave through it?]

31

When you pulse something, and you pulse it through a coil like this, a magnetic field builds and collapses. It's the collapsing, building and collapsing, off and on that spins the electrons. The magnetic field on, and then the magnetic field off, and they'll spin back to where they were. So you've got the coin flipping with the electromagnetic component flipping. Once you've got it flipping then you're in business. [Question. So that's the whole trick to this?] Yeah, once you've got it flipping then you've got to have something to collect it with. Then there's the rest of this is the collector and storage part. [Question. So you have just a couple of capacitors there?] No, there's 8,000 volts worth of capacitors there [four 2000V 8uf capacitors]. [Question. Are they special made?] No. [Question. What's the little tube on the end there?]

32

Which one, the yellow one? That has a double meaning. In this case it has no meaning, because I'm already charging the battery from over here. [Question. Oh, so that used to charge the battery?] But the energy coming from here before it goes out to the transformer, is passing through this one. That's building a magnetic field. This one is not connected to this one in any way, except through a magnetic field. So this becomes a generator of electricity and you can charge batteries from that. [Question. That's great, how much energy would you say this little system can produce?] This one, what's the output on it? [Yeah.] It's about 16 kVA. [Great.] Which is a lot of energy. [Yeah.] It could go higher than that. That's with this thing turned down on low. [Question. Wow, does it run itself out?]

33

No, there's nothing here to wear out. The capacitors are self-healing. [Question. This is it. You got a closed loop system that charges itself and kicks out enough energy...] Once it's going it's on its own. This is sort of an intermediate model. There's a larger one sitting over here. A more commercial type. This one in the suitcase calculates out at about–I put in the book a little article I gave you, I think it was 27KVA. It's about half the size of this. So this is capable of probably around 50 KVA. [Question. Now is this something you give away as far as selling plans to people, or showing people how to build one?] Well the plans for some of these things are in the publications.

34

[Question. So this is not any great secret, this is out there?] No, it's looking people in the face.

...commercial model [page 166], and the absolute output level on it I don't know. But some of the components in there would probably limit it to about 35,000 volts. Probably about 200 amperes. That is an enormous amount of electricity. You can see that it has the same components that the other device has, they're just scaled up some what. [Question. Have you used this one here?] I started it. I've tested it. But I haven't ran this particular one for any extended period of time, except to just start it. [Question. So this was just a sort of experiment to see if you could build a commercial model?] Well, basically it was just to see how the components fit on the board, that's what it was.

35

Because I did the schematics, and passed them on to other people, and they're building them.

How to build your own device!

[Question. Use that on this...] Actually I'm going to show you how to make a simple one. Most of the things that are in my devices which I demonstrate were put there because people expect to see them, not because they need to be there. So this device [Bentonee NPS-12D8], which I had turned upside down, is the high voltage module here. So we come out and we need a L1 coil here. There's going to be a L1 or input coil.

36

Since this is alternating current, it's neither plus or minus. What it's going to be doing is coming over here, and into here, and coming back, just like that. So we've got our high voltage device, and we've got an L1 coil and it's all working just fine. At this point is running at 35,000 cycles per second. The method I'm teaching you here, you don't have to know about tuning. The length of the wires that go on this coil here are basically irrelevant, because it's being pumped by the high voltage device here. That's going to set the frequency. The length of the wires are not going to have anything to do with it. So you don't have to know how to tune it to do this.

37

The next thing is, that we're going to have an L2 coil here. It's going to have coils on it, and it's going to be out. This is going to be your minus, and this is going to be your plus down here at the bottom. Electricity, or electrical field, always moves from the highest of concentration to the lowest. But in this case what's happening is, each coil there has a capacitive and inductive characteristic. So it's magnifying. For example, say that this is 3000 volts here. We have ten turns here. So this is volts. We have ten turns. We divide 10 into 3,000, and that means that each turn is going to equal 300 volts.

38

So each turn on a L2 coil here, each turn on there will have 300 volts on it. So if you have 30 turns you've got 300 times 30 which would be 9000. So, they're going to have 9000V on your L2 coil. It's going to do that regardless of the amount of turns or the frequency or whatever.

39

What we're doing by forcing the 35,000 cycles per second onto this first coil, we're forcing that second coil to duplicate the same thing, as long as it's an even division of, or multiple of the wire length in the first coil. So we've got at this point 9000 volts out. We probably will want to put a spark gap of some sort in here. This goes to your earth ground. This center part of the spark-gap will go to earth ground. This is going to limit the voltage so that it's within the range which you can handle it with whatever device you're going to load this with. So you can go directly from this into your transformer out here.

40

It will be an isolation transformer. At this point what you have to do since you've got a frequency which the isolation transformer cannot use, you have to correct the frequency. So in order to correct the frequency, you put a resistor across the two poles there. In order to determine what resistor size you use, you go to either the American Radio Relay League chart [pages 96-97] showing a capacitance and resistance and other types of things, which will tell you what approximate frequency you'll end up with. Now in the case of American Radio Relay League you're going to be off the end of the chart, and you're going to have to take a little piece of paper, and make the grid, and extend the chart over about an inch or so. An inch or so down in order to get say 60 cycles or 120 or whatever it is you want.

41

But you can do that real easy. We've got through our 12 volts, and none of the electrons from the 12 volt battery have left the high-voltage module. The high-voltage module has excited electrons and are separate, just like your transformer here, where you have the two sides are separated from each other. So that none of the electricity on this side actually goes through the system. What it does through induction, it excites the coil on the other side, and your electricity over here is coming from the earth grounding. Your load is going to be in between the earth grounding and the other pole of that transformer.

42

You're going to have to also put some sort of voltage control on this thing here. That will be a varactor [or varistor]. What that is, like on your computer where you have a voltage control thing that keeps the spikes from coming in. It looks like a little capacitor but it's not. It basically will limit the voltage to some level, which is okay for the device that you're trying to run. What I have shown you here, you don't have to know how to tune it. That takes care of itself for you. You have to make only one correction in the frequency. Any coil and resistor, or any coil and capacitor, or any combination of those will give you a particular frequency.

43

Now RadioShack has a book on electronic tables and such and you can use the charts in it. [Question. Is this the smaller simplified version?] No, there's some solid-state versions of this where all of this is thrown out. But I'm not at liberty to talk about it just now. [Question. But this can give somebody an idea that you don't actually have...] You don't have to know what you're doing at all.

Build Your Own Device

From The Office Video 4

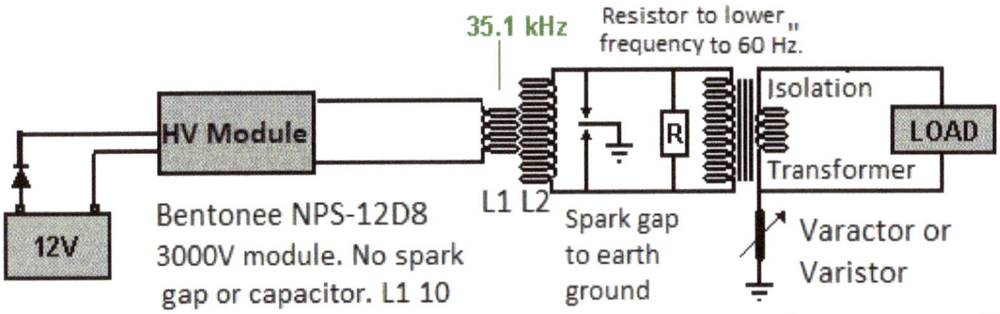

Bentonee NPS-12D8 3000V module. No spark gap or capacitor. L1 10 turns so 300V each turn. Each turn on L2 coil will have 300V on in. With 30 turns you have 9000V.

None of the electricity on this [output] side actually goes through the system. What it does through induction, it excites the coil on the other side, and your electricity over here [output secondary side] is coming from the earth grounding. Your load is going to be inbetween the earth grounding and the other pole of that transformer."

1. 12V small battery to power module.

2. Diode (1N4007) on + from battery to module.

3. Cut battery wire 1/4 wavelength of the L1 frequency (into an even division of that).

4. Use 3000V module to do this exact setup.

5. Cut length of L1 wire so that it is 1/4 size of L2 (includes wire ends). "What we're doing by forcing the 35,000 cycles per second onto this first coil, we're forcing that second coil to duplicate the same thing, as long as it's an even division of or multiple of the wire length in the first coil."

6. Make 10 winds on L1 and 30 winds on L2. L1 can be inside or outside L2.

7. Add a 0.0025" spark gap on L2 ends. "This is going to limit the voltage so that it's within the range which you can handle it, with whatever device you're going to load this with."

8. "You can go directly from this into your isolation transformer."

9. Add resistor. "At this point what you have to do since you've got a frequency which the isolation transformer cannot use, you have to correct the frequency. So in order to correct the frequency, you put a resistor across the two poles there. In order to determine what resistor size you use, you go to either the American Radio Relay League chart [page 96-97] showing a capacitance and resistance and other types of things, which will tell you what approximate frequency you'll end up with." You're bringing this down to 60Hz from much higher. You need a frequency meter or an oscilloscope to know the L2 frequency and an LCR meter to see the primary of the transformer's inductance and capacitance. Capacitance can be added across the primary just as the resistor. But most resitors also have capacitance and inductance that have to be factored in. This is the hardest part to get used to doing.

10. You're going to have to also put some sort of **voltage control** on this thing here. That will be a varactor [or varistor]

Video 5, 2001

And once you upset the balance you trigger off a lot of things that are happening. At that point if you have a device for collecting the energy you've got a generator. Because the generators don't make electricity. By the coils and the magnets moving in relation to each other they upset the background balance. When that happens, well then you've got energy available. If you have a way of collecting it, you've got energy. It comes from the ambient background. How do we know that? Of course from here to here there's not much going on, because both places have the same ambient background.

1

But if you look at how that ambient background out there. The earth rotates at a certain speed and has a certain weight, so we can take that. We can calculate the amount of energy that it takes to make the earth turn. Then we can convert that into watts, and we know exactly what the ambient background is. Which is enormous. This device here off course is a–you see the field out here. I can go quite some distance from it. Any collector system I put inside this magnetic field will generate just like an ordinary generator. [Question. Magnetic field or B field?]

2

This is used for tracing wires and things, but the magnetic flux in the area is what's running it. You can get these at [?] or just about anywhere else. You know something we didn't bring in? This one, the batteries run down on it, that's why it's not working. This one is working just fine. My briefcase? I need it. Where did I put it? Or did I even bring it in? It may still be sitting out there in the automobile. It's going to be interesting to do the speech without my overheads.

3

We do have one overhead which we can put on there if you want. I think I left my briefcase in the automobil. Suppose somebody handed you this particular slide that you're looking at, and you would wonder what's going on there. How many of you know are familiar with this? No one familiar with it? That came out of the Australian deal.

4

Unfortunately when I got out of the car, and got the things out, I had an enormous number of real nice slides that I had made to show you. I really don't have any of them. So I'm going to have to do this without the slides or other things.

5

What we can do, I'm sure that there are plenty of questions out here and you're familiar enough with the types of things that we'll just kind of start, and take you one at a time and we'll do your questions since we don't have any slides. Who would like to start it off? We start over here. [Question. Can you give us an explication of the unit?] You wanna know about these devices? [Question. Yes] Early on when I started first studying electricity, the first device I'd like to show you will be this one [page 153].

6

If people say that you can't get more out than you put in, this is a good example of it. It runs off a 7 amphour 12-volt battery. This is a Tesla coil here with the L1 and the L2. It puts out about 30,000 volts, this central Tesla coil here. That magnetic field that we were talking about over here, it's all over the area. As a matter of fact, you can go anywhere in the neighborhood and if you have another coil that's tuned to that same frequency, it will generate the same amount electricity as the one that is in the center here. If you notice the little neon two bulbs here from Radio Shack.

7

What they do is when you tune these other coils to match this one–due the fact they're slightly different you have to tune them. When you do that those little neon tube things light up. I'm going to have to use a handkerchief. I had a several strokes and I don't have good control of my muscles in my mouth. You have basically three receivers here, and you have one that's producing. Each of these will duplicate exactly what's here. So we've got three times more energy. Well actually we've got this one, and then we've got these three, so we've got basically four times what we're putting in, coming out. It's undeniable.

8

If you can build one Tesla coil you can build three more just like it, or 100 more like it. Or you can fill the whole room up with the other ones that match same frequency. That should answer the question whether you can get more out than you put in. But actually you're not getting something for nothing because it was all there to begin with. All you're doing, you have a generating system, you have a collection system, and you've disturbed the magnetic field around it, which is enormous and it covers a large area. Anywhere in that magnetic field it's going to equal whatever is coming out of this one. If you want to build something which is relatively simple, which answers the question as to whether you can get more back, than you originally put in it, this would blow anybody away that might not like what you're doing.

9

This I built about 15 years ago, and it survived up to now. So you're getting to see it. We have a lot more to talk about and we have time to cover. Does this [page 167 bottom image] look familiar in anyway? This is Tom Bearden's device about 10 years before Tom Bearden knew that he could do it. This is the one that was in the black suitcase in a 1996 presentation at the Tesla convention in Colorado Springs.

10

There were engineers in the audience, that had their instruments with them, and they tested it as producing 120 volts at 80 amps. This is kind of small for something that's coming out of a seven amphour 12 volt battery. These things become self-sufficient. When that happens, because all you have to do is have an extra tap off of the thing there. That will produce more electricity than is required to run the device. Once you get that initial ignition going it's self-sustaining. There's no moving parts here. If you can stay within the tolerance of the parts it will probably last for a very long time. [Question. I think that one wire is ...]

11

You see the one, that extra wire that's in there. That's something that Tom Beardon didn't have. How that happened was, I used the wire that you used to hook up the speaker systems. The reason for it was the fact that they were multi-stranded. I thought well I'll just have to pull these apart in order to get a single wire so that the thing will work. So I decided to leave the the second wire in there, and turn it into an antenna system... [Don get's shock from device when touching.] It's running right now off of that [plasma column]. Anybody care to come up and touch it. It's kind of like touching a thing you poke cows with. [Question. Am I right if I think the one wire is the positive and the other wire is negative?]

12

What it does, it boosts the ambient background that it's exposed to. Sort of superchargers it. This thing puts out,.. Bearden was getting a 1.8 I think on return on it. Well this one will get maybe about twenty or thirty times that. It's probably due to that little antenna that I just accidentally touched. It's high frequency that's why it didn't kill me. [Question. Is that a battery?] It doesn't have a battery on it right now and it's running. [Question. It's a neon sign transformer.] It's running at full force, let me tell you. It's like getting hit with a cattle prod thing. No battery.

13

Anybody cares to test me on that? [Question. Can you explain the components?] How this came about,.. I won't be able to hear you very well, so you'll have to get me later with your questions. But how this came about, you see the toroid that's in there? The United States government bought a lot of those things in Alamogordo, New Mexico. I paid about ten thousand dollars a piece for them. I found the whole box full of them in a surplus place. I didn't have the slightest idea of what they were, Or what they were used for. I didn't know what Metglas was at that time. No one else did either because it was highly classified. When [?] and the other people declassified it, that's the first time I realized what I actually had.

14

Because what this is, it's a toroid. You make it with metal tape, silicon steel type thing. You wrap it around the form, and you get any size, or shape, or form, whether round, or square, or oblong or whatever. That's how you get the toroids. I have some of these that are 14 inches in diameter. They probably cost the government probably half a million dollars a piece, and I've got six of them. In my website, you'll see a large round toroid that's a little push cart [page 157], and it's got a Tesla coil built on that thing.

15

What happens there, when you do that, it's a cyclotron type of deal. Once you get the electrical field circulating in that closed system there, you can put other coils around it, and however many as you want. You're not taking electricity. None of the electricity that started out in the system ever gets through it. It's the point A and point B thing where you get electrons jumping up and down at point A. They do not go from point A to point B. At point B they jump up and down in resonance to the ones at point A. That's your resonant energy type thing. Once you get the resonant energy type thing going, and you get the magnetic fields going, and that sort of thing. Well you can tap off of that system, without taking away from it.

16

You can take off as many other coils as you can get into the area or onto the system. You can take that much off. It's not affecting what's being put in in any respect. It's highly obvious and it. Also people will tell you, you can't do this sort of thing, but you've already seen one example here when I got shocked just a minute ago. Those that are skeptical about it, I invite you to come in and touch this thing here, and tell us that it's not working.

18

Working from the photographs there, the pancake type deal, is that what you want to ask about? You want some general statements on it, as to the why it works, and how it works, and that sort of thing? Whenever you take take a coil, and instead of being a normal... Let me put this down, before I touch it again. In a coil system like in a normal Tesla coil, they're normally high and skinny and that sort of thing. That's the wrong way to go if you're want amps and stuff, because the length is volts and the diameter is amps. If you get the right combination of the length versus diameter, than you've got plenty of amps, and plenty of volts. But also depending on how you tap onto the coil, will determine whether you get amperage.

19

But there's the same amount of amps there as there are volts. Because in the magnetic system that's driving it, it has to be equal and opposite that's your Lenz law. You know that equal and opposite. That means that one of those is your magnetic field and the other is the electrical field. Magnetic field is amps. [Question. What about bifilar coil?] How is it made, what's the ratio? The center of the pancake coil is certain size, and the other ring is a certain size. Then the L1 coil is maybe a couple of turns on the outside of it.

20

The ratio of the diameter of the inside versus the outside turn on it is the improvement, or the advantage that you gain by doing it that way. In the one that's in the picture in the book there the ratio is one to five. So not only do you get the L1 and L2 gain, you get five times that. That's the advantage of the pancake coil. [Question. One wire or..?] The the L1 one wire should be an even division of whatever the length of the L2 is, that way they are able to talk to each other. The frequency, if you worry about the frequency, don't worry about that. Just build the thing so it looks nice and make an even division.

21

Even multiple or even division of the lengths of the wires if you want to talk to each other, and they'll be talking to each other. [Question. What's the ratio?] The ratio in the case of the one that's in the book, is one to five. So it's five times, plus the fact that it's a normal Tesla coil , so normal gain times five. [Question. Do you ever do quarter of the wave length, like Tesla did?] They will be able to talk to each other. You can calculate the wavelength of those, you take the speed of light and divide that by a certain factor and that will tell you how long the wire will be for a certain frequency. What I did on that particular one [see pages 100-101], I used about two hundred megahertz which meant the wire is going to be about eighteen to twenty feet long. I think that's eighteen something.

22

Once I had the main frequency set up, then I made the other coil. As long as it's a even division or multiple of, you know, which ever way you go, they're gonna talk to each other, and that's all you need. There are a lot of different things. This [page 154] thing's a little bit heavy, and I see it's got a lot of sawdust on it because it was sitting.

23

You notice that this main coil there, has the center part of it cut out? The two cut things are brought over as your ground deal. The electrons are spinning one direction, going this way, and spin the other direction going that way. So you've got electricity and you've got amps, and when they come together it continues on. You've put the two together and you've got more amps then you know what to do with, and more electricity you know what to do with. So it has to be cut back. This is a diode bridge here. You go in and out of these inverter systems since this thing has to run on alternating current at 120 volts. In order to get to that from the battery I had to use an inverter. People make fun of me because I use RadioShack inverters.

24

But in this case the high voltage thing there has to run off with 120 volts at 60 cycles. That's how I get to there. I've got a rheostat here where I can raise and lower the input so that over here it comes out to a correct amount that I want. Plus, this is 8,000, well they are 2,000 volts each so it ends up being like 8,000 volts [2000V 8uf each so 2000V 32uf total]. I have to be careful not to get beyond that. They are 8 micro farads each. I have some 50 farad capacitors which are newly available on the market, and they actually go up to well beyond that.

25

The way they manufacture them today, they are down in the voltage level which is too low. So about paralleling them, and other ways, you get up in the voltage range that you want. [Question. So you use super capacitor?] Yeah, I have several of them and I use them in some of my larger units. I may tell you, one like this, if you stick a screw-driver in there while it's running about half the screwdriver evaporates. I know it makes a noise. It'll unnerve you and the whole neighbourhood.

26

You'll have everybody in the neighborhood for walks around running outside to see where the explosion was. You'll end up with a screwdriver with a handle with just a little bit of metal left on it. [Question. Why do you have the white cap on one side?] That one on there? When I got to that point, it was out of tune. I had to do some tuning on it, to keep it going on. The reason for this is, it's also a tuning mechanism. But it turned out when I got built, it didn't really need very much tuning. I never really had much of.. This is your old slide radio thing, or you're tuning antennas. The L1 is the red that's on the inside there, and this is your L2 system. But it's a double one. Set up so that the amps and volts and everything are in there in a proper order. This one I built in 1994.

27

There are over 200 of these devices in existence right now. The more recent ones are subject to business deals and things, so I can't show you them. The idea is to try to fix it so that you can get the idea, and you can build your own. I had something in my briefcase that I wanted to give you, because everyone wants schematics and this particular one was so simple that it was ridiculous. What I'll have to do is give it to Bruce here, and let him post it. I actually ran out of ink printing the darn things up to bring over to you. Then I walked off and left them. I had a lot of help this morning in getting here, and that's why I was late.

28

Let me tell you about these. People were asking me about them, and they're not something that you can buy. They're not in a catalog anywhere. When I call Cornell Dubilier and ask them, do you have this? They are very high voltage, if I remember correctly, they're about 10,000 volts [4000V]. They're 0.5, if I remember correctly, 0.5 micro farads.

29

They're impossible to find is what I'm leading up to. Cornell Dubilier asked me what I wanted them for, and I told them I need them for research and development. They said, well how many can you use? I said, maybe a dozen or so. They said, well there, I think there were 350 dollars each, and they said well fine, you'll be hearing from us. I forgot about it, and about 6 months later they showed up in the mailbox. For free, compliments of Cornell Dubilier. I found out how to work that system now, so I've got a lot of things which don't really on the market exist, they exist in people's minds, but there's not a demand for them. That'll give you some idea about it.

30

These diodes here, again like I said on the Tesla coil, the length is voltage and the diameter is the amperage. The same thing holds true on diodes. It probably holds true on a lot of other things. These diodes are long, and these are 25,000 volt diodes. They give you some idea. Everything that's in here, is probably an over-kill, but the results are real interesting. It does work. [Question. The device right beside the caps?]

31

A lot of your electronic devices have things which control the voltage, you know, like spark gaps. That is a radio frequency spark gap. It's used for lightning arrestors on computers and things that are sensitive to voltage. If lightning hits, anything over a certain amount, will go to ground. That's the ground close on that high-voltage thing there. I'm grounding it back into itself. When it's running, that flashes, you see the flashes in it all the time. That's like my little neon things that I used to check the circuit with. A lot of times I'll put them in different places, and I know up to that point, that everything is okay. Because when it's running they light up.

32

That's getting pretty heavy about now, isn't it? This device that you see on the slide up here–and all of these things–you have a dipole type field where you have minus and plus and so on and so on. If you have a situation like that, and you interfere with that with some sort of collection device, then you've got a generator. You'll notice that there's no input on this, except the ambient around it. If you go back to Faraday, in the very early days of the Faraday generators.

33

They were basically magnets with some sort of something that would interfere with, and that caused them to kind of go off and on. In this case the plus and minus are on the poles of the magnet and you're interfering with them in a certain way, so therefore you're getting 700 volts. This is straight out of the era with no–you see the little antenna things like that second wire that I have on my other device? That winding up there simply is taking from the air, the ambient background, which is enormous if you do the calculations.

Serway's University textbook for professionals and college physics, SERWAY. He's actually the editor of the book.

34

Each chapter would serve–many many chapters there are done by some worldwide expert on that particular subject. The part on electricity and such is done by a certain professor or certain university. Serway is the editor of the book, and it's a collection of experts from all over the world. It is used as a professional physics in the university. That's the one where they get their start. It's the one that has the statement about the earth or the weight of the earth, and this and that. Since we can calculate the ambient background from the movement of the device–current or movement is always in the direction of the current movement. If we take this energy coming in from outer space, we think the Sun as being the major source of energy. It's just a tiny bit of it.

35

If you look at the electromagnetic spectrum you have from a few cycles per second, into the multi thousands of gigahertz, I guess. The Sun is kind of in the middle. What this matters is, that when you have a swing, current swing like that, from plus to minus, that's like the speed on a generator, if you run it say 700 cycles per second, you get a certain amount. You run it at 2100 cycles per second you get more. It's a rate of disturbance that's going on there of this ambient background. The generator is not generating electricity at all, it's actually disturbing the ambient background, and it's collecting the energy out of it.

36

All of these devices are doing the same thing, except they're doing it with radio type devices. If you want to see that, the computer programs for schematics, if you get one that's with nonlinear system, which quite a few of them are, and a open system, which is quite a few of them are, you can actually do the schematics on these things. These programs have oscilloscope, and all your meters and things are virtual. Its built into it. You can build your device with schematics, and you could test it and see how it's working. Then you can play by switching out pieces, until you get the right combination.

37

I can tell you that these things which are called over unity show up absolutely the way they are. Some of them may be rigged, but there's several of them that I know about. The first one I picked up off the internet was out of Slovenia. They had a lot of examples. They had done dozens of examples of schematics where the output was thousands of times greater than the input, and that sort of thing. They gave the schematics that they did, and showed what they looked like on the virtual instruments. It was highly obvious that there was nothing terribly unusual about getting what people might call over unity. The over unity has an albatross hung on it in order to make the person look ridiculous that thinks that it might be there.

38

It's intentional, and driven into all the books. This all started back when Tesla confronted some people early in the century. They found out that there were things out there which might be disturbing to their monetary interest. Since they contributed to the universities, and this and that, well they started making sure that when the books were published, that all of that was written out of them. And they still do. There are a few people that are bold enough to write books that are in universities. But in generally-speaking, they don't last very long. University is there for the system, and it's there to make a rubber stamp out of you.

39

If you don't like being a rubber stamp, you're in big trouble. Apparently there are enough people with that renegade attitude that it's beginning to change. I also forgot to show you this, it's called a electro sensor and it's does the same thing as the other deals I have except that it's measuring in Gauss. This thing is greater than 30 Gauss. If you convert that into electricity, based on the original definition of Gauss, if you do it the way that textbooks tell you, you'll end up with something that's not significant. But, one Gauss is equal to 100 hundred volts RMS, or 100 watts rather.

39

When you get into your coulombs, and your watts, and all this and that, the definition is the same for all of them if you notice. There one volt for one second at such and such. It ends up a Coulomb or Watt and so on. There are about four of those things, and there all actually the same whenever you know the value. With Gauss–Hans Christian Oertsed and Gauss were professors over in Europe, and quite some time ago they were trying to measure the electrical value of the Earth's electrical system.

41

Off course one of them did it at night and one of them at day time. One of them got 50,000 volts, and this was an imperfect measurement they made because they did it by flying a kite or you know like a friend that likes to fly kites. The night time value is about half of what it is in the day time. This comes from the fact that in the Earth's rotation, when the earth is spinning, it's spinning into the cosmos so to speak. On that spin inside it has a greater electrical system then on the side that's away from it.

42

That's why the two are different. A Gauss is like 500,000, the way they were measuring it, equivalent to 500,000 volts And the other I think is about 1/5 of that, if I remember correctly. When you get my age here, you don't understand what's going on there. Anyhow, it's there, and if you've got the right shovel you can get it. That's kind of what you're working on whenever you're attempting to intercept it or intercede. Of course in order for it to work you have to have more of something, and there has to be less somewhere else. The leg in between the more and the less is the active system which you're going to be using to get your useful energy.

43

As these things are working right now the energy is not useful. You have to do certain things to them. You can measure what's going on here, but there's three or four different components that have to be in there in order to bring this down to where you can actually do something with it. You can work it either from the back side into it, if you know what you want. You can back engineer it through the system, or you can come the other way and put controls on it so that you can draw output on the other end. You need instruments that will work with radio frequency. The reason for radio frequency is that the more of these up and down things you have, it's whatever it changes direction there, that it throws off either a magnetic or electrical wave.

44

On the electrical side you're pretty well boxed in and all the rules and the books and things, you're in Ohm's law and all that sort of thing. If you're in a closed system, and you're boxed in, Ohm's law works. That's on your electrical side. But when you move over onto the counter part of the magnetic side, none of those laws apply. It's a whole different system, and most of the critical information they're is missing, intentionally. In effect, what you have on the magnetic side, you can make as many copies of it as you want, without depleting the source. That's where your units that you're interested in why they work.

45

If you know that, then you can just turn your nose up at those people that are trying to make it on the electrical side. Because they require very low temperatures. They're in a very complicated situation. It's got you where you can't really do what you need to do. If you're intent on doing it, stay off of the electrical side, as much as you can. Let's see how we're doing? We can stay here till this time next month and will not get the subject covered. If we can somehow maybe find out what's bothering someone here, and see if we can untangle it for him.

46

[Question. When you see when the big power transmission systems going through a right of way, some people have taken coils and put them where they can get some of that energy.] They're not taking away from the company at all. If the power company says you're stealing their electricity they're exposing their ignorance. Because it's just like the magnetic field around this thing. If I took something from it–you know put coils in it and produce electricity–it's not affecting the output at all. [Question. Their statement that they can find those things by the power drain is just...] They can probably locate it because the magnetic field is repeating itself or reflecting out of the thing which...

47

For example, a common case there is where someone has a fence that's not grounded, that's running parallel to the power system. It can be thousands of feet away, and that fence will have a charge on it. It will kill you if you're not careful. But usually the humidity and other kinds of things ground it, so it's fairly safe. What really happens is the power the company we–you have neither on matter or energy, it's neither destroyed or consumed, everything is always there. It just simply flips over into some other temporary arrangement, and then it will end up going back to where it came from. So nothing has been used.

48

So therefore, when the power company sends you an electric bill for electricity that you use, you did not use any. It just simply flowed through the system and went on out and went back into the system. It's not used or worn out, or degraded, that's not the end of it at all. It's just like at your house when you turn the light switch on, and ground that system into the earth ground. You haven't destroyed the electricity. Nor when a machine heats up and there's heat involved, and this and that. All of that electricity is still very much there, it just passed through resistive point and therefore you get the heat, but the heat is something in addition of what was already there. Just like if you put other coils in the system you get repeat of energy in them. That heat is just something which occurs as a side effect.

49

It in no way has degraded the electricity. It just flows on through the system and goes back into it. That's your ambient electrical system. I bet we've got another interesting question out there? [Question. I got couple from a buddy of mine and then I want to ask a couple of mine. You always talk about you have your business interest, your business buddies and stuff. I know you got certain things you have your contract obligations. Will you be able to lease your units anytime soon?] I myself, I'm going to stay out of that. The various business interests are the ones that are going to be involved in that.

50

I wouldn't sell you a unit if you offered me cash. I've had people offer me like five million dollars for one of these devices, and I turned them down. [Question. Who would we go talk to, you know?] I would say, kind of wait and watch and we're not looking for publicity, so if you ask me that question again one day soon, I might point you to somebody. [Question. The other one he had was then, have you solved the problem needing a ground connection, if not, have you looked at the Hendershot circuit, which uses the same principle?] Grounding works in several different ways. People think of earth grounding as being something that goes in the earth. The Earth's atmosphere is also part of the earth, so you can you can have air grounding, or you can have a ground grounding.

51

In citizen band radios, and the ones that are in automobiles moving around and things like that, they frequently are grounded into the metal body of the automobile. Or if they're not that, then actuality that antennae they're sticking up there, is grounding into the earth's electrical system. That's why you can pick up the radio station a long way off. Because it's interfering with the magnetic system and sending out these waves that are different than what's normally there. So it's upset the electrical or magnetic balance, and that's why your receiver up there is receiving it. [Question. Are you working on anything to produce a fuelless aircraft? But that's back to the ground. But if you don't have a grounding can you use the car as the ground?]

52

When you're flying on an airplane they have electrical systems right? There grounded into the metal parts of the airplane. [Question. Right, to drain off the static.] It radiates off of that. Grounding is a relative sort of thing. It's that there's more of something here, and less of something here, so it's going to move between the more to the less of something. It's that leg that's in between there, there's your useful energy. [Question. My personal question for you When we have the simple schematic right here like this?] Yeah, you'd be surprised that it actually works. But there have been a lot of people that try to build it and it they weren't successful at it, and I can't guess as to why it wouldn't work. Because I try to make it as simple as I could.

53

I think what happens is people change the lengths of the wires, or something is a little bit changed there. That is problematic because depending on that power source there some of them are self-grounded. That's where part of the confusion may come in, because sometimes when you hook that thing up if you're not careful, everything you got, goes there. That's the end of it, so nothing happens out here. As a result of that it didn't work and I was wrong. [Question. On the capacitors, that's a great story you know, I love it. But how can I get some?] You find something they don't have and you just call them up, and you get into a nice conversation. Then they ask you what do you want them for. If you say research and development, well they say we have a department that takes care of research and development problems and how many of them do you need? Then of course you tell them, and then six months later, after you've forgotten all about it, they turn up.

54

[Question. You had some handout you want to do, and it was real simple. Could you draw it on this? Or could we just copy it?] Yeah, I can draw it, I remember the schematics.

55

[Question. You got resistors with millions of ohms so you have millions of different values, I want specifics.] The problem with that is, that everyone when they buy their wire, and other things, those things are not going to match with what's on the list. All the things I do are generalities, and if you get one of those electronic schematic deals you can fiddle with it all, all you want to. You might get better results then what I would give you.

56

[Question. I want something, when I walk away here today, I can build it, and put out, you know, something to run a house. A house, that's like 10,000 watts, that will cost me too much in parts, but like on a house where I have a circuit breaker, I break it down...] Any of these things you get are you're gonna produce more power than the system can handle. What you do inside the circuit, you put these deals where circuit limiters, like you have on your computers and telephone to protect them from spikes.

You put those those all through there, and you set the only amount that can come out of it, as what you want to see at the end. So most of it's going to go out of the system, but you will end up with what you wanted at the end of the system.

57

[Question. I see it as a giant breaker box. Here you got the primary. Okay, this is the living room. This is a bedroom...] You would only need on that deal there that one coil in the center is more than you ever could use. [Question ... but I can build a couple of smaller units the smaller coils or smaller circuits that are just dirt cheap.] What I recommend is you take the end product of what you want and work it backwards. [Question. Give me specifics that's what I want.] We'll see what goes on here.

58

[See pages 80-81 for the following.] This is a piece of material like–same kind of chemical composition say of Metglas, or piezo-electrical type material or what, maybe a rod of any size. It may be this size, or this size, or whatever, it depends on the amount of energy that you want out of the thing. I'm working backwards on this. In order for the system to work, you have to have off and on. It has to turn off and on. What we're going to do, is put a–this is not exactly the symbol for it, but it's a NPN transistor. This is the base. I forget which way the arrows point here, so I'm going to leave the arrows off. What happens–this is your input here, and this is your output here. It comes over, and this is not going to be very pretty. Then this comes back here. On the base down here we have a resistor which is our timing device, which opens and closes the gate on this transistor. What's coming through here is gonna be a certain frequency. You're setting the frequency there. This is the first coil. There's a second coil or a number of other coils which are up here.

1:00

What's happening is, the magnetic flux in this piece of metal here, is set up such that any amount of coils you put up here, will duplicate whatever is being put in here. We have our input. I think it comes like this. We have this one there's another deal here. This is going to be your negative side. No, this is a negative right here, this is plus and this is plus here. So this negative one here is coming back over here. The output, this is your output here.

1:01

This is your main input here. [Question. I have a question, in this you have these two shored together is that right?] This one goes around that one, and this one, let's see. [Question. I just want to make sure, this looks like. Well, it's interesting that this is shorted together up here.] I almost swear that this is 100% correct, but the one that you will see later is [page 81].

I think it–it may look illogical there in a place or two but I think that's pretty close to correct but you will know later.

1:02

[Question. Did you came up with any new devices other then..?] I come out with three or four a day. I don't often have time to put them together, but I can say that once you get off onto something like this, that every one of them relates to something else. There's no such thing as one of them. It's kind of like a cat that wanders up at your house, there's no such thing as one of them. At this point someone had some questions a little bit ago that wanted to come up and ask them. Do we have someone else that would like to come up?

1:03

[Question. This part is L1 and L2?] The smaller one wound around is the L1, the inner coil is the L2. That's just for convenience. I named them so that you know which one is which. It's very possible that this thing is not tuned in to that [the plasma column that had powered the device that Don was shocked from]. That's the reason why it's not acting up. The one like this one acting up was this silicon steel toroid there. [Question. Is there a magnet inside?] No, the toroid in there, it's a toroid that's been flattened. Instead of being round, it's kind of rectangular.

1:04

But how you make those, it's made that shape because it will fit into a particular component in some device, if it's big and round it's going to take up more space. By making it shaped like that, it will fit into... They will bite you, at least a while ago, I don't know whether there doing it now, but it definitely bit me. [Question. What is the frequency of this device?] Frequency of it? Of this particular would I think it's 55,000 cycles per second. Frequently, when you build something like this you end up with a frequency that you hadn't planned on. When you work it out on the electronic programs it will frequently turn out something different than what the program says. They do kind of haunt you.

1:05

They're nice to have. [Question. I have one myself. I'm really curious about this since you showed it. This wire here?] What I've done there, two of these wires here, normally come off to another system. So you're not seeing the complete system. You're seeing the main part of it. I've built parts of these systems on different little blocks and then when I hook them up, I've got the whole system there.

1:06

Now this has all of the different components on it, which sometimes there's more over here, and this part, and this part, will be on different blocks. Because I can use them interchangeably with other things I'm doing. [Question. Is there a reason for this load being disconnected?] You notice those two are loose, that's probably why it quit sparking there. This thing was–when I set it down a certain way, it closes the system. I noticed when a fellow came up just now, it wasn't doing that apparently. But when I tested it, it did. [Question. These windings on the inside..] If you look close you can see that it's like metal tape, it comes on a roll.

1:07

They roll it around a form to get the shape they want. Then they dip it in shellac, or something, and it stays put. [Question. These coils, are they mirror to each other?] They're layers just like scotch tape. When they dipped it in the lacquer stuff, well they put it in a form and it keep it there until all that dries. When it's dried, well that's ready for use. [Question. This is something similar to this right here?] They all do the same thing. They look very different, but basically it's your bucket for dipping the water.

[Question. You indicated you use software to model these sort of like Spice. A circuit analysis system. I guess you were talking about software that you used that would have displays. I assume that something like spice. You know the Berkeley Spice.]

1:08

On the internet, if you do a search on the internet you come up with a great many different programs. One of the best ones, it's about a 9000 dollar program is from the University of California Berkeley. They quit updating it maybe ten or fifteen years ago, and the University over Slovenia picked it up. They've been updating it, and then the University of Barcelona got in on it. You can download it from the physics department at University of Barcelona. It's a program that you buy from a commercial source, that quality is about 9000 dollars. You can download it for free.

1:09

But you have to know a little bit about it, because the different components that do different things have been upgraded by different people, and you have to know how to put those components together so that they complement each other and become one program. A lot of your computer programs are composed of other computer programs. So it goes on all the time. It's just in this case they give you the instructions on how to do it. You have to be able to understand what they're telling you in order to get a working copy of it. The others I'll put on the internet for you on the discussion list. I'll give you a list where you can just snap and go right to these particular things.

1:10

[Question. The other question I had is, do these systems show that effect that you produced, the over unity? Those software, do they reproduce in software the effects?] Yes, it has a virtual oscilloscope, a virtual–every type of instrument that you would spend thousands of dollars for it. It has it built right into the program so that you end up with oscilloscope screen with everything going on, and it reads just like an ordinary–it looks like an oscilloscope and it does all the things that are there. [Question. This is an effect that violates Maxwell laws?] If you put this particular thing on there you get very startling results. [Question. So it's nothing that would violate like Kirchhoff's current and voltage laws, because spice is built on that.]

1:11

What you do, you pick out your components and you slide and drag them over onto the board, and put them where you want. [Question. I know what I'm doing in spice. The question I'm having can Spice show this effects? What I'm trying to make is, you have Spice simulator. Can you take a Spice simulator and show these effects, or is this effect something that is not part of standard Spice, or a field simulator?] You can build your own components and you can tell it what you wanted to do so you can get those facts out. That's the reason why you have it, and the reason you run it is to see where the defects are, and where things are happening that you don't want. Then by playing with the components, that's how you take those problems out of it.

1:12

[Question. Spice will show this effect?] Yes.

Not all spice now, it has to be a non-lineair open system. A lot of those are closed systems which put you right back into Ohm's law. You're stuck with what they want you to know, not what you'r about to find out. More and more University people are using the open systems because–open system things are hard to defend. The university people like things that you could do mathematically, and you always end up with the same answer regardless. That way they look good if they're competing with something that's not going to give the answers that they want to be repeated every time.

1:13

Then from university standpoint they're in danger of needing another job. The University is not interested in people that can't repeat what they're doing. [Question. Can you draw up there the URL for this group thing you were talking about, that's on the Internet, where you have people join, or read ask you questions?] You want the Internet thing. I've got a lot of cards but I'll write it on here here for you. The email is: donsm1@earthlink.net

1:14

The website is: www.altenergy-pro.com [See pages 145 to 161 for the complete website.] The best place to find it, is go to Lycos.

1:15

But actually it's on all the major ones. You find it on Altavista, Lycos, and it's actually on over two thousand servers worldwide. You can get it whether your in Russia, or wherever. The students at the University of Moscow in St. Petersburg literally wear me out asking questions. There's a lot of people in Russia think that it's a real great thing. I got one real nice response out of Finland. I've had probably 20 or 30 out of Finland. But one the other day had to do with gravity and new approaches to electromagnetics as such.

1:16

It was all stuff which does not show up in any of the textbooks. I think you all have seen it, it showed up at Oli Tina. First name is Oli, and last name is Tina. He's a professor of physics. I can't say the name, it's down in Finland. There was a little tit-tat going on, on a question-and-answer thing on yahoo there. If you go back two to three days, or a week, you'll see it. He had the most appropriate idea about what goes on in electricity and gravity, things like that.

1:17

It was totally out of the box, but this is somebody highly educated, and he speaks and writes absolutely perfect English without a glitch anywhere in it. If you want to correspond with him, I'm sure he will be flattered, if you would pay attention. [Question. Could you write his email address?] Of Oli Tina? I'll put it on the package of things so that you get it. I'll send you an email on it. We need some more thoughts here. [Question. The alternate energyProCom, that's where your discussion list is at?] No, that's on Yahoo. [Question. Can you put that, that's what I want to go to?]

1:18

[Nuenergy at Yahoo groups.com You can go to my alternate internet resource page off my main website and you can sign up for that discussion group.]

1:19

[Before you can use it you have to sign onto the group.]

1:21
[Question. If people are getting capacitors of large capacity from salvaged sources they're frequently vastly mislabeled as to what the value is.] Oh sure, you have to check them against other things. I've had pretty good luck with it. [Question. I want to let you know there's a way to do this with large capacitors where you do not have to use a capacitance bridge, because capacitance bridge will not give you an accurate answer. The way that it can be done is by measuring the time constant by applying a voltage and measuring the time constant, to bring up to a certain value, this will give you a very accurate answers.] Yes sir, a charge or something which you could put that on. [Question. There's a particular number for it to come up to a full capacity. Let's say it's coming up to 10 volts.]
1:22
[Then there's a time for it to come up to like 6.7 volts, and that gives you the time constant.] This is your percent accuracy thing that they put on there. Because they will vary considerably. [Question. The reason they're mismatched or they're mislabeled is because they're using a capacitance bridge to give you the measurement, and it's not accurate.] If you have control of the situation, and you don't mind spending a lot of money, you can get where they will swear to whatever values they have, and actually give you a certificate on it, won't they? [Question. But you can do it with an oscilloscope, all you have to do is measure how long does it take to value come up to that time constant value, and then you know exactly how much it is. That was a patent that Boeing put together, and I was part of that project, that's how I know about that.]
1:23
It would be nice if you put that patent number on the discussion group. [Question. I'd have to go look it up and find it. But it works very well, and especially for these real large capacitors like a Farad or 3 Farads, or whatever.] Using the powdered carbon bits, you can get an enormous capacity in a small packages now. You can go up to over five million micro Farad's with the present-day technology. And that, everyday, is moving out a little farther. At five million you might not be much bigger than two shoeboxes. That used to be the size of this motel. The things are really changed, the carbon particles take enormous charges. That's what makes the difference.
1:24
[Question. Has anyone ever seen an electron?] Yes. Actually, when Thompson said it was out there–and two Thompson's. There's Lord Kelvin is one, and the Thompson that actually postulated the electron. Not too long ago, they actually can photograph them now. They know what they look like. It turns out they look like just what they said they were. [Question. It might help if we could see them and see what they doing.] Well you kind of sneak up on them. [Question. Going back to the circuit here, can you make a comment on the number of turns used?] The output will be a ratio of the number of turns on the first one versus the other one.
1:25
If you had like two or three on the first one, and you put back ten or twenty on the other one, the output improvement would be the

ratio of the first one versus the second one. [Question. There's no magical number, say one to five that works better?] The thing is you want to make as few of them as you have to, because it's real hard work you know winding those things. Not only that, you want to do it in such a way that you slip that rod out and use it in another device if that one doesn't live up your expectations. After you make the second device, you've got two points to work off of, so you know which way to go for the third one–and the third one you should get pretty close to your target.

1:26

[Question. You'd mentioned that program that's at the University of Barcelona in Spain, what is the name of the program?] I may need some help here, because I'm having trouble turning corners right now, but it's the original name for Spice. Spice something or other. It's the original one that started it all pretty much. It's out of the University of California Berkeley, physics department. It's the one that all the others basically hinged off of. It's spice free something.

1:27

It's got two, three other letters in there. [Question. And that's the University..] The one out of Barcelona, basically I don't know that any one has given it a name. But they tell you the history of it where it came from. That's what it is. So they may have named it, but I don't think on the computer program. It's just they use the original name and the other components which they fit on to it have their name. You've got something there that's got maybe four different components. One of those components will be your virtual test instruments, that sort of thing. Then there will be other ones for building your own devices and fitting them into this library of things which can be used.

1:28

You can build your own devices like transistors and coils, and you know various unusual things, which may or may not exist. They have over 70,000 different components in the library. You probably can add some to it. [Question. Since you know what goes where..] Actually, you're running a risk of frying this because.. These two things right here, and I've ruined a number of test instruments.

1:29

[Question. Can you measure something without blowing my meter?] We don't know that. [Question. I really want to do it, but I do not want to blow this meter.] I'm no trying to talk you out of it, I had disasterous experiences that involve a lot of meters and things that were absolutely nice, that I was real proud of, and all of a sudden one day they quit working. Mostly it was just when they where sitting close to that thing there, when it was running. [Question. You can measure it, without touching?] [Question. Everything is extremely high voltage right?] It's high frequency and high voltage and your meter is probably, unless you got a frequency meter. [Question. I've got a frequency button.] If he's got a frequency thing on there, just simply turn it on, and what you got there you can measure the frequency standing where you're at.

1:30

[Question. I'm reading 43.9 now does this need to be further apart or what?]

[Question. What's the scale, kilo..] Actually, that decimal point needs to be about three places. [Question. Do these need to be further apart?] For some reason, you're reading a component of the thing that's there, you're not actually reading a frequency. I don't know why that is but it's running at about 50,000 cycles a second, that I do know.

1:31

[Question. When you stepped away it changed, come on back over here. 43 now back up, it jumped to 80 there for a second. No still 43 almost 44.] Actually, it needs about three decimal places on it. [Question. If I go to build something, this is the meter I'm going to be using.] You're in trouble. [Question. What kind of meter do I need to have?] You need radio type meters and devices. [Question. Give me some part numbers.] What you need to do is go on the internet and do a search for radio device test instruments, and you get thousands of them. [Question. I wanna know what you use, that's the way I can duplicate.] You would have to just simply come out and look because I've got a lot of different ones. I've got Hewlett Packard high-voltage high-frequency meter, and they cost a bundle.

1:32

[Question. I want to get into something low but easy to make, just like the one transistor thing.] The low end test instruments are things like this, and this is like seven dollars. [Question. If I build the circuit that you showed there with the one transistor, we can't use this?] You can use it to a certain extent. What you do, you can set your own frequency on that transistor by the base there. The timer that you use on the base deal, it's opening and closing the gate there, and you can pick whatever frequency you want. So you can run it at a frequency which is within your test instrument capability. You don't have to go out and buy radio type test instrument. [Question. Just lower it down?] Yeah, just lower the frequency into the range as you can handle and you're in business.

1:33

[Question. I can handle 60 Hertz.] You got it. Anybody know what's this thing is all about? You want me to do it for you? You know if I drop this thing it's going to go just like a stone. So what I'm gonna do I'm gonna put it through this copper tube–and this is a neodymium magnet–and I'm gonna spin it. Can we do that again? [Question. Yes, how do you explain the phenomena?] Well, let me show you what happens when you drop it. See how fast it goes? It's heavy. When you put it inside this, it's none magnetic material, copper, and you spin it.

1:34

In case anybody thinks there's some obstruction in there, there's not. This is from Bill Beaty in California. It's one of his favorite toys. It demonstrates the effect of gravity, nullification of gravity. [Question. Don't spin it and drop it?] The reason you're spinning is because it enforces what's going on and if you just simply drop it will still.. It just takes a little longer when you spin it, and that's why I did that. Sort of emphasize why it was slowing things down.

1:35

You can look down in here, if any of you went to Dennis Lee's show in [?] the other day, he had a big one like this with a mirror setup. It had a large copper thing, with the mirror set here, so that you could see that it was going down through there. It was projected on a screen, so that you could see it in the big scale. If this had more magnets on it, it would slow it down more also. [Question. It's Faraday drag isn't it? Like on the Faraday disk the drag that's being created.] The faster you spin it, it also adds some to the slowdown. You can make it to work with the larger magnets, and a longer tube where you could just sit here half a day waiting for it to fall through there.

1:36

Actually probably three or four minutes sometimes maybe, where it would normally take less than a second. I'm sure you could set up a chart on that, and you could calculate how much or what it takes to neutralize gravity. You know build a graph off of it. Experimenting with different sizes, and spins and so, you could actually project into something that might be real interesting. [Question. You got different materials that you're using to do the coils, which one did you find that was the best to do?] I tell you whenever it comes to choosing what, basically he's saying well there's different kinds of coils, and different kinds of capacitors, different kinds of wire, and the different devices have different things on them.

1:37

These devices occurred sometimes over very long periods of time, and the basic rule of thumb, is that building this type of device you want multi-stranded wire, which is basically the wire used in DC things normally. The more of those little strands are there, the better the deal. Because it gives a surface for the electrons to dance around on, and by having more of the strands, you have more circular mils. Which means that the total surface area is increased by quite a bit, by using the multi stranded wire. That works in your favor, so you won't use what is normally probably called DC multi-stranded wire.

1:38

The larger the diameter the better, but the bigger ones are hard to work with. You pick some size which is convenient for whatever you're trying to do. That's why when you're building these things you will fiddle to a certain extent, and you find that you're getting different results. Once you figure it out which is the better way to go, then as you go along the new ones have the changes built in. [Question. Is it better when it's naked, or covered with something?] I've got on the small one here–what is asking–actually I have the thing disconnected. A couple of those things are actually touching when I was fiddling with it and got shocked. I had them pulled apart so that it's not supposed to be doing anything.

1:39

Somehow, by moving it around, while the wires ended up touching each other, and that's why I got shocked. Some of the other people that tried it, and didn't get shocked. I move the wires after that happened. It was putting out enough that it'll get your attention, that was about equivalent of a cattle probe. Or the thing that police use with the electrical shock stick. It's kind of that equivalent to that. I moved the wires that were touching and that's why it quit working, at least for the time being. [Question. Is there a possible relationship between

these devices and gravity reversal devices?]
1:40
On my website I describe what gravity is. Everyone else is taking a potshot at it, so I figured I'll see if I couldn't come up with something. What I came up with is, that movement in electrical devices is in the direction of current movement. All of the huge amount of electrical background–the earth is like a big black hole. Anytime you have a solid or accumulation of something that's because the matter is kind of hanging on to itself there. It's affecting other things that are similar to it. That's why you end up with blobs like the earth, and the planets, and solar systems, the Sun, and all this and that. For that stuff to move in, it has to be affected by the gravitational effect somehow.
1:41
What's happening there, the energy and electricity–power and energy and matter are all in physics basically the same thing. This movement of the energy out of the cosmos into the solar system, and into the planets, and such like that, is the power which is spinning the earth, which is anormous. This movement into this mass which we call Earth, the energy coming in, is the equivalent of–it's a friction factor of this energy being pulled into the earth.
1:42
That's why when you fall you fall towards the earth. In electrical devices movement is in the direction of the current movement. So gravity is a friction factor of the energy moving into the earth. It's a simple thing, and it's undeniable if the person will think clearly. Gravity is a component of the movement of energy into the earth. It's a friction factor, related to movement. When some motor is spinning, or the magnets are pushing each other away from each other, that's an electrical reaction there, and that effect between them is the gravitational field, either pushing them apart, or pulling them together, depending. Gravity is not something magical. It's not something which can't be described. It's something that you can see the effects of. You can see the effects of it, and you it describe it because it does this or that.
1:43
I like my defenition a lot better then the one that's in the physics books. [Question. Theoretically you could take a flyback circuit and reverse the charge that goes to a TV, which is positive, make it negative, and it would by controlling the rate of charge, it would control the rate of levitation so to speak?] Actually by these quick swings of alternating current, and that sort of thing, that flux–the magnetic field switching off and on–and they're actually into that right now. By using square waves, and pulsing them, they get the very noticable change in weight of any metal device, that they feed that into. But they're actually getting heavy pieces of metal lifting up and moving around.
1:44
[Question. On your circuit, the base of the transistor, were you driving that with a square wave?] It depends on the transistor. There's about nine thousand different ones. But they will give the characteristics of that particular transistor. So whenever you look at a chart of them, you'll be able to do that. An ordinary NPN will work. [Question. Was that a square wave you're driving with?]

Yeah, actually a square wave is just simply this, with the chopped off. It's actually part of the energy there is being chopped off of it. So it is still this whether it's a square here or square here and so.

1:45

[Question. You got a part number for the transistor?] NPN is all you really need to know. You go to Radio Shack and that's a generic description of it. The N means something in each case. The P means something. NPN means that it simply will work with alternating current. [Question. How about the resistor value?] As long as it will work within the voltage which you're going to propel it with–and it says on the package is good for 25 volts or 100 volts or 3000 volts, or whatever. As the voltage goes up, the price goes up. You can get them for any amount, and that sort of thing.

1:46

I know many of you came from long distances and I appreciate very much. I'm very flattered by your coming. I'll do my best to answer other questions if you take advantage of the Internet there.

1:47

Those of you that want books, I have CD ROMs that have a lot of stuff on it related to what we're talking about here. Those that want books I'll make up however many we need here.

1:50

I can tell you that there's more than 30,000 copies of the book that are out there. The book has been translated into every major language around the world. In one day I get like 200 emails from all over the world. About 20 of them are from here in the United States, and they come from very interesting people. Sometimes these are people which you hear about that are in the news, books. You never know who's going to turn up on the doorstep.

1:51

I had people from China calling me and they wanted to make a deal. I kept trying to put them off. So one day when they called, they asked, well how much do we need to make this thing work? I hadn't the slightest idea, so the number I pulled out I thought would stop it. I said 300 million dollars. This was like on Wednesday. Now on Saturday morning I get a call in the airport and these people from Dow Chi Ping's office in China were at Intercontinental Airport and they wanted to know if I could come out and pick them up. So I did, and they had Halliburton briefcases with three hundred million dollars US in cash. So I got them checked in to the motel, and got them a car, and every morning before breakfast they'd show up at my house. It was like nine o'clock at night when they go home, and we would sit at the breakfast table and they'd opened up these briefcases, "please take."

1:52

When they were gone later in the evening I called friends of mine who were in international business deals, and things like that, to find out what do you do in a case like this? They said well don't. They said what will happen will be that I totally take the money and use it however wanted, and that they would take devices from me and charge them against it. Where they made one fatal mistake was they asked me how much do they cost to build these things? I said well they're not particularly expensive, maybe a couple hundred bucks or so if you do them in mass quantities. Otherwise it's probably three or four thousand dollars or more or if you do just one at a time.

1:53

So anyhow they decided that they were going to buy them from me at that price and credit that against the 300 million, and I had no intention of doing that. I was the one to decide what the price was. They were not going to decide. The only way I ever got them to leave was to tell them that whenever I got a factory I'd been touch. I was finally able to get them to close their briefcases and go back to the airport. I took them back to the airport and they went. They're out of Vancouver. Dow Chi Ping had two daughters, and they both married to Americans. This particular one was an electrical engineer in Vancouver British Columbia. The Chinese government had put him on to me, and he still calls about once a month. So they're still interested. But this was about three or four years ago.

1:54

I have things like that happening all the time. But what would have happened if I had taken the money? I would have so many problems I wouldn't know what to do with. I did not have proper facilities or anything else So I used that as a way to get out of it. But if I had taken that money I would have been in big trouble probably. Not that I really didn't want to. Anyhow, that's one way to stay poor if you know what I mean.

1:55

It's great you came and I'm really ticked to see you, and I wish you all the luck in the world. If something doesn't work that doesn't mean that it doesn't. You know if you get the right combination–and I've tried to keep it as simple as I can, so that the options of screwing up are limited. So just do it by keying in on it. Do one and what it's not doing, well then do another one. Once you get several points here well then you can find a point there where it does exactly what you wanted it to. So thank you again very much.

Audio, 2003

[My next speaker is Mr. Don Smith. Now strain situation. His doctor is not allowing him to travel at this point. However he will be speaking with us through the speaker-phone, and any questions that you have can be addressed to the speaker phone. So this may be a personal conference. You know you can't get here but the speaker phone could be just as good. I think it ought to be very interesting. Let me try to get him on the phone here. Where's he from Bruce? Texas. He suffered a major stroke. I'll probably let him explain to you what happened. But a very strong person. He had only a 5% chance of making through. He's a strong bugger.]

1

Hello? [Hello Don, how you doing? This is Bruce from Inventor's Weekend.] Yes good morning. [Question. How's your health today?] Quite good. [Question. Maybe...How your health is at at this point?] Actually I'm on the mend.

2

I had major stroke about a year ago and it pretty well wiped everything out including a speech. But I got the speech back, and except for my left hand, right hand rather, I'm in pretty good shape. [Question. That's really good to hear Don. Do you have anything you want to start off with?] Yes I'm sure there's some fellas there with a lot of questions. There's some general things which we agreed to talk about, like my dipole transformer generator I believe. And why it works, and that sort of thing.

3

If you take basic system it would be like a north-south deal like in your Tesla coil where you have a North South componant, and take a right angle off of that which would be similar to the rings of Saturn. And off that you take two plates, one aluminum and one copper, and put spacers between them. You can copy your energy directly from your Tesla coil in large amounts. [Question. Don, when you say large amounts..?]

4

Of the experimental ones I did here [page 165] I used plasma tubes for my north-south component, and I was getting 65 thousand watts off of that. There was milliwatts in and and 65 thousand watts out. [Question. Wow, that sounds really incredible.] Yeah it's basically unbelievable. But what people don't allow for is the fact that in your electrical system it's dead on arrival, so to speak. It's so amazing that the magnetic flux elements are perpetuate basically forever in great quantity except when you have lower frequencies.

5

So you have to run the frequency up like a Tesla coil maybe, and then you bring it down into a direct current from the system with the rings on it that I was talking about. [Question. so the output is high frequency at first, and what you're doing is converting it into direct current and then you could probably put an inverter on there?] Well the north-south component is electrical and the east-west component is your magnetic flux which when captured on those rings changes back to electrical. [Question. It's based on the dipole system? Could you explain maybe the transformer.]

6

Well the invention is called a dipole transformer generator. The dipole is a North-South component and the generator part is your Tesla coil deal. The other part–you have a generator and you have a receiver kind of deal, so that your generator receiver. [Question. Does anybody have any questions for Don here? I'm asking the members here they have any questions for you.] Yes I'd be happy to answer any question. Remember this is extemporaneous. [Question. How you doing? My name is Mark Whitford.

7

I'm from Columbia Maryland. My question is regarding the tools that we used to measure the energy.] Whether the measurement systems are correct? [Question. I'm concerned about two things. One is whether or not the the tools that you use to measure the amount of energy that was input and output.] Actually I have all kinds of fancy gadgets to do that with and I'm not one bit wrong. I don't know what your concern is that you feel that I've seen something which is not really there. [Question. no I didn't say that. I'm curious about how.] I know what people think. Go ahead. [Question. The question really boils down to what is the accuracy of your instrumentation? In other words, if we're talking about an accuracy of a small percentage, but you're able to compare the input and output ...]

8

When you're doing something like this on an experimental basis there's plenty of room for error. But the preponderance of evidence is there that overwhelms the [?] points one way or another. [Question. Okay so in other words you feel that the results are overwhelmingly clear?] Yes to someone who knows what they're looking at. [Question. Okay have you published those results, and have they been published?] I'm not really hearing or what you're saying. [Question. Have the results been published?] Published, is that what you're asking?

9

[Question. Yes sir, have they been published?] They are in the system of the patent office. But I haven't publicly published any of them. [Question. Okay. And you have any designs or drawings that are available? I mean did you submit a patent application?] Basically as simple as I've described it here you don't really need anything else. It sort of explains itself when you put it together.

10

Do a separate Tesla coil and at right angles like the rings of Saturn place the plates on the ring-like deal there and separate them so that they're not shorting out, and then measure between the two plates. [Question. Okay, and when you measure the total energy output, or what you claim was the power output was 50 kilowatts, what did you use to absorb that? Did you absorb it into like a dummy load for example?] First of all you have to change your radio frequency, and you change it to pulsating DC. Once you've done that then you can use it just like ordinary electricity. [Question. Thank you very much, that's all my questions.]

11

[Question. How are you Don? My name is Greg Dun, and I'm from near Baltimore Maryland. I appreciate the opportunity to chat with us here a little bit this morning. Can you hear me okay?] Yes I'm fine. I understand you guys are having a good time there. [Question. Yes. We have a beautiful hotel, and beautiful weather, and a great conference. We got a bunch of people here all perking up and sitting on the edge of their chairs here listening to you, and we thank you for this opportunity, and hope everything goes well with you. I have read your book that I got a copy of, I guess a year or two ago, and looked at that video that was produced some time ago that had demonstrated. And we've studied your your drawings, and your works, and your explanations quite a bit. I'd like to ask what is the function, and what can you describe what really happens in the spark?] What is happening and why it works?

12

[Question. Well specifically, I want to know about the spark gap.] The spark gap is a very important thing. That's why the Tesla coil thing can be part of it. You have to have the of and on component present in any situation where you're switching between magnetic and electrical flux. What your not switching there you're already getting the electrical component which is used up almost immediately, the magnetic component has no bounds on it. It's an open system where the energy can–depending on the frequency–can travel throughout the universe.

13

At that point you need a system to harvest the useful energy which would be the magnetic flux which, when it passes an electron it flips back and you have electrical components [..] to you and you can have as many of these sets, duplicate sets of the energy as you want. There's no limit to them. [Question. So your concept is that the electron kind of flips between two states or?] Yes flips between the magnetic and electrical impulse, and it sends out waves, and the magnetic waves are not quenchable like the electrical. The electrical component is dead almost immediately absorbed. But the magnetic component is not absorbed.

14

It's either reflected or deflected. At that point when the magnetic component is passing it accepts additional electrons which would go back to the electrical deal. That way you have a endless output from the magnetic resonance. That's why I call my devices the resonance energy methods. [Question. Yeah I can understand that to get many many iterations of this you need to have it in resonance. I understand that part.]

15

Yes, also let me tell you something else. I have copied my book in an Adobe system. It will be available sometime shortly, and working on getting copies of it the CD-ROM copies. [Question. Where will we find that?] Adventures unlimited will probably end up with it. Are you familiar with who they are? [Question. Yes I have seen that. That was in your book.] Yes they're going to have several more books and things tied into the other stuff. [Question. Okay, great. We look forward to that because I think communication is the key to the answer here.] Yeah they actually have the patent work on the dipole transformer generator. It's the actual patent documents are published as part of it.

16

[Question. Okay, Do you have a patent

application number that you will share with us yet? The patent, I decided I didn't like the United States Patent Office so I [...] It's coming along nicely but I don't want to talk about it very much except for that. It's okay for me to talk about what I do. But I don't ordinarily discussed an actual patent just yet. [Question. Okay certainly we can understand that. To get back to your device that was in the book that use the L1 and L2 coils. Are we talking about different device now?] I have dozens of different devices. Most of which you've probably never seen.

17

The dipole transformer thing I'm talking about you've never seen it most likely. [Question. Okay. So that's very different than what you had published?] Yeah they all work on the same principle, however. [Question. Okay and that's based on this understanding of the electron and how it really functions?] Yes, and what the magnetic flux [?] differs from electrical flux. Magnetic flux is not quenchable. It doesn't die out. It basically is unstoppable. Each time it passes another electron it flips the energy from–all of those electron flips added together ends up being a huge amount greater than what the electrical plug would provide.

18

[Question. Now when the electron is sort of excited into the electrical mode, does it have to be near a wire or something to collect that?] Let me say this. Most people don't realize that every electrical device that you have, regardless of what it is, or where it is–that sort of thing–is giving out a huge amount of the magnetic flux along with the electrical flux. Most people think that its inconsequential and it doesn't really mean anything. But if you have a Gauss meter, and you map the electrical flux area, and then do this right angle component on it, like I'm talking about, you could turn any device into a generator of electricity.

19

[Question. Now does this magnetic flux have any relationship to what Bearden calls the heavside current?] It's resembles what Bearden is doing. There's another fella from over in Switzerland that has a deal that it also resembles. It's not put in clear terms so you can understand it. When I described it for the patent thing it's in very clear terms. It's basically the same thing. You know why Bearden can get things complicated. But that's the problem there. [Question. Yeah we're all agreed with that. Very much so. Takes him seventy pages to say something.] He's and absolute genius. We own him a great deal. I wouldn't want to distract from you in any bit.

20

He actually got his thinking on this system, like I'm talking about, from the guy over in Switzerland.[Question. What is that fellas name.] The man in Switzerland? [Question. Yes sir.] Methernitha group. Yes I guess I don't know exactly how to say it. [Question. Okay. When do you anticipate this device that you have of kind of hitting the commercial routine?] Off and on I've had an aweful lot of people approach me on it, and different times I've had things which almost [?]. At the present time most of my connections are overseas, not here in the United States. Intentionally avoiding the United States.

21

So commercial to answer your question I don't really know. It's there. It can happen any day, except I don't know the exact sequence yet. My health situation kind of screwed things up. Because last September I was at the point where I had everything working exactly right, and it had already been commercial. [?] last November now. So I'm still working on it and things look very promising and there's nothing at all wrong with the system that I've tried to do. It checks out in all respects. [Question. And physically, how big would this machine be for comparable, or you know a particular size outfit?

22

Well, if you pick a number in terms of, you know, kilowatt output?] Kilowatts worth of potential? It depends on the electrical output of your dipole system like we agree that you could use a Tesla coil. They come in all sizes. Question. So if you have, say a 10 kilowatt generator, how big would that be?] It wouldn't be that large at all. It probably would sit on an ordinary desk with a lot of room to spare. Question. Now in your book you said that there's kind of a magic number of 20,000 Hertz, above which things seem to change in terms of the way electricity and magnetism seem to work. Would you explain that a little bit?]

23

Okay, the higher the frequency the more often your off and on component is coming on. So each off and on component is add to the composite of the energy that's present. So if you're at 60 cycles, you're not getting very much. If you have four hundred thousand cycles you're really beginning to add up. When you go up about a million the curve on that type of thing is about 90 degrees slop I guess.

23

When you get up to 200,000, and go above that it's basically almost straight up. Infinite from then on. [Question. Is this obeying the square laws as you said in your book? Is the output proportional to the square of the frequency? Is that the concept that you're talking about here?] Yes, whenever the frequency goes up to a certain level the energy is infinite. You're basically plugged into the universe, I guess you might say. [Question. Now is it necessary that this high frequency be at a very high voltage, or does it not?] The voltage of course is a variable.

25

But once you get up into the higher deals, in order for it to be useful you have to bring it back down to pulsating DC, and then you can change it to 60 cycles or whatever else you want. Because if the high frequency and high voltage, which is the basic idea, you don't have good control of it and you had to change it to something which you could control better. [Question. Also in your book you mentioned about this magnetostrictive rod at one point.] Yes, magnetostrictive rods. Okay, the curious thing about electricity is the fact that whenever you have a magnetostrictive materials the resistivity becomes enormous.

26

It's playing a trick on the materials and it's exact opposite of anything that would conduct electricity. [Question. When you put a field around one of these rods?] The fact that doesn't conduct electricity it becomes a perfect conductor for magnetic fields. [Question. So you put a magnetic field around one of these rods and it actually grows longer, right, by a little small dimension? When you put a field around one of these magnetostrictive rods?]

27

[?] put the magnetic field, and they talk about getting zero resistance or even more so. Well in the case of the magnetic stuff you have have this way beyond zero. It goes into the millions of times on the zero side. Where if you did that electrically you couldn't do it. [Question. So the magnetic component?] Works just the opposite to the electrical. [Question. So instead of having as we would think of ohms resistance in something, you have actually an enhancement of the magnetic traveling through this rod?] Yes the idea is to make a magnetic field work for you. [Question. Now when the rod sort of, as I picture this rod it goes through a cycle every time you change the field around it. It goes from.]

28

That's your off and on component your asking about. You have to have something–it's breaking a circuit and alternately. That's necessary in order to start the action. [Question. These rods can accept a very rapidly changing magnetic field?] Yes they have enormous magnetic fields. That's the reason for using them. [Question. So when you pulse this rod and make it go through one cycle, it takes energy in and then it gives it back to you in a different form?] The idea is taking something out, which has some resistance to it, and your putting it into a system where the negative resistance is almost infinite.

29

At that point you're swinging a system around so that you're taking advantage the negative resistance. [Question. I'm beginning to piece this together a little bit here in my head.] By the way those drawings are to give the general idea. They're not the absolute way to do it. But that was a simplest way I could demonstrate it without getting overly complicated. [Question. That's part of the art of this whole process.] Yes everything I do I try to simplify it as best I can. So it may look like I'm a fool but actually I've been very careful to try to make it so people can understand it. If people don't understand, it doesn't mean anything. [Question. Are there other devices besides these magnetostrictive rods that makes sense?]

30

Actually I have a system that I don't want to talk about very much. But I can take the magnetic field and convert it directly into the energy without going through all these gyrations we were talking about. [Question. Can a simple air coil do that?] I'm trying to figure out and tell you what happens without giving the secrets away. I take an [?] magnet, and just like it's a battery. I will take the ambient directly from it.

31

The idea there is that the magnetic flux, magnetic sphere is embedded into the magnet, and when you take it out it resupplies itself. [?] it always maintains the same level regardless of how much energy you take directly from it. One of my inventions which you don't know about is a device which uses a magnet directly without any complications such. It takes the energy directly from it in any amount you might want. [Question. Well, I've used up all of my questions I don't have any more at this moment.]

32.

[Question. How about an address for the video and book that he was talking about from adventures unlimited.] [He want's an address where the book can be purchased, Adventures unlimited?]

33

[Question. Hi this is John Drawer. I'm from Ohio. My question is for the experimenters that are here today. Do you have anything that you can offer up to us. Maybe something that we could build that would give us proof of concept?]

34

[Question. I was asking if you have anything that the experimenters here today could build that would show proof of concept of your device.] I understand your concern and I understand that this is a novel and healthy question. I appreciate a great deal. All along in my experiments I have been irreligious as far as proof of concept and other things. That sort of gets in the way when you have a lot of people wanting to argue about things. So in order to get the work done, and get things accomplished I've not intended for it to be disrespectful in any way.

35

I do have all the backup that anyone who knows what they're doing would have. I have tonnes of it. I just haven't published any of it. Or I haven't been inclined to want to get in arguments with people over it. Hello? Did I answer your question?

37

[Question. I heard you had a representative in Brazil.]

38

I have business connections there. Senator [?] is one of the major Senator's in the southern part of Brazil. He's my business partner along with the Russian government.

39

I've been in Brazil and have done presentations for the Brazilian energy department and other people.

40

I was with [?Petrovos?] for a number of years out of Rio Diginaro. I was the person that [?] when they read their reports for petroleum engineers graduate I was the person took them on and taught them the oil buisness.

42

[Question. Good morning Don, my name is Warren Opheim. I'm from Minnesota. We really appreciate. My question is for your magnetostriction rod. If I remember the diagram it is similar to an autotransformer in resonance.]

43

I'll tell you what that is. That's the basic thing that's used by the submarines for their sonar system. It's identical materials with wiring diagram and everything else. When you have a submarine and your listening for like fifty miles away and it comes back and you pick up the energy and you can what's out there. It's the same energy system as the submarines use. [Question. What frequency works better?]

44

It's magnetorestritive material which I would gather the fequency is probably about forty-fifty thousand cycles per second depending on the chemical composition of the particular material of the rod. I might tell you that stainless steel rod works very [?] [Question. Would this be a hollow stainless steel rod then or solid?] Stainless steel has a dielectric quality in it, and its very sensitive to magnetic things but not very sensitive to electrical. [Question. In the latest issue of Faraday Labs magazine they do a time machine experiment, and they use magnetostrictive material and I just thought that.] Basically, depending on the size of their [?] and how you're going about it it works just like a big Tesla coil so to speak.

45

What's happening is related to one coil and two coil of a Tesla coil. [Question. I guess that's all the questions I have. Thank you very much.]

Video 6, 2005

[Video starts cutting into sentence. Picture shown on page 160] ... consider is electro and magnetic and they're very closely related to various methods. The electro dies a heat death very shortly the magnetic flux field continues onward. It's actually what makes possible radio waves and all kinds of energy applications that are not possible in electricity. So with that in mind I've worked on about 200 or so different devices, and a large number of them actually show evidence which is contrary to what we were taught in physics.

1

My experience with that was when my son was quite young, and in university level, I had the opportunity to see his physics books and I realized that something is badly amiss as far as energy was concerned. With that in mind for about 20 years I've been pursuing the subject in great detail. I like killing snakes. I tend to overdo things so I've spent a great deal of time and very intense effort in learning about energy in the various aspects of it. In regards to that these devices which are going to be seen on a screen here, I think the first one is the eight unit device. That is manufactured in Russia now.

2

The Russians sent over a group of technicians to Houston and we spent about six months working out the details on it. It was replicated and about four four times the size of what you see in a photograph there. It was highly successful and did everything that I said it would do. This type of device normally has the enemy of heat. If the heat is involved in the process it kills the magnet system off. We had to have a device which would not have that heat killing aspect in order for the magnets to continue on for long periods of time.

3

In this system, the way it's constructed the magnets will last basically forever from a human standpoint without needing to be replenished. But what most people are not aware of is the fact that these magnets have hundreds of Gauss in them, and a Gauss is an energy equivalent of about a hundred thousand watts of electricity. That's on a basis of a temporary exposure. When the system rotates it has a shielding thing inside it which breaks the magnetic contact in part. By constructing it in the way I do, it doesn't kill off the magnetic system.

4

The Russians have been particularly successful in developing this device. The Russian Academy of Science, and a whole slew of other people in Russia, followed me very closely and were in contact almost on a daily basis. They have a number of other devices which I have. They have actually patent some through the European Patent Office. So I have a number of real important landmark patents which have come down the line courtesy of the Russian Academy of Science. I think I didn't have to pay for it. They supplied the funds and everything and took care of it automatically from my descriptions which I supplied them. They considered that it's a way to go.

5

We're going to see something that is very

simple later on. It's the pièce de résistance after we get through with this talk and near the end of it. There's two things coming up. There's on electron paring. There's someone here that has developed and designed on this electron pair sequence, which is very clever and helps illustrate the nature of electrons. We'll have him come up and demonstrate that for us before we go into the final part of the show.

6

At this point if we go on to the next device. I'm going to have to be able to partially turned in order to be able to discuss this with you. We can continue with the continuous photographs. There's quite a few of them. As a matter of fact 116 of these photographs that I wanted to show you earlier, and was able to edit it out to about 76 I think. So we're going to have quite a ride here. So hand onto your seat and let's go. We'll take questions later, so if you have a question, make a note of it and we have time for that. Just so you can see this first device from several angles.

7

This is when I was developing the shielding device. I had to make it myself there because it hadn't ... what we have there. Go ahead. There's a sheet of plastic, and we take the cover off, the paper cover off the sheet there and we spray shielding material–a powdered metal, the shielding material–over the thing with after spraying it with rubber cement. I did several layers of that. The part of it is going to be when you pull up the mass there, the center part which you see right now, with the paper will disappear so that the shielding material.

8

There are openings inside the mask there that allow for the magnetic flux to pass through it. Seeing the same thing up closer and there's going to be a point there where we can pull some of the shielding material off and you will see how that it works. This [page 166] is a different device. Will skip over it for the time being. That's a thirty five kilowatt device we just saw, and it works like gangbusters. This [page 152] is one of my early coil deals. I had one now which is built [?] for me in California which I go to visit him after I leave here.

9

I'll be there for July 4th and [?] such annual presentation. The July the 4th the huge coils that he builds. All the news media from all over the world are present for that. I get a front-row seat. But he's building a very special Tesla coil for me, and it's actually three layer Tesla coil with one energy source running three coils. It's 40 inches in diameter and the wire is a much larger than your thumb.

10

A type of welding cable that is multi-stranded. The reason for multi-stranded is that it causes the magnate flux to multiply in such a way that it is very beneficial. The diameter times the number of strands that are available is a multiplication factor in the coil set. So if you put a half a million watts into it the magnetic flux will end up being–say it has a 30 turns or something like that–you'll be having approximately 30 times the output in usable form or magnified to 30 degree.

11

So this is a very early one I built to demonstrate that. This is about twenty years ago. At that point I knew that was getting a lot more energy out than in. It was very easy to see and prove. This energy has to come from somewhere since it's neither created or destroyed. It simply changes form. This energy has to come from somewhere so the only possible place it can come from is in ambient background. Actually, you see it when the Earth rotates or turns it's a lick an electric motor.

12

Something has to cause it to spin. So you take the weight of the earth, and rate of rotation, and you can calculate the ambient energy present in the Earth's electrical system. This is true of all the celestial bodies. What makes them rotate and the degree to which energy is present in order for that celestial body to be rotating. This at all it gets back into the electron thing. Electrons are cause of it all, and we'll have something on that a little bit later.

13

This is a diode bridge that goes with the flat coil we just saw. Since the the frequency which this thing is running down within an easy range of what diodes can do, that sort of thing, you can actually extract the energy very easy–to see the multiplication factor. These are just the different views of the same thing. You can see it's not very many turns but it was very effective.

14

This [page 163] is something else. We had a group of people that were able to take ... When you have pressure on devices and they generated electricity that's what this is all about. This is a pressure-sensitive type device which when we put pressure on it generates an electrical field.

15

With that in mind, most people don't realize that electrical current, or alternating current coming into something like this, is the same pressure type deal, so you're generating electricity with a piece of solid state device... put it together and there it is complete. The idea was I put the energy into the center part there and it will [?] energy out on each side there–a great deal more than what was put in. For example 5 volts was put in and I had a ten thousand volts out, that sort of thing. So you got two sides there, and times ten thousand, that's twenty thousand volts coming out of that thing.

16

What really happens there is that volts and amps of course are very different things, but amps only occur when you have a grounding. So in order to turn this into useful energy it's the way in which you ground it will determine whether it becomes a practical source of useful energy. But it turned out very successful and a very high rate of amplification. We look at some more of the parts when I'm putting it together. This [page 165] is a diode bridge which I designed myself for my plasma tube experiments. In plasma if you have a way to measure the magnetic flux, which there's a number of ways, plasma tubes put out such huge amounts of energy you can't believe.

17

So you put maybe twenty volts in, or ten volts in or whatever, and out of the plasma tube–the energy coming out is maybe fifty thousand or more units of energy than going in. A great deal more than that actually. So if you have a way to capture the energy from the plasma tube in a practical practical way, and you work it down in the frequency zone that the diodes work for you, instead of against you, you can get a lot of useful energy directly for the plasma tubes. This is a sort of a stranger that got in between. This is one of the 200 or so devices told you about.

18

Go ahead. This [page 154] is the inside of one of the more sophisticated designs. That thing in the center there with the coils on it was intended to correct the frequency so that you can control it properly. It turned out that I have calculated everything such that it didn't need that part of the device. This is where the energy passes through one side of it there. The magnetic flux feeds on the secondary coil there, which is the feedback coil. Which means that the device, once it's started it can run on it's own and generate electricity basically forever.

19

That's the details on some of the capacitors and things that are there. This is a device which you'll recognize for measuring DC current, DC flux. It's something you used in combination with several other instruments. This is a high voltage high frequency device for measuring energy. This is one of my early devices with the shielding material on it which keeps the flux inside the box.

20

This is one of the plasma tube devices [page 155] with the diode bridge thing in tact there. See that pyramid there? The people in Egypt there and in the pyramids back several thousands years before Christ had energy devices. What you're seeing there is a photograph of the energy flux in the area of the pyramid, and it's enormous. They [?] or shape is what contributes to that. The base material there would be one type of mineral which had electrical characteristics, say of one type, and the pyramid itself was made it was another kind material which has a different electrical component.

21

Between the sand, which it's sitting on, and the pyramid material, which is a different material, and the triangular shape of it gives us this energy flux. Mountains do this all the time. If you're in the mountainous area you can see the electrical display coming out from the mountains. So the triangular type system relates to the energy which you see in mountain changes such. In central and South America you can see that part good at night. It lights up a type of lighting which comes from nature.

22

Von Dankin built some of the electric lightbulbs that were shown in the pyramids and they worked. It's documented in all of their information sources. He actually shows photographs of electric light bulbs which he duplicated which looks exactly like the ones that Egyptians were building three thousand years ago. Electricity has been around in useful form for quite a while, and we just didn't understand it. We're getting to the point where we're beginning to understand it. When you leave here today you'll have a much better view of what's out there. Because you're not getting something for nothing, and you're not creating anything.

You're just simply changing it from one form to another, and in that change process from one form to another you have useful energy.

23

If the energy is static and not moving, or it's mass, or other forms of energy, you don't really see anything useful. It's only when it's going from one form to another that it becomes useful, and you capture it just like the water running down from the mountain back to the sea level. In between you put a paddle wheel there, and you've captured the energy. This is true of all forms of energy, whether it's electrical or whatever. This [page 159] is one of many. This is a 14 each Tesla coil which I built and engineered. Unfortunately when I got it built, if you don't have the place the coils.

24

Tesla coils have a lot of useful energy and in this particular case it was totally engineered from a computer before I did it. What increases the energy available later would be the diameter of the coil, and amount of turns on the coil, and a number of other other things which indicate it will tell you what your amps and volts and things are going to be. This thing came out when I finished it I got a surprise because the computer program was very wrong. It was very wrong in way which meant that I would never be able to use the coil because there was so much and Amps that thing would burn up like a flash bulb. So actually calculated everything, and checked it before I turned it on.

25

I found out that I built a device which would be like a flash bulb if I turn it. So I actually never got to use it. A big disappointment. This is one side of the pancake coil again. Some of these are out of sequence, so we just skip by them as we go. This is the set of light bulbs which had 10 one 100 watt bulbs and which I demonstrated at the Tesla Convention back, I think it was in figures 79 [1996] or somewhere along through there. It caused a quite a stir because they had the thing turned on and running at full force for a better part several hours.

26

Based on the amount of energy there it would have only lasted a few minutes if it was not something unusual. It caused a [?flap?]. To this day there are people who still talk about it. This [page 168] is one of the couple of hundred devices I've done. This one on the blue part down there is where you plug into. The input is over here. See it's 12 volts. Here's the output. It's at 120 volts at however many amps you want.

27

It was good for about four or five kilowatts continuous from the 12 volt battery. So that must tell us something. It's not something for nothing either. It actually happened. Someone bought this from me by the way and payed me good money for it. They took it home with them. You can see every part of it. There's no need to tear it up to find out why it works. But they did tear it up, and when they couldn't put it back together and make it work they got very upset with me. So I told them if they would give it back to me the way they have received it I'd be more than happy to give them their money back. To this day I've never heard from them.

28

That was quite some time ago. Notice the five hundred microfarad coil there, and another coil set. There is just some of the things that I have worked with at different times. This [page 166] is a capacitor set. This is 35 kilowatt device we're looking at here. You see it from several angles. This [page 162] is a 50 kilowatt capacitor. I have this same size–this is a shoe box size by the way. I have these from several manufacturers, and they go up to over 400 farad.

29

One farad is equal to a kilowatt. So each farad is equal to 1000 watts of energy continuous. The airplanes, and other things that we're involved with, use this type of thing and in order to run run electric airplanes, and various other devices, and very successful with it.

30

A very special device we're being introduced to here [page 172]. See the metal rod in the center there? You can't see the metal rod because it's inside the plastic. We will see some more photographs of it a little bit, but it's from the submarines where you have a resonant device, a resonant rod. In this case it's Terfanol-D. These things were $3,000 each. I had four of them that was playing with it at the time. This is one of the devices I built with it. Basically, what it does: those are capacitor plates in the center there.

31

A right angle component of the magnetic flux is useful energy. Those rods are set at angles to the magnetic flux put out by the Terfanol-D rod. This is at the frequency it's working at and everything we can use a smaller cable which might surprise you that the energy present there is about 50 kilowatts. Your input is maybe about 20 or 30 milliamps with 50 kilowatt output. That should get somebody's attention.

32

The device does not exist. I have a patenting thing in process, and I work through the European Patent Office, and Russian Patent Office. One through China and several other places. I avoid the United States usually because I spent millions of my own dollars trying to work with the Department of Energy and the other people only to be severely disappointed. I guess I was pretty dense because it took me about 15 years to figure out what was going on. Once I realized what the government is doing to me to stop me, I decided to go somewhere else.

33

That's when I started–my website is in more than 200 web servers worldwide in every language available. I get daily emails from the ends of the earth in many languages. I'm able to work this because of the fact that we have translation software and things like. So I have a direct contact with so many people that you can't believe. I found out that in Eastern Europe, those people are about the most open-minded people I've ran into. They're not brain washed like the people here in the United States when it comes to physics. There were several other areas but I found the Chinese are not brainwashed in regards to physics.

34

So I like an audience like that because I like to talk to people who have an open mind. They apparently realized the same thing because I'm told by the Chinese government that they consider me a national treasure. The same thing happened with Japan and Saudi Arabia in several other areas. These are more views of that same device which doesn't exist worldwide. It doesn't exist anywhere because I do my own patent search. I Google worldwide patent search in all major languages before I even do the first device. This one does not exist anymore except where I sit so to speak. It's a new technique. It's a new system completely, and it works magnificently.

35

It appears to violate a lot of laws of physics. But it was based on the fact that Terfanol-D was what these submarines use when they make their sonic signals that go 50 miles out and come 50 miles back. Sound is a form of magnetic energy. It's just like radio signals and a lot of other things. It's just a different wavelength. If you notice the devices here that tell you that there's magnetic flux present. I use those. I have very sophisticated instruments also. I follow the magnetic flux. Anywhere there's a large amount of magnetic flux I consider a source of usable energy.

36

This is something you're going me use in a little bit. When we get through the speech they're going to demonstrate the ambient energy background, and the amount of it out there. Where you can visibly see what I'm talking about when I say that all this energy is everywhere, all the time, and is available to you. It's a very simple process to collect it. I'm going to show you how that happens. That's coming at very end of the show.

37

This [page 170] is one of the 200 or so devices that I did. This was a very unusual transformer there in the center. Unlike coil like transformers this is heavy sheets of copper, and is special made in India for me. It has a very high degree of energy output that wasn't present before. There's three or four points in this device where energy is gained above and beyond what's put in it. You can see that the input is 12 volts. This is a 12 volt battery there. The output down here is–there's different taps there.

38

You can get 110, 220 and I think it's 440. You can get your choice of any of those. You get it continuously. Basically, forever because there's a feedback system here, which you can't really see here. There's a feedback system that works with it. This thing is self-perpetuating. Once it's going it can feed itself and have output in addition to it. This is about twelve twelve years ago I guess. I've been having pretty good luck with this sort of thing for quite a while. Also notice this is one of those 50 farad capacitors. Present day those capacitors have over 400 farad.

39

That's what makes it possible. The electric airplanes and other things which we're going to talk about after a while. This is a closer up of the multi taps on it–special coil. This is at 120 volts, more or less. It's the output measured while the thing is running and generating electricity. This [page 165] is one my plasma tubes. Devices which I seem to specialize in. But you see that metal plate there? There's a copper plate and an aluminum plate looking at you.

40

There's copper plate underneath there that

you don't see. These points here are where they contact the different plates. The aluminum plate is this side. The copper plate is the bottom side. Separated by a plastic sheet. This device here has a milliwatt input on it, and has 65 thousand watt output. It's at a frequency which your diode bridges work, and you can tap it directly. Again, the reason why this energy is there is because of the ambient background which is everywhere–all places, at all times–is part of the earth's resonant background.

41

So you're not getting something for nothing, you're simply tapping into what's already there and using a catalyst in order to arrange the tapping sequence. More views of the same thing. A closer up view. This [page 152] was one of the first devices that built. The object of building the device was the fact that each one of those coils there–one of them is the output coil. The center one. See this one here? This one is the center one and it's the output coil, and whatever energy you put into this thing shows up in however many other coils you might put around there.

42

These coils are independently separate from the other. They're not electrically connected anyway so it's magnetic flux which is controlling them. You can have thousands of these other coils and each one of them duplicate what's there from the other thing. So that tells you something about the fact that you can get from the same source of energy the same amount of output–you can get thousands of copies of it. Because here we're getting three copies of the same thing. If you notice real close here, there's a little light emitting diodes [actually neon bulbs]on these. It's part of the tuning process here. So it's real easy to tune the thing so that the lights light up and that tells you that that coil is active and producing the same energy as is coming from the other one, from the output.

43

That was about 20 years ago that we're looking at there. It was the first device that bill because I was told that you couldn't do it. That's a mistake when somebody tells I can't do something. I'm going to find out whether you can or not. More often than not I'm right. This [page 166] is looking at a 35 kilowatt deal. This is a neon tube transformer here. You see the heavy diodes here and you see them coming together in the [?]. So you've got the output here, from the capacitor, to the two output points. Here's another view of something that we've seen before.

44

This is the high-voltage, high-frequency testing device. One of them. This [page 156] is the project which I did for the Japanese. You notice the Coke machine over here? What I did–once I demonstrated the thing for them I served cold drinks. This is one of the devices we'll see that closer in a little bit. But this device was made for like the 2kW, and continuous, and self generating. Well there was a coke machine again.

45

There's the large coil which I never got to use. There's he coke machine. I rented a Coke machine full of drinks, and had a large number of people there to see it demonstrated. These are some of my friends from

Japan. The man on the right here is the CEO of a large shipping company in Japan. the largest manufacturer of ships in the world. This fellow here his family owned a large number of vending machines. In Japan vending machines–everything exists including Auto Parts, fresh meat, fresh fish, vegetables, and everything else.

46

When you're in the subway and you get out going home, on the way home, you pass this huge number of vending machines which you can get your supper, whatever it is going to be, in front of the vending machines. There are millions of them in Japan, and it's very important. They were spending so much on electricity running their vending machines that they were upset. So they commissioned me to find a way to get rid of all the energy cost associated with the vending machines. That's where that Coke machine came from. It was a device I did for them. This is a very interesting device. The black thing here, again, this is a material–when you put an electrical current on it it generates a lot more electricity than is going into it. It's piezoelectric material.

47

They put the cables under the street and highways so they do counting of autos passing by, that sort of thing. Pressure-sensitive material. This is a closer look at one of the plasma globes I use. Of course another view of the transformer. A lot of these views here will be just duplicates. This [page 157] is a different device, which, if you notice it it's a Tesla coil wound around a toroid.

48

The benefit of this is once you get any energy into the toroidal system, and it's circulating, you can have however many. You can see it very good, there's a tap off of it. You can have as many taps off of that as you want, and each one of them duplicate what this Tesla coil is putting into the... Here is a neon tube transformer running the Tesla coil. The red thing is the L1 coil. The L2 coil is the larger coil on the outside. That was one of my inventions that ran off of a large toird. These toroids were from the Los Alamos laboratory. I have all together about six or eight of them that I got military surplus.

49

It's very special metal. It's nickel with another mineral which I can't think of just now. But it has very interesting characteristics which tend towards the overunity type of results. Another view. This device was built for the market in Brazil. They approached me about redoing their entire infrastructure or energy. We had a deal, written contracts, and everything with the Brazilian government and several of the Senators there.

50

They were to raise 200 million dollars to get this thing checked off. The problem was that when they got their money raised the exchange rate was fluctuating in the wrong direction so fast that the amount of money that was left over for an exchange to the new level wiped out anything they had done. So we raised 200 million dollars about a half dozen times, and every time the thing fell through. Because when we tried to change it back to dollars, which I had to have some dollars, because I had been under write the thing myself and pay for all the costs and expenses. They would pay me with their money and that just wouldn't work when the exchange rate was so unfriendly.

51

So that thing felt fell through. It's still very much there, and they keep coming back at me about once every year or so trying to redo it again. That included redoing the set of electrical trains for going across to the Pacific. They don't have a Pacific outlet for the all the goods and things that are going from Brazil to the Pacific. They have a large contact with people in China and Japan. All the goods that they were sending them had to go around the earth in order to get there. So they had a direct route through Brazil over the mountains. One of the projects I was to do electrical training system for Brazil.

52

This is still very much alive and still working if we ever get the details worked out. That's been going on now for more than 10 years. They have seen my devices and I was familiar with them. I had their engineers working with me and know what I'm doing and they're very pleased with what I do if they could just access the thing. This [page 167] is the close-up of the Japanese energy device for the coke machine if you want to see what it looks like. The input part is over here. This is a milliwatt high voltage Tesla coil type device and it's like I said milliwatts.

53

It's coming through this diode bridge, and it's coming out into a capacitor here, that has the certain level of energy in it. From that it goes into this system here of which you notice these are heat shields, and a very special type of diode here. What this does, it makes the energy into pulsating DC, which in turn goes through this toroid here, which smooths it out and makes it into alternating current, which could run the Coke machine. And to the Coke machine, after that toroid, it's almost perfect sine wave.

54

Which means that the Coke machine doesn't know the difference. It's still pulsating DC. The the Coke machine seems to think it's alternating current and it runs on it just fine. So here we are with milliwatts in and we got 2kW out actually, continuous. Is there anyone of these devices that anyone would like to see or know more about?

55

There's a phenomenon called electron pairing. I'm going to as a friend to come demonstrate for you. The fella who is going to do this is very much in the physics and I think it's one of the Eastern the universities at high level.

56

It's Denis Lee, and he's going to bring his– this is a patented device demonstrating electron pairing. So Denis, please come forward. Dennis is going to take over. This interesting system which demonstrates electron pairing and how it comes into the energy situation. [Denis: One way of explaining where the energy comes from is. The sun emits electron pairs and they're everywhere.

57

In subatomic theory the electron shells, they always talk about the up and down electrons, or left-hand and right hand spin electrons. Well when they're in a pair it's undetectable. But you can use weak magnetic fields to separate them. This sort of demonstrates an electron pair, and how they may combine together. There is a unique mechanical property which keeps the–I called concentric toroid–which keeps the small toroid centered in the large toroid. You can see they can come out. You can use a weak magnetic field and separate the two, and this becomes positive electricity and negative electricity.

58

An interesting thing about this is that when you roll it on a tube–I don't know, this could be a flux line–and I don't if you can see this but it roles and it spins, and that's a mathematically correct fourth dimensional rotation.] Denis is much more sophisticated than you might suspect here. At the University level of MIT and that sort of thing. He really knows what he's talking about. These electrons, like he says, in the ambient background, is everywhere, in all places, at all times throughout the universe [?].

59

So you can't go anywhere without running into them. So if you have energy devices, where you can turn this energy into propulsion and various other uses. Like the in between I was talking about. It's when the energy is moving from one happening to another happening it becomes available.

1:01

In the case of capacitors and what they do in regard to energy, this is part of the secret of this [?] We can utilize the energy to our benefit. For example, a capacitor is a blocking device. Energy cannot pass from one side of the plates to the other side of the plate. But the second side of the plate, due to the van der Waals Casimir forces, duplicate what's on the first side of the plate, even though there's no way for that energy to get past the plate. This duplication is due to the ambient energy background.

1:02

With that in mind the second capacitor plate which, is not part of the system, as I'm going to be showing you very shortly. But if you have a diode bridge for example, and you have energy coming out of the bridge you put the charge on to one of the poles of the capacitor. The second pole will duplicate what's on that first–the capacitor without having any obvious reason to do that. Then as the ambient energy, which is available through the second capacitor deal, and if you ground that properly, it becomes useful energy almost immediately. So we're going to actually demonstrate a case here so you can see it working with your own eyes. It's not very complicated. It's much simpler than anyone would have ever had imagined.

1:03

The energy which we are using is the ambient energy background which spins from the earth ground. The handout which I gave you is telling you basically what's going on there. In nature all things are simple and uncomplicated if you truly understand them. The energy source is available to you in unlimited amounts with a device as presented. The energy available would be equal to the amount of the energy put into to it whether it's a major volts five volts. So basically you design the output device so that it runs at the rate which you want to see at the collector point.

1:04

So if you're running a scooter, or running

airplane, or whatever you happen to be running, an electric motor, or automobile motor, or ship, or any other type of energy device. The amount of energy you want is an amount equal to the energy on the plate, of the first plate there. With that in mind we have a device here to demonstrate that. If you will set it up we'll let you see it with your own eyes.

1:05

These are two capacitor plates. This is a dielectric tester device which is used for certifying the electrical durability of films and other kinds of things. We're going to use it for a purpose which it was never intended to be used for. I'm going to put 8,000 volts on one side of the plate, and you're going to see the same 8,000 volts show up on the other side without any input to it. With that the mind the grounding that you do from that second plate depends on how much energy you want in the device to be running.

1:06

What I'm going to do is put 8000 volts on this plate here. See those sparks there?

1:07

We have a thing [plug] with the earth ground only present on it. The green wire is the earth ground. You should be seeing the sparks on the other side. See how long and how warm they are? There's no connection there between the energy on the one side and the other side. So the second side has to be your ambient energy background which is present everywhere at all times. In this case it is alternating current.

1:08

It turns into pulsating DC on the other side. See how long those sparks are? How many volts is that? 1/4 inch is about 10,000 volts. This thing is upwards one inch. So between the two plates we're developing tremendous background about 40,000 thousand volts.

1:09

[Question. Why is it not alternating on the other side?] The ambient background energy is more related to the DC.

1:11

[Would it matter if you had copper on both sides?] No, There's just a slight electrical difference between aluminum and copper which makes it more [?].

1:12

This is a forever source of energy.

1:13

[Question. How do I ground it if I'm doing it on an airplane?] In the case of air grounding you have different degrees of air grounding.

1: 21

[Question. Where can I buy one?] Actually these have all been for own personal entertainment more than anything else. I spent several million dollars of my own money playing with these things over a long period of time. A surprising number of them have over unity type characteristics. Do you want to ask another question? [Question. You haven't answered that one, where can I buy one?] I just told you that they're not ready for sale. I ocassionally made the mistake of selling one of them and the people that get them–you can see they're out in the open. There's nothing concealed. It's fairly easy to understand.

1:22

There's no reason to destruct the device that say the person gets. I've tried to sell you a few and I found out that what happens to us.

The person who got tried to tear it up and see why it worked. They could [?] they tore it up they wanted their money back. I told them if that's the case then please put it back like it was and I'll give you your money back. [Question. Where can we get directions to build one?] Actually you can figure it out yourself, I told you just what you need to know. [Mark Smith: If I could interject real quick. This, it comes up all the time. Please tell us exactly how to do this, step by step and piece by piece. But there's a problem with that. My father is trying to partner with industry partners to construct devices like this.

1:23

They don't like it when he comes and does these conventions. They talk about, 'well what are you going to say?' 'How much you going to give away?' Because the reality is if he stands up here and gives a workshop, and says now take coil A, and insert it into slot B, those industry partners will disappear like mist. People then say, 'We're not going to buy these things.' Well the reality is, most of the industry partners, in fact all of them to this point, that have partnered with my father, and he developed something for them, and they take it in-house. They're not turning around and selling Donald Smith technologies overunity devices. A friend of mine, who is in fact a CEO of one of these companies, who actually develops filament technologies for light bulbs, once told me, 'You can't make me stab my company in the chest.' I might have a 90 year light bulb, I'm gonna sell you a 25 year light bulb until the market saturated. Then I'll sell you a 30 year light bulb.

1:24

Saturation can take 15 to 20 years. So the reality is, they take his device. They improve an existing process, and you'll see, you might see a new whatever widget X come out that's ten percent more efficient than it was a year ago, that may be based on things my father has done. But you will not see which widget X that will power itself forever. So my father is trapped between two worlds. He needs industry partners to take his devices worldwide, and to try to make an impact. By the same token he'd like to just unleash this to everyone, to the world. So he comes and says as much as he can at these conventions so that those of you who want to try to follow along with this, and do this on your own, can pick up from what he's talking about. But don't ever expect him to give you the whole nine yards, because even if he wanted to, I wouldn't let him. Because the reality is, we have had enough people come to the table at our house and try to steal his intellectual property, and I'm sick of that.

1:25

To be frank, after he had his heart attack and his stroke that left him half paralyzed, I offered to move him with my money, if he wouldn't use his money, to New Zealand to the country. Stop doing this, it's too much of a headache. But he's still here, and he's still presenting to you guys, because he would like to see the world benefit from what he's doing and it not just become a closet technology that someone sits on.] Okay Denis, let's take your question. [Question. In device 9 you use an acrylic plastic sheet?] Yeah. [And powdered metal shielding?] The powdered metal–the metals come in a molding. Metal projects today, most of the metal molds are made from powdered metal.

1:26

They have a cylinder that reaches under pressure and they heat it up, and when the pressure is a certain level the particles all are dispersed black powder. And they cool it down in that state so they end up with powdered metals in any quantity you want. If you want to build a solid device, you take this powder and put it in a mold and you center it. You bring it up to the point where the metal begins to–wear touches against the [?] each other and you have a solid device. [Question. Was this out of a spray can?] Yeah. When you get the [?] it's already been made into a powder. There's companies with large industry where they supply unlimited quantities of all kinds of metals, rare and unusual. [Question. What type of metal was that?]

1:27

It is something that magnetic flux doesn't like. [Question. That airplane capacitor at 400 farads, what voltage was that rated at?] It usually is at 24 volts or something like that. [Question. How did you reduce the output of the Tesla coil to be able to accommodate only 24?] Well actually, what we're seeing here is an extreme case of what happens when you do this, you can do to all voltage levels. So a 24 volt output at say 200 amps or something, is that what's accessible, because the amps don't show up until you do a grounding thing.

1:28

If you dead ground there you don't have anything left. If you grounded it properly you can bring an amount of amps or volts in that you want, custom make it. [Question. In the plasma tube device you said the copper and aluminum plates are in the frequency range of a diode bridge.] Well the frequency range will be whatever the plasma tube itself is putting out, which is from say twenty to forty thousand volts, for example. The frequency will also be about fifteen or twenty thousand cycles per second. [Question. Diodes are rated for about 25 kilohertz.] Well they go into the megahertz.

1:29

If you pick your battles so you can win them, you try to make the energy source at a level which is easily accessible, so they are off-the-shelf diodes. [Question. The toroid is a military surplus material? Is there a name for that?] Nickel-iron alloy. There's a name for it but I can't think of it at the moment. [Question. And the last question is on the Coke machine. You have a diode device that's on the output on the first set of capacitors, and could you specify a little bit more on what that is? Could you specify a little bit more of what that diode device on that output.] Any diode bridge type thing will set it up so that it will work. [Question. Oh so it's pulsing DC and it goes into the toroid?]

1:30

Yes. We're using pulsing DC and it comes through the toroid deal. When it goes through the toroid it smooths out the thing so that it comes out almost perfect sine-wave, which is a very simplified version of an inverter. Inverters are [?] complex, but they don't have to be. They could be [?] simple. That's what I've done. I've simplified the inverter so they had very few pieces. You can see the Coke machine running beautifully, so it must work. [Thank you.] [Question. I just want to add a comment here, to you, and that is, in your present model of getting this technology out this assumes that it's business as usual on the planet.

1:31

If you read about the peak oil issue for example, that there may be not very much time. Several years.] It's a lot less than that. [?], worldwide. My background is in petroleum engineering. [Question. So you must be following the peak oil?] I know exactly what's going on and it's a lot sooner than you think. About a year and half from now. [Question. I'm going to quote Tom Bearden, most people know about Tom Bearden here. But I read a briefing for the United States Senate, senator from our state here a few years ago and in Washington, and Tom Bearden was one of the people I had there at that time, which was the year 2000. Tom made the comment that if we aren't making these devices–anybody knows Tom those that hes got a [?] from Louisiana.

1:32

He said if we aren't making these little things like sausages out of a machine, for the first quarter of 2004 we might as well go fishing because it was too late.] That's very true. [And that was his assessment five years ago about where we should be, and right now we aint there. We're there and we aren't making a like little sausages.] What's going to happen here, and we're going to have, when you disrupt the system for food and energy and that sort of thing, you basically bring the world back to the back to the 1800s. And you've done it overnight. So what's going to happen, you're going to have hundreds of millions of people starving to death, killing each other for food and all signs of types of things like that. Take my word for it, I know exactly what I'm talking about.

1:33

We're about a year and a half from that. I'm very knowledgeable on the subject. If anybody wants to debate it I can handle myself quite well. [Question. I don't want to debate it for a minute but what I do want to do is pose something for us to think about. Here we are talking together about the existence of abundance for all of humanity being a possibility. But we're living in a social climate of fear, power, and control domination, and violence.] Very true. [We, by changing the way we're thinking, and helping each other to get to abundance for everyone, can change. How do we do that?]

1:34

Unfortunately human nature is constructed in such a way that it's gonna be very difficult. And although I think what I'm doing is for the benefit of humanity in the long run it's going to be very difficult to implement. But my connections to China and Russia are well on the way right now. So the real thing is already beginning to happen. I'm considered a national treasure in China. Also the Russian or the intellectual part of the Russian... [Question. That sounds good. Let's hope that there're people here that may be connected to the mainstream institutions will understand that the US will become a backwater unless we...]

1:35

That's my plan, I intended for it to be delivered. Because I tried so hard to give it to the Energy Department and the other people. I spent an enormous amount of money over a long period of time, with the Energy Department thumbing their nose at me, and so I decided myself that the United States was going to become a third world county.

[Question. I notice you have brochures and handouts for scooters and airplanes. Are those companies that you would allow us to buy from?] I'm not the output source of it.

1:36

[Mark Smith: You'd have to contact the company that we've dealt with and see if they would be willing to sell you what he sold them. Which they almost certainly wouldn't. They're restricted to only selling in the spheres, like if it's a part for a plane, they can put it in a plain. It is unlikely, but it's up to them, whether they'll put his full device on the plane or not.] Let me tell you how the airplane came about. You know guys there was an air show a little over two years ago. The device was presented there and the Chinese government happened to be present. They got a contract for twenty four twenty meter aircrafts using that system on the spot.

1:37

So the Chinese government was what the broke the ice on that. Believe it or not General Electric here in the United States ended up making a contract there. And the first 24 of them are out running now. They fly basically forever without a visible source of energy. They fly several hundred miles an hour. They're very effective. 48 more of them are contracted recently, so there will be 24 and 48. And there will be that many of them before the years over here. Actually out physically running, so chances are you may have actually seen one of them.

1:38

You may have actually seen one of them and didn't know it. They don't make any noise [?]. Basically they run forever without any apparent source of energy, which actually as we know is the ambient energy background which we were demonstrating on this system. [Question. The question is about the demonstration we had that previously with two plates. Do you need the electrical arc to have the thing work, or you can just put the high voltage on one side?] Actually what we were seeing here is totally unnecessary.

1:39

Because if you have any energy source which you can put on a capacitor plate, which is not going anywhere, when it hits that capacitor plate that's the end of the transit for that energy. But the other side duplicates, through the van der waals and Casimir effect, the first side. The second side is an open circuit totally independent from the first side, and the degree of Earth grounding you put on at a second side depends on the amount of useful energy present. So you adjust that so it matches your requirements. [Question. Could they actually share the same ground from the input to the output?] They don't share. The amount of energy actually physically used is very tiny.

1:40

It's in the milliwatts or less. You could put a feedback thing on there which will make up for the difference there, and it would basically go forever. I said it was so simple that it's ridiculous. It's not complicated.

That's my kind of device. I like things that are simple and easy. [Question. You don't have go at 10,000 volts to make the thing?] No. You could work it on 10 volts or 5 volts or whatever. It depends on if you have a capacitor [?] you got a huge amount. Or if you have a small capacitor it will work just the same.

1:41

[Question. To continue his discussion about the voltage is necessary, you can buy these 1 farad capacitor for 4 or $5. They're quite accessible for five or six dollars, or whatever they are. They're about that big. They're about that thick. One farad but they hold voltages.] One farad is equal to a kilowatt. [Question. But they only take four, five, six volts or whatever. That's okay. If your final output is going to be 5 volts, 10 volts, or whatever, you just put more. So that's how you run your motor scooter.

1:42

If you want to build a motor scooter that does this just look at the capacitor that's already in that motor scooter and figure out that the wiring slightly different, you've got a motor scooter that will run forever. Aren't you glad you thought of that? [Question. I got one more question. If you don't mind, tell us a little bit more about the material that would put on the road. You say you use it in some of your machines, and it's some sort of material?] Yeah, the one that knocks the magnetic flux out? [Question. Yeah, they put it on the road and count the cars, and you use in your machine?]

1:43

[Mark. Piezoelectric material.] It's available off the shelf. It's commonly used for counting automobiles and things like that. I think everyone has it in their minds, it's pretty clear what happens. You're still expecting things to be complicated or difficult, which it's not. [Bruce. I don't if you're still in contact with Jimmy, but he built one of your devices and sent me a video, and lighting up a light bulb with it.]

1:44

You can light up however many light bulbs you want to light up, basically forever. [Bruce. He threw it all together with spare parts. So he's having a problem with one of the components in there, and he just hasn't replaced it. But a new component in there will last forever.] Actually, you could build that light to last forever. I have some of the original light bulbs that were built back in 1989 [1889] or something, and there absolutely as good as they were the day they were built. Some of them have been used ever since they were built, originally. They don't wear out. Building a light bulb that doesn't wear out is not a complicated thing at all. [Bruce. All Don's circuits are on my website. All you have do is download it and actually get hands dirty and build the circuits, and make mistakes, and then eventually you'll have working circuit.

1:45

But you have to get your hands dirty. You know, Don's practically handed it to you on a plate. The schematics have been up there for a few years and there have been some people that have duplicated it and contacted me. It works. It's from my newsgroup on Yahoo. Nuenergy from yahoo groups, the discussion list. It's in my archives on the file.]

1:46

I can tell you this that in eastern Europe there are probably four or five hundred people that have duplicated my devices with real good results. One of the most interesting ones was they wanted to know in making a coil, like a Tesla coil deals. They used stainless steel tubes on the Tesla coils when they wound them, and the net result was that the whole countryside for about ten miles around lit up at night and they couldn't turn the lights off. So that was one case. I don't know. There were quite a few more but it seems that people in the eastern European block are more open-minded than they are anywhere else that I've have seen. I have a lot of communication with them and I know they've duplicated.

1:47

Most people don't tell me when they've duplicated, it suddenly becomes a secret and I'm not supposed to know about it. But occasionally I'll get someone that is very friendly and talkative and they'll tell me what's going on. So I know that my devices have been duplicated in many, many parts of the world already. [Question. Don, on your very first device, the picture you showed us with the coil and the three coils around it right. Those smaller coils would suggest a very high operating frequency.] Yes and no. You can run it at quite a number of frequencies. The frequency depends on how fast the thing is spinning because it's got an off and on device there where the magnetic flux has been cut off, and it's turned on. Cut off turned on.

1:48

Each time part of the hole passes there that is open. See it blocks the magnetic flux normally, and there are holes that are cut in the thing. When they pass the magnet the magnetic flux comes back, and it causes an energy surge. It's not like Denis Lee's. The Denis Lee I'm talking about is not the one that's here with me today. But the devices that Denis Lee did came directly from me. But he refused to use a system which I asked for and a net result those devices are killed by the heat factor when they run. The magnets will last a very short period time, and that's why you won't find him demonstrating the devices over a long period of time.

1:49

[Question. A question about the frequency. The input frequency you said was very high in the example you gave us, is that critical, does that matter?] The natural frequency of magnets–they have, just like everything else, every substance, it has a natural frequency which is built into the systems so to speak. So the magnets themselves can be very low frequency or very high frequency. But they're more inclined to be backwards. It's more likely to be out of the range it's like your megahertz.

1:50

If you see the energy spectrum from the electrical display that is available, I have to tell you about it on the internet. But all substances or materials that are solid or otherwise, human or whatever, everything has it's own natural frequency. Most of the frequencies that we're running into on magnet type stuff are actually in the gigahertz, or something along that line. If you know what those frequencies are you can actually make an energy device direct, without having any

moving parts.
1:51
One gauss is equal to 100,000 volts worth of electricity. So you can see that the magnet with a hundred gauss in it represents a considerable amount of energy. And the magnet itself may be the size of a peanut. So if you can access that energy properly, correctly, you have the equivalent to that amount of energy accessible to you. I've done it, and it works beautifully. It doesn't deplete the magnet because the magnet works like a sponge. When you take energy out, it immediately rebuilds itself.
1:52
So the energy is coming directly at you, and you come out with unended amounts without depleating the magnet. One of my inventions does this. I think it's very interesting because it means [?] in a little tiny flashlight battery, or a battery like in your hearing aid here, you can have a million watts of energy direct from the source without having other complications. That's one of those 200 devices that I built. [Question. Have you patented that?] No I haven't patented it yet. But I dare anyone else to do it.
1:53
It is a very interesting source of energy, and it's highly obvious that it's there. It's what to do about that matters. If you know what to do about it, it's accessible for you.

VIDEO 6 DEMONSTRATION OF AMBIENT ENERGY GENERATOR DON SMITH EFFECT

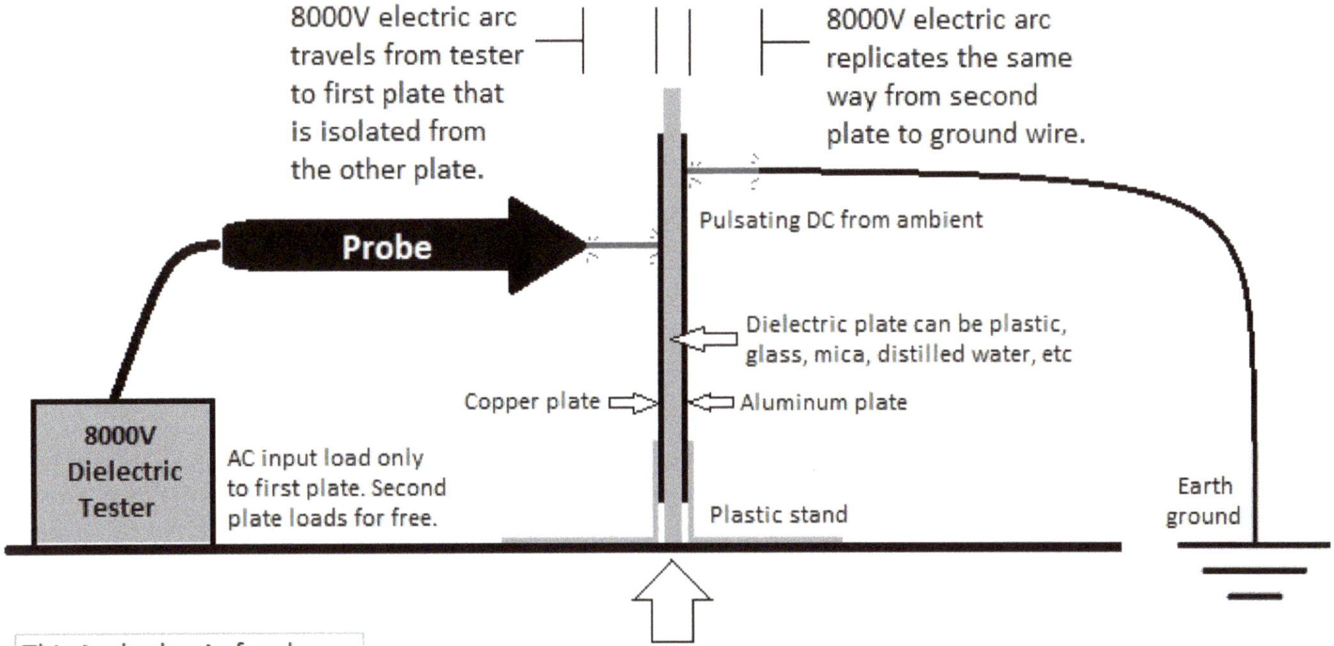

This is the basis for the DSE and Ambient Energy Generator on pages 76-77.

A capacitor is a blocking device.

Energy cannot pass from one plate to the other plate. But the second plate, due to the *van der Waals Casimir forces*, duplicate what's on the first side plate, even though there's no way for that energy to get past the plate. This duplication is due to the ambient energy background. If you ground that properly, it becomes useful energy almost immediately.

The energy which we are using is the ambient energy background which spins from the earth ground. You design the output device so that it runs at the rate which you want to see at the collector point. Alternating current on the left side. It results in pulsating DC on the other side. The ambient background energy is more related to the DC.

Tesla Impedance Circuit

otherwise known asTesla Hairpin Circuit

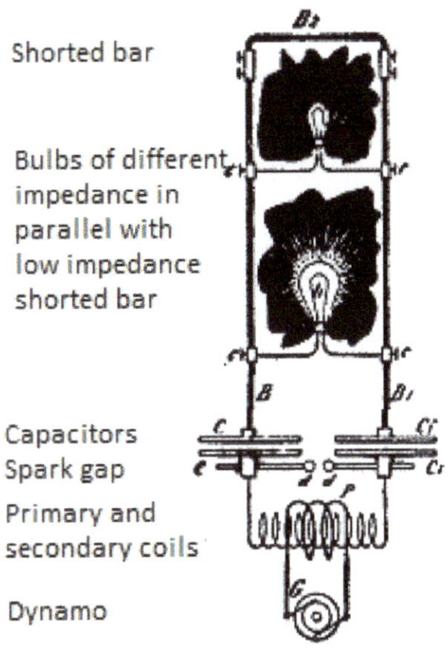

The *Tesla Hairpin Circuit* was named such because it resembled a hairpin. But it was meant to show the *impedance* phenomena. The two capacitors isolating the load side are the same sort of thing Don is doing here.

Ambient Energy Generator

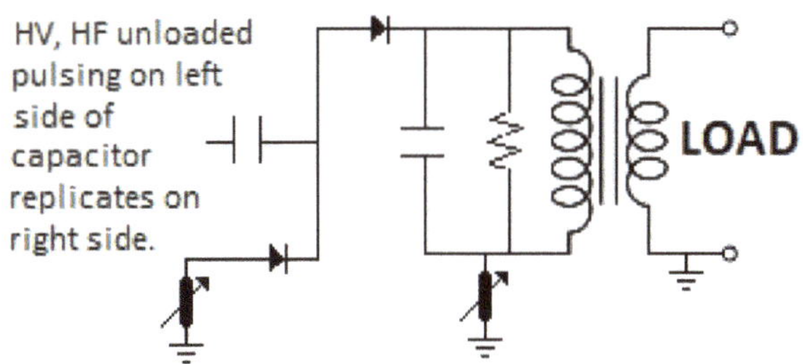

These are applications of the demonstrated system in video 6. This provides the means for electrons to be drawn into a high potential and cycle through a load on their return back to ambient ground in a controlled way.

Video 7, 2005

[Introduction. Back around 1990 I believe. He's always had things that make everybody think. He's going to do the same thing today. He builds power transformers. He spoke first about 93 or 94, at I think it was an extraordinary science conference in 93. At that time we were doing in odd numbered years conferences that focused on technical building information. In the even-numbered years we'd go more into theory. Don has always been very practical. He's always built stuff that's real, that you can get your hands on. From that we prove and disprove all kinds of theories. He promised us there'll be a lot of sparks flying. His basic thrust is explaining why the term ambient energy is better than zero-point energy. The ambient background is a better term than aether.

1

He's got a university degree in petroleum engineering and geology from the early 50's. He was first employed at Gulf oil, later Phillips, and many many others, including some independent companies. He says oil is not a necessity anymore with the system that he's building. Here's Don Smith.] Thank you very much. Nice to see you. We're going to hop along the bunny trail here and see some things. We can start with number one. The following video is a presentation from [Don Smith company video: The following video is a presentation from Resonant Energy Systems. This material contained in this video is confidential and for the intended recipients eyes only.

2

Viewing by the personnel or persons is unlawful. If you were not the intended recipient of this, please turn this presentation off now. What we're looking at is the development of Resonant Energy Systems, for using the Casimir effect to produce more electricity out than we actually require in. Briefly stated, what we're looking at here is the scope output of the readings taken in the various places in the system. The top, or amber, is the output sine wave as coming out through the regulated transformer, inverted AC 120 volts. The next group, or sine wave, is the intermediate inverter sine wave.

3

The violet trace, or magenta trace, is the DC voltage at the input, and back to the battery. The bottom sine wave is the resonant back feed energy sine wave. Currently for this test, we're using a Tektronix model TDS3034B. See if I can get this. For fine measurements, so that there is no question about the integrity of the measurements of our equipment. This scope is commonly available in the US. It may not be available abroad. Also currently being used are fluke commonly available model 177 and model 189 meters.

4

The meter on the left, reading 115 volts, is currently coming from the output transformer. The model 189 on the right is measuring current battery voltage and return voltage to the battery from the system. This transformer in front of the screen currently has the capacity to produce 120, 240 and 480V off the different taps. This particular meter here is the load. The voltage control for the transformer in the foreground is the variac as in the back in the foreground of your picture presently. The whole unit is completely encased in this cabinet presently.

5

Using this special capacitance inversion circuit [Editor: 50F banks, where two capacitors are independently in series with each input and output terminal.] we're utilizing what's called the Casimir effect to produce AC 120V, and DC 12, AC 140 and AC 480 as needed for commercial, industrial or residential use. Just a brief walk around the cabinet. This is a 12-volt DC load, 2 x 1 million candle watt power lights in the foreground, we're going to switch those on. Those are 1 million candle watt power lights. Here we see the backside of the transformer, the 120V, 240V, 480V transformer and the variac for controlling it.

6

These lamps are aviation landing lights, they are 1 million candle watt power each. The load that is placed on them. We are going to also transfer the load from the lighting that we are using for our output here, to the lighting for,.. we're loading the system. This is a small system. Right now the landing lights are putting on about 60 amp DC load. It's about 65 amp DC load per lamp. We have a 500 watt lamp to my left here, that we're currently using. The sine wave is remaining stable.

7

On the scopemeter we have stable voltage, without variation and stable sine wave. To be able to represent and capture all of the data, we've compressed four separate channels into one screen. Hence the reason why we're dealing with very compressed data there. We're compressing quite a bit into a small space. This is a 20 amp AC space heater that we're currently loading with. The voltage is currently remaining fairly stable at 114V 115V.

8

There's our DC voltage on return. We're pulling 65 amps aside from those lights out the top. Once again, this system can be closed. This is a brief review and presentation of Resonant Energy Systems, energy generation product. Thank you.] We're going to look at some slides of some devices that I have done and I'll talk in detail about those things as they come along. We have about 70 slides of devices which I have which are working the way they're supposed to.

9

Doing exactually what you would hope they would do.

10

We have a pancake device on the top there. This is a small pancake device and it was one of about 4 dozen different devices that I've done. With the pancake device, you get an amplification of the energy such that you actually physically can generate through inductance and other things. There's an energy gain. The energy gain is coming from the ambient background energy. It's not getting something for nothing.

11

We'll go into that in more detail later. When I first started doing these things I had one of Nikola Tesla books by Thomas Commerford Martin and his official authorized biography of Tesla. From that there are a number of experiments which he shows in the book. I went through every one of those experiments in great detail. That was my learning process for understanding what Tesla had done.

12

At the end of it, I realized that technology had advanced quite a bit and a lot of new things, which were available, which were not available the time Tesla did his work. All things I've done are extensions off of Tesla. There beyond in many cases where Tesla had arrived at when he had been completed his work. I like to play with plasma tubes and that sort of thing. This is one of my early experiments where I took the ambient background energy using the test instruments they use for plasma and it worked beautifully.

13

This is something which I brought to the 1989 [1996] Tesla convention. It was my black box that I brought, and it caused quite an upset and quite an uproar. Basically it's a very simple device. It was working completely off of the ambient background energy. It could have basically ran forever if anyone wanted to look into it. This is one of the other devices. This one generated about 35 kilowatts. That seems impossible with a small device like that, but it actually did.

14

We're seeing some of the other views of that particular device up close. So we'll skip along, until we get through that. What we see here, if you can read the things, this is a 35 kilowatt system. This is where I put a Tesla coil onto the toroid there and use the Tesla coil to generate additional energy,which was not present there originally. All this stuff that comes from the outside or the ambient background is added into what you have. It appears to be that you're getting energy from nowhere, which is not correct at all. You do not get energy from nowhere, it has to all come from somewhere. That is also a device I made especially for the Brazilian government, and project that I was doing for them. I was supposed to renew and remodel their whole power system for them.

16

We had the written agreement and everything was boxed up. The problem was every time they came up with the money to do the project, the value of the Brazilian money had decreased to where it was basically worthless. The net result after about 10 years of fooling with the project, it never got done. We're still very good friends and maybe someday we'll do it. They wanted to pay me with their money and I couldn't. I had to pay for all the things that are necessary to make it happen with my money. But they were going to paye me with there money and I couldn't very well do that. I didn't have quite enough funds to suplement the government of Brazil.

17

We've gone to another device here [page 160]]. This is one which is currently being manufactured in Russia successfully in large quantity. It's one of the devices that appear to get quite a bit of attention. There's eight different power sources there. They're not intermingled. So each one of those power sources will do about 50 kilowatts. This [page 83] is something which I have a patent on, which relates to getting a huge amount of energy out when you put a few milliamps of energy in.

18

You end up with like 100 kilowatts of energy out. That won't make sense to anyone who thinks like the electrical system. You have to think in terms of the magnetic flux. If you're using electrical flux everything is dying a heat death, and you lose the energy almost immediately in electrical flux. But in magnetic flux you can make as many copies of the original as you want without depleting the original at all.

19

If you're chasing electrical flux you're chasing the wrong rabbit. You need to be chasing the magnetic resonance flux. This [page 162] is a early stage of the 8 cylinder thing you saw just a moment ago, where it was demonstrated how simple it was, and how well it worked. This particular thing has been duplicated all over the world in many countries successfully. There have been people here tried to duplicate it, and done so successfully. There are quite a few people apparently had frustrations with this. We need to talk about that on a one-to-one basis.

20

If you can focus that thing in, where everyone can read it properly [page 83]. [John: Question. Did you want to explain it, or do want to let them read it?] You explain it. You're familiar with it. Take it a little bit at a time where everybody can see what's going on. This is actually part of patent documents. You're seeing things firsthand before it was ever in a final form. [John: This invention relates to the dipole antenna system and their electromagnetic radiation. When used with a transformer and appropriate energy collection system it becomes a transformer generator.

21

The invention collects and converts ambient energy which with conventional devices is radiated and wasted. This is like the Moray valve or the Tesla valve. It's in concert with that similar technology. When tuned to the proper frequency, you can pick up a strong ambient resonant energy that's available in your area. The frequencies vary from place to place. A system that is assembled at your house may not work ten miles away as well. What you have to do is tune it to match to work in your area. There's a gentleman in New York that has a system that four blocks away, when he moved from one house to another almost was unfunctional, until he retuned it. Basically it's picking picking up ambient resonant energy.

22

Some individuals would tend to call that well I'm depleting the ambient RF that's produced by my radio station. I'm sucking a black hole of the RF in, which you can do. In fact we've run tests where we've set up everything from a regular monopole to dipole to a ground plane and yagi, and with the signal strength meters located just a thousand yards away, we've figured out how to suck down a specific frequency so that we can convert it to usable electro-motive force to charge batteries and run things. We've watched the signal strength deplete just a thousand yards away. When Tesla ran his 1931 pierce-arrow there were very few if any commercial radio stations in some of the areas where he drove that, and drove just fine. Which proves that there's still plenty of magnetic resonant energy out there to be had.

23

The dipole basically uses that. Sometimes you need to excite. I'm going to pass through some of it because I explained it. Sometimes you need to excite the ambient magnetic field. You can excite it, or disturb it, like water in the pond. As the ripples pass through, say a wave generator, you can, if you have a small wave generator there, that just the float on the surface of a pond with a magnet in it, floating up and down over a coil, you can generate electricity just from some child jumping into the water. All you need is a rectifier or transformer circuit for each one of those that are in the water. You can disturb the magnetic field in different ways. If you can disturb the magnetic field in such a way that you're using little or no current, or reduced wattage to do so, while disturbing the magnetic field.

24

Anytime you disturb it, you create a flux or a disturbance in the flux field. When you disturb the flux field, 10,000 lines of flux over conductor constitutes 1 volt ampere or 1 Watt you have usable electro-motive force. The dipole turns out to be one of the more efficient methods that Don's invented and work with. He invented applications that use the dipole that some of them he's not even shared with more than two or three people. Sometimes I've visited him and he's been thinking and explained the inventions that haven't even been put to paper yet. What you'll see here is that the dipole creates a disturbance with a coil on the bottom.

25

We create a disturbance in the magnetic field in these plates, or our inductors, that we take through the transformer and rectifiers, that make it usable. The dipole generator here, you can see the plates. The top plate is an inductor, the bottom plate is an inductor, the middle plate is an insulator. There's your dipole circuit with your exciter coil. Much like the eight coil device that he was explaining earlier. The disc in the middle, all it has is just voids with a material like bismuth, neodymium or there's actually another material that even works better, that are sprayed on something as simple as a Plexiglas disc or a phonograph record. One of those prototypes that he still has working, there's still a phonograph record in it.

26

This is an assimilation of the plasma with the tube. All it's doing is disturbing the magnetic resonant field. You have your inductor coils, or inductor plates in this case, that also are qualified as a plate capacitor to 17 where your terminal block, where you go to your transformer rectifier. There's many different applications. The one that is now on your screen, let me see if I can shrink that down a little bit. We're gonna reduce that, so you can see it. At this one, 21 and 20 there, is a series of plates that are assembled. You'll see in a moment that on Don Smith's website, that he has a product gallery that he numbers many of them one through nine.

27

This one isn't on the website I don't think Don. But www.altenergy-pro.com, in a little bit we'll put it up. This one right here is similar to the plasma tube. Number one is a dipole, with north facing up. I told myself I wasn't going to do this, I was thinking about getting out that funny green laser. Don has a laser pointer there, well I'll just use the pointer on the mouse here. That's the North Pole, down here is the South Pole. There's an exciter coil. All we're doing with this coil is using that dipole magnet, we're creating a disturbance.

28

We're not producing a generation of electro-motive force, by sheer mechanical effort in, and magnets passing over induction coils. This is a different approach, it's creating a resonance that's allowing in this particular case 20 and 21 constitute a group of 11 plates with 20 insulators plates between them. Don feel free to jump in, should we just keep going? What was the question please?

29

What you've just seen there is something which European Patent has already been issued on. I had tried for years and years to work with the United States Patent Office and the Department of Energy and various other people connected to the US government, and they never could prove what I was trying to do was wrong, they just simply sand bagged me and causing to spend several million dollars of my own money trying to work with them. Running my biological clock, out among other things. I decided I had enough of that, and so I went to the European Patent Office.

30

I actually have a very good relationship with the Russian Academy of Sciences and a number of other people in China and other places. I have no difficulty getting a patent through the European Patent Office these days. I have completely given up on the United States government because of the dishonesty and there methods of handling situations which they don't understand. This device is one which, if you don't have any good working knowledge of resonant energy, that you want to purchase. It costs about 50 bucks, and it's supplied through one of the services. You can get it through several science services. What it amounts to is that you have a transmitter and a receiver. Of the transmitter and receiver you can have, like in the radio station, or television station, you can have hundreds of thousands of people tuning in, or thousands of people tuning in, and it doesn't deplete the system at all. In magnetic resonance energy, the output and the receptor things is a multiplying method, which works very well. Almost all of my devices depend on that system in order to get the extra energy, which apparently was not there originally.

32

This is a pitch for my resonant energy systems. I don't know whether you can see it well enough to make it out or not? Are we ready to go to the ultimate presentation here? Where are we in time-wise?

33

[Assistent: That was a letter from Don of a recommendation for a submarine power system. It uses a very unique material called Terfenol-D, which also happens to be the.., it's extremely expensive, but the most optimal material. That's that one device, that you looked at earlier, that had the eight coils on it. It's probably as good as diamagnetic that you can obtain. In the picture before you, there's an example of the one device that he talked about earlier that.

34

The disc in the middle if you use Terfenol-D and just make slots in the disc so it interrupts the field between the magnetic coils, you deflate and inflate the magnetic field between them. The field inflating and deflating constitutes, or a magnetic field flux field passing over the conductors and produces the electro-motive force. In this case here, the rod that you're looking at is made of Terfenol-D which is extremely expensive. You can use the inductors in the ambient magnetic flux field that is not stagnant or stable on its own, induces a voltage and a current in these coils. Therefore provides usable electro-motive force.]

35

This thing you're looking at here is an offshoot from the military use of sonic methods for submarines. In that case the energy goes out about 50 miles and can come back about 50 miles, and you can read what's going on with the instruments. In this particular case we were using a rod Terfenol-D which is similar to what the military use would have been. The Terfenol-D is very highly titanium zirkinate in case you want to know exactly what it is. It has very unusual electrical characteristics which you can use to your advantage in building these devices like I like to build.

36

It's also a little bit expensive. When I originally started on this, those rods you're looking at there, were about $3000 each. There's other types of material you can use that will react somewhat similarly. Certain types of stainless steel have some of those same characteristics. You can pick and choose, and get something which does not cost so much maybe. [John: We apologize for this one being on its side. This is actually the schematic that was inside the black box [Editor: page 69. He must be referring to only the second half of this circuit as the first part Don said in 2001 video 5 was the MEG-like device in the shoe box.]. If you look up at the top right hand corner at number 1. That's the lead that came out of it, that was actually the ground lead.

37

Much like the crystal radio set, or the Tesla valve, or the Moray valve, you had to have a ground lead. In this case, it's not quite the same as the Moray valve or development. What happens is, it produces electricity the same way, it gets there by a different route. Two of the devices that we saw earlier, that we saw the wire induction coils are basic forerunners of, or similar to, this system. The suitcase ran and ran and ran, and the suitcase is still somewhere today. I don't have permission to tell you where it's at, but it is somewhere today. Do you have any comment on that one Don?] Well there's several different devices represented here.

38

Here's the alternator part over here, and of course this is your bridge system here. This is the input part back here. This is basically your L1 coil right here. This would be the electrical system of input into the L1. This is a timing device. It adjusts the timing so that you're getting a frequency that you want on your L1. This is basically the L2 system here, it's split up at two different parts for reasons of getting a better signal into the deal.

39

This of course we mentioned was a diode bridge where you're taking the alternating current system here, and changing it to DC so that you can reform the waveform here, through the electrical system here. Of course this is the output transformer here. When you end up with this thing, you end up with a waveform that you can use in any system that's out there, including your computer or whatever else.

39

What we're seeing here [page 40] is the right angle componant under the different factors involved in an electrical system. The magnetic effect and gravitational seem to be related in some respect. The electrical thing is on it's own. This is explaining the separation of the block wall in regards to the electrical component and the magnetic component. If you're going to go the electrical route, you lose the battle before you get there. If you go the magnetic resonance route, the world is your oyster.

41

This [page 65] is explaining how the different parts of the Tesla coil is separate so that different parts of it perform different functions. The ideal Tesla coil is one that eliminates the magnetic part and dwells on the electrical end of the system. That's a mistake, if you really want to get in business, you have to go to the resonant magnetic flux. This [page 70] is explaining the amplification factor, which actually in fact is the Fibonacci series showing how things increase with resonant energy.

42

We have this same resonant energy showing up in the growth habits of seashells and all sorts of things. The energy is amplifying as the system multiplies. The output is much greater than the original input. These are simply some mathematical notations in regards to various things you run into in the electrical and magnetic system.

43

That is sort of self-explanatory I do believe. What you're seeing here is part of a book that I did. The book is on a cd-rom and I used to sell copies of it. I sold over 40 thousand copies of it worldwide. Here in the United States not too many bought it, but overseas there was quite a following. With that in mind, I decided if I was going to be successful, I should be chasing my rabbits where I was most likely to be successful. That's why I moved out of the United States and started doing things overseas.

44

This [page 50] is showing how energy is transferred from one media to another and how the electrons are reacting to that. These [page 51] are some early experiments I did, which show how simple this sort of device can be. There are several different versions of it.

45

This is showing the electrical and magnetic interplay and how you go about achieving certain frequencies that you might be interested in incorporating into your device. That address, I might tell you, is no longer applicable. After about 30 years at that address, we recently moved to North Dallas area. We're not in Houston anymore.

46

Some of these statements here are simply self explanatory I think.

46

These are various devices that have lots of amperage. You run into the problem where people think you have lots of volts but no amps that sort of thing. That's not necessarily true at all. If you know the right way to construct your Tesla coils, then you'll have amperage equal to if not greater then the voltage. This is in this case the point. This [page 134] is showing the system's break down. The power out of the plant here might be like 5,000 or 50,000 volts.

48

As this comes along, where it starts coming down the system, you start with transformers reducing the power. When you finally get down to your house, it's maybe a couple of thousand volts in this general area. This is all using an electrical resonant system. Anytime you have alternating current, you have a resonant system. If you use a magnetic flux component of it, you've got the full force effect of it. You can multiply it, and do things with it that you cannot do with electrical flux. You're wasting your time, if you're going the electrical flux route. It's not going to take you anywhere, except to heat death. Which is a normal system, which you are taught in physics in the university. They don't bother to teach you about the resonant magnetic flux for obvious reasons.

49

[John: I think that's about everything in the PDF there. This was just a quick note of the website. This [page 160] is one of the devices. There's a neodymium magnet, one each in each of the PVC cores there, that have a counterpart on the opposite side. The rotating disc inside this particular one, happens to be a phonograph record with holes drilled in it, so that the field is disrupted. Much like the drawing earlier. The small motor turns on 28 mW at 12 volts and just rotates the phonograph record. It creates a disrupter, interrupter. Terfenol-D is the most advantageous material, although bismuth or powdered neodymium will work.

50

Spray adhesive on an old phonograph record and sprinkle it with, even the neodymium dust is enough to disrupt it. If you have the holes equally spaced there's little or no load on the phonograph record while it rotates disrupting the field. This is Don's website here. This is one of the showcases that you can look at a lot of the material. You can see what's on his website. Here's the front page. You can see the contact point right here. If you want to, that's the best way to email him. Remember Don.. like some of us get more than one or two emails a day, Don's average is about 300. I average about 1100 myself.

51

Make sure that you when you email him that you put some large block letters, like Tesla tech or something, so he knows that it's your email, and not one of 250 spam mails that come with his regular 300. Don, how about your demonstration?] We're going into an actual demonstration here of some of the things that relate to what you're interested in. I'm going to also need someone to come and help me. Let me tell you what these devices are here.

52

This is a Tesla coil with the output on it. This is a dielectric testing machine that we see here. What we're basically going to do, in electronic gadgets like capacitors and resistors, and all this and that, they have their own characteristics. This particular case here, a capacitor, is a blocking device. I don't know whether you're fully aware of that or not. When something goes into one side of the capacitor, it does not pass through that capacitor at all. What does pass through there is the reflection of it, which is from the ambient energy background, caused by the Casimir and Van der Waals forces.

53

What I'm going to do is put 8,000 volts on this plate here, and let the fellow with the earth ground over there play with the other side. With that in mind, see if I can turn it on here. We've got 8,000 volts here, and you should be seeing sparking up there very noticeable. When this sparking, what's happening on the other side,... move it away from it as much as you can. None of the energy from the side that I'm on is passing through the other side.

54

The other side is reflecting ambient background energy, which some people call zero-point energy or whatever. You can see that 8,000 volts is putting up quite a fuss on the other side. We have a sheet of plastic between the two plates here, so that none of the energy from the side that I'm on, can go to the other side. If you don't believe in ambient background energy, which some people call zero-point energy, but I consider that an albatross. Ambient background energy, is a much better term. [John: Many of you are familiar with the fact that the AC sine wave passes through capacitance plate like this. What about DC, DC voltage doesn't pass through the capacitor.

55

In this case you're looking at, if we were using pulsed DC you can excite a field on the opposite side with very little. And just like Tesla did, put a load on the opposite side. This doesn't make a difference which way the anode or cathode is. On this particular, you can set it up so that if you're putting a pulse DC signal into it, you can extract a pulse DC signal out and put a load between the ground and your anode or cathode, whichever you've chosen to use. If you buy the regular like Mallory or one of the other DC capacitors, those are very anode, cathode specific you have to be careful which one you use. Right now we're using the Casimir effect. That's basically how all capacitors work.

56

Casimir did some excellent work in defining, it wasn't that he was the first one, but he did some actually work on defining what it is, and they called it the Casimir effect now. This is not a closed circuit to ground. The power source to run the exciter is actually coming from the little pack in the back there. We are using earth ground through the grid house system here, because they have in one building over, have a 1 inch diameter 8 foot nice copper ground stake driven in. There's no way you can see that as developing a closed-loop system. On DC, you can actually put in a load. You can cut this white wire and put in a load. We've had a situation where we've had an 180 volt DC motor, 3hp with 26 milliamps exciting, spinning–no load–but it did spin at full 3450 rpm.]

57

The commercial version of this is already being used to power electric airplanes, scooters and all sorts of things. [Question. Is that hooked up to a battery?] [John: The power source to run it is coming from the battery, yes, the light up pack. So it's not coming from grid, there's no closed loop there. It was done specifically to ensure and demonstrate that there was no closed loop.] [Question. Do you use Casimir cavities, or maybe 50 nanometers wide, to extract energy from the ambient energy?]

58

The Casimir and Van der Waals forces are very much apparent in this particular type of device. Very little is known about them. There have been relatively few publications. Most of the people who published, add to the confusion rather than help. It's very much in evidence. We have commercial versions of devices which run on the system. They're very obviously there, and they'll be there a long time. Next, Patrick Bailey. [Question. I think it's wonderful that you've gotten patents in other countries and I really like the fact that your marketing to Saudi Arabia, because they want things to replace oil.

59

My question is really to John, who has seen your stuff for years and I guess has really helped you out a lot, and John knows the proper way to measure fancy devices like this, John is this really working?] John and I are relative new acquaintances, about six months or maybe a year ago we didn't know each other. [John: Actually, Don and I have just gotten to know each other, we first met only 19 months ago, in fact 20 months ago, in another week. We've kind of only gotten to know each other well over about the last year. It turns out that a lot of the research we've done is identical, and so we kind of fell in together.

1:00

[Question. I'm not interested about your relationship, I'm interested *does this work?*] It very definitely does, and it's been proven by the Russians and the Russian Academy of Sciences and a number of other people, that I'm connected with, have actually proved–we actually have commercial devices running. There was an order placed by the Chinese government for 24 or 20 engine electrical aircraft, that run on this system, and it does not require fuel cells or anything else. Once you've started, you can run on its own. The aircraft that I'm talking about, the first 24 were delivered. There are 48 more in progress right now.

1:01

They're twin engines like the ones that are New York Airport in New York or areas where you see the community aircraft. [John: Don would you like to show off a few pictures from one of them?] One thing you have to take your mind here is that the contractor on this system we are talking about was none other than General Electric right here in the United States. If you're trying to find an evil side to it, the United States government and General Electric are the biggest culprit.

We've got about three or four more minutes to take questions.

1:02

[Question. John, on the demonstration, if you had the second side plate twice as big as the first one, would you get twice as much reflected ambient energy into your circuit?] In this type of system as you noted earlier I showed you a demonstration where there was one transmitter, with multiple receivers. You can do this with the resonant energy, and you can have as many receivers as you want. They all have the same output as the original had, in addition to the original, and you won't deplete the original in any way, in any sense. [Question. Could one of these be purchased to power my home?] Can it be purchased by someone today? [Right here in the United States?

1:03

The financial control over it is out of my hand, I can influence the possible purchase but the rights are in someone else's cotrol at this time. [Question. I'm not quite clear on the contact point and how much the PDF costs?]

1:04

How to get a copy of Don's PDF there, the website address there, If you look at my website, you'll have the whole story there looking at you. Most of his common developments are on the website. One thing, I get about a thousand emails a day from the ends of the earth in every language imaginable. You can imagine trying to sort through something like that and what's important and what's not important and that sort of thing. I'm covered in all countries and all languages all over the world. The website has been translated into all these different languages.

1:05

Depending on how serious you are about it I've got some calling cards over here. I will have them available for anyone who might want to take issue on anything that I said or done. There's a contact button here on the bottom left, it's hard to see there. The website addres is www.altenergy-pro.com You'll find out, if you go to google information and try to look this up, that out of a million, two hundred thousand people that have devices and such related to resonant energy, that one two three slot belongs to me.

1:06

I didn't do that myself, it was done by Google. [John: Thank you very much Don Smith.

Relevant Emails from NuEnergy Yahoo Forum

These emails from Don Smith are from the NuEnergy email group. Relevant emails from others that Don is responding to are also included within parenthesis (" "). Emails from Don that have no content, or are unrelated to his systems or technology, such as political rants, are not included. No editing of spelling or punctuation have been made.

Oct 13, 2001 **Dennis Lee's OU motor Generator**
Here is a recent photograph of a unit that works Correctly Notice that the charges do not state that the Unit does not perform.
This is because the Attornery General has knowledge that it does. The charges by the Kentucky Attorney General are pure Elephant Pooh Bureaucracy.

["Don, looking at the pictures, it looks similar to the device on your website listed as device 20 and 21. Is this because it works on the same principle as yours? Dave"]
Oct 13, 2001 **Dennis Lee's OU motor Generator**
They do look alike. They are very different in construction and method of function. The motor/generator of Lee requires a 40 Hp electric motor to make go and yields about 180% return. Mine uses a milliwat motor and each pair of coils gets 1000 Volts at 50 Amperes. Lee's is about 3 feet high and mine is 14 inches. The inventor of the Lee Device lives at Houma, Louisiana. He is an Electrical Engineer with all credentials.
My device has 8 set of coils, each set is a separate generator (eight sets). The coil sets come together at a buss bar for a composite output which is regulated by a speed control. Mine runs off of a 12 Volt self recharging battery. Lee's starts from a conventional electrical source.

Oct 17, 2001 **Dennis Lee's OU motor Generator**
John, no am not a dealer. There are a half dozen or so Devices out there that work as stated. I have several of my own which are simpler than the others. I do not sell directly, only through companies that are already long established in a related industry.

Oct 29, 2001 **Magnetostrictive Oscillator (Energy from Ambient Background)**
All, here is the Ambient Energy Collector Circuit promised (Don Smith Device - one of many). The magnetostrictive rod collects energy from the ambient energy background. The piezioelectric characteristics of high magnetic permeability material performs the equivalent functional counterpart of negative resistance. This permits ambient energy present everywhere in the Universe to enter the system. Chasing the electrical side of the equation assures less than 1 COP. Magnetic permeability extends into the millions. Negative resistance is found only in resonate circuits courtesy of magnetic side of the equation.

[Question: "In addition to drawing in the energy from the vacuum in a cirduit, it would be very nice if one could refrain from killing the dipole with our closed circuit wiring. That is probably where Tesla was trying to go with his one wire circuits. That way one wouldn't have to exert so much effort to make the dipole last. Any ideas in this department?"]
"Hi Greg,
[Dave answers and Don replies: "Excellent!"]
"The short answer is that the dipole does not come into play in a one wire system.
"Some people think the one wire system is supposed to work like an antenna, but this is not the case. Although the one wire system does use tuned resonance whereby the inductance of the

wire and its capacitance to ground determine the tank circuit. But the object is not to radiate energy into the environment in a one wire system.

"The object is to maintain resonance in the wire itself and within its capacitance to ground. This means that if a one wire system is tuned for the first floor of a tall building, then it will have to be retuned to use the circuit on the top floor (unless the building is used as ground reference.) Tesla invented many novel approaches to variable lighting in his laboratories. He would use a high frequency generator and output to a bar of copper. The frequency of the generator and the resonant frequency of the copper bar would be in tune. Then he made a sliding contact that traveled over the copper bar such that a wire connected to the contact would receive varying voltages depending on its position relative to the voltage nodes and loops on the bar. Moving the contact along the bar would vary the voltage to the lamp.

"In the one wire system just described, AC current is used. Pulsed DC current from a MEG could be used and converted to AC for the purpose of one wire electrical transmission. Resonance is used to store energy in the wire and it takes a relatively small amount of input energy to maintain the resonance. The dipole is no longer an electrostatic charge at the opposite ends of a wire, but now dipole is converted into a standing wave.

"I have puchased a copper bar about one meter long and hope to use it in a one wire experiment this winter. I'm curious as to whether it is possible to tap a positive voltage loop and a negative voltage loop to get DC current. If it is possible, this would be an interesting variable DC power supply. Dave"

Nov 2, 2001 **Re: Negative Resistance**

Excellent! Open circuits and none linear where ohm's law does not dominate. External advantage: falling water, wind blowing, supper steam, radiant energy of any kind including ambient energy. Mechanical advantage: levers, pulley systems, hydraulic jacks, and various others.

Physics teachers dare not mention the real nature of this. In electrical systems as in coils length = volts, diameter = amperage. Multi stranded also adds to amperage as does a central core. Vortex energy is a combination. Ohm's law is king in closed circuits. Magnetic Resonance Power Systems such as the schematic recently posted is a good example being outside ohms law.

Nov 2, 2001 **Re: info MAgnetic resonance power system follow uo**

Mikael please go to http://www.islandlogix.com/ where You can download a copy of a nonelinear electronic's schematics program with which You can switch and play with componets and see the results on vertial instruments included as a part of the program.

Nov 3, 2001

Magnetic Resonance Power System (MRPS)

From Schematic previously provided: Construct such that the rod and other componets can easily be intechanged, particularly the rod. As a trigger some trace element such as thorium, cobalt or other should be present. You may want to sinter powders of certain minerals such as tourmaline, certain garnets and substances like those found in early coherer devices. If properly tuned physically the rod should access the earth's radiation belts frequency ambient energy background. This is diurnal and the rod behaves like a loaded CB antenna (see ARRL annual). When all is correct, insert the rod and You will have power, just as Tesla.

Enjoy

4 Nov, 2001

Harlan Sanders icognito innominatus

Site moved to http://mixmaster.nullify.org/remailer OPEN MESSAGE FOR DECKER, YORK AND ASSOCS.

See: http://www.nuenergy.org/iw/2001

Under Don Smith, Photo Row 2 at right end is the Device You question.

The Device was built by me in mid 1980's from German and Canadian Patents which Tom Bearden recently copied and is patenting. The difference is that I demonstrated mine at the 1996 Tesla Convention. The Device ran during the 3 1/2 hour presentation. At the end the audience was invited to bring their test instruments and check it out. They did so as witnessed by the VCR Taped Presentation. The results were as You know from a 7 1/2 Amp Hr., 12 volt battery was 120 Volts at 80 Amperes. Hundreds of People are witness to this. Among other items ran was a bank of 10 100 Watt Light Bulbs. This was the small black suitcase.

Heinrich Kunel of Germany, Patent # DE 3024814 Issued 28 January 1982.

View @ http://12.espacenet.com/dips/viewer?PN=DE3024814&CY=ep&LG=en&DB=EPD English translation at: http://hometown.aol.com/DMBoss1021/Kunelpatent.html

First Paragraph from that Patent:—concerns procedures and devices for energy production, which converts magnetic flux without rotary components—a Large modification of the induced flux electric current—without torque such a manner that the electric current is strengthened to the n-th number multiplication.

The Canadian Patent issued prior to the German Patent is more direct and a better one than the German. At the moment it is misplaced. You now know where Tom Beardens Device came from.

I use Electronic Schematics Simulations Software to check all of My 200 plus devices before I built them. The nonlinear schematic programs here listed clearly show greater output than input when present.

1. VisualSpice Pro V 5.10.OR
 http://islandlogix.com/products2.html
2. MICROCAP 7 Ver. 7,
 http://www.spectrum-soft.com/index.shtm
3. CircuitMaker V 6.2c
 telephone # 801-224-0423 Also on Internet.
4. Win VLSI a $5,000. program for free.
 Undated Berkeley University Spice.
 From University of Barcelona, Spain
 http://www.cnn.es/~pserra/winvlsi/

More than 40,000 copies of My book "Resonant Magnetic Energy Methods" are in circulation, being translated into all major languages. It has been republished by others and sold world wide. There are 7 different editions out there. More than 12 separate VCR Tapes exist and are sold by numerous unaccounted groups and persons. News Film groups have been here from China, Japan, Australia, Europe, Saudi Arabia, Brazil and others. They sometimes stay for days filming. Sometimes they are only 3-5 hours of filming, then return home. They are always pleased at what they see and have trouble sleeping for weeks after the visit. See my recent response to Michael Couch message 17856 at jlnlabs. http://groups.yahoo.com/group/jln/messages for how locals are handled. Because I don't roll over and sit when You bark does not mean any disrespect towards You and therefore doesn't merit the numerous unfounded and meritless diatribes which You hourly bestow upon me. I am extremely busy and do not go around blabbing business secrets. Have a Great Day, Don Smith donsm1@... t See Lycos and then altenergy-pro.com (2000 Servers World Wide). also on Lycos TransWorld Energy.

Nov 6, 2001

Re: Harlan Sanders icognito innominatus

Would love to know more about Joe Flynn's work! Take Care, Don Smith

Nov 8, 2001 **Geoffery Spence OU Patent, US Patent: 4,772,816**
United States Patent 4,772,816 Spence
September 20, 1988 Energy conversion system
Abstract:
The apparatus uses a magnetic field (80) to accelerate a charged particle radially towards a target electrode (10). The increased kinetic energy of the particles enables the particle to give up more electrical energy to the target electrode (10) than was initially given to it. This charges the target electrode (10), and the increased energy is extracted from the apparatus by connecting an electrical load between the target electrode and a point of lower or higher potential. [Rest of the patent not quoted.
https://patents.google.com/patent/US4772816]

[Question:]
Nov 8, 2001 **info Magnetic Resonance Power System follow up.**
"It seem to be difficult to create the type of coil in this software, how do you succed to enter the stainless steel core and the special coil in the design? Michel"

Nov 8, 2001 **info Magnetic Resonance Power System follow up.**
Mikael, there are three basic types of devices, inductors, capacitors and resistors. Then there are variations from these. Of the thousands of devices in the software inventory including solid state and other the physical activity is part of the computers knowledge base. Experimentally build several of the specilized cores with the windings and measure the inductance, capacitance and resistance with various windings and frequencies to determine the information which you will add to coil system information upon importing Your Coil system into the software as a new or not previously cataloged component. Hope this helps. Remember the metal core must be high magnetic permeability which always has extreme resistance to electrical flow. Remember You are doing the magic on the magnetic side of the equation. Good Luck.

Nov 9, 2001 **Relates to Bearden Patent, CIPO - Patent - 2172240**
[Important Canadian patent predating the MEG
https://patents.google.com/patent/CA2172240A1/en
Actual pdf appears to be here:
http://www.tuks.nl/pdf/Reference_Material/Magnetic_Flux_Switching/CA2172240A1.pdf]

[The following are from two forwarded postings from Don from someone on Keelynet:]
Nov 19, 2001 **Energy_Experiment**
Sunday b History
Inventor Don Smith has created numerous devices that are related to creation of excess energy. Little is known of any of these devices by the transmutation team, but for reference an experiment was conducted with the device shown on the following URL.
http://altenergy-pro.com/id29.htm
Mr. Smith has reportedly observed numerous observations and anomalies from his experimentation in the area of free energy.
Mission In a total blind study to see the output energy could be increased or changed on demand.
Experimental Conditions
The device is located at the following coordinates:
8110 Bent Oak Lane Spring TX 77379
Device is a ~14 inch in diameter toroid that is energized by a small battery. The battery drives an inverter that drives a fluorescent tube transformer This transformer connects to the toroid and the output of the toroid is connected to a bank of capacitors.
The output of the toroid is alternating current (60 Hz).
The experiment was witnessed in Spring, Texas by the following:

Don Smith
Roland F. Harvey
Steve C. Johnson
Experiment

It was reported that this device had not been operated for over a year and the battery had only a 2.9 volt potential. When the circuit was activated the AC output to the capacitor bank was ~220 volts.

When the ADAM technology was activated stabilized. The voltage was further raised by the ADAM technology to 290 volts then to 350 volts. The observers reported that the only power to the device was the one battery.

Accepting the assumption that the ADAM technology was operating the equipment disconnecting the battery voltage fell to ~170 V (AC) and remained stable. Upon reconnection of the battery times.

Upon disengaging of the ADAM technology to the original 220 VAC and with the ADAM technology disconnected was no output voltage when the battery was removed from the circuit.

An important note were no wires, connections or associated apparatuses of the technology located in vicinity of Mr. Smithb The entire experiment was controlled from 1,200 miles away.

Conclusion

Anyone reviewing this report must accept one of the following as to the reality of this experiment.

o' The test was not accomplished and the parties involved conceived a very interesting tale to share with you.

The events observed and reported by Mr. Smith and his observers were events that would have occurred without the application of the ADAM technology.

The events that were observed were in fact the relationship of the ADAM technology with the device that Mr. Smith constructed. This report is factual. However, it is up to the reviewer to develop their opinion.

J

History

Inventor Don Smith has created numerous devices that are related to creation of excess energy. Little is known of any of these devices by the transmutation team, but for reference an experiment was conducted with the device shown on the following URL.

http://altenergy-pro.com/id29.htm

Mr. Smith has reportedly observed numerous observations and anomalies from his experimentation in the area of free energy.

Mission

In a total blind study to see the output energy could be increased or changed on demand.

Experimental Conditions

The device is located at the following coordinates:

8110 Bent Oak Lane Spring TX 77379

Device is a ~14 inch in diameter toroid that is energized by a small battery. The battery drives an inverter that drives a fluorescent tube transformer This transformer connects to the toroid and the output of the toroid is connected to a bank of capacitors The output of the toroid is alternating current (60 Hz)

The experiment was witnessed in Spring, Texas by the following:

Don Smith
Roland F. Harvey
Steve C. Johnson

Experiment

It was reported that this device had not been operated for over a year and the battery had only a 2.9 volt potential. When the circuit was activated the AC output to the capacitor bank was ~220 volts.

When the ADAM technology was activated stabilized. The voltage was further raised by the ADAM technology to 290 volts then to 350 volts. The observers reported that the only power to

the device was the one battery.
Accepting the assumption that the ADAM technology was operating the equipment disconnecting the battery voltage fell to ~170 V (AC) and remained stable. Upon reconnection of the battery times.
Upon disengaging of the ADAM technology to the original 220 VAC and with the ADAM technology disconnected was no output voltage when the battery was removed from the circuit. An important note were no wires, connections or associated apparatuses of the technology located in vicinity of Mr. Smith The entire experiment was controlled from 1,200 miles away.

Conclusion
Anyone reviewing this report must accept one of the following as to the reality of this experiment.
The test was not accomplished and the parties involved conceived a very interesting tale to share with you.
The events observed and reported by Mr. Smith and his observers were events that would have occurred without the application of the ADAM technology.
The events that were observed were in fact the relationship of the ADAM technology with the device that Mr. Smith constructed. This report is factual. However, it is up to the reviewer to develop their opinion.
J

Nov 19, 2001 **Energy_Experiment**
Joe Champion submitted this. ADAM technology is something the public does not yet know about. It's a way to interfere with the Time Dimensions and remotely change things. Joe Champion (Electronics Engineer) is retired from NASA and works out of Kingman, Az, USA. He is best recognized for commercializing the Transmutation of various elements into precious metals. Electrical Engineers were present and all facts are given. Nothing is amiss. No University Professors here to screw things up. The whole thing is too honest for them to understand.

Nov 19, 2001
24 AWG Magnet Wire, New, Available on eBay
Dave magnet wire is single stranded which works against the circular mil's approach. I have found that multistranded Amp King high amperage DC wire is all around best. AWG #'s 8 and 10 are the one's I prefer. It's very flexible and easy to use. GateWay Electronics in St Louis, Denver and Los Angeles sell it. otherwise near impossible to locate. They are on the Internet.

Nov 19, 2001
24 AWG Magnet Wire, New, Available on eBay
Dave a couple of other thoughts. In Coils Systems length helps the Voltage and diameter helps the inductance-amperage. If You want functionally useful energy stay away from long and skinny L-2's. Try something like say L-1 being 10 inches D and L-2 being 8" D. Say 40 turns of # 8 multistranded Amp King wire for the L-2. Once wound onto the Coil remove it and measure it. Then cut # 10 AWG for the L-1 to one forth the length of the L-2 wire. Think of the wound Coils as antennas so wound for convenience since cut in this manner they are very close to tuned upon construction. Mount the L-1 coil so that it covers the approximate center of the L-2 Coil. This gives a better shot at amperage in the output. The coil will be deadly so be very careful. Tesla coils are normally built to mostly eliminate amperage which based on electron spin right (= magnetic-amperage) and left (= voltage-potential). They should be equal. The magnetic is radiated and mostly wasted, therefore an under unity device. It's the magnetic flux which converts back into volts. This flux is not diminished and therefore can replicate many times and not be diminished. Any energy spun off from the electron does not diminish the electron in any way. The energy You

collect from this action is from the Earth's Ambient Background and is free except for the collecting Device. The method of collecting Energy is a function of Human ignorance. Otherwise the Energy is Free just like the rain and the fish in the stream. Hint: the Coil works just fine laying on it's side. It's also easier to manage.

Nov 19, 2001 **Energy_Experiment**
Robin no evasiveness, some people are doers and some are explainers. Until better documented I am still adjusting to the happenings. My plate is extremely full and I have hundreds of my inventions that need explaining. My thinking is highly original and explaining any one of these Devices would be a lifetime project for the Explainer Type. One has to decide which is more valuable, doing or explaining. In each case all the details for construction are in my head before the Device is made. It's extremely rare for me to miss my target, small adjustments take care of the problem. Since neither of Us is going away we should try helping each other.
Regards, Don Smith

Nov 20, 2001
24 AWG Magnet Wire, New, Available on eBay
Andrew it's similar to Litz but no individual wrapping on each multistrand. It's flexible, plastic coated normally used for high amperage DC, but works beautifully with High frequency AC. I assumed You want a practical Unit which produces Energy which becomes useful upon passing through a diode bridge into a capacitor bank and then through an inverter to useful load. Regular Tesla Coil Transformers don't produce usable amperage. You can up the amperage by putting ferromagnetic (soft magnet material) as the core of the L-2.
Enjoy, Don Smith

Nov 22, 2001 **Bearden Patent Query**
Stephen, there are two patents which closely resemble the Bearden Patent Application and work using the same principal. Boday has a Canadian Patent among others. I had previously named the Kunel, German Patent and did not at the time remember the Canadian patent number. I constructed more than ten years ago a Device which is an identical twin to the Bearden Device. It produced 120 V at 80 Amps continuously when demonstrated at the International Tesla Society Convention at Colorado Springs, USA before a large audience in 1996. I tried for a US patent, spending lots of money with zero results. At that time the US Patent Office as I were totally ignorant of the German and Canadian Patents. They were fully aware of the fact that My Device worked as stated, but used bureaucratic maneuvering to frustrate My efforts. The Nuenergy "Inventor's Week End" post has a photograph of some of My Devices including the one of interest.

Nov 23, 2001 **Real Power Generator RePost**
Len I went back and checked and printed out each message which I sent You. The message listed below does not make any sense and has been tampered with. jlnlabs@yahoogroups.com messages are presently in the 2000 to 3000 range. Sorry! Accept my apology? This is a clear cut case where some one changed the information I sent in order to keep it from being posted. Will go back and repost.
[The following posting was claimed by Don above that it was tampered with and chenged:]
> From: donsm1@e...
> </group/jlnlabs/ >
>
> Date: Sat Nov 17, 2001 5:03 pm
> Subject: Re: Real Power Generator Information
>
>Some help in understanding the altenergy-pro file referenced below. The PVC Coils have Amp King # 10 AWG, multistranded wire for 50

Amperes. The cylinders upon which the wire is wound have two 400 Gauss magnets, one on each side as a pair. The winding is continous for each pair. The magnets when seperated have a neutral zone where electrons are not spun right or left. This allows a disk of shielding material with cut outs to pass without back pressure between the magnet pair. Carl Gauss's original definition of one gauss is 100,000 volts worth of magnetic flux. Shielding material available ranges from 1 percent to greater than 90 percent effective. If one percent shielding effectiveness,then 1000 volts of useful electricity results. Therefor a milliwatt input (for the disk drive motor) easily produces 1000 volts at 50 amperes. With eight sets of these coils feeding the composit output into a busbar is —??

>Using higher amperage wiring and greater percent shielding the results are very interesting. The Device referenced is 14 inches high, 16 inches long and 10 inches wide.

Nov 26, 2001 **your book**

Dave the Book is called "Magnetic Resonance Energy Methods". It is $30. which includes the PO 2 day mailer (probably one week). Address is 8110 Bent Oak Lane, Spring, TX 77379. Am working on a CD ROM with lots of goodies, colored photos and assorted helpful material. When finished in about 2 weeks, will also be $30.

Dec 4, 2001 **Depressing Memory etc**

Reference to magnetostrictive rod material: some forms of stainless steel will work just fine. Have one party who get's from each watt in an output of 1,000. watts from a stainless steel tube - open ended. This makes the Device very dangerous. Take Care, Don Smith

Jan 1, 2002

UPDATE: Motionless Battery Shock Charger

Robin, you are correct MetGlas is a maximum insulator. I use all the time, you don't have MetGlas.

Happy New Year, Don Smith The bulbs I presently have are 100 V at 55 cycles for the Tokyo market. Factory puts out one million per month and they sell quickly at equivalent of US $60. Don't worry, the ones which will be sent to You are free. Supposed to have some which You can use probably mid January.

Jan 1, 2002 **Motionless Battery Shock Charger**

MetGlas has extreame permeability to the magnetic field, not DC. That's why it's special uses make it attractive. General Electric bought the Patents from Allied-Signal about 2 years ago. They apparently relized that it could change the market as relates to energy. All my MetGlas is prior to GE's takeover. I have copies of the original patents which confirm what I say. The ones on file at the Patent Office probably have been tampered with since GE is one of the Governments Industry Partners who has complete run of all the confidential files, including removing them from the Patent Office and losing them if necessary. I know this happens, I have had the Patent Office do it to me. (see the file listed below, for the material here discussed)

Happy new Year, Don Smith

Jan 1, 2002 **Honeywell Metglas**

Metglas® amorphous metals were designed by Honeywell (formerly Allied Signal) at their research facilities located in Morristown, New Jersey. Honeywell first began research on amorphous metals in 1970.

Metglas High Frequency Cores Catalog Honeywell manufactures four different types of high frequency magnetic cores with Metglas amorphous metal alloy ribbon. They are Magnaperm®, Powerlite®, Magamps, Microlite™ and Microlite XP™. These cores utilize the unique characteristics of Metglas amorphous alloys and are used in a wide variety of applications.

Metglas Amorphous Metals Overview:
Metglas amorphous metals are metal alloys with noncrystalline atomic structures. The metal is cooled and cast quickly in order to produce this unique atomic structure. This structure is the key to the incredibly low power losses exhibited with Metglas. Other unique properties of Metglas amorphous metals are effortless magnetization and demagnetization, and low temperature rise at high frequencies.

Jan 2, 2002 **Honeywell Metglas**
They put a metal band around the core to hold it together, which is not the metglas material. The insulator seperator material also makes the core a form of capacitor. Half the energy is usually thrown out as waste which asures under unity. With the (C = Joules) and (I = Amperes) both retained You are on Your way. See
http://www.makisoft.net/quantronix
first paragraph for the formula. The pump speed rate (CPS) then becomes (times spin reversed) becomes very important. That why what You are using makes a great difference.

Jan 7, 2002 **Gravity a la Grebennikov**
NATURAL PHENOMENA OF BIOLOGICAL ANTI-GRAVITATION ASSOCIATED WITH INVISIBILITY IN INSECTS & GREBENNIKOV'S CAVITY STRUCTURAL EFFECT ("Flight"- V chapter from book V.S. Grebennikova "MY WORLD")

Jan 7, 2002 **Antigravity Research to date and the men who did it**
[http://maddsci.tripod.com/id13.html]

Jan 9, 2002 **Something to think about!**
Dave, the Flying Thing is about 6 month's to one year to commercial production. Projected cost about the same as a Corvette. The Power System will probably come from Me. No explosive or dangerous fuel, fully self powered without external requirements. Don't have a setup for credit cards yet. The book "Magnetic Resonance Energy Systems" and the CD "Don's Note Book" are available at $30. each which includes S & H with Priority 2 day Post so they say. The CD has among other things the updated version of X-Spice with manual and numerous other goodies.

Jan 11, 2002 **Problems reading the CD**
Warren, there are about 200 of the CD's out and only two people have had problems, once understood that many different files from from all over the World are included, the potential for problems exist. The Copy Program I used is Easy CD Creator V-5, Platinum which is the Rolls Royce of Copy programs (Look up on the Internet). The Computers involved are Pentium II's and II's, Athelon 333 Mhz up to 1.5 Ghz. The CD-RW used for the Disk is HP 9600 series. Multiple file readers such as Paperport, Paint Brush, and others are helpful. A large % of the files require a PDF reader up and running, then accessing the files. All the files are quite readeable to a knowing person. For You to duplicate the effort required to put the disk out, would require Your siting at the computer, 18 hrs per day for months. The Disk is a goodwill effort at sharing information which I consider relevent, as is the Book. The Book is largely My Theories going through maturation and the Disk relates to the same developing process. Together they represent a life time of thinking by someone who is in the know. Money collected for these items requires subsidizing by me. For example one set of ink cartages for one printer cost about $80. per set and requires replacing about 3 times per week. There are three of these printers. Priorty mail is about $200. plus per week. Add in about 80 hrs. of none paid work This is on top of a very crowded schedule which is none comprimising. Include responding to about 200 e-mails per day and other. You keep the disk and if You think appropriate, I will

refund the $30. which I am shure You need more than I.
Best Wishes, Don Smith

Jan 11, 2002 **Cosmic Ray Study**
Cosmic Rays as in any other radiation, as in Radio Broadcast and et al - Fields - Waves - Rays. Rays are simply very high frequency waves. Therefore action at a distance applies. Example a space satellite going past Pluto, with it's 9 V battery is sending signals, clearly arriving at the surface of the Earth. Magnetic action at a distance allows this reception. Physically the electrical component becomes part of the Earth's ambient background energy upon impacting it's atmosphere. However the magnetic component of this high energy system directly impacts the surface of the Earth and as in all electromagnetic systems can change from magnetic to electrical. It's part of the electron spin mechanism present everywhere.
Have a Nice Day, Don Smith

Byron, Global Trading Company out of Tokyo of which I am one fourth owner already is making Light Bulbs that have a negative ion generator and runs at frequencies known to destroy pathogens, but not harm humans. The Tokyo Light Bulb is in advanced patent procedures with the World Intellectual Property Organization out of Geneva. Looks like we might have been barking up the wrong tree, don't You think! The Tokyo Bulb is already on the market and selling about five million per month and that's with little or no effort. It is high quality and built to last beyond 100,000. hrs. The US is included in the patent, so will be marketing them here most anytime. The Bulb uses 20 Watts and puts out greater Lumens than the standard 100 Watt US counterpart.
Take Care, Don Smith

Jan 15, 2002 **Response to negative review on Inventor's Weekend**
To whom it may concern, I have viewed the nuscam post below, and am not amused.
It's a fabrication from a very sick mind (JWD). Please view evidence, photos and other material. I don't think You appreciate having some Jerk telling You what to think. Not to My knowledge was the audience abused or mistreated. There were many Devices presented and discussed. It's not My Job to be a Baby sitter or Butt Wiper for anyone. The Devices Presented were a goodwill effort on My Part to share information and techniques which might be useful to persons having common interest. The Person made mysterious and threatning was Bollenger who's photo is at the bottom of the photos posted from the Meeting. He is a nice guy and I am sure he does not appreciate being portrayed as a Spook. He was very helpful in asking correct questions of the Speakers which he felt were appropriate. One of the Devices shown by me was the MEG of Tom Bearden fame (Photo in the Box Device shown on Bruce's Web Site). It preceded Beardens by more than 10 years and had an output ratio of 1000 to one of input ratio. Take a look at the strong resemblance. I had a sampling of other Devices and interesting experiences to share. Anyone Who has been present at more than three dozen Presentations which I have done including the Willard Hotel across the street from the White House in Washington, D C., will immediately recognize the malicious nature of these fabrications by JWD. I was the main Speaker at the Washington meeting and several thousand people were in attendance including a majority from the European Elite. You should let JWD and his Scam Machine (H. Sanders) know that You appreciate his unbiased opinion in which He claims to represent and save You from the evil persons in the world. Appreciate very much Your Interest, Don Smith

Jan 17, 2002 **Fwd: Scientists' breakthrough shows objects fall in mini steps**

This nodular movement means that gravity has a known frequency. Determine it and You have broken the Code.

"An experiment has shown for the first time that objects falling under gravity do so in tiny lurching steps. Scientists in France have seen the very small quantum forces involved in gravity. They did it by monitoring tiny, slow-moving, uncharged particles in a specially-designed detector. Full story:

http://www.ananova.com/yournews/story/sm_497419.html

Jan 18, 2002 **Re: tetra coil**

Is it possible that Tetra (four) is two Bifilier wires, possibly used in a coil for amperage amplification. I use multiplier (single-wire) amperage type for the same effect.

Jan 19, 2002 **Self Sustainable Devices**

Information from " Drake's Radio Cyclopedia " 1929 [https://www.americanradiohistory.com/Archive-Early-Radio/Call-Book-1926-Amateur.pdf]

By: H.P. Manly

Pub: Drake and Co., Chicago, USA

See this System in use at http://www.altenergy-pro.com/device [Device 2 in this book.] 02.htm http://www.altenergy-pro.com/device03-2.htm

The 02 Device in the 1600's Chair has three separate coils tuned to the same frequency as the Tesla Coil from the center. A limitless number of separate coils, each duplicate the energy present in the Tesla Coil. This does not draw down or limit in any way the original energy source, however it duplicates fully the source. Magnetic Waves which spin the remote electrons duplicates the original energy source without reducing in any way the source. (Physics as prescribed by the establishment is 100% wrong) Feedback from the remote Coils supplies excess energy, some of which can make the Device self sustaining.

The 03-2 Device uses the same system for feedback. See upper left end of Device for the dual Coils on a ceramic rod center mount.

Have a Great Day, Don Smith

Jan 19, 2002 **Self Sustainable Devices**

This is the Device from the small Black Suitcase (Briefcase Size) presented at the 1996 International Tesla Convention at Colorado Springs, Colorado. Input was 12 V DC at 300 milliamperes and the Output measured by an Electrical Engineer, picked by the audience, was a continuous 120 V at 80 Amperes as Documented in the VCR Tape of the Presentation. You may see a photo of the Device taken at "Inventor's Week End" and shown to the Audience. On the Web [page 167]

http://www.nuenergy.org/iw/2001/photos/don8.jpg

This is a close relative of Tom Bearden's Device. However I used a different Coupling System. In the Attached Snip from "Drake's Radio Cyclopedia (1929)", [see page 167] the coupling on the left is Bearden's and the one on the right is Don Smith's. It's obvious from the energy produced as to which is the better (Don Smith's). Also, the Smith Device precedes Bearden by more than 8 years. The Smith Device is self sustainable if desired.

Jan 21, 2002 **EM Simulator Is Free Of Charge**

Voltaire XL Nonlinear Circuit Simulator - Applied Wave Research Jan 21, 2002 for Microsoft Windows 98/Me/NT4/2000/XP

ELECTROMAGNETIC (EM) simulators have become well-established recently as useful stand-alone analysis tools and as invaluable companion tools for linear and nonlinear circuit simulators. raph program allows users to plot their response data from em (and other simulation tools) using standard engineering formats, such as Smith charts or Cartesian graphs. Internet: http://www.sonnetusa.com.

Jan 21, 2002 **EM Simulator Is Free Of Charge**
You can simulate anything that exists in Spice. You build a profile for the part and enter it into the catalog of devices, where it becomes just another part. Something exotic may require breaking it down into smaller units each of which has it's own function. Several functions from one device tends to screw the system up.
Take Care, Don Smith

[Question: "Can we use trifilar secondarty coil in the magnetic resonant system? and bring out 3 phase?]
Jan 21, 2002 **EM Simulator Is Free Of Charge**
Would require knowledge of solid state switching devices. I am not sure that it's practical. Making two phase out of single phase is not all that difficult, a small cigarette package sized solid state device is required and can cost $200 to $300. if not more, depends on amount of amperage required. What You are asking about is a rare bird and probably not in stock, so would have to come direct from manufacture and would be very expensive in small quantities.
Take Care, Don Smith

Jan 21, 2002
Negative Resistance Amplifier Device
That's what I refer to as self sustainable, which is a fact. Slight changes in the coupling on the toroid and the skys the limit. We have been there for more than ten years based on My personal knowledge.
Take Care, Don Smith

Jan 22, 2002 **The Mystery Crystal Set**
Basics of so called Free Energy Devices are present in Early Radio. For example take the examples of the Crystal Sets shown in the schematics herein.
1. Replace the antenna portion with a high voltage source. Tesla coil derivative - -
2. Upgrade Diode Detector to Volts and Amperes. Diode changes signal to pulsating DC.
3. Replace Capacitor across the Ear Phones so as to set to the desired frequency. This can be a resistor due to built in capacitance of Transformer.
4. Replace the Ear Phones with appropriate size (desired output) Transformer. Presto, You have the Basic Space Energy Device. This Device sucks in energy from the Ambient Background if properly adjusted. The Tesla Coil Derivative requires some input to trigger the row of dominoes, so to speak.
Have a Great Day, Don Smith

Jan 22, 2002 **ByteMark's BookShop**
BYTEMARK'S BOOKSHOP
Amidon Tech Data Handbook
Amidon 148 pages, including all the technical information on this web site, with additional Q-curves, formulas, charts and specifications for most Amidon parts.
Transmission Line Transformers
https://ia902709.us.archive.org/6/items/fe_Transmission_Line_Transformers/Transmission_Line_Transformers_text.pdf
by Jerry Sevick, W2FMI This hardcover book is a practical design manual covering the use of transmission line transformers for both commercial and amateur applications. It covers types of windings, core materials, fractional-ratio windings, efficiencies, multiwindings, series transformers, baluns, ununs, limitations at high impedance levels and test equipment. 272 pages, 15 chapters.
Transmission Line Transformers Design Handbook
http://www.introni.it/pdf/Amidon%20-%20Transmission%20Line%20Transformers%20Handbook.pdf
by Jerry Sevick, W2FMI Describes 48 improved designs covering every impedance ratio with data sheet - circuit diagrams, types of cores to use, numbers of windings, sizes of wires to use, wire spacing and photographs of wound units

for baluns, ununs and multi match. All the information you need to construct your own.

Building and Using Baluns and Ununs: Practical Designs for the Experimenter

by Jerry Sevick, W2FMI The all new definitive source for his latest practical information and designs. A unique way to learn about the application of baluns and ununs for dipoles, yagis, log periodics, beverages, and antenna tuners with countless examples.

RFI BOOK replaces - Radio Frequency Interference: How to Find It and Fix It

Over 200 pages, 16 chapters. Written by RFI experts, it's filled with proven ways to solve common - and not so common - RFI or EMI problems, whatever their source. In addition, you'll learn how to build a cooperative environment with neighbors and how to contact skilled volunteers who can assist with those tricky situations.

Hints & Kinks for the Radio Amateur

http://www.n5dux.com/ham/files/pdf/Hints%20and%20Kinks%20for%20the%20Radio%20Amateur-1978.pdf

http://www.n5dux.com/ham/files/pdf/Hints%20and%20Kinks%20for%20the%20Radio%20Amateur-1992.pdf

ARRL, 15th Edition This latest edition includes hundreds of good ideas, projects and techniques that will help you solve problems, improve your operating and have more fun on the air. From the popular QST column, this book covers all ham subjects, from hand-held transceivers to antennas, batteries to computers and digital modes to mikes.

Solid State Design for the Radio Amateur

https://www.scribd.com/document/352658368/Solid-State-Design-For-The-Radio-Amateur-1986

by Wes Hayward, W7ZOI A Classic in its field: circuit design and applications, descriptions of receivers, transmitters, power supplies and test equipment. Much of the data cannot be found elsewhere. Essential for every technical library.

Jan 24, 2002 **New file uploaded to nuenergy**

Zip Files # 2, 6 and 7 have numerous color photos of a New Device based on the two metal rods inserted by Tesla into his Electric Automobile Device. [Energy Receiver file]

Have a Great Day, Don Smith

Jan 26, 2002 **Does Don have any working devices for sale, if not devices**

Yes there is in circulation such! As to eliminating all the guess work, that's a near impossible job due to the differences with which the human mind functions. Humans have a way of making simple things impossible depending on their motivation or lack of it. The Magnetic Resonance Power System is a spin-off of the basic sonar system used by submarines, from which the magnetic field emitted allows viewing a 50 mile radius. In it's simplest form I published "Magnetic Resonance Power System" for water systems. The rod shown in practice is one of the materials listed with the Report. However those are somewhat of a problem.

The Terfenol Rods cost > than $5000. each, I have several of them. In Eastern Europe they have been getting COP's greater than 1000. with stainless steel rods and tubing. problem with that is there are many many kinds of stainless steel. Which one, well they don't know, since they are using scrap. A chemical analysis should resolve that. I am getting excellent results and I refuse to get into a Turkey fight with the populous, therefore will not further discuss it. Will be more than happy to see others duplicate the results I have already achieved. Any one that gets bad results needs to examine their mental qualifications and motivations, Jerry Decker included.

This is one of several hundred working Devices which I have built. I have developed my own theory as to electromagnetics and it works 100%. My Web Site altenergy-pro.com very generously in simple terms explains why My Systems always work. It's written such that the understanding level requires very little effort on

the part of the reader. This makes me look unscholarly to those who believe in muddying the water so they look intellectually superior. I have news for them, this battle they lost a long time ago.
Have a Great Day, Don Smith

Jan 27, 2002 **NOT SO......most oil is formed just like other mineral deposites**
Reply to Robin: My sources of information? :-P : Several - In National Geographic - Pieces of Amber with air bubbles has 35 % oxygen from Biblical times. Secondly, US Government research in 1900 has 20 %. Thirdly, Carl Sagen's book " The Blue Planet" 1984 shows 18 %. Fourth, US Government Research shows 15 % for early 1990's. Cities of 20 million or greater, examples Mexico City, Tokyo and a dozen others from NASA sensors and photographs-chemical identity show 7 % or less oxygen. This is confirmed by actual ground level test in all of these Places. A hand held oxygen meter, like the one used in buildings, offices, schools or anywhere a large number of people congregate is a simple way to actually know. The meter I am talking about is the size of a packet of cigarettes and universally available. Take note, in Mexico City and Tokyo with which I am personally acquainted they have numerous Oxygen dispensers throughout the City, in Office Buildings where You put Your Quarter for a couple of sniffs. Jakarta, Indonesia, Several Chinese Cities, Sao Paulo, Brazil, Indian Cities, Buenos Aires and more than a dozen others of which I have personal knowledge have 7% or less. All the Cities mentioned are 20 million are greater population. Several are 22-24 Million range. If You use a depletion chart, oxygen verses time, human mortality along with all air breathing creatures becoming rapidly extinct is quite obvious. Those who have self sustainable energy sources can put oxygen back into limited space and therefore survive perhaps until nature rebuilds the system. Would much appreciate Your Response.
Have a Great Day, Don Smith

Jan 27, 2002 **NOT SO......most oil is formed just like other mineral deposites**
Robin, oxygen studies taken from ice cores in the polar caps which are even more remote, are at odds with the Tasmanian Study. Any mechanism which converts oxygen to a form which is not accessible to air breathers damages their environment. We have here what is called red tide (a micro organism from the sea) which covers enormous areas. When it becomes active all the fish in the infected area die from lack of oxygen. One of the Biological warfare studies uses a form of micro plasma for infecting masses of people who develop hypoxia and die from oxygen deprivation. This was used extensively by the US Government Black Projects people in various parts of the world including Canada. It's contiguous and may be dormant for very long periods in the person, until something else triggers it.

Jan 27, 2002 **NOT SO......**
Robin, will try to respond to Your questions, did not plan on a legal brief. All the documents mentioned are located in any large library. I use the US Library of Congress, the City Library of New York and several dozen other world class University Libraries. This has been going on for more than 60 years. at 83 I still read more than half dozen deeply technical books per week and don't miss a beat with a killing work load of > than 18 hrs., 7 days a week. I intentionally avoid getting caught up in harangues which I violated this once.
Carl Sagen's book mentioned, can sometimes be found at used book stores or in a University Collection. The book was not a best seller, so very copies.
NASA photographs of flumes over and around

Cities have been around for more than 30 years. These are current and and ongoing projects. Suggest You contact NASA.

That Oxygen Depletion Chart You point to would get laughed out of existence. It's obviously from some self appointed authority with credentials, but not much intelligence.

During World War II My military duty was being in charge of a Petrochemical Laboratory where all sorts of problems related to the US Government's purchase, transport and quality assurance of hydrocarbons of all sorts were monitored. My Laboratory was located inside the World's Largest Complex of Petroleum and Petrochemicals here at Houston. This Laboratory performed ASTEM (American Society of Testing Materials) test. Which includes all known materials and is the Bible on the Subject, World Wide. When I tell You something related to Hydrocarbons You had better believe me. Pollution and other related statements are covered by the ASTM test.

I did carefully read the documents which You referred to and will reserve my comments since I have no time for Turkey Fights. Some of the Information is correct, not all. I am well acquainted with oxygen meters for almost 70 years now. I also carry one with me when I travel, which includes into and out of all those large cities mentioned. Bruce Perrault's Web Site has a brief Bibliographic Skit on me which will give information, which You are obviously not aware. I do not normally quote authorities, since they only point to others and so and so. The authority game and publishing worthless papers is for University Professors who have plenty of time to waste.

Have a Great Day, :-) Don Smith

Jan 28, 2002 **NOT SO......**

In turkey fights things get disjointed sometimes. Reputable is something that in today's world does not have the meaning which the dictionary attributes. It depends on who is doing the defining, which very often depends upon some special interest in the background. A good example is the blaming of Ozone, when it's actually a pollution index, not the culprit. Hugh headlines, horrible Ozone is back. It's natures house keeper and becomes obvious only when the air has been poisoned by other chemicals. The perpetrators of this hoax have considerable influence over the national lab which is a casualty of special interest. It happens here and all over the world. The chart mentioned is one such happening, and there are many. The information sent this time is great. In Chemical Laboratories here, there are all kinds of charts related to poisons and tolerances Putting in focus what I have been saying, in large population areas, from high above an anomalous change in the atmosphere is clearly seen. This flume when chemically analyzed has components which are clearly not desirable. Special film sensitive to the differing components tells the % 's of this and that. Now the percentages mentioned which appear to upset You are from this source. This information is publicly available from NASA. I do not keep sets of this information, but periodically take a peak, next time will send You a copy. When they sweep the bodies from the street in Calcutta each morning, one wonders how many are dead from unclean air.

Feb 3, 2002 **NOT SO......most oil is formed just like other mineral deposites**

Michel, the L-2 coil was ruined when the guy scratched the enamel off the copper wire. You can get the same measurement with A Gauss meter (reading milligauss scale with out coming in direct contact with the coil.

:-)

Feb 3, 2002 **NOT SO......most oil is formed just like other mineral deposites**

Michel, if You smell ozone it's sparking, You just

don't see it. Normally quenching refers to a spark gap that has several stages as with very high voltage. A spark plug is only one stage and I don't think quenching properly applies.
:-P

Feb 5, 2002 **precison**
High frequency changed to lower frequency requires going to DC first then to 60 CPS for instance. DC can be changed directly to any frequency. With AC the options are multiples or division of starting frequency. All components have both inductance and capacitance. This often screw's up the desired effect. Sometimes have to use trial and error to approximate the desired results. Both tend to increase as the length of wire and number of turns increase. Everything is resonate at some frequency.
:-)

Feb 5, 2002 **precison**
Michel, will try to understand Your question. Are You asking why I call the diode bridge a frequency adjuster. If so, then HV and Radio Frequency both at the same time, for ease of changing to a much lower frequency (60 CPS at 120 Volts out) is simplified by first changing to pulsating DC by the diode bridge. Numbers 11, 12 and 13 are voltage limiters for protecting the # 15 circuit which yields 60 CPS into the isolation transformer.
:-P :-) :-\

Feb 5, 2002
Last Post to The Radioactive Rock List (or not)
Dave, in the case of the flat coil, one L-1 can cause multiple stacked L-2's to produce from the same input without diminishing the input. The inside circumference length divided into the outside circumference length gives the ratio of gain greater than input for the L-2. The magnetic waves produced from the L-1 replicates the output in as many duplicate set's of coils as You may desire. See the Device # 2 on My web site, three receiving coils with one Tesla Coil power source, the spiral configuration works even better. There is no draw down on the L-1 when using extra coils as reactors.
:-)

Feb 6, 2002 **precison**
Michel, the center tap You refer to is not fully understood. It is probably the broken dipole referred to by Bearden, which allows the background energy to enter the system. The center tap and Bloch wall are closely related in function. Coils, antennas, capacitors and such are energy pumps reacting to external forces, therefore nonlinear and have the potential to collect energy from the ambient background, when activated by external action. An earth and or air grounding is reguired to complete the energy system. These are essential to magnetic resonance energy sustems.
:-)

Feb 6, 2002 **precison**
Michel, the metal structure, negative pole of battery and such become the ground in airplanes, boats and such. There is no positive if negative is not present. It's the equal and opposite concept such that every thing must equal zero, therefore everything equals nothing concept built into all mathematical and physics concepts. That's why I switched to disturbing the ambient as the source of all useful energy. The disturbed leg in the process of returning to ambient becomes useful energy. Rain on the mountain, returning to sea level turns the water wheel and such.
:-)

Feb 8, 2002 **precison**
Michel, the wire is DC high amperage AWG # 8 single wire, consisting of multistrands. It works

beautifully with high frequency, each of the many strands radiates and multiplies the energy present by both inductance and capacitance.
:-)

Feb 8, 2002 **precison**
Michel, only the wire is DC type, however it works better than the wire commonly used for AC especially at high frequencies. Input into the L-1 is HV and HF not DC. High Frequencies tend to move on the surface of wires. Many small wires in a single bundle has greater circular mils than a single strand of the same size. The magnetic field is more intense this way and the over unity can then take place.
:-)

Feb 15, 2002 **Tate power ambient circuit update**
Geoff, the reason the power is so low is the capacitors being used. If larger capacitors such as used for power audio systems or the newer super capacitors such as the 50-500 Farad units You will begin to see real power. The schematics shows a great deal of naivety on the part of the Inventor and the issuing Patent Office. The basic idea is 1000% It relates to My Earth Electrical System II from 15 years ago.
Don Smith :-)

Feb 19, 2002 **Dr Smith equation/fogal Chip**
Michel, on page three of the "The Truth Behind Charged Barrier Technology" Paper there is an equivalent circuit to the Fogel Transistor. The Charged Barrier shows in the circuit as a timing circuit, which normally consist of any two of 1. Capacitor, 2. Resistor 3. Inductor. The Capacitor is in Coulombs, Resistor is Ohms, Inductor is in Farads. All these units equal one volt at one amp for one second. The physical size of the Fogel device is much less than one fourth of an inch in diameter, so the surface area is very limited. The measurement You asked about is to the right of the decimal point by more than 12 places. The Fogel Transistor is about the size of a grain of rice and sits on a small circle of metal. The equivalent circuit can be built to any size for more power.
Have a Nice :-), Don Smith

March 5, 2002 **Gauss's Law**
The total of the electric flux out of a closed surface is equal to the charge enclosed divided by the permittivity.
The electric flux through an area is defined as the electric field multiplied by the area of the surface projected in a plane perpendicular to the field. Gauss's Law is a general law applying to any closed surface. It is an important tool since it permits the assessment of the amount of enclosed charge by mapping the field on a surface outside the charge distribution. For geometries of sufficient symmetry, it simplifies the calculation of the electric field.

Mar 7, 2002 **CD I mailed you**
Warren, Your new CD is on the way by First Class Mail, could get in a week or so.
I like to work with battery powered 12 Volt, neon tube transformers. Their operating frequency is 20-30,000 cps. This happens to be in the frequency range which Hulga Clark uses for much of her work. It's in a range which is mostly harmless to humans and deadly to disease mechanisms including most viral which includes many so called incurables. My new fluorescent light bulb with the negative ion device built-in uses this same system and the Clark Frequencies (not mentioned in the packaging) as a bonus. In about one to two months these bulbs will be available for about $49, each at Home Depot, Radio Shack and many other places all at once. The bulbs give 100 watts worth of light for 20 watts of electricity and last 100,000 hours. My Tokyo Office is selling one million bulbs per month in Japan for the equivalent of $60. US. The factory which we own is stepping

production to 5 million bulbs per month so that other markets can be serviced. Probably will also have a factory in Mexico. The Mexican Patent is pending. Japanese, US, WIPO and EPO Patents are Pending. The Bulb is called-marked "E-On light". The factory is in Korea. The Japanese and Korean counterparts came to Houston for the month of February for strategy planning for World Wide Distribution and My Input.

Mar 7, 2002 **Star Trek warp drive 'impossible' says scientist - Your News from Ananova**

This is the opposite of what is common knowledge. The negative charge on the front or foreword moving end of an object such as ballistics or a super sonic aircraft allows it to travel faster and farther with no increase of energy required. This is several times the distance at a higher speed. Have a nice day, Don Smith

March 7, 2002 **Re: Preliminary Radio Intermolecular Generator (corrected info)**

If You pay attention to the source, My Devices obtain their energy from the ambient background which is the source of the present conventional Devices. They are simply a more intelligent way of obtaining the same energy at much higher levels.. They are much higher effency and from copies of the source through magnetic induction allow feed back and self sustainable Devices. This magnetic resonate energy is normally radiated back into the ambient as wasted energy. Would like to know what iswrong with this approach. It works for me and is fully described on My Web Site.

Have a Nice Day, Don Smith

March 11, 2002 **Re: Resonance Energy**

Colin, thank You for the nice note. In order to put this in focus and see where I am coming from, You must obtain a couple of simple test instruments. The ones I recommend are the none contact Voltage detector (used for locating and tracing wires from their magnetic component.) One manufactured by GREENLEE, Catalog No. 1110 for voltage range 50-600 AC, Cost here at Hardware Stores is about $12.00. and a simple Gauss Meter called Electrosensor (electromagnetic field sensor) ($35.) calibrated in milligauss. Both are on the internet. The object being that when an electric field is present there is always an equal and opposite magnetic field present. Therefore as a none contact and non obstruction way of measuring, one knowing the other is present tells Us immediately that 50% of the energy is radiated as wasted from conventional Devices. Due to resistivity causing heat loss and limiting the electrical component results in only a small part of the electrical system being productive, less than an estimated 2 or 3 %. This is known by those who plan the conventional system, so that waste and inefficiency will be the captor. The sensors listed above are an el cheapo way to see this in action. When any electrical component or Device is operational it is radiating as wasted the magnetic component and especially the MilliGauss Meter can tell You the amount. One Milligauss is the magnetic flux equivalent of approximately 100 Volts of active energy. From a 120 Volt Device in action, You will see readings like between say 2 and 30 Milligauss. At this point if a method of capture is used, the wasted energy is many times greater than that being used. The result is a Device which has put back into the System Energy many times greater than what is thought to be all there is. This energy is derived from disturbing the Earth's Ambient electromagnetic background which is also where conventional electrical devices get their energy. Generators do not make electricity, they only harvest that which is disturbed by the rotating magnet and coils arrangement. My devices simply utilize in a more correct way so more of the energy is available, which is many times greater than that available through the Stupid

Devices Pandered by the conventional systems. Hope this might clear some of the fog intentionally put there by the establishment.
Have a Nice Day, Dr. Smith

Mar 11, 2002 **Fwd: Re: Resonance Energy**
Don't mix bananas, oranges and cactus. The statement was the energy present, not the energy being used. Obviously 2 gauss is the magnetic flux equivalent of 200 plus watts. (volt amperes). Think of the milligauss as being Volt Amperes with VAR being an artificial number related to all the components of the system in action and their composite contribution. It's obvious then that the energy present and being actively used are two separate items.
Take Care, Don Smith

Mar 11, 2002 **Response**
It's best to think in terms of One Volt Ampere, One Joule, One Watt, One Coulomb, One Farad. All at one second are the same unit of Power. Therefore if You want (10 Amps at 100 Volts) You need to pump the farad a certain number of times per second to maintain the energy level You require. Since most of the capacitors You will encounter are milli, micro, pico, nano and billy goat. You first must place the decimal in relation to the storage value of the capacitor. For example You mentioned 5 mfd, that's three zeros worth a 5 inserted or.005 of a farad. being one volt over.005. Therefore.005 devided into whatever Volt Amperes will tell you the required CPS or pump strokes per second to obtain the energy level requested. the little m is three zero places. Obviously the operating range listed on the capacitor should be several times the planed usage rate. The Capacitor is a DC Storage Device. Therefore You need to put a diode before it for two reasons, one a diode yields pulsating DC and two as a kickback one way protector. Now if You wan't to do AC, the capacitor has to be an electrolytic one which has a plus and a negative side. Put the kickback diode on the plus input side of the High Voltage module. That.005 capacitor requires 5000 cycles per second for one volt ampere. Therefore if You wan't a continous 120 watts multiple that times 5000 for the required frequency of operation to produce the energy You desire. See there is nothing to It when You know where You are going. To answer Your other Question, it can be any wattage range in the book, it's users choice. You are the Boss.
Take Care, Don Smith

Mar 15, 2002 **interesting presentation**
See Answers inserted below - - - - - -
"M:
http://www.geocities.com/agravity/ANTIGRAVITY.htm
"B wrote: "Michel,
"This is the type of coil that T.H. Moray was using in his RE device. This allows the device to absorb energy like a sponge."
"Michel wrote:
"If we look at this Tesla discussion, it seem to be that current is appropriate and EM field is lost energy."

ANS: Electromagnetic Field, Lost Energy!!. Electro and Magnetic are the equal and opposite of Lenz's Law. Without one the other does not exist. The System is a Dipole and when something interferes with the Symmetry, it is referred to as broken. At that point useful energy can be extracted to the extent that the dipole is interfered with. The magnetic side which passes through almost every thing without significant loss will replicate as many times as You wish, which in turn is easily changed-flipped into useful electrical energy. The electrons present in the flipping area provide energy not previously present. This is seen as free energy. It's the same energy provided by the normal conventional devices. The difference is a more intelligent way of getting it.

"I just came across this tonight. Below is an

excerpt from NIKOLA TESLA ON HIS WORK WITH ALTERNATING CURRENTS. This is very relevant to what I'm trying to prove in the way of longitudinal waves. Here Tesla clearly speaks of the difference between EM and current,"

ANS: Current is movement. With out the off and on switching seen in resonance, spark gaps and various timing combinations, normally two of the three, capacitance, inductance and resistance, no current-movement of energy can occur. There is an overlap between current and inductance. The unit of measurement for movement is amps.

Mar 15, 2002 **interesting presentation**
Michel, spark gaps and lightening arrestors, spark plugs included when I use them are not used as a or frequency or device timing. I use them for spike, overload and lightning protection. I use high quality capacitors, resistors and inductors.

Have a Nice Day, Don Smith
Mar 15, 2002 **interesting presentation**
Michel, try this - - - - - The best is to plan the construction, so no tuning is required, if so, very little, then the pica and nano farad capacitors can be used. I think where the confusion is surfacing is that capacitors are also used as temporary storage prior to continuing on into the load. This is the tricky part because it must supply ample coulombs (volt amperes) to the requirements. Best bet is to build the separate basic ciruits used seperate boards so they can be switched in and out until You settle on the particular Device that matches Your changing requirements. There are several circuits involved which with time and knowledge may change. keep the high voltage module separate, keep the temporary storage and inverter modules separate. as well as the final usage module. Set them up such that a particular board can be used in many ways for latter experiments. Saves buying a lot of possibly expensive components many times. The high voltage modules are already made in many and varied outputs. Do not go for the highest or biggest or fancyist. One to two thousand volts works just fine and preferably with a low amperage, 12 volt DC is just fine. The Idea being with a known input and latter a known output makes it easy to assess the results. I would recommend the Device photo # 2 on My Website with the small Tesla Coil output with the 3 separate Coils which duplicate the energy field from the center output. The ready made Coils shown B and W Coils (They are on the Internet) Each of the separate receiving Coils have a tunable mica capacitor and a tunable resistor in the circuit Neon 2 lights with the resistor on one leg are from Radio Shack and are placed with the resistor leg attached to the circuit of each of the receiving coils. This simplifies seeing the energy present when the coils are tuned to the sending Coil in the center. Start out with something that's manageable until You get the concept and then get complicated if You wish. Otherwise frustration will win.
Take Care, Don Smith

[Question. Don: > Michel, You must be doing something right, 500 ma X 12 volts is 6 watts and 3 amps X 4000 volts is 12,000. watts out.
[snip]
"Even this need not imply OU, if the input is DC and the output is sharply spiked."

Mar 21, 2002 **Dr smith**
Answer to Robins Question - - - - - In all cases, output comes from a capacitor bank which has arrestors attached such that spikes if present are earth grounded and never become a part of measured energy. This has never been a problem. The capacitors I use are 50 Farad or greater.

March 21, 2002 **Don Smith**
Use a diode bridge to change to pulsating DC so that the capacitor can handle it. AC requires electrolytic capacitors. Capacitors are constructed for all voltage ranges. The dielectric must allow for high voltage. Any voltage You can imagine can be used if dielectric is made of proper material.

Mar 21, 2002 **Don Smith**
ANS: Below
M wrote:
"When you talk about the 50 farad capacitor, do you mean on the L1 or the L2 at the exit?"
ANS: Remember I use separate modules for differing functions. Exiting from the L-2 the energy goes into a module which changes the High Frequency and High Voltage to pulsating DC which goes to the next module. This module takes the High Voltage DC and temporarily stores it in a super capacitor and special battery combination, which yields instantaneous power foreword to a heavy duty converter module which shapes and prepares the energy for end use requirements. Depending on the switching devices used this energy can be very substantial. In addition I add in other modules which use the energy radiated and normally wasted, such that feedback from this wasted energy results in self sustainable energy devices which require no external power once running.
"You say one farad one volt, then if we want to reach 120 volts, so what do we do with the amperage?"
ANS: Equal and opposite of electrical systems which the establishment insist upon tells You that magnetic and electrical are equal. Since magnetic and amperage are such that any current movement creates a magnetic field. Therefore from a practical stand point Voltage and Amperage (movement) are required for useful energy. Now when High Voltage exist it does not exist alone. Therefore in transformers when changing from high voltage to lower voltage, the necessary lowering of the voltage is compensated by an increase in amperage. In short in lowering from High Voltage to 120 Volts, a great deal of current movement which can temporarily be stored (dammed up for use) Capacitors store energy as Coulombs which is Joules per second or Watts which is Volts X Amperes. That's how You get the required amperage to go with the 120 volts.

Mar 22, 2002 **Dr smith**
Check Your history on the reason for and origin of the terms in question.
This is the kind of confusion that science is famous for. If I remember correctly it was an English Brewmaster who had a problem relating the various types of energy, who finally put things into a useful condition. In general, as in the field of science, nothing is defined without passing the buck to some other term which intern repeats and so and so - - - - - - - - - Don't let the terminology game get in the way, just do it and see for Yourself.
Have a Nice Day, Don Smith.

Mar 26, 2002 **Negantropy from Common Dipole**
I have constructed and tested several Devices which are in the Patent process using this System. It's the basic approach that I have used from day one. Bearden was nice enough to organize and put it in more acceptable form. I physically have all the rare books which Bearden quotes. Some I bought, some came to me through interlibrary (Library of Congress) loan which I Xeroxed prior to returning. I also inherited several large personal libraries from friends that died. These friends were students of Albert Einstein during the 1920's through the 1940's. They were PhD's in Engineering and Physics. Princeton, MIT and about a dozen other major Universities are represented in the Electrical Engineering and Early Physics. During

their Lifetime My sideways approach to electricity was a source of amusement which obviously ended up in My receiving several hundred very rare books on the subject. Just maybe they did not think I was chasing the wrong rabbit. The fact that Bearden surfaced in the Rabbit hunt I consider a very High Complement.
bearden1.pdf

Mar 28, 2002 **Hubbard Matrix a possible answer to his free energy invention design**
Using the magnetic flux field to make duplicate copies of the energy present without depleting the source, which My Web Site discusses seems to have been around at least since Hubbard. I know Tesla used it with his coil systems. The magnetic flux passes through the collectors without degrading. Electrical flux degrades rapidly and follows ohms law just as conventional thinking likes. Most of the energy present radiates away from the energy source and is not used. Simply put collectors where the flux passes through, preferably at right angles to the direction of magnetic flux movement. Magnetic and electrical flux are at 90 degrees of each other.
Have a Nice Day, Don Smith

Mar 28, 2002 **Beardens Patent**
A preliminary look at the Patent and with comments:
1. There are hard magnetics and soft magnetics. It's magnetic permeability which is important here. Using hard magnetics as they do is 100 % wrong and those connected with the project all are aware of this. To me it's highly obvious that considerable fornicating with the facts has occurred. Bearden's health places him at Death's Door and in My opinion a Patent Fabricated by the Patent Office with the Help of Special Interest (US Government Goons) has been switched in place of the Patent as submitted. The intent was to send a clear signal to others who have similar ideas as to their future. It would take me several days to list all the tampering with that is present in this Patent.
Have a Nice Day, Don Smith

[Editor: The following is only the response of a heated exchange on Keelynet. Only Don's words are quoted below. The full exchange can be viewed online.]
April 11, 2002
KeelyNet: Re: missing two from the pile
A RESPONSE TO THE ATTACHED ITEM:
Out of Context information, being incomplete should always be stated, in My Opinion, not as God's word. Casting stones should never become a full time occupation, since nature allows two way trips. This is something that the Middle East and Others have not yet become aware of. Each side assumes they are the chosen and therefore the stones are allowed only one way trips. Should the stone be returned they become highly agitated.
The single item that defines a criminal mind is their choice between constructive and destructive behavior.
My advise is, unless one knows correctly, what he is talking about, he should always say in My Opinion at let the World judge the appropriateness.
In regards to Jerry Decker's Opinions of Individuals, they frequently are intentional fornication's of a sick mind. This tells Us that Decker has a very low self esteem which requires putting down others as a means of self gratification.
It is appropriate to lift some one who has stumbled and sit them on track, with a helpful, not malicious intent. Lifting people who do not stumble is folly and can result in a black eye.
"tapped into Van Allen Belt radiation" This is My Beam Energy System which works exactly like the down beaming being done by NASA and Japan. The difference being I use an ionized upward beam which much provides any desired amount of useful electrical energy. It uses the

energy gradient going upward towards the Radiation belts.
"Sells them for about $2000" The answer here is an absolute no! The devices are licensed to large international companies in exchange for between 25 and 50 % of the net income derived from the contract. In order to maintain the license certain requirements, in part being commercial viability must be maintained with minimum periodic income pre stated. Also 50 % of the controlling shares goes into My International Trust.
The Devices (many) are presently hidden inside other Devices such refrigerators and other appliances, other power Systems which are considered as commonplace. Automobiles are in progress as well as many other items.
At present time factories which I control are located in Japan, Korea and several large Cities in Mexico. The object is to make My Devices commonplace before most of the World finds that It has been using them all along.
For Your information I have a special light bulb which produces negative ions and uses the Hulda Clark Frequencies. There are many large user's. The largest being Japan. About one million bulbs per month are sold there at the equivalent of $60. US each. My factory in Korea will top out at about 5 million bulbs per month.
Am presently d dealing with the Mexican Government for several factories to be located in the nicer parts of that Country. These bulbs are constructed to last 100,000 hrs each and use 80% less energy.
Stay out of things of which You have no comprehension and continue that which You do best and the World will be a better place.
Have a Nice Day, Dr. Smith

Apr 26, 2002 **HUBBARD back to life??**
The Device on My Web Site that is close to Hubbard Device is the one in the 1600's French Hall Chair. It was originally built based on the Tesla Photograph showing one large coil with multiple smaller coils empowered from the flux field of the central coil. As it turns out that is almost exactly what Hubbard's Device does. The theory about multiple copies from the same flux source without any draw down on the original source allows self sustainable Devices is correct. The magic is in the magnetic waves as a catalyst, result in duplicate copies of energy present. Since electrons present at point A (source) do not travel to point B (duplicate output) additional energy from the ambient background is accessed. This is also the case with conventional magnets and coil systems. All access the Ambient background, which is the source of electrical energy. The Devices do not make the energy, they collect it in all cases. That Milk originates from a bottle mentality of the Physics Community demonstrates their intentional ignorance when understanding energy sources.
Have a Nice Day, Don Smith

April 28, 2002 **Answers to Your Questions?**
Please Post and Repost to all Discussion Groups!

1. I happen to know personally the person called Wizard, not the impostor.
2. There are large numbers of people worldwide who's interest is not offended by information that affects all.
3. Publishing by Me is for the purpose of stimulating interest in related matters. My theory in case You have not noticed is very different than others which are influenced by coat tail hanger one's and name droppers. What is presented as theory moved from that stage more than 20 years ago. To people who fumble the ball it's still theory. One does not quote oneself and academia has yet to produce anyone worth quoting in this subject matter. Chemistry Text's define Free Energy as that Energy available for use after a reaction. Why not then for Physics, have Devices which disturb the ambient

background, collecting available energy for use after the reaction (also Free Energy). Has anyone noticed that this is exactly what happens with conventional coils and magnet Devices. The Device does not create energy, it only collects it after disturbing the Ambient. (End Lesson 1) Second step as to origin of this background energy, please see altenergy-pro.com. All the Devices (more than 200) built by me, simply disturb the ambient (background) and collect from the unbalance ambient the longer leg of either electro or magnetic. Once the field (or waves) are established then an appropriate receptor collects any desired amount of energy. There are several helpful doors which allow access to ambient energy. Coils and magnet Systems obey ohms law. The magnetic 1/2 which is ignored does not bother with ohm's Law and when resonate (and regenerative) builds (amplifies on it's own). Since as any physics book will tell you, magnetic and electrical radiation (waves) are at right angles to each other then the collector of the desired (magnetic) should be placed as such. Magnetic is then the catalyst which accesses the ambient background and if properly managed is an open door to the energy of the universe. To see this in action one needs a few cheep insturments (such as a Gauss Meter, ElectroSensor from the web is one such). It measures from 1.5 to 30 mG. One Gauss from original definition is the magnetic flux equivalent of 100,000. Volts-Watts. You will immediately see that all Devices such as motors, fluorescent Lights and any other such are radiating enormous amounts of energy back into the ambient. This is wasted free energy. Now who possibly could be the winner in such a JWD scheme.

4. Lesson # 2. Other doors include any gaseous tube, spark gaps and such. Gaseous tubes, depending on gas used can yield magnetic waves thousands of times greater than the input. Lord Raleigh's experiment with Nitrogen in a glass tube with spark gap is an example.
http://www.rexresearch.com/articles/activen.htm
Plasma Globes, Plasma Tubes, Mercury Vapor and others are common examples. These are examples of magnetic dipoles as in Tom Bearden's (Giant Negathropy). Collectors placed at right angles to the Dipole have access to the ambient background. See Bearden's statement on Heaviside Current as relates to Dipole. Locate the magnetic radiation from the magnetic Dipole and place the collector at the proper angle and the rest is history.

5. Just because You are ignorant does not mean that everyone else is more so.

Have A Nice Day, Dr. Smith

Apr 29, 2002

Radio Frequency Triggered Nuclear Energy (was Cosmic Ray News) - patents.htm

Here is where a good working knowledge of Chemistry will save Your Life. Two of the elements listed when used together constitute an extremely dangerous explosive device. It would evaporate several square miles with ease. Phosphorus will self ignite in the presence of any oxygen. Cobalt is extremely unstable and requires special handling. It's a practical substitute for uranium in high powered reactions. It's My opinion that the Patent Examiner had no working knowledge of Chemistry. It is also My opinion that when they actually tried to build the device that it exploded and did much damage. If One must build it substitute other elements such as thorium for cobalt and yet to be determined for the phosphorus. I have a better idea, use a glass tube with gaseous mixture xenon, radon, krypton, users choice. The basic device without the dangerous material mentioned. One of the Devices which I built is a 14" florescent light bulb with a 2" Diameter by 10" Length B and W coil, 10 TPI, placed over the light bulb and L-1 of 5 turns centered with a 9000 HV module input to L-1 all ran by a 12 V at

7 Ah Gel Cell Battery. This is a convenient and easy way to get into the gaseous tube electrical generators. A very strong Electrical Dipole Develops which substitutes for a very large battery. The setup requires storage such as capacitors and an inverter to be useful.
Have a Nice Day, Don Smith

May 3, 2002 **chart**
Please Post and Repost to All interested Groups
Several People in Eastern Europe (who I told about this) got better than 1000 to one greater energy output (Watts) than input by wrapping coils on the stainless steel tubes (thin wall). If You raise the frequency You can avoid the awful sound. Sparking like a large welding torch is to be expected with huge amounts of energy output. The combination is deadly so be very careful and keep Your distance. That's the main reason for measuring the magnetic flux as an indication of the energy out put present which is enormous. My invention using the Terfenol-D Rod System is what started this. The Electrical Engineers and University Professors in Eastern Europe ask Me if they could substitute thin walled stainless steel tubing and I told them Yes, not as good but Yes. Instead of > 1000 it goes beyond 10,000 times when input with Terfenol-D. This is My Piezoelectric Generator System. About 8 years ago Norm Wootan at a Tesla Conference has some washer shaped Piezio units which He was selling and I bought some not having the slightest idea as what to do with them. Since they have a capacitive reaction and amplify magnetic energy applied I ended up making a small Cockcroft-Walten System which equaled a huge one. That led to other experiments and the System above described.
This is one of several open doors to the Earth's Ambient Energy as described in My web site altenergy-pro.com In Chemistry energy left over from a reaction is called Free Energy. So if You insist in Physics, Energy resulting from a Reaction should properly be called Free Energy. After all Coils and Magnets when reacting with each other produce energy from the Earth's Ambient Background. Problem here is that only the Electrical one half of electromagnetic is acknowledged and this obeys Ohm's Law.
Using the magnetic side You can thumb Your nose at Ohm's Law. Magnetic Permeability provides for magnetics the same benefit as negative resistance. Why so much intentional stupidity exist in the Physics and Science Departments of the Universities with regards to this is very disturbing.
Have a Nice Day, Dr. Smith

[Question: Don, I have finally assembled most of what I need to start building. I have a question about the frequency and wavelength. When you use a neon light transformer that operates at 30khz that coresponds to a wavelength of 9,993 meters. What do you typically divide that by to get a resonable length to make a coil?
Thanks, Dave]
May 3, 2002 **frequency question for Don Smith**
Use an even division or multiple of and allow an extra inch or so. The length includes the rat tails used for connecting. 30,000 divided into the speed of light will give the wave length. Two things, there is a considerable argument as to the speed of light and a special conference was held and they simply voted what they thought might be the speed of light. There are great variations so it's a shoot in the dark. Once built You can locate the nodes where resonate points occur and then trim the coil ends to match.
Physics Books show an experiment using a single wire of about one meter in length with a small LED which is moved along which lights at the resonate nodes. The Neon-2 bean sized diodes with the resistor on one leg that Radio Shack sells work just fine. You will be working harmonics of the frequency since a wire length mentioned is impractical. Radio Type Devices do

this all the time. Good Luck.
Have a Nice Day, Don Smith

May 3, 2002 **frequency question for Don Smith**
From the magnetic field reading You can get an approximation of the energy level present. At that point select the appropriate diodes and switches. If possible allow about twice the required amount for the present power. This should give longevity to your Device. The Bridge You build is to reduce AC to pulsating DC such that temporary energy storage can be had in a capacitor bank and or batteries. Then You need an inverter circuit to regulate the output, such that the load can properly operate.
Don Smith

[Question: Michel Bisson wrote:
Do you think Ring shape rod of Terfeno-D can make the job or we need a full rod?]
May 3, 2002 **frequency question for Don Smith**
Tubular shaped would be Ok. To see the effect try stainless steel. Terfenol-D is a giant overkill. Besides that it's very expensive unless You luck out. Try 1 inch or larger diameter and about 14 inches long. I have several and they were $5500. each.
Don Smith

May 22, 2002 **Free Energy Device**
http://www.cheniere.org/misc/static%20poynting %20gen.htm

May 29, 2002
A question to dr. Smith about resonance
ANSWERS BELOW AFTER THE QUESTION:
On Wed, 29 May 2002 wrote:
"Z,
"Yes, resonance is very important if you are going to obtain energy from an external source. However, when we are trying to get energy from the cosmos the wavelengths are much shorter than what the average experimenter is used to seeing in their experiments."
ANSWER: Very true, very true, however this energy is washed into the Earth's Ambient Background as a part of the whole. Going for it directly requires specilized knowledge. In part this energy is the energy seen in magnets or temporary magnet systems. Their degree of permeability acts as an antenna type device. A constant level of energy is cycling through this material such that this continues possibly forever. It's radiation is detected when wires or coils of wire conductors pass through it's area of influence. Simple way is to indirectly intercept and capture as useful energy. Useful energy is that which is temporarly disturbed and is on the return to ambient. It can be greater than Ambient through resonnance or less than ambient, the difference being useful.
"I have a question for you. Are you getting better yields now with your radioactive dual detector?
-B"
"bihari26 wrote:
"Dr. Smith, I have read an article about a russian free energy researcher named Melnitchenko or Melnitchenko, who is working with resonance based power multiplying free energy devices."
ANSWER: Thanks for the information. The problem and the solution is the same where ever encountered. When You know the details they will be very simular to the things I do.
"(http://www.keelynet.com/energy/rpe.htm)
"Your devices, as I can see mostly resonant circuits, too. I see if"
ANSWER: Correct, there is no other way unless other types of radiating energy is used. In either case the magnetic and electrical components are there, when captured become useful energy.
"I drive a parallel resonant tank circuit with a small power exiting source oscillator, than in the tank circuit MUCH larger currents building up. And the source is just a source of voltage, not current."
ANSWER: Potential is one thing, current occures

only when disipating into an earth or air grounding.

"The current is minimal, only needed to maintain the resonance."

ANSWER: It's the magnetic field/waves that You then need to capture. When they pass through a conductor material the electron cloud at it's surface, (electrons flip) changing to the electrical radiation component. In this action which has been going on for more than 4 billion years has not in any way dimenished the electrons presently yielding useful energy. When in the magnetic wave/field stage the conservation of energy laws do not apply. To see this, a radio or TV Station when radiating their magnetic wave siginal has no drawdown, wheather one customer or millions. Magnetic waves freely pass through most conductive materials as undimenished. They induce energy collection and movement, which can be used. Otherwise they deflect, reflect or absorb. When passing through the Earth they absorbe becoming heat. This in part is the infrared heat seen inside Your automobile on a sunny day.

"But how to tap this high power tank circuit? Parallel to the coil and the caps, or put the load into the tank circuit?"

ANSWER: Use an inexpensive Gauss Meter such as "Electrosensor" to determine the flux background which is available (This is available on the internet for about $35. plus or minus). Then a cheep Frequency Meter ($65.) from Radio Shack to determine the frequency. Then build as many seperate coil sets as needed for the output required. They have to be tuned to the frequency of the radiating magnetic waves.

"Please help us. Now I experimenting with low voltage oscillators,"

ANSWER: Low voltage, no problem, it's the frequency which flips the electrons, producing at the same time 50% electrical and 50% magnetic energy which when impactig switches to the opposit form of energy.

"and noticed a real power amplifying effect, but this is not really"

ANSWER: If on the additional coil sets You make the receiving coil the L-1 and the output as the L-2, the potential can become megawatts very quickly. High Voltage is only useful when properly disipated into a proper grounding. When You flip the on switch at Your House You are doing just that on a small scale.

"usable. High voltages in the tank circuit gives more power? Many"

ANSWER: Power is only seen when energy is dissipated. This is a very different animal. Any time You have High Potential, power is the controlled dissipation process making it useful.

Take Care, Dr. Smith

May 31, 2002 **yet another replication**

RESPONSE: This louse picking contest is silly and counter productive. In any endevor, intellectuals improve on what is available without destroying that which is established. Contribute to the effort already extended, don't hang up in the distruct mode.

Here is a New Toy for You! Take the Weight of the Earth which is a known and the rate at which it spins and calculate the required horsepower required. This equals to the Ambient Electrical Energy available for use when plugged into. Next, build the Device which allows one to use this. This has existed since 1820's for those who are in the know. Slow learners still do not have a clue.

Have a Nice Day, Dr. Smith

Jun 4, 2002

A question to dr. Smith about resonance

Thank You for the nice note. The answer is Yes, 100%. The answer is that I am much advanced beyound the person in Russia. I get thousands of times more out than in. Typically one of the Devices which I am presently working on uses radiation from Plasma Tubes and the Heaviside

Current described by Bearden. With small sized energy collectors with 24 watts in, it's 61 KW out. I can do much better with properly configured collectors. The amerage is 134 for the device described. So, when someone tells You it's not possible, just smile. Jerry Decker and associates know this, but insist on being Jerks. Rotten is as Rotten does! The net effect is very damaging to their longevity.
Have a Nice Day, Dr. Smith

June 5, 2002 **Earth's Charge**
Bruce, the general mechanics mentioned in this article are more or less correct. As to lightening one strike is 10 trillion Volts at more than 200,000 amperes and there are more than 20,000 strikes per day world wide. Does that sound like microamps and insignificant amounts of electricity. This electrical process has been going on for more than 4 billion years for the Earth. It's easy to count the strikes since where ever they are they reflect off the ionosfere. The National Weather Center at Denver keeps score on them. The Earth's atmosphere contains enough electrical charge to spin the Earth, exactly as in a large motor. The weight of the Earth is a known, and the rotation rate known. Therefore the energy required for the motor action tells us the active electrical energy present, which is enormous. You can calculate it for Yourself. It's the poney brake approach since You know the force that must be overcome. This is something which I have a very large amount of information and can be rightfully considered authoritative. The people at Lockhed sound like a 6th grade science class, copying from an outdated Encyclopedia Please put this with the file in question and questions are invited! Robin might like a copy of this.
Have A nice Day, Don Smith

Jun 5, 2002 **Ball Lightning**
Personal observations on ball lightning during the 1930's might provide some clues. We lived on a farm in north Texas near the Oklahoma border.
Summer storms especially at night frequently looked like fourth of July in NY and WWI rolled into one. During day time it's not so highly visible. The ball lightening there was frequently automobile size. It bounced and frequently moved in several directions. It was bright like a welding torch, tinted with blue and yellow. Based on information described in the attached article, I would hazard several guesses. One is that since they don't immediatly go only upwards that they are not hot plasma, but negatively charged plasma being cool which being heaver than air keeps them near the surface. The glow cause is simular to that which occures with flourescent tubes, bulbs etc. It's probably atomized phosphorus and a lightning bolt that hatched but fizzled.
The particular area in north Texas which I speak of when a display is in progress, several to more than ten Balls going off at a time. This at different times lasted longer than an hour.
At first it's very scary, but shortly becomes so fascinating that danger gets lost. Have seen them (smaller ones) pass through the house, bounce off the floor and travel along a display shelf passing through a beautiful lead crystal butter dish, not breaking it but a quarter size clean hole upon entry and one on leaving. I never did time any of them, but it seems forever before some of them flashed out. They also traveled normally several hundred feet to maybe 1/4 of a mile. Also much later when I lived on the east coast of central Africa I got a good look at them a second time. We lived on Victoria Island near Lagos, Nigeria. Frequently at night time high wind and blow in, could set You clock by their arrival, mostly 1:30 AM. The electrical wiring in the houses was 240 V and finger sized wires. The storming coming from the ocean was related to Daytime and Nighttime temperature

changes, although only a couple of degrees. Anyhow with storm in progress, all of the plug boxes throughout the house spued ball lightning which was like a pool game gone mad. Did not dare to get out of bed, just watch in horror till it would go away. Needless to say our small dog would have a mental breakdown. They would cross the marble floors some times 3 or more times before going out. Was also an engineer for an Atomic Energy Project, immediately next door to Area 51 during the time that Lazar was doing his thing. During snow storms there was a lot of lightning and ball lightning. In this case light comming through snow filled air changes colors, all colors of the rainbow from one flash, whith many at varying distances going on at the same time. It's a real psycodelic experience. There are a lot of other experiences including ball lightning inside a TWA Jet around the World flight when landing at Heathrow Airport at London. Enough boring news!
Take Care, Dr. Smith

Jun 6, 2002 **Lightning and Energy**
Reminds Me of the hillbilly song looking for love in all the wrong places!
It very obvious that canning lightning bolts is not very productive. Let Us consider other closely related possibilities.
1. Electrical approach is restricted to in the box logic, closed system and underunity.
2. Remember one half the equation is magnetic, which is an open system, not restricted to the normal in the box logic. Proof of this is that magnetic waves travel unimpeded great distances. When they pass through a conductor, flip the local electrons, producing useful electricity. They may pass through many such conductors, repeating the electron flip and continue onward undiminished in any way.
3. The next step is identifying the frequency and occurrence such that one or many resonance units can change the the magnetic waves into useful electricity. Note that the magnetic wave and it's source is not depleted in any way, but produces useful electrical energy. This is Energy, free for the taking, just as wind turns the windmill.
4. In order to see how this all comes to gather we need first to look at what we already know and then tie this into a neat bundle.
Fist from Physics, conventional style, there is a laboratory experiment wherein a meter length of antenna wire is mounted on a board. A radio frequency, times shorter than the one meter is induced. An LED is used holding one end and touching the other and moved along the antenna wire. As each node, intensification area is touched the LED lights. The energy intensification is the resonance of the particular wave.
5. Now that We know about resonance, how does that relate to collecting naturally occurring energy sources?
6. The several ionization layers that the Earth has corresponds to the energy intensification described in # 4 above. These layers are the result of disturbance by shear and slippage as the Earth spins. Their frequency is fingerprinted by the distance they are separated from each other and the distance above the Earth's surface. Their frequency changes with day and night conditions, but is in an envelope which is consistent. The resonate magnetic waves being emitted work just as all magnetic waves and therefore when passing through conductors which respond to these frequencies flip the local electrons, yielding useful electricity.
7. This huge unending source of resonate magnetic energy translates into a simple source of electricity when properly accessed. The information herein is repeated many times in My Web Site and Lectures and Books. The Magnetic Resonance Wave Energy from lightning is collectively warped into the Energy Present in the Ionization Layers and is accessed as stated herein. Have a Nice Day, Dr. Smith

Jul 1, 2002 **Don Smith in Laymans Terms**
Technology of New Energy:
Developments in understanding of Electricity and Materials not previously available allows construction of Devices which collect energy from the Earth's Ambient Electrical Background in large quantities. This Energy is naturally occurring, is environmentally benign and available everywhere. It's available where ever and whenever required. The New Devices use Resonate Magnetic Waves which replicate upon spinning the locally present electrons providing multiple duplicate copies of the Energy Present. Each electron when spun yields both magnetic and electric waves in equal proportion. The electrical component is a closed system limited by Ohms Law. The magnetic component is an open system not limited and replicates multiple copies of the energy present. Special materials of recent development allow the magnetic energy to reproduce through resonance, unlimited duplicate copies acquired from the ambient background. The Device harvests the energy that has and always is present universally. Conventional methods consist of coils and magnets systems. Upon moving past each other, the magnetic flux field disturbs electrons which yield electricity, which is collected by the coils system. This is accomplished electronically with the new technology, with out moving parts with the energy multiplied such that the Device becomes self sustaining once started.
Dr. Smith

Jul 17, 2002
Don Smith's Rotating Seperator Coil // OK
Try several thin layers of magnetostictive material alternating with thin layers of electrically conductive material. For example stainless steel and copper and or aluminum. This is not yet an exact science and requires a skill level. You may get some weird results at first. It requires mixing electrically conductive with none conductive which scatters-disorganizes the magnetic waves.
Dr. Smith

May 27, 2003 **PLASMA ENERGY**
Have been flowing the Babel on Plasma and among the knit pickers not one has mentioned how to access this energy. Obviously someone needs to point the way. The PLASMA forms a very strong - and large MAGNETIC DIPOLE. The right angle component of this DIPOLE converts to useful energy. In conventual's power generation this thrown away. There many thousands of times the energy present by conventual's methods. HAVE A NICE DAY, DON SMITH

Nov 17, 2003 **Re: free energy source?**
Norman, when the forces are balanced as in Henry Ford,s Device it becomes self propelled. A lawn mower will do the same. Dieseling by diesel engines is a mechancle form of this. Bedeni knows this. The moderator has deleted me from the free-energy group.
Regards, Don Smith

Jan 15, 2004 **Atmospheric Electricity**
Time to intrude, this directly relates to My Earth Electrical System, spent several millions of $'s and ten years trying to get DOE and Patent Office to act. The System is Scientifically Sound with out reservation. The Establishment simply ignores that which it can not ridicule or explained away. Regards, Don Smith

Feb 19, 2004 **Radiant Energy—**
Wireless Transformer of High Power Lines?
I did this about 20 years ago minus the power lines. I called it in the patent application "Electromagnetic Radiant Energy Conversion System" It's a winner. All it takes is an open mind. Regards, Don Smith PS it was a part of My resentation to the Tesla Society at Colorado Springs, ask JW McGinnis.
Regards, Don Smith

Emails

Relevant Emails from Free-Energy Yahoo Forum

These emails from Don Smith are from the the Free-Energy email group. Relevant emails from others that Don is responding to are also included within parenthesis (" "). Emails from Don that have no content, or are unrelated to his systems or technology, such as political rants, are not included. No editing of spelling or punctuation have been made.

Oct 6, 2003 **Don Smith Devices**
Your like of SUCCESS tells me that You did not duplicate the Devices.

Oct 7, 2003 **Re: even with amplitude modulation applied device didn't work**
TRY THIS ONE. IT,S A SINGLE UNIT FROM DEVICE # 9. IT,S NOT A FULL DISCLOSURE FOR PATENT IN PROGRESS. REGARDS DON

[Question: "Hi Don,
Nice drawing, but the principle is patented already under the name of Ecklin if i remember correctly? Garry"]
Oct 8, 2003 **Re: even with amplitude modulation applied device didn't work**
Look closely they are not they are not the same Device. REGARDS Don Smith

[Question: "Device #2 at altenergy-pro.com, I tried to duplicate using b&w coils and mica capacitors and variable resistors. Well it turns out that the only mica capacitors i could find quite possably could not take 9kv resulting in arcing between plates. So until i can find a variable capacitor of apropriate sturdiness i can't accuse Don's #2 device of not working (which is what i recently did). My results were a retrieval of 70% of source between all three coils. Its possable that when energy levels reached 2Kv or so arcing occured effectively wiping out OU. I wasn't thinking of this until Mr. Smith suggested using commercial radio products which may be imposable to purchase. Of course your sugges-tions are welcome."]
Oct 9, 2003 **Re: even with amplitude modulation applied device didn't work**
You are closer than you think. If you look real close you will see separate coil sets each have small light emitting diodes attached. They are Radio Shack 2 for dollar. They have a resistor on one leg. When tuned the extra coil sets. light up. However many coil sets at resonance, they each each duplicate the source with out reducing the source. REGARDS Don Smith

Oct 9, 2003 **Re: even with amplitude modulation applied device didn't work**
With respect the Devices have not failed. All the possibilities mentioned was water under the bridge more than 20 years ago. What you see on the Web Site is mostly outdated by about 10 years. The information presented is based on known facts not conjecture. REGARDS Don Smith

Oct 10, 2003 **Re: My Bold Overunity Prediction**
It,s already happened. DEVICES 1 AND 2 ON MY WEBSITE CASES WHERE TESLA DID JUST THAT. Device #1 the Flat Coil.s gain is the ratio of the outer diameter verses the inner diameter. For the flat coil Device # 1 on my website it,s a gain of 5. For Device # 2 the gain is how ever many sets of resonate devices added. I intentionally started my website using these examples. How obvious can one get? REGARDS Don Smith

[Question: "At the 96 International Tesla sym-posium Don says the device with only one b&w coil and red primary coil are resonant at 218 (my guess 218khz) but i can't see coils of that size

and caps that size being more than low mhz."]
Oct 13, 2003 **Re: Don Smith's ion valve resonance at 218mhz???**
218 MHrtz is correct.

[Question: "I don't believe resonance has anything to do with OU or FE, but that's just my opinion."]
Oct 16, 2003 **Re: Don Smith ideas**
When resonate the source is duplicating it,s self without depleting it,s source. The magnets used in a generator is one of many examples. REGARDS Don Smith

Oct 21, 2003 **Re: Don's post**
Notice the small input verses the the larger output.. It ignores Lenz,s Law. REGARDS Don

Oct 21, 2003 **Re: Don´s Anti-Lenz rotor, "Don is Sitting on the Answer"**
When running at full speed it quickly melts the # 10 multistranded DC copper wire. I use infrared heat detectors for that reason. REGARDS Don Smith

Oct 22, 2003 **Re: Device Test**
I spent years trying get past the US Patent Office without success. A crew was sent from Kiev Russia to Houston to test Devise # 10 from my website. The results were excellent and they set up a factory here to build MEGAWATT Units. Once started the Devices are self powered. In case You have not noticed the single unit is one from the set of eight. Megawatt units are about four feet high by two feet wide. REGARDS Don Smith
[Next email:] Sorry about that, I am parallelized and have trouble responding. Try Devices # 9, with <the 8 coil set >. REGARDS Don Smith

Oct 23, 2003 **Re: Farad capacitors**
Something is very wrong with this line of thinking. One Farad equals 1,000. Watts. Regards Don

Nov 6, 2003 **Re: Don´s snake oil site...**
The testing was done about 5 years ago with positive results. See Device 9. REGARDS, Don Smith

Nov 6, 2003 Re: **Don´s snake oil site...**
You are twisting things around out of context. There are eight separate coil sets with eight separate energy systems. The composite output is as stated. If you look carefully you will see. The nice Slide showing a single unit, you recently viewed. REGARDS, Don Smith

Nov 6, 2003 **Re: Essential Parts**
Hi Don, Is it possible that you vary the speed of the motor to find/make the two coils and the two Caps. all in series (series LC circuit) resonant thereby increasing their efficiency/output? It would seem that a full wave bridge diode would be necessary to keep polarity correct into the battery/load if DC wanted. Ben

[Question: "Hi Don, Is it possible that you vary the speed of the motor to find/make the two coils and the two Caps. all in series (series LC circuit) resonant thereby increasing their efficiency/output? It would seem that a full wave bridge diode would be necessary to keep polarity correct into the battery/load if DC wanted. Ben"]
Nov 6, 2003 **Re: Essential Parts**
You are correct. The variable speed switch is from an automobile dimmer. The drawing is the simplest possible and is therefore incomplete. The basic idea is clearly shown. REGARDS, Don Smith
[Question: "You seem to be using the rotating wheel simply as a magnetic field blocker. Correct? What materials are you using as the blocker? Is it possible to use a flux carrier like METGLASS instead of a flux blocker?"]
Nov 7, 2003 **Re: Essential Parts**
My "Rotating Separator Coil" one of a dozen

related projects I did. It's my improvement over the Henry Ford Generator Starter Motor. The Aircraft Industry has starter generators in mass production which produce excess energy and with minor change are self sustainable " 60 Kilowatts and upwards of excess energy". Thanks for the comment. REGARDS, Don Smith

Nov 7, 2003 **Re: Essential Parts**
The small millwatt motor shown powers only the rotating disk. The output is from the magnetic variation resulting from the shielding. Which is quite large. REGARDS, Don Smith

Nov 10, 2003 **Re: Get off the keyboard, out of the books and into the workshop.**
Expert Couch Potatoes beware! Accessing the none Lenzing Option with my Rotating Disc requires use of the Block Neutral Zone for maximum success. REGARDS, Don Smith

Nov 11, 2003 **Re: Get off the keyboard, out of the books and into the workshop.**
It was not an Heart Attack, rather a stroke massive, caused by several blood clots in the right side of the brain. Heart checks out is perfect. I was 100 percent paralyzed for three months. Had to learn to talk and body movements from scratch. One year latter am almost back to normal except for the right hand. The blood clots are still there and could move at any time resulting in a total wipe out. That is whey the Doctor objected to my traveling to Seattle. I spend a lot of time in Physical Therapy sessions. It present have only the left hand operational. As you can see I can partially manage. REGARDS, Don Smith

Nov 11, 2003 **Re: Get off the keyboard, out of the books and into the workshop.**
An enormomous amount of water under the bridge prior to now. Those caught in the middle are confused and I try to help as best I can which is difficult not knowing the exact problem. On the book sales - - - Internationally averages 10 sets per day, Domestic averages 3 sets per day. That,s with no visible effort on my part. Regards, Don Smith

Nov 11, 2003 **Re: Get off the keyboard, out of the books and into the workshop.**
I am financilly independet in the upper 1% of 1% bracket and my needs are well taken care of. Smart Allics and Jerks present have no such option. Big Mouth, Small Brain syndrome I guess. Regards, Don Smith

Nov 11, 2003 **Re: Get off the keyboard, out of the books and into the workshop.**
The term " neutral zone blocker " is a malicious misstatement. Please notice I said BLOCH,which is man,s name. The neutral zone of electron spin was named in his honor. REGARDS, Don Smith

Nov 16, 2003 **Re: EMF, electrical signal, current, electron flow, voltage**
Samantha and All, what is numerical relationship between the unit gauss, oersted and amperes, coulombs. This is necessary using a gauss meter when avoiding physically touching high voltage output surceases. Regards, Don Smith

Nov 21, 2003 Re: **Flux control**
This an improvement closely related to Bearden,s Device. Regards, Don Smith

Nov 22, 2003
Re: Here is the FE, I need help to connect!
Richard, It may be easier than You think. a custom transformer made to my specifications does just that "made in India". One of the pages "Tesla Coil Geometry" from my book was posted to the group which explains how to correct the amperage problem. It was totally unnoticed. Regards, Don Smith

Dec 2, 2003 **Re Tuned-base Armstrong oscillator**
Only that energy present in the closed system in availabe. Magnetic Resonance is open to the Univsrse. It takes slow starters for ever to catch the ball. Regards. Don Smith

Dec 7, 2003
Re: Another hidden dimension in free energy
Gary in a polite society we are sure of our facts before we pass judgment. I happens that you are very wrong, many times over. I try very hard to help people understand. It happens that I am in the business and have obligations such that I can not freely discuss many things. Such as, self propelled electric vehicles. small magnets that work same as a huge battery, simple and easy limitless useful energy from the universe and several hundred more such Devices. Wishing You the Best, Don Smith

Jan 19, 2004 **Re: peer review process — your comments invited**
Time for a reality check! The intent is great, however reality in badly flawed. Physics is a Basket Case. It's a Gift Horse with one leg and a horribly deformed swayed back. The peer review process is fraudulent in concept in that it is representing it's self as a race horse. Even a trolley and a wooden leg can't fix that. An imperferfect System passing judgment equals What ! Regards, Don Smith

Jan 19, 2004 **Re: Ion valve by balash ahmedov**
Jay, I have been using 50 Farad capacitors for more than 4 years. They are about the size of shoe box. One Farad equals approximately 1,000. Watts. They are common in BOOM BOX sound systems. Search the Web. There is also a special battery that comes with the system which works better than any other available. Regards, Don Smith

Jan 20, 2004 **Re: 30.4 j (300uf 450v) cap and how to deliver 100k watts or more**
Garry, glad You asked. If You will give he your private e-mail address. The yahoo system blocks photos is the reason I have-not done it before. Actually you have the photograph already in Bruce's last September program. The original is from 5 years ago. I use several 50 Farad Capacitors. One Farad equals approximately 1,000. Watts. They are shoe box size. You can then figure out how to get around the Yahoo block and send the Group a Copy. Happy to send a replacement! Please excuse the garble, I do not have the use of my hands.

Jan 20, 2004 **Re: capacitors,and there importance to gaining OU**
Nice sense of humor. But wrong, try impedance matching as in antenna and radio type circuits. For example, every thing has resonance. This includes all types of batteries. For the antenna system to be open to the ambient energy of the Universe the impedance must be matched, Once this happens all the energy in the World is Your Oyster. This is what Bruce has been trying to tell You. REGARDS, Don Smith

Jan 21, 2004 **Re: Up the Hill Backwards**
Bill, being a geologist and engineer, it's meteor-ite buried in the area composed of magnetic shielding material. This also explains their random distribution. Have a nice day, Don Smith

[Comment: "The heated air will cause rapid air movement, similar to an airplane wing, creating a low pressure area. Within this area materials will tend to be sucked together. Expanding air is somewhat irrelevant to this phenomena."]
Jan 22, 2004 **Re: re.004BIG STORY:GRAVITY!-2**
Try this with Gravity Shielding equivalents and You have a Flying Saucer. It's how You control it. REGARDS, Don Smith

[Comment: "Hi Art I'm surprised you can say this when we have Don Smith visiting this site from time to time (has seven degrees) who's credential are as impressive as they get tells us free energy is possable via magnetic flux duplication through resonance. But we talk and talk and talk when all we have to do is listen accutely. I'll tell ya that its so easy to accomplish even i did it with a basics electronics book in hand and no degrees and many hours of experimenting with coils and capacitors within a year. Art, if you need some insight into resonance i'll hook you up with a 30$ video info but most people for some reason have a hard time conceiving its use. There's probably a good thousand copies in circulation but only a hand full of people actually benefited."]

Jan 22, 2004 **Re: Let's put it all together**

I would be pleased to put Your mind at ease. If approach in a You a respectfully manor. I know why the group has difficulty an what to do about it. will take some patience and self control on Your part. Is it a deal ! Don Smith

Jan 23, 2004 **Re: healthier perspective**

Reality Check: About 1,000. Copies in the US, balance world wide, largest concentration is Eastern Europe, Includes Croatia [Tesla's home]. Several dozen successful duplications of My Devices have occurred there. About 1,000. copies were gifts to Key People in various Universities and Educational Groups World Wide. Regards, Don Smith

Jan 24, 2004 **Re: Quantum Electro-Gravitics**

Howard and Rayburn, this has already happened, about 20 years ago. I should know, it,s the early stages of a flying saucer. It's much simpler than the approach presented. Extraneous PHD type approach. In realty the system in simple terms the system works very well. The PEACOCK APPROACH IS NOT REQUIRED. This is one of many patents for which I filed. Regards, Don S

[Comment: "miniscule current no loss!Need i say more?probably"]

Jan 24, 2004 **Re: pure potential**

This is the bird on the high voltage line, not much current but lot's of potential. This is the background energy of the universe, With a proper grounding system it's fully aces source of usable energy. This is what I am all about, ambient resonate energy systems. Until resonated and with controlled grounding is required.

Jan 24, 2004 **Re: reality check**

I take this as invitation to reply. The miss under standing in all in the eye of the beholder. A mentally mature person would see it as a part of the learning process, nothing else. Learning is a full time, life time experience for some, for others it's only populated as a desperate measure. The most BANG FOR THE BUCK comes from third world countries, that's why my efforts have been concentrated there. As to $50's being a problem, if someone will volunteer as moderator,I will as thousand of previous cases donate the files. Basket Case third world types can then access them. In stressful situations is best to conserve ones energy, as an octogenarian I find this necessary. My intent is to be helpful by nature is to requirement, any perceived evasion indicates an abusesive person who is up to no good. I do the best possible when permitted. My apologies, Don Smith

Jan 30, 2004 **Re: oil runout article**

True Very true, will totally deplete the OXYGEN IN ABOUT 3 YEARS, then nothing else matters. This has been My motivation for inventing energy devices all along. By natural causes I don'r expect to be around so what next ! REGARDS, Don Smith

[Question: "Don, This is the first I've heard of an impending oxygen shortage - where's all the oxygen disappearing to?"]

Jan 30, 2004 **Re: oil runout article**
The Oxygen situation is easy to see, for example: The percent oxygen present in a bubble of amber deposited during Biblical time tested 40 %, pre internal combustion period tested by US Govt. tested 30 % Oxygen, large population areas to day test about 6 %. Plot the Depletion Curve on this. This is one of many in My book. The Human immune system FREAK'S OUT BELOW 15 %, This explains cancer in the developed countries. It's a metabolism of Oxygen problem, plus the absence of germanium and selenium in the Food Chain. Germanium is not poisonous, selenium should use caution. It is widely used in Japan with a cure rate approaching 100 %. Cancer is a very BIG CASH COW so do not expect the AMA OR YOUR GOVERNMENT to recognize it anytime soon. As far as they are concerned 10 million cancer deaths per year is ok. Bet your boots it's they know all the details. REGARDS, Don Smith

Jan 31, 2004 **Re: oil runout article**
Check it out for Yourself, as I have done. Oxygen meters are not expensive. Do a web search on the subject. In Mexico City [population > 20 millions], Tokyo and dozens large cities of others oxygen meters are placed on the streets for public use. More on this as required. Regards, Don Smith

[Question: "In postulating a model for how Permanent Magnet Motors work, I've wondered if they may be tapping into either some cosmic magnetic force or the earth's magnetic field."]
Feb 16, 2004 **Re: Q. PMM disruption to planetary mag field?**
My response would be that magnets are a magnetic antenna system reflecting the Earth's Geomagnetic Structure. A form of Corerer. Regards, Don Smith

Feb 16, 2004 **Re: Q. PMM disruption to planetary mag field?**
When the temperature reaches a certain level the magnetism ceases, this is about 50 miles down inside the Earth's crust. There is a name for the event which I do not presently remember. Regards, Don Smith

Mar 5, 2004 **Re: Coriolis force**
Keep going, the weight of the Earth and the force in watts to spin the motor. At that point You have the Ambient Background of the Earth's Electrical Background. The "Chemical Rubber Publishing Company" states the weight of the Earth, so answer please, anyone ! Regards, Don Smith

[Comment: "Apparently, the fellow who attacked us was certain we could not have a room temperature superconductor."]
Mar 8, 2004 **Re: Standing by Mark Goldes**
For those with very limited knowledge, the same occurs in magnetic resonance at room temperature all the time. It's the dunder heads stuck in electrical syndrome that cause all the trouble. Regards, Don Smith

Mar 24, 2004 **Re: UTILIZING AMBIENT RADIO FREQUENCY** Norman, the statement about light not getting tired is profound, develop that line of thinking. Physics including Einstein ignore that which might be difficult to explain. Best Regards, Don Smith

Mar 24, 2004
Re: UTILIZING AMBIENT RADIO FREQUENCY
Kateb, as Paul Harvey the news broadcaster used to say, now the rest of the story. Ford and Tesla's self propelled automobile, Tesla later developed his 3 phase electric motor using the same idea. Regards, Don Smith

Mar 24, 2004

Re: UTILIZING AMBIENT RADIO FREQUENCY

Art, You presented me with as opportunity to interfere and make a statement, thank You. The ignition system of the 1920's Maytag washing machine, early tractors and others when balanced out by a fly wheel had a strong tendency to become self propelled. This is where Tesla and Ford's self propelled idea came from. The modern day equivalent is the starter / generator used by the aircraft industry, being mass produced. A simple computer chip change and it's self propelled. My web site at altenergy-pro.com mentions this. Regards, Don Smith

[Question: "Did anyone actually look at Einstein's "original" work?"]

Mar 24, 2004 **Re: re A Santa Cruz wager**

The source of it all was Oliver Hevysides including the E=MSquared. Einstein borrowed it and forgot to tell. Have a Good Day, DS

Mar 24, 2004

Re: Mag driven Generator is it Overunity?

Gentlemen, a nudge here and there. This is something I did about 4 years ago, so I can help. For now go to Goggle Search Engine and see " Pancake Motor Generators " I personally built 3 different motor sets, all preformed exactly as planned. My website on the first page mentions them. They were based on the 3 phase pancake Tesla rendition. Best Regards. DS

Mar 25, 2004

Re: Free energy motor-generators and coils

Sir, It or what ever, or Your information a full size self propelled fully functional airplane exist and I would be happy to supply a photo. Regards, DS

Mar 26, 2004

Re: Free energy motor-generators and coils

Art, We already tried that and it did not work. I will tell You that the Airplane is in commercial production in China. GE has a contract for 24 Aircraft. The original device is in storage at an undisclosed location. The Chinese came here, liked what they saw and ordered some. Disjointed and rotten logic of the mind control system [group] makes precautionary measures a necessity. Heights Regards, DS

Mar 27, 2004 **Re: Homemade solar cells!**

What's missing here is that all the energy incoming from the Cosmos is magnetic resonance as I have been saying for more than 20 years now. This type of energy can not be filtered or stopped, it can only be reflected, deflected or absorbed. Normally it passes through all the electrons present, flipping them to the electrical position where the energy produced dies a heat death. Not to worry about the electrons, they have been doing this since the beginning of time and have not been diminished at all. See My Web Site altenergy-pro.com Regards, DS

Mar 27, 2004 **Re: Self powered airplane & technology exporting**

Please do a Goggle Search for electric airplanes and You will see things which You obviously not familiar. You will also find a complete disclosure. All the people You are threatening my with are a part of it so my advice is dig a hole, climb in and cover up as fast as possible, Regards, DS

Apr 7, 2004 **Re: Ambient power module uses**

Mark, the limiting factor here is the capacitor and it's Coulombs. Use of the new super capacitors multiplies the out put many thousands of times what You are getting. You are in business. Good Luck and Best of Regards, Don Smith

Apr 7, 2004 **Re: Neutrino rods videos posted**

Robin,

I have business interest in Japan. The first ones I physically had contact with were at a Dennis Lee presentation at Austin Texas 3 years ago. The source is from Patent issued and assigned to the US Government. The Rod business is a decoy to hide the real function, which is a replacement for solar cells. The difference being captures the 24 hour a day radiant energy, and each cell puts out 35 watts of useful energy which is several times that of a solar cell. It's a ceramic material and not difficult to make. An American from NASA quit his job, went to the Philippines, got a Japanese partner and took a Patent there. They opened up business there and sold a very large amount of stock. Yes it works as presented except for the NEUTRINO GIMMICK. Let Me point out that no single cell battery or solar cell can equal this. Not only that, they do not deplete. Several of them together and You are in business. Best Regards, Don Smith

[Question: "what brand solar salt are you using? Maybe it has some special isotopes in there that are a bit radioactive and thus enhance your output?"]

Apr 17, 2004

Re: Re: Walt´s alloy batteries, solar salt brand?

How about Iodine potassium.

Apr 17, 2004 **Re: renewable energy push**

It took me years to come to the same conclusion. A lot of time and money wasted based on good faith on the US Govt. Special Interest decides and mind controls every thing including all publications and University Science Programs. Highest Regards, Don Smith

Apr 17, 2004 **Report**

I have 3 hard drives on this computer with about 300 GByts about 90 % full. Their are 4 different Virus catchers, Norton, MacAfee, Penicillin and Spy Catcher. The show 22,120. different attempts at messing up this computer during the last 2 years. It appears someone is really intent on screwing things up. It's enough to give one indigestion. It has a complete history which I will look at very close with some expert help to track down those responsible. I almost forgot but both in and out are checked. Highest Regards, Don Smith

Apr 18, 2004

Who on this site said OU was easy/simple

Dear All, I put a trail balloon out when last I started to explain a fully working and tested Device. I only got 1/3 the way there and got shouted down. This ominous development cause a change in plans. Two major publishers, one in Honkong and one in Germany approached me with offers to do my book. The Books will now cost $250. per copy verses for free due to bad behavior of the group. Regards, DS

Apr 19, 2004 **Dear All**

W ith My Inventions a corporation is formed, I furnish the technology and the investor funds the endeavor. I take shares which only become valuable as the shares become of value. I am open to two to three American Investors. There is a finders fee equal to several years salary available upon successful completion for each new venture. I have done his overseas and presently have more than 500 Million Shares, includes Worlds largest ship builder and Mitsubishi and several others. The prospective investor is required to have large resources and with an aggressive approach. For Your information those shares are presently worth $50. each. Have a nice day, DS

Apr 19, 2004 **Clear the air**

It's generally accepted that matter is forever and that it only changes shape. It's also generally accepted that when matter changes from one form to another that excess energy there from this is named free energy. This is electron spin energy [electro magnetic] [ambient background] and does not degrade in any way the system.

The Establishment uses shifty naming systems to degrade and confuse these issues. Most notable is the misleading term ZERO POINT ENERGY which is intentionally misleading. The publish or perish crowd love since it makes them appear to know more than someone else. Have a nice day, DS

[Question: "I have been to your web site and the volts / Amps claimed are not possible through the big red wire you seem so fond of. Also, your ability to create multiple design prototypes which produce the exact same 8000 volt 50 AMP rating is amazing...Perhaps you have me confused with a moron. I assure you of this my friend, I am not confused and I am not a moron."]

Apr 23, 2004 **something to ponder**

Your understanding of electricity is limited to 60 cycle understanding. High frequencies are a different world. See Tesla's Book "My Inventions, Researches and Writings ". Any and all of My Devices Work exactly as I state. I can not be responsible the shot comings of others. Regards, Don Smith

Apr 23, 2004

Obfuscation and lies by certain members

Such things are impossible only if Your understanding does not allow You to comprehend the facts. All the Devices in question are high frequency until point of use and therefore do not require large wires. Your description is an exercise in Ignorance. Regards, DS

Apr 26, 2004 **radiant_energy_diatribe**

Radiant Energy DiatribeMy two cents! By altering the ground plane You can get several millon times improvment. Govner to be Jimmy Davis, hustory proffsor at LSU in baton Rough, new at politics was up against the Hueghy P. Long maffia out of New Orleans. To make it short all the media including radio and news papers included. One postage stamp sized local transmitter only from a friend allowed Davis to advertise. The 50 wqtts available could only reach the city limits. I advised him to soak the earth grounding with salt water, All points world wide then had greater signal strenth than million watt stations then available. This was how Davis won the election. Something to think about, Don Smith

Apr 26, 2004 **Fw: DON SMITH basic circuit**

Stefan, this is more or less item #'s 9 and 10 on my website. Where it shows a battery, place a high powered diode in and replace the battery with one of Maxwell's new super capacitors. You can also place in series an out put transformer for experimental load. Have a nice day, DS

Apr 26, 2004 **Fw: DON SMITH basic circuit**

The off - on disturbs the electrons, any method including a plane spark gap will do. An input diode can do the job. The capacitor determines the out put. Whenever pressure put on the input plate it duplicates itself from the ambient background. This then feeds the output to earth ground. Being a blocking device no energy passes through capacitor, The output in series into the earth grounding provides an endless energy source. No energy leaves the input capacitor. Therefore all energy from the action in excess of the input and is named free energy based Gibb's. The capacitor used determines the power output. Therefore any and all of my Devices work exactly as I say. USE YOUR BRAIN FOR GOSH SAKES. Regards, Don Smith

Apr 30, 2004 **organic radical Battery recharged only in 30 seconds!**

By organic they mean carbon, probably Bucky Balls. Any of the newer super capacitors do this. Regards. DS

May 5, 2004 **Re: Bargain CF from Nigeria!**
Norman, what do You think? The sharks circling convinced me that this kind of information is not good for ones health. I am working with several of my regular clients presently. There is a bonus of 3 years salary for a contact that results in a buyer contract, I have done this several times before. I take shares in a new company which is formed, who's value depends it's increase in value when marketed. Small upfront and 51 % to the technology. Validating the license requires successful on the market within one year. Certain minimums are also required. I have several hundred other similar inventions, all have worldwide patent searches and nowhere exist. Any how would appreciate Your comments. DS

May 10, 2004
Re: Don Smith succeeders to organize
Chris, for starting go to rexresearch.com then continue down the list to Frank Wyatt Prentice's Electrical Power Accumulator, this Device works as stated. When properly constructed it works exactly as stated. The 3 Kwatts is not to be sneezed at. My devices are a vast improvement Prentice's. If and when You get Your Device up and running I will supply You with additional information which will move You in as a big winner. Regards, Don Smith

May 10, 2004
Re: Don Smith succeeders to organize
Mark, thank you very kindly. Don't believe the gossip mill. It's very clear and as simple as it gets. He states that he uses telluric [earth] currents. They are every where present, but there are several things about them that You need to know. There are special maps used by the Geologist which show these energy sources in detail. What part of the World are You located. I am a Professional Engineer and work with these things. If I have your personal e-mail I will send You a nice map showing what I have told You about. You can then post to the group. Regards, Don Smith

May 14, 2004 **Emailing: GibbsFreeEnergy**
Gibbs Free Energy — from Eric Weisstein's World of Physics

May 14, 2004 **Emailing: HelmholtzFreeEnergy**
Helmholtz Free Energy — from Eric Weisstein's World of Physics

May 14, 2004 **Don Smith on Free Energy**
That excess of energy remaining after a reaction ! That is why resonate magnetic flux is the most obvious source in that it makes multiple duplicate copies of the same source which become additive when properly harvested. DS

May 16, 2004 **Re: build your own gaussmeter**
Howard, magnetic resonance flux which is every where present in huge amounts having a Gauss Meter is very important. One milli Gauss is 100 Volts of resonace flux. If PRENTICE or MORAY had known about this rest would have been HISTORY. Using a magnetometer in placing the magnetic dipoles, instead of 3 Kilowatts for PRENTICE would have been MEGAWATTS. This very correct as I use Gauss Meters Professionally for more than 60 years. Don Smith

May 16, 2004 **Paul-Brown-12-05-86.pdf**
Paul Brown's Over Unity Device experience is very similar to many which the same which I have I have had. However I use extensive electronic simulators so I got it right the first go around. Several hundred very different Devices have resulted. Money not being a problem every thing was first class. DS

June 3, 2004
Subscribe: beyond_thermodynamics
Resonate magnetic flux is not subject to ohms or Lenz's laws. DS

July 21, 2004 **Subscribe: Re: powertrain sites**
You All ! And the train that could Poo, a recent Mexican Patent Filing related to rotating iron wheels of locomotives. The resonate electromagnetic energy produced by the rotating iron wheels is enormous and requires special grounding to prevent a disaster. You will have to get Your own copy. The grounding if feed through a transformer system is in the high MEGAWATT RANGE, Regards, Don Smith

July 22, 2004 **Re: MEGII, COP>12**
Also note: the use of magnetic resonance resulting in negative resistance which amplifies the power out. The stainless steel plates do this, and are very noisy. Regards, Don Smith

[Question: "Don, are you saying we should use stainless steel for the core in MEG2?
Please explain."]
July 22, 2004 **Re: MEGII, COP>12**
Yes that is correct. This is based upon an experiment using L-2 Coils when winding upon stainless steel tubing, when in use the electric lights for miles around light without being turned on. Regards, Don Smith

July 22, 2004 **Re: MEGII, COP>12**
Steve, thanks for the nice response. As an introduction http://www.altenergy-pro.com. More than 2000 webservers worldwide carry my website which is strongly represented in Eastern Europe. One of these persons located in Croatia duplicated one of my projects and without telling me, wound his Tesla Coil on a stainless steel tube [4 in dia] unexpected results were extremely noisy and horrific energy display, which I previously mentioned. It has commercial possibilities and there are no plans which I am aware of for development. So go for but be very careful. There are many kinds of stainless steel, Regards Don Smith

[Question: "What about copper plated stainless steel laminated as the core?"]
July 23, 2004 **Re: MEGII, COP>12**
The copper will probably cause sparking. DS

Aug 8, 2004 **Re: Building a Free Energy Car**
Nicolas, please see altenergy-pro.com, I have a self propelled electric airplane in commercial production, a full sized automobile is a peace of cake. Some of the Devices I built 20 years ago are shown on my web site. Have a nice day, Don Smith

Aug 8, 2004 **Re: Building a Free Energy Car**
Jared, Perendev technology is highly out dated by technology presently in the patent process. I should know since I am the source. Regards, Don Smith

Aug 17, 2004 **Re: Where can I find info on this**
A couple of clicks and - - - - See attachment? Do a PSPTO search for the two words Fabrizo and Pinto and You will get the 5 Patents issued to him, #' 'S ' 6665167, 6661576, 6650257, 659566 and 6477028 Regards, DS

Oct 1, 2004 **Fwd: Perendev Update**
Incoming Resonate Magnetic Flux propels - rotates the Earth at 1000. miles per hour. The weight of the Earth X the speed of rotation tells us the Magnetic Flux Energy Present in the Ambient Background. See my website altenergy-pro.com Regards, Don Smith

Oct 1, 2004 **Fwd: info**
Ted, this is the energy source demonstrated by my before a large audience of energy professionals on 1908, being described in detail in the

document sent You earlier today. When and if, You and your crew get tired of confidently playing ignorant, let me know, we have serious business that needs prompt attention. Regards, Don Smith
My Patent filing earlier this year. Regards, D.S.

Oct 1, 2004 **Quote**
Gravity is the Right Angle component of the Dipole of the Earth's rate of Spin and Mass, which increases with it's depth into the Earth. Don Smith Would like to know Your thoughts.

[Question: "Explain why Venus which hardly rotates at all has plenty of gravity"]
Nov 8, 2004 **Re: Quote**
Mass is dominate, rotation is second ! D.S.

[Question: "I just talked to the people at GMC Holding and they say the motor/generator is closed loop with no batteries or external power source!"]
Nov 17, 2004 **Re: You won't believe this!**
Dear All, an Aircraft Starter - Generators with the circuitry rearranged can easily do this, I discuss it and the implications in My web site of 9 years ago. http://altenergy-pro.com Regards, Don Smith

Feb 19, 2005 **Re: VCR Bert Info and Don Bashing**
Dear All, two ways of saying the same thing. The electrical side of the equation immediately dies a heat death, the magnetic side replicated and added to such that from the ambient electrons present from magnetic resonance provide greater output than input. Go to Goggle Search Engine and call up [magnetic resonance energy] it shows 1,360,000. hits. My site is # 1 and 2, see all those below me and their competency level and compare with Yours. Case Rested. Highest Regards, D.S.

Feb 21, 2005 **Response**
Gentleman, challenging is the name of the game. There is much work to be done if humanity is to survive the impending energy disaster caused by special interest and which is less than 2 years from realization. Are You up to the task. I have tried several times in the past, only to be harassed by the special interest and Government Agents here on this very web group. Thank You very kindly, Don Smith

[Question: "Mr. Smith: You have a provocative website. The devices and their claims are thought provoking. What I and perhaps other members of this group would like is nothing more than to understand how to replicate your work, and acheive the results that you have claimed. Having said that, I have not found enough information that would allow me to replicate your work. I am an empirical sort of person, not a theoretical person and believe that if someone has built something great then I needn't reinvent the wheel. I am unclear what motivates you. If I knew this then I would try to provide it in trade for the ability to test your device and learn the techniques needed to replicate it. What would you require to allow someone such as myself to do this? If the answer is that you do not wish to share your knowledge, what is the purpose of your website, and posting in this group? I can understand your frustration when people attack you or your ideas, but as eating banquets in front of the starving masses have caused revolutions, displaying devices that seem to have the ability to answer many needs, and not providing the device, or its benefits will similarly cause attacks. This is a simple request for information. Nothing in it is intended to raise your ire, and if I did so, I apologize in advance. Paul"]
Feb 22, 2005 **Re: Open letter to Don Smith**
Paul, thanks for the kind words. The sole purpose of the website is awaken people to the

fact that there are ways survive the enslavement of special interest. After many years of trying to cooperate with the US agencies it became apparent that the were the problem which resulted in my activities being relocated offshore. I have a huge following world wide with immediate success. Many people realize the necessity for what I am doing and most want results only if some one else hands it to them on a platter with no effort on their part. This causes a vitriolic reaction which is difficult to manage. Again Thanks, Don Smith

Mar 15, 2005 **Re: Devices using perm magnets**
Dear All, using none technical terms will try to explain what You are having problems with. User description is advise when encountering the half full - half empty syndrome - debate. The magnetic intensity level available is set in the activation process. The sensitation level reflects - activates a portion of the ambient resonate background energy - everywhere present in humongous amounts. Various Devices collect this energy which replaces it's self without diminishing it's source. Let my know if this is helpful. Regards, D.S.

Emails

SYSTEMATIC INDEX

This Index quotes Don Smith according to the subject under each heading in the order they appear in this book as indicated by the page numbers that follow each paragraph. This gives the reader the ability to study in one place what Don wrote or said about a certain subject. There was a lot of repetition so it is not exhaustive.

General Historical Development 388

Don Smith Historical Development 389

Don Smith Electron Theory 391

Types of Over Unity 397

Example Proofs of Over Unity 398

Illucidations of Over Unity 400

Sectional Processes in Over Unity Systems 403

1. *Agitation, Imbalance, Disruption of Ambient Equilibrium* 404
2. *Asymmetrical Engineering the Imbalance* 405
3. *Gathering and Collection of the Potential in Capacitor, Inductor, Ground* 407
4. *Waveshaping Conversion to Make Useful* 410
5. *Inverting Energy as Needed.* 412

Devices and Materials Mentioned 414

Using Tools and Making Calculations 415

Benefits of Magnetic Resonance 417

Consequences of Under Unity Systems 417

Where Conventional Theory is Mistaken 418

Hindrances to Over Unity and Why Replicator's Fail 419

General Historical Development

Pre Civil War electrical devices used Earth as a battery, and Radiotelegraphy, 14

Early Radio Devices could not be explained until after Thompson around 1911, 15

Related Devices began to appear in numbers in the 20's and 30's, 15

Cohere of Tesla, precursor of diode and solid state devices helped radio, 15

Farnsworth (TV) built upon coher and radio tube, 15

Peer Review principal weapon of the establishment in preventing destablilizing to self, 16

Near the turn of the 1900's the French Patent Office declared all inventions having greater output than input a violation of the laws of nature and therefore would not be considered. Edison through grants and other favors managed to control any and all information dispersed by publications, text books and sources of higher learning, 20

The Field of Science proflgates by notation, that which does not, is not, 22

Isaac Newton (1642-1727) with the help of Charles Coulomb (1736-1806) provide us with the square laws which are based on an open circuit, 23

Heaveyside, Oliver (1850-1925) provided us with Energy equals mass times the square of it's acceleration. Albert Einstein (being an insider) took credit for Heavyside's work, 23

Leonardo Fibonacci (1170-1230) of Pisa demonstrated very clearly that, which directly translates from laws of nature, an orderly form of progression, 23

Scientific Notation showing overunity, is and has been available prior to Newton, 26

Maxwell, J.C. (1891), active electron field gives rise to magnetic, 28

Benjamin Frankilin demonstrates ectrical flow, 28

Gauss, C.F. (1777-1855) and Oerstead, H.C. (1777-1851) define Earth's electrical field, measuring magnetic field flux as an indicator of negative electron energy active, 29

The electron as a particle was postulated by professor J. Thompson in early 1900's, 60

Prior to Tesla, there was a large group of persons in Europe who were building resonate coil systems related to medical usage, 61

A short list of those active (1860-1880 onward) in resonate high frequency coil systems include the Curies, Roentgen, Ruhmkoff, Oudin, Hertz, Levassor, Dumont, D'Arsonval, 61

Peugeot, Panhard-Lavassor, Bollee, Renault had successful electric automobiles using AC motors, 61

Various electrically powered airships like the Dirigible France were in service, 61

D'Arsonval, Professor of Experimental Medicine at the College of France, invented the electrocardiograph, oscilloscope, am and volt meters, thermography, etc., 61

Electric vehicles dominated until the 1920's when the electric start motor made the internal combustion engine practical, 61

Joule was a English brewmaster who had all kinds of machinery that ran on steam, and some electrical, and this and that. He had to come up with a system which would make translating one system of energy to another more universal, 202

In the 1920s people discovered that in studying the Earth's crust that it had electrical characteristics, 271

Don Smith Historical Development

"Do you need to put it in a particular geological area?" Well *it started out that way* because of these high concentrations of negative electricity in certain areas the problem in the very beginning was: You've got these two areas of different potential so how do you turn that into something useful? It became obvious that you are not going to be able to put a cable between the two because the cable of any size regardless of how large was going to vaporize like a flash bulb whenever you grounded it.

1998, Energy as it relates to Chemistry and Physics Formula: Free Energy (Δ G) = Ambeient Electrons (mass) times CPS (cubed not squared), 18

I had a neighbor who was an electrical engineer who knew what I was doing, and he and his son I switched out their microwave oven, and they were curious as to what was inside. After they tore it apart and found this thing they came bouncing over to my house and said, 'well we think you need this.' This was about six months ago [1994] and I didn't know that I needed it. But whenever I got to the point to where I knew that I needed to do something like that well I realized I had this thing out in the garage, and rather than go out and spend $400 buying one, I already had one, 186

In my experience with this device, when I hook it up to it, it turns out that the L1 coil on this device has a four point something feet length of wire. It turns out that the wire which hooks onto the battery here has that same length. We're dealing with a radio device so we've got an L1 there which is oscillating at the same frequency that this wire here that goes to the battery would oscillate at. So guess what? When I hook it on to this battery here it's going to be generating electricity, and feeding back into the device, and it's not going to be using the electricity out of this. How I know that? These devices have been ran for as much as a month at a time with no lowering of the capacity of the battery, 188

1995 video: I'm a geologist by background and work with electromagnetic Maps and locating all deposits, mineral deposits, and other sorts of things. Of course that's a direct reflection of the electrical characteristic of the earth and our portions of the earth. So that got me involved in electricity initially, and I've had a life-long interest in it. About seven or eight years ago when the petroleum industry went sideways, I had a lot of extra time and I was asked by the local school system to help out in their science courses. Of course when you look at the physics books fifty years after you saw them the first time, or you had them in college, you see things which you probably didn't see the first go around. You've had some life experiences and you begin to see how these things all fit together. So it became quite obvious to me that there were some real big holes in the way that electricity was being presented. Now one of the early on things which I found was this. And it's from a science supply. The initial device that I ran into was from a science supply house and it turns out that this company is still around and still makes these things and I'll show you the actual device, [Editor: We have reproduced this kit with extensive additions.] 197

As a matter of fact it was one of the first

things that I knew about. I'm a geologist actually. In about ten years ago when the petroleum industry went sideways, friends of mine try to get me to assist in the high school science department in this large high school that was close by. I did and of course I ended up teaching physics and chemistry and a bunch of other things. This was one of the devices that I used to teach radio characteristics in physics. That was the first time that I was aware of the fact that this overunity type thing was a possibility. That was more than seven years ago. I was very much into it, far advanced, before I knew about Tesla, I knew him, but I didn't really remember much about him, 219

Basically what I have done is pick up where Nikola Tesla had reached, and I've put myself inside his mind. I've expanded and extended what Nikola Tesla was doing. [Question. Do you think you actually figured out what he was doing?] Yeah, I can read his mind almost perfect. Actually, I did not know about Tesla until I was into this thing more than five years, I guess. [Really?] Yes, and I knew the name, but I had no idea exactly the significance of it. People started saying, well you're doing what Tesla did. Building the same kinds of coils, and doing all the same types of things, 270

About ten years ago I had reached the level in the petroleum industry where my salary was too much, and I had too much experience. The people that were in the company all had less, and they felt uncomfortable around me. So suddenly oneday I found myself without a job, 272

In Eastern Europe there are probably four or five hundred people that have duplicated my devices with real good results, 326

He's got a university degree in petroleum engineering and geology from the early 50's. He was first employed at Gulf oil, later Phillips, and many many others, including some independent companies, 330

When I first started doing these things I had one of Nikola Tesla books by Thomas Commerford Martin and his official authorized biography of Tesla. From that there are a number of experiments which he shows in the book. I went through every one of those experiments in great detail. That was my learning process for understanding what Tesla had done, 331

12

At the end of it, I realized that technology had advanced quite a bit and a lot of new things, which were available, which were not available the time Tesla did his work. All things I've done are extensions off of Tesla.

Don Smith Electron Theory

All electrical Power Systems have only the magnetic flux move from point A to point B, 3

No electrons pass through from the input side of the isolation transformer to the output side, 4

With the down stream system in place, all that is now needed is a potential system. Any flux system will be OK, 4

Any input system wherein a lessor amount of electrons disturbs a larger amount of electrons, resulting in the output being greater than the input is desirable, 4

Non-Ionic electron doublets in vast quantities in universe. Origin is from the emanation of Solar Plasma, 5

When pushed apart by agitation one spins left, yielding (Volts-potential) electricity and other spins right, yielding (amperage-magnetic) energy. One being more negative than the other, 5

Impacted emit visiable Light and heat, 5

Never deminish, 5

The sea of Continuum "Ether", governed by the Law of Squares is from it's electrical and magnetic character composed largely of right and left-handed electrons, 5

Controlling the flux rate controls the electron (potential) activity, 8

Solar Plasma is 65% None Ionic Electrons, 9

Early Radio Devices could not be explained until after Thompso's electron theory around 1911, 15

Radio glass tubes, electromagnetic stimulation dislodges electrons which avalanche in dislodging many-many more electrons. The excess of electrons at the Cathode magnify and react with the electrons present at the anode, resulting in more energy out than input, 15

Daylight caused by electron enrichment of upper limit of Earth's Atmosphere in the radiation belt (shear zones), 15

Recovery spin right is magnetic - amperage, 16

In chemistry physical change, from one form to another results in one of three actions: 1. Mixing only, physical only, no reaction. 2. Endothermic, loss of energy by absorption. 3. Exothermic, additional energy not previously available. This is referred to in all chemistry Texts as Free Energy or (Δ G). These changes occur at elemental and or molecular level. At particle size, as in Electrons there is an overlap into Physics, 17

Free Energy Harvesting Requirements, 1. Electron acceleration or deceleration must vary from ambient. 2. Spin rate CPS fluxing provides electrical with magnetic energy. 3. At Resonance (negative resistance), all electrons move in the same direction without interfering with each other. 4. Electrons must Avalanche, pulsing magnetic waves provide three dimensional and nonlinear energy at each cycle, electrons presently available disturb and bring alone many more electrons, resulting in gain, 19

Electromagnetic stimulation disturbs electrons by avalanche, dislodging adjacent electrons, resulting in more energy out than in, 21

Electrons, when disturbed, first produce magnetic flux and then electrical flux when they spin back to their normal position. Therefore, any electron movement produces above ambient energy, being over unity, 24

The basic unit of electricity (the electron)

upon encountering a moving magnetic field (or wave) spins, giving off an electric impulse. When this impulse collapses, it spins back to it's natural position, giving off a magnetic impulse. Therefore, magnetic and electric are two sides of the same coin. When the magnetic side is pulsed, it yields electricity and conversely, pulsing of the electrical side yields a magnetic field. Moving one in relation to the other produces useful energy. When done consecutively, each cycle pushes (current) forward, while pulling electrons into the system...in much the same way as a water pump moves water. These electrons are obtained from Earth and air grounding, 27

70% of the Earth's crustal surface silicone related non-metals, Electron donors source of electrical energy, 27

Weak and strong forces rather than positive and negative, composed of electrical magnetic and gravitationals fields and waves. Energy from electrons represents he weak force. Energy inside the atom represents the strong force. Controlled resonant induction of any two of the three, changes into the third and is the motor that runs the Universe, 28

More negative charges move to the less negative, unequal batteries comparison, 28

Each cycling of this resonant induction system pulls in additional electrons from the Earth's electrical field, 30

Functions, become active when placed inside the critical distance allowed by their negativity. Active electrons provide: Electricity, Magnetics, Gravitational thrust, The source of Visiable Light, and it's charge is Negaive, 34

Move in a closed infinity symbol loop, half of loop is a magnetic impulse, and return half electrical impulse like sine wave, 34

Flash of light occurs when two electrons too close, 34

Electron travel: In Electrical Systems, Electrons active at point "A" are not the same Electrons active at point "B". The Electons activated at the Electrical Station are not the ones used at your house. When you ground your system by flipping the wall switch, you use your own electrons. In closed energy systems, electrons communicate with and replicate the activity of the overbalanced potential, when provided with Earth and or Air Groundings, 36

Electrons which cycle through this system, after being used, are returned intact to their former state for future usage, 39

Electron spin causes electrical current and magnetic lines of force, 39

The effect of current results from the unequal distribution of negativity (electrons), 39

Magnetic imbalance causes the gravitational effect. This is evidenced in electric motors by magnet-gravitational displacement of mass which causes the motor to rotate, 39

Diagram of Energy Acquired by Magnetic and Electrical Impulsing, Alternating Resonate Electrical Power), 42

With alternating electrical current, electrons do not move from point "A" to point "B" as commonly envisioned! Electrical potential (oscillating electrons) at point "A" results in harmonic electron activity at point "B", when the grounding switch (circuit) is closed. Point "B" supplies it's own electrons and mirrors the activity of point "A".

Impulsing (turbulence) by magnetic induction causes electrons to be pulled into the system, which then oscillates. When the magnetic field collapses (becomes absent) the electrical potential returns to it's natural background, 47

Each time the Hz's. are doubled, the effectiveness of induction is squared. At about 20,000 Hz, when radio frequency is achieved, the electrons begin spinning free, outside of the inductor. They becom increasingly free of the inverse relationship of volts-amperes, 47

When Electrons are disturbed by turbulence-movement of any sort they become active and try to regain equilibrium. This movement towards equilibrium results in magnetic displacement, which causes the motor to rotate. Forward thrust of Electrons results in Magnetic Energy and backwards thrust results in Electrical Energy, 54

The electron as a particle was postulated by professor J. Thompson in early 1900's. Physics as it exists, cannot explain why the elctron remains intact and is not diminished by the energy it releases. A volts worth of electrons, when cycled yields a volts worth of electricity, 60

Since magnetic and amperage are one package this suggests that electrons in a nutural non-ionic state exist as doublets. When pushed apart by agitation one spins and supplies electricity and the other spins and provides magnetic (amperage) energy. When they reunite, we have Volts X Amperage = Watts, 60-61

Bloch Wall (space), area of Electron Spin Separation, 63

The Source of these Electrons being from the Solar Plasma, are non-ionic and occupy all Free Space. They are commonly obtained from Earth and Air Groundings. They exist in Doublet Pairs, one being more negative than the other. The more negative one has a Left Hand Spin. The less negative one has a Right Hand Spin, 63

The Electrical Circuit equivalent of the Bloch Wall is a Diode. Electron Doublets once separated must recouple, resulting in useful energy, 64

I = Current = Amperes = Right Hand Spin. V = Electrons = Volts = Left Hand Spin, 64

Resonate induction transfer from one power system allows other resonate induction systems to duplicate the original source, which in no way diminishes the orignal source. The number of radio devices or televisions concurrently running in no way diminishes the output source, 66

Resonate induction disturbs a large number of adjacent electrons which were not a part of the original source. The Pulsating (pumping effect) then incorporates the newly available extra electrons into the ongoing energy generation source-system. A near unity energy system of reonate air core coils and extra aquired electron-energy source constitutes an over unity system, 66

The amount of Energy transferred relates directly to the density of the lines of flux present. Each unit of increase causes a squaring of the flux lines present. One volt, Two Volts = Four, Four Volts = Sixteen. In resonate air core coil energy transfer the increase in flux lines present disturbes more electrons than previous, resulting in over unity energy being present and available, 70

You can see where the envelope is set up

and where the thing has passed outside the envelope there, and it continues even greater distance outside the envelope. As it goes along and it gets out to a certain point here and it collapses and then it comes back and it starts the whole thing over again. This is a particular resonant circuit that you find depending on how you work your shunt on the resonant circuit, 175

But in the third one is a particular arrangement of a grounding shunt, which means that its ability to communicate with the electron source, which would be your grounding. The resistance there is like a dam on a large body of water and it lets the electrons through and they get caught behind this dam. This whole device is like a pump. All it's doing is like the dominoes we were talking about, its tripping the electrons here and that electrons trip in the next one, and this whole bunch of electrons in all directions are being tripped. Each time one of those would be tripped it gives off a magnetic impulse first, 175

We have two kinds of kinetic energy. We have expanding magnetic energy and it's dominate for that. It gives you your electrons moving apart, pressure decreasing. It gives a cooling effect. That explains why it's two hundred and seventy-six degrees minus out in an open space away from the earth. When the electrons are moving apart or expanding it's cooling, and sometimes very cooling. There's less scattering of energy in an expanding situation. In addition to that you have negative resistance as dominant. If we go into the contraction stage of energy or mass we get electrons moving together. Pressure is increasing, heating effect dominates and you have a scattering of energy, and positive resistance results. You're under unity people work primarily and in that particular set of circumstances, 177

The movement it's not a circular movement in the manner which most of the books describe it but it's sort of a closed lazy eight, sort of thing, that's compressed in on itself and this also happens to be the symbol for infinity, 178

Forces of nature to operate: Composed essentially of a weak force and a strong force. Each of these forms of energy occur as fields and waves depending on what the resonant frequency is at a particular level, 178

Basically electrical energy is achieved when an electron density at point A becomes greater than electron density at point B. This is your potential difference between two points. If that doesn't exist you don't have any energy, regardless of how how many electrons are there. It's just like the high pressure and low pressure areas of the weather system. Things are going to move away from the high pressure area into the low pressure area. The electrons density acts the same way, 180

One is wound clockwise and one is wound counterclockwise. That as part of the slowing down of the electron so that you end up with a waveform that you want, 187

If we have ambient, which is background, which where you're sitting is perhaps two hundred thousand volts per square meter. Anything above or below that is going to be useful energy because the electron is just like the snowball that you put on top of the hill, and you roll it down the hill. It's trying to get back to gravitational level which it was before you put it up on top of the hill. Well if you raise the electrical potential above the background it's always going to try to go back to the background level. It's the in-between part there that's useful to you and it can either be below ambient or above ambient, 197

For each line of electrical flux we have a line of magnetic flux., 206

When two electrons get too close together they push apart. When that happens you get a flash of light. That's where your daylight comes from. It's the electrons that are coming in from the solar plasma that impact on the side of the earth that's spinning into the solar plasma, 210

Since the electrons are in an awkward position, they will try to get back to ambient. In the process of getting back to ambient, they're going to be doing the things that you want them to do, 241

Two types of electrons. There's one that's a little bit smaller and weaker then the other. When it spins its direction it becomes the amperage part. The stronger of the two electrons spins and makes the voltage part. In nature they are coupled together and that's the reason why they're inactive and floating about in space without anything happening. It's whenever you disturb them or knock them apart that they try to get back together and that's where you get your useful energy, 252

This business of having electrons coupling and being disturbed and pulled apart. And they have to fight to go back together, and when they come back together, you have volts times amps equal watts. That is an extension of JJ Thompson's description of the electron originally. It's expanding onto it, and there's some other things there, 265

You cycle say at a million cycles per second, the curve on the increase in the amount of magnetic and electrical field there is doing this. When it does this, you capture at a higher point, you've got a lot more energy from the same amount of electrons, then you have at the lower point, 271

I knew that there was an electromagnetic. You can't have one without the other, and it flips. When it flips, at that flipping point is where you have your useful energy. The more you flip it the more you have useful energy, 271

As it turns out there's good evidence that some electrons have more negative charge than others. And being more negative or less negative that that's the same as being plus or minus. So therefore the electron with a weaker charge will attract one with a stronger charge, and they will be orbiting in each other. They're relatively easy to pull apart. So the electrical and magnetic charge of the electron, of course, when you spin them, spin them right you've got a magnetic field. When you spin them left you get an electrical field. So when they spin apart you're getting a benefit of both of those. And then whenever they come spinning back together you're getting the volt's and the magnetic field coupling again, which means your amps, amps times volts equal watts, or you've got useful energy, 272

Denis Lee demonstration for Don: The sun emits electron pairs and they're everywhere. In subatomic theory the electron shells, they always talk about the up and down electrons, or left-hand and right hand spin electrons. Well when they're in a pair it's undetectable. But you can use weak magnetic fields to separate them. This sort of demonstrates an electron pair, and how they may combine together. There is a unique mechanical property which keeps the —I called concentric toroid—which keeps the small toroid centered in the large toroid. You can see they can come out. You can use a weak magnetic field and separate the two, and this becomes positive electricity and negative electricity. An interesting thing about this is that when you roll it on a tube —I don't know, this could be a flux line—and I don't if you can see this but it roles and it spins, and that's a mathematically correct fourth dimensional rotation, 319

It's the magnetic field/waves that You then need to capture. When they pass through a conductor material the electron cloud at it's surface, (electrons flip) changing to the electrical radiation component. In this action which has been going on for more than 4 billion years has not in any way dimenished the electrons presetnly yielding useful energy. When in the magnetic wave/field stage the conservation of energy laws do not apply. To see this, a radio or TV Station when radiating their magnetic wave siginal has no drawdown, wheather one customer or millions. Magnetic waves freely pass through most conductive materials as undimenished. They induce energy collection and movement, which can be used. Otherwise they deflect, reflect or absorb. When passing through the Earth they absorbe becoming heat. This in part is the infrared heat seen inside Your automobile on a sunny day, 368

Types of Over Unity

Electromagnetic Energy Harvesting methods include a catalyst, a collector, and a pump. Catalysts include such things as Air and Earth grounding, sensitization trhough doping with certian elements. Collectors include temporary storage as in capacitors, coils, and transformers. The pump system includes movement onward to point of use. The rotating coils ang magnet systems activate electrons present such that action at a distance can occur and therefore is an energy activation pump, 93

Nonlinear systems, 2

In chemistry physical change, from one form to another results in one of three actions: 1. Mixing only, physical only, no reaction. 2. Endothermic, loss of energy by absorption. 3. Exothermic, additional energy not previously available. This is referred to in all chemistry Texts as Free Energy or (Δ G). These changes occur at elemental and or molecular level. At particle size, as in Electrons there is an overlap into Physics, 17

Useful Electrical Energy is obtained directly from electron spin induced by incoming magnetic waves or indirectly through mechanical exchange as in dynamo type devices. Nature for free provides grand scale magnetic wave induction throughout the universe. As an accretion mass, the Earth is an Energy Sink, getting it's energy from elsewhere, being Cosmic, Galactic and Solar, 57

Indirectly: Electron Stimulation—Induced Electron Spin Systems, Electron Avalanche Pumping Systems. Direct Conversion, Magnetic Wave Open Systems: Ambient Sources, Air Core Coil Systems, Gaseous tube Systems, Solid State Marx Generator Avalanche Tye Systems, Leyden Bottle Capacitor Types inserted in Lakes, Magnet Systems, Electron Beam Antenna Systems, 58

Example Proofs of Over Unity

Piano provides a greater sound level than the input, 2

Lighting worldwide daily exceeds 57,600,000 million volts and 1,152,000 million amperes of electromagnetic flux, 2

DC from converting Sun's electromagnetic flux, 2

Dynode in cameras is self-powerd shutter, 3

Electrons as a source of energy have been moving and yielding magnetic and electrical flux since the beginning of time. They are never diminished and are an excellent example of an over-unity source of energy, 5

Negative resistance, at resonance Impedance is zero. Adjacent electrons, not a part of the orignal are pulled into the composite, becoming greater energy out than in, 9

In a diode, the excess of electrons at the Cathode magnify and react with the electrons present at the anode, resulting in more energy out than input, 15

Earth Electrical System II, use of naturally occurring electrical differences in the Earth's Crust and Radio Device in combination for Transmitting of the Energy to Points of Use, 16

Motor-Generator Coils and Magnet System, 16

Gaseous Tube Electrical Power System, Tesla Coil, Plasma Tube, and Laser Systems, 17

Propulsion-Star System, uses Spin System seen in Star Systems, Solar Systems, Tornados, Cyclones and related to Rail Guns, Gyroscopes, and Negative Gravitational Systems, 17

Power System, Piezoelectric Solid State Marx Generator, 17

Solar Power Power System relating to Magnetic Waves where domains aligned equates with resonance, 17

Electromagnetic waves incoming from point "A", the Sun, reacting with point "B", a solar cell which spins the local electrongs sideways, yeilding electricity, 21

In Physics all Matter relates too, and is Energy. Electrical Energy at its denominator is the acceleration or deceleration of electrons. Ambient or background being Earth Ground. Variations from Earth Ground provides useful electrical energy. By accelerating (disturbing) electrons as in Resonate Coils and Magnet Systems, or Converting Mechanical (dynamo) into other than ambient, results in useful energy. The faster the cycling of electrons the more energy present. Energy greater than ambient, based on the established definitions from the above Gas Laws is then referred to as Free Energy. 18

Kystrons, and all radio type devices, 19

Explosives, 19

Brownian movement, 19

PSI tire pressure, 20

Living cells multiply, 20

Avalanches start small and increase enormously, 20

Lightning starts with one electron, 20

Electrons, when disturbed, first produce magnetic flux and then electrical flux when they spin back to their normal position. Therefore, any electron movement produces above ambient energy, being over unity, 24

Dominos, energy required to flip #1 must be added with thousands more, 36

The number of Radios and Televisions running at any one time do not diminish, in any way the electrical output of the source station, 36

The flipping device (reactor L-1 Coil) is pulsed which then provides a resonant induction pulse, in turn this flips the electrons present at the (reactant L-2) Coil. The energy input in L-1 is divided by the number of turns present. The induced magnetic pulsing in turn flipps the electrons in each turn of L-2. If more turns are present in L-2 than L-1, there is a net gain in Energy present, as demonstrated by the dominos, 36

Resonate induction circuits in conventional radio tubes (high plate voltage), 38

Negative feedback systems in Op-Amps, 38

Abundant Energy Everywhere Present, User's Guide, 76

Don Smith's pending patents: Magnetic Radiant Energy Direct Conversion System, Motor Generator, Generator Gaseous Tube Electrical Power System, Propulsion System, Magnetic, Power System, Piezoelectric Solid State Marx Generator, Power System, Super magnet, Power System, Direct Energy Conversion, 94

If you use a small amount of energy to extract a larger amount of work, regardless of what the books say, haven't you've got something back more than what you put into it? 179

Everything that is useful is above ambient so everything that is useful is overunity, 214

It's already beginning to behave as an over energy device [NST], 241

The television tube itself is an over unity device., 273

Illucidations of Over Unity

Piano resonance activates side strings for greater sound level, 2

USPTO has already issued hundreds of Patents related to Light Amplification, all of which are overunity. Dynode used to adjust the self-powered shutter in your camera receives magnetic flux from light which dislodges electrons from the cathode, reflecting electrons through the dynode bridge to the anode, resulting in billions of more electrons out than in, 3

Useful Circuits, Suggestions for Building an Operational Unit, 6

Resonant Electromagnetic Power System Schematic, 10

Suitcase Model demonstrated (1996), 11

Brownian movement, Maxwell, Boltzmann and Clausius "the molecules of a gas carry out ceaseless transitional movement in all temperatures above absolute zero." 19

Air Coil Resonant Induction System, 36

The system consists of three seperate modules: 1. *HV Induction Transformer Module* with a connection allowing the high voltage output to pass onward through the induction coil L-1 and then to it's grounding. 2. *An Air Core Induction Coil Transformer Module* with Reactor coil L-1 with hv radio frequency capactior between it and it's grounding and reactant coil L-2. Input into the L-1 is divided by the number of turns. The magnetic flux field provided from each turn of L-1 replicates itself as an electrical potential in each turn of L-2. L-2 can have hundreds of turns. Net gain depends on number of turns, and output from L-2 is in V.A.R. where volts and amperes are the same till work (resistivity) is introduced. 3. *The Inverter Module* Inverting to DC or AC for designated usage, 48-49

If the plus connector wire's length is a wave length component of the L-6 or L-7's frequency the battery will recharge itself at the same time it is being used. To see this substitute two opposing LEDs. They show the direction the electrons are moving. DC electrons out and radio frequency electrons into the battery, 69

Each turn of L-2 is exposed to the electric flux of the lowered voltage of L-1. Each turn at the bottom end of L-2 acquires the same volts. The flux lines are squared and are additive as the voltage progresses towards the top end of L-2's many turns. A huge amount of flux lines not previously present occur at the top end of L-2. These flux lines excite the electrons nearby in it's earth and air and groundings. This high level of excitement above the abmient causes a large amount of electrons not previously a part of the energy present to become available. At this point overunity is present in large amounts, 73

Ambient Energy Generator, 76

Dipole Antenna Systems, 83

Dynamo is a combination collector and pump of energy acquired from the ambient background, 93

The system can use magnetometer studies for locating areas of greater electron density, or the fact that one has a greater density than another one and event that you want to do your electricity between the two points. For example if you put a contact on top of the Cheyenne Mountain out here, and you put one down at the bottom, and you ran a wire between the two, and you left a place there where you could plug your device into it, you would have over five million volts of usable electricity there that would essentially have the same as five million amperes because your resistivity would not come in, or your amps would not come in, there after the fact until you put a load on it, 180

In my experience with this device, when I hook it up to it, it turns out that the L1 coil on this device has a four point something feet length of wire. It turns out that the wire which hooks onto the battery here has that same length. We're dealing with a radio device so we've got an L1 there which is oscillating at the same frequency that this wire here that goes to the battery would oscillate at. So guess what? When I hook it on to this battery here it's going to be generating electricity, and feeding back into the device, and it's not going to be using the electricity out of this. How I know that? These devices have been ran for as much as a month at a time with no lowering of the capacity of the battery, 188

Why over unity is possible. These are three different conditions under which you get resonant circuits. There are other conditions. The bottom one there is the one what you're really looking for. You could see it on your oscilloscope, and you can measure it. I'm going to show you a more detailed breakdown on it next. You can see that it's building energy. It builds it up to a certain level. Then it cuts off, and then it cycles back. The resistor shunt in the middle there, is basically what helps cause this cycling effect, 248

That goes back to the circuit I showed you over the overunity thing awhile ago, where you have a resistivity in there. But when this thing becomes resonant, of course the resistivity all becomes negative. Therefore you've got energy which is greater than what you have to begin with, 249

Basics of so called Free Energy Devices are present in Early Radio. For example take the examples of the Crystal Sets shown in the schematics herein. 1. Replace the antenna portion with a high voltage source. Tesla coil derivative - - 2. Upgrade Diode Detector to Volts and Amperes. Diode changes signal to pulsating DC. 3. Replace Capacitor across the Ear Phones so as to set to the desired frequency. This can be a resistor due to built in capacitance of Transformer. 4. Replace the Ear Phones with appropriate size (desired output) Transformer. Presto, You have the Basic Space Energy Device. This Device sucks in energy from the Ambient Background if properly adjusted, 353

A 14" florescent light bulb with a 2" Diameter by 10" Length B and W coil, 10 TPI, placed over the light bulb and L-1 of 5 turns centered with a 9000 HV module input to L-1 all ran by a 12 V at 7 Ah Gel Cell Battery. This is a convenient and easy way to get into the gaseous tube electrical generators. A very strong Electrical Dipole Develops which substitutes for a very large battery. The setup requires storage such as capacitors and an inverter to be useful, 366

Take a charged capacitor, and lay it on top of a permanent magnet so that the E-field of the capacitor is at right angles to the H-field of the magnet. Then the standard Poynting flow S is given by S = ExH, which in this case is maximized for a 90-degree angle between E and H. In fact, the magnitude S of S is just the product of the two magnitudes E and H. The direction of S is at right angles to both E and H, and given by the usual right hand rule.

Well, even by orthodox theory, that is an actual Poynting energy generator. It just sits there and pours out free energy, directly extracting it from the vacuum. There are two dipoles — one electrical and one magnetic — continuously serving as an asymmetry in the fierce vacuum flux. Once you pay to make the permanent magnet and charge the capacitor (or use an electret), that simple gadget will extract energy from the vacuum and pour it out indefinitely.

http://www.cheniere.org/misc/static%20poynting%20gen.htm

Use an inexpensive Gauss Meter such as "Electrosensor" to determine the flux background which is available (This is available on the internet for about $35. plus or minus). Then a cheep Frequency Meter ($65.) from Radio Shack to determine the frequency. Then build as many seperate coil sets as needed for the output required. They have to be tuned to the frequency of the radiating magnetic waves, 368

Sectional Processes in Don Smith Over Unity Systems

These are quotations arraged generally as five different possible sections of Don Smith systems that are organized in the way of functions or processes more than parts on a module board. Don talked about making 200 or so prototype models over the years. Many of these were not complete systems but were mostly various module boards that performed one or more essential functions. These modules could be changed around with different systems so that it was unecessary for him to duplicate many parts in designing new systems. Don thought in terms of modules and sections rather than in complete systems. Because replicators tend to look at a particular model as consisting of all the parts as one they are unable to properly recognize the different stages and functions of each individual part and process. Related to this is what Don mentioned in their being several separate sections where only magnetic flux transfers from one side to another (resulting in action at a distance).

In contemplating how to group his words about his systems into various modules it became evident that they really could not be easily grouped into specific modules but rather into five or so general processes. These can overlap each other. For years I have said that over unity systems have three or four stages: 1. Disruption, 2. Collection, 3. Transformation, and/or 4. Utalization. I have attempted to organize various quotes in this way so that the reader can better understand what happens at each point. These generally overlap the particular modules. Diodes and transformers generally separate modules in a system. A homework project could be to separate the schematics in this book into module sections, or classify the different parts into the below categories. The quotations below are meant to help you do that. This is the most important purpose of this book.

1. Agitation, Imbalance, Disruption of Ambient Equilibrium. 404

2. Asymmetrical Engineering the Imbalance. 405

3. Gathering and Collection in Capacitor, Inductor, Ground. 407

4. Waveshaping Conversion to Make Useful. 410

5. Inverting Energy as Needed. 412

1. Agitation, Imbalance, Disruption of Ambient Equilibrium.

Piano string is impacted, 2

Turbulence upsets ambient and we see electrical displays, 2

Earth's Mass and speed through space gives it an enormouse electrical potential, 2

The Intensity/CPS of the resonant-frequency-flux rate sets the available energy, 3

Methods of disturbing electrons and ambient includes moving coils past magnets, or vice versa and pulsing (resonant induction) with magnetic fields and waves near coils, 5

Electron acceleration or deceleration must vary from ambient, 19

Once you get the magnetic flux going, you can have coils that are not tuned to it, that are in that magnetic flux, and they will oscillate at the resonant frequency that they are meant to resonate at. So you can go from one frequency to another and still have your electricity there because actually the electrons are not going through your system, they're just being disturbed by the magnetic flux lines of another electrical system, 234

2. Asymmetrical Engineering the Imbalance.

Piano strings are tuned to be in resonance, and connected to frame so that they experience sympathetic resonance, 2

Negative resistance, at resonance Impedance is zero. Adjacent electrons, not a part of the orignal are pulled into the composite, becoming greater energy out than in, 9

Resistor R-1 used to set electron pump rate, frequency of the capacitor. Maintains the desired voltage level required to operate the system, 10, 19

The farads and henrys of the resonant system provide the resonant frequency when pulsed by and external energy system, 36

Two coils, one being a reactor coil (L-1) and the second coil (L-2) reactant. Magnetic field fluxing (on-off of the electrical source) causes inductive reactance of L-1 which replicates by induction in L-2. Pulsing of the magnetic field (from L-1) in the presence of L-2, generates electrical potential. L-1 10 turns, 1.2kv A.C. Each turn of L-1 120V. This induced magnetic field then replicates itself in each turn of the L-2 coil, 47

Air Core Induction Coil Builder's Guide, 78

Output frequency is going to be determined by the length of this wire. The highest frequency that you have represented in any of the components is going to be the dominant frequency, 182

What you want to do on any turns, or whatever you put on your second coil, they should be a relationship such that it will be either a multiple or a division of the same frequency. The second coil is going to run at a different frequency because the wire length is different, 183

Once you've got this thing divided down, and you've got two thousand volts on each turn here, then you're ready to go to your L2 coil which will be the second one. In many cases that is going to be a great many more turns than the L1 coil. But it may be one turn. If you want two thousand volts instead of a great deal more, then you make just one turn, and make it so that it's a division or multiple of from the length of the other coil so that they will react from a frequency standpoint properly, 183

L2 has to be grounded in order to pick up the maximum effect of electrons. Anytime you add a capacitor, or any other device, or a resistor, or anything else into this circuit here it's going to lower the frequency. So the highest it can be will be based on the length of the wire, 183

We've got two coils and they may be very slightly out of tune, not precisely tuned, so you need a variable device here, which can be a variable blocking device, which can be a variable capacitor. Or it could be a variable resistor or a number of other things. You have to do this by experiment and don't put any real heavy resistors in there. I mean any high resistant deals. Put something say like down below 500 Ohms, maybe around a 100 Ohms or something like that. This is going to be that shunt that's in the resonant circuit that we were talking about a little bit earlier. It's going to be the dam that's going to hold the water behind it so to speak, 183 *Continuing:*

We've got one turn here, and we've got it grounded with a variable device on it so that we can fine tune it. At that point we've got an output out here which is in an oscillating alternating current type situation at radio frequency. [Question.] Yes that is the shunt in a resonant radio frequency configuration that produces the dam that keeps the energy. Once it gets to oscillating inside the system, it can come in but it can't get out unless it is hooked into another system. Yeah well it's simply a blocking device which keeps your resonant coil system, 184

The reason for that is that you want one end to be delayed just slightly more than the other so that you're going to have an offset on those spikes, 185

Lightning arrestors, used in place of spark gaps, 191

One of the problems that people run into in building these things is getting the 2 coils to react with each other. You've got to have them so that they can talk to each other. So if you use a quarter wave system your shortest wave will be the longest length of wire and your shortest length of wire will be your higher frequency, 198

You're actually tuning this thing by pushing the L1 backwards and forwards in there, till it actually matches up with the tuning deal. Then you've eliminated the possibility of having to put capacitors and a number of other things. Because the capacitance is already built into the wire that's part of the coils. So you've simplified and you've taken out some stages there, 220

Without that center tap, you wouldn't have the amperage available to you, 240

As you lower the voltage [on the NST dimmer], the amperage goes up on this thing. So if you run it at less than 110 volts on this inverter here, and then you back it off again with the variable deal, you're going to end up in some cases with more amperage than the device can handle, 240

I calculate the length of the wire from the battery to the input into the device so it's one quarter wavelength of the L1 coil, and for that reason it picks up the radio frequency. With the kick back diode, it forces the high frequency electricity back into the battery. The high frequency electricity is charging all the DC that's coming out, 242

Use an even division or multiple of and allow an extra inch or so. The length includes the rat tails used for connecting. 30,000 divided into the speed of light will give the wave length... Once built You can locate the nodes where resonate points occur and then trim the coil ends to match, 367

[Question: "Hi Don, Is it possible that you vary the speed of the motor to find/make the two coils and the two Caps. all in series (series LC circuit) resonant thereby increasing their efficiency/output? It would seem that a full wave bridge diode would be necessary to keep polarity correct into the battery/load if DC wanted. Ben"]

You are correct. The variable speed switch is from an automobile dimmer. The drawing is the simplest possible and is therefore incomplete, 374

3. Gathering and Collection in Capacitor, Inductor, and Ground.

Piano strings store resonant energy, 2

Coil crown on top of Plasma Globe resonant air core coil system, 6

Power factor correction capacitors should be as many microfarads as possible to allow lower to match primary side of isolation transformer, 8

Electrons must Avalanche, pulsing magnetic waves provide three dimensional and nonlinear energy at each cycle, electrons prensently available distrube and bring along many more electrons, resulting in greater amound of energy, 19

Each turn of L-2 is exposed to the electric flux of the lowered voltage of L-1. Each turn at the bottom end of L-2 acquires the same volts. The flux lines are squared and are additive as the voltage progresses towards the top end of L-2's many turns. A huge amount of flux lines not previously present occur at the top end of L-2. These flux lines excite the electrons nearby in it's earth and air and groundings. This high level of excitement above the abmient causes a large amount of electrons not previously a part of the energy present to become available. At this point overunity is present in large amounts, 73

At this point we're going to have to go through a diode-bridge. We're going in with alternating current, and we're coming out to DC. At this point we've got something which we can take through one or more steps and we're where we want to be. We're very close to where we want to be. We take off from a the DC here and we come out like so. We're going to inject a coil in here which is actually two Tesla coils "bottom to bottom." This is the middle here. This would be a base of a small Tesla coil, and this would be a base of a small Tesla coil. We've got DC coming out here so we've got to pulse that in order for this coil to work. What we do we put a capacitor in parallel there, which gives a pulsing effect, or an off-and-on effect, which will cause the magnetic fields to be, or induction to pick up in these two other coils, 184

What we're going to end up with is basically a spike. A DC of course, if you've got your reference point here. You've got alternating current which is doing that sort of thing. In DC you're gonna be on just one side of that reference point which would be the top side, or the bottom side, and you're going to be on just one side. At this point the waveform coming out of here is going to be a DC spike and we're going to change that when we go through this thing right here that's going to change. When we go through this system here we're gonna have a DC spike above and below. Only it's not going to be DC anymore, it's going to be alternating current. It's going to be double the voltage that you had here. So we've gone through a voltage doubling maneuver here. This double Tesla coil thing here has a variable blocking device on it, and it's grounded. Two bases of the Tesla coil are grounded, 185

The ground in this case it's inside the module here and it's coming back through one of these hookup wires, 193

At resonance you have a tremendous intensity of magnetic flux lines which will transfer beautifully over to another electrical system. That's what's happening between your L1 and your L2 coils. Once you get your L2 coil going it's going to be moving from the base of the coil towards the top of the coil. As the magnetic flux starts building on each turn there it's transferring over to the next turn and it's carrying with it what it had and it's multiplying by the flux lines that are being active there. So you're getting an amplification of the magnetic flux lines. These magnetic flux lines are building in intensity as they go towards the top of the coil and they reach their maximum at or near the top of the coil, 201

Whenever we talk about ambient and that sort of thing we get into a situation where as the frequency increases, the magnetic lines increase. It's those magnetic lines of force that disturb the electrons which cause more electricity become available to you. That's basically what's happening on your L2 coil as you start from the bottom of the L2 coil and go towards the top of it. Each winding there has capacitance and other types of things. Those things start squaring as you go up towards the top end of the L2 coil. The net result is that you have a tremendous magnetic flux field going on at the top of that coil there. That also presents you with some real interesting possibilities. Because that's the end that you really want to get your flux off of in order to go ahead to your next stage, 212

You put a timing device on that ground there so that these two coils are activated for 60 cycles, 222

Well remember now I said that on the grounding wire, that center one in the middle there, it has to go to an earth ground, and you have to put a timing on that earth ground which will pulse across the two coils that you see there, 223

I calculated the L1 coil which is on this sliding 2-inch PVC tube inside the 3-inch coil. It was calculated to be one-fourth the length of the wire that's involved in this L2 coil, or the outer coil. The reason for that is that coils have both capacitance inductance and resistance. All those things that play part in the tuning of the thing. But since you started out with about one quarter length of the longer coil, you're pretty much tuned. You're almost tuned immediately. As it turned out—of course, you have to put a capacitor on one of the two coils. The easier one to work with is the L1 coil over here, 239

In order to have useful energy, you can't go directly from this coil system and use it. You have to have some way of storing it. The storing options are basically either battery or capacitors. And it turns out capacitors are—we'll come back to that later. Capacitors are probably the best option, because they can be built to where they're self healing. They can be built to where they have over 40 years of useful life that's known, 239

An earth ground is where your amperage comes in, 240

The electrons are moving this way, and then the tail back here, that hooks the two together, it goes to an earth ground, is where your amperage comes in. So whenever you take your amps and volts apart and you start bringing them back together, you get volts times amps equal watts. So that's your useful energy, and you do this as you go into the capacitor bank, 241

Question. If the ground cable is removed from your suitcase will the unit still operate?] Yes. You notice the screws on the top here, that's where the thing is bolted in. It actually takes up just the top half of the suitcase. [Question. But if you pull the actual ground cable off that, that small block cable, will it still operate?] The ground cable is actually necessary for it to operate correctly, if you take it off, it starts building up heat and other things, 265

If you take a light emitting diode—electricity can only flow in one direction in a light emiting diode. Take two of them, and reverse them. Make a pigtail and put them in the line that comes from the battery to the device. Depending on which ones lighting up you know which way the electricity is flowing. When they're both lit up you're taking out and you're putting in at the same time. The reason you can do that is that they are different frequencies, 275

4. Waveshaping Conversion to Make Useful.

Piano frame amplifies sound, 2

Solar Cells convert Sun's magnetic flux, 3

Power correcting factor system, being a capacitor bank maintains an even flow of flux. These same capacitors, when inserted with a coil system (transformer) become a frequency-timing system. Therefore the inductance of the input side of the transformer, when combined with the capacitor bank, provides the required fluxing in activating the proper style of electrical energy (CPS), 4

Voltage - Amperage Limiting resistors required for adjusting output level and desired CPS, 6 Time constant (frequency) information in chapter one of Handbook of Electric Tables and Formulas, 7

Isolation transformer needs to operate at 60 CPS, that's 60 up and 60 down, 120 cycles, 7

Earth grounding is desirable as a voltage limiter and transient spike control. Two necessary, one at capacitor, other at input side of transformer, 8

Limiters, Voltage, Amperage, V-1, 8

V-1 limits the voltage as required for the Output Transformer L-2, 10

Correcting the output frequency with R-2 (120 total), 10

Controlling the flux rate controls the electron (potential) activity, 8

Correct frequency at primary of isolation transformer if needed, 8

Transformer comes alive (amperage) when pulsed, 8

System shunt in the resonate circuit sets the continment level for energy potential, 36

Once we're off out here then we have another wave-shaping manoeuvre that's going to occur. At this point what we want to do is round those spikes off so that we get what we're looking for. Out here is a normal sine wave thing. We need to delay the time a little bit there so that when it's up at the top there that it's going to spread out in this top area. When we do that we end up with this type of thing. Now we're at the wave form where we can do something. We've got a hundred volts and we've got the proper wave form. We're at the point now where we're going to set the frequency. There's going to be another maneuver at this point, 185-6

There's a capacitor that goes with the output transformer. It has a certain amount of induction, and there's a capacitor that you put with it that forms your resonant circuit, 186

You have to go through the wave shaping and set things up so that it matches the devices that you want to run on it, 193

Since the electrons are in an awkward position, they will try to get back to ambient. In the process of getting back to ambient, they're going to be doing the things that you want them to do, 241

Any of these things you get are you're gonna produce more power than the system can handle. What you do inside the circuit, you put these deals where circuit limiters, like you have on your computers and telephone to protect them from spikes. You put those those all through there, and you set the only amount that can come out of it, as what you want to see at the end. So most of its going to go out of the system, but you will end up with what you wanted at the end of the system, 290

High frequency changed to lower frequency requires going to DC first then to 60 CPS for instance. DC can be changed directly to any frequency. With AC the options are multiples or division of starting frequency. All components have both inductance and capacitance. This often screw's up the desired effect. Sometimes have to use trial and error to approximate the desired results. Both tend to increase as the length of wire and number of turns increase. Everything is resonate at some frequency, 357

I use separate modules for differing functions. Exiting from the L-2 the energy goes into a module which changes the High Frequency and High Voltage to pulsating DC which goes to the next module. This module takes the High Voltage DC and temporarily stores it in a super capacitor and special battery combination, which yields instantaneous power foreword to a heavy duty converter module which shapes and prepares the energy for end use requirements. Depending on the switching devices used this energy can be very substantial. In addition I add in other modules which use the energy radiated and normally wasted, such that feedback from this wasted energy results in self sustainable energy devices which require no external power once running, 362

5. Inverting Energy as Needed.

No electrons pass through from the input side of the isolation transformer to the output side. Therefore we only need to flux the transformer to have output, 4

We're going into a resonant iron core transformer system, and this is where we're going to end up with all the amps and other things that the electrical people would like to see maybe. Say that we go into an iron core transformer here, and we're working off of it. This device is going to—it's resonant transformer. It's the self resonant isolated transformer which means that the other side of it is not connected in any wires. The only connection is through the induction. So we're running it through this thing that's going to set the 60 cycles at this point. Then we're going to take the one which we're going to come off of, and here's where a load is going to be added. This is where the wall plug in the house is. So we have a magnetic flux field on this one, which is driving this one. And the self resonance of that particular transformer is 60 cycles per second. ou can either step it up or step it down at this point. Depending on what kind of output you want, if you want 440, or 6 whatever, or 220, or that sort of thing. If the two units are equal you will end up with equal amount of say a 100V. You see this has a certain amount of induction, and there's a capacitor that you put with it that forms your resonant circuit, 186

Well before you can put a load on them you have to put it into a capacitor bank and get your thing built up so that you have something to put a load against. If you try to put a load directly onto it it's going to detuning it and you'll end up with nothing, 218

Once you've taken measurements on number 11 [output transformer primary] there on the input side, you can calculate what number 10 [capacitor] will be, 250

That's why your capacitors—even though they don't appear to have a lot of electricity in them, based on the fact that say they're 8 µF or whatever—if you repeat that enough times in a second it becomes acumulative, and all of a sudden you have hundreds of thousands of volts that are available to you. They're available outside of the system from which you were to begin with. The amperage at that point depends on the degree at which you ground that capacitor to the earth. If you should touch it, and touch the earth ground at the same time, it will cook you on the spot. You don't really want it to unload at a very fast rate. So what you do, you choke it back with dividers, so that only a limited amount will pass through the load which you're actually wanting to do the work for you. At that point you have useful energy, 254

Question: "You say one farad one volt, then if we want to reach 120 volts, so what do we do with the amperage?"

ANS: Equal and opposite of electrical systems which the establishment insist upon tells You that magnetic and electrical are equal. Since magnetic and amperage are such that any current movement creates a magnetic field. Therefore from a practical stand point Voltage and Amperage (movement) are required for useful energy. Now when High Voltage exist it does not exist alone. Therefore in transformers when changing from high voltage to lower voltage, the necessary lowering of the voltage is compensated by an increase in amperage. In short in lowering from High Voltage to 120 Volts, a great deal of current movement which can temporarily be stored (dammed up for use) Capacitors store energy as Coulombs which is Joules per second or Watts which is Volts X Amperes. That's how You get the required amperage to go with the 120 volts, 362

Devices and Materials Mentioned

Plasma Globe emitting 400 gamma/40,000 volts, 3

Solar Cells convert Sun's magnetic flux, 3

Composite core transformers, correctly designed, run cool, 4

Plasma Globe Radio Shack's Illunna-Storm, 6

12V DC neon tube transformer oscillates at about 30k CPS, 8

High voltage sources: ignition coils, TV fly-back transformers, laser pringer modules, 8

Bertonee, NPS-12D8 Neon Tube transformer, 10

Kystron, 19

Magnetostrictive material, 80, 81

Terfenol-D, 80

Neon Tube Transformer, 85

The equivalent circuit of a diode is basically a capacitor with a resistor shunt on it, 208

How to identify a diode that works at high voltage, the length of the diode has to do with the voltage, and the diameter has to do with the amperage, 208

The dielectric constant for distilled water is about two thousand, 226

It's already beginning to behave as an over energy device [NST], 241

The thing is actually presently a solid-state device which is very small and does all the things that these things do but it's a solid-state device, 259

You want multi-stranded wire, which is basically the wire used in DC things normally. The more of those little strands are there, the better the deal. Because it gives a surface for the electrons to dance around on, and by having more of the strands, you have more circular mils. Which means that the total surface area is increased by quite a bit, by using the multi stranded wire. That works in your favor, so you won't use what is normally probably called DC multi-stranded wire. The larger the diameter the better, but the bigger ones are hard to work with 298

Terfenol-D rod, 315

Using Tools and Making Calculations

Gammer meter can read magnetic flux, 3

Units of Measure, 5

One milligauss of magnetic induction equal to 100 volts, 6

Reactance charts makes approximating the required three variables, capacitance, inductance and resistance, much easier. If two of the variables are known, the required missing one is obtained, 7

Timing formulas, T = RC, T= L/R, 10

Measure the electromagnetic radiation bleeding out from motors, transformers, power lines and any other electrical device and you will see much more energy is wasted than is ever used, 19

Remote flux reading requires the squaring of the distance x the remote reading, 24

$E=MC^2$ electrically: $E=CV^2$ and $E=LA^2$, 24

Electrostatic voltmeters and magnetic flux methods, can safely measure when based on the inverse square law, 24

Gauss (one unit = 100,000 volts), 29

Oersted (one unit = 50,000V), 29

Tesla (one unit = 10,000 Guass), 29

Gamma (one unit = 1/10,000 of a Gauss, 10V), 29

Volts and Ampere are equal (V.A.R.) until work (load) is introduced, 30

Volts = Energy Available (Potential), 35

Ohm = Scattering, dissipation of Energy (Load), (after the fact) 35

Ampere = the rate of dissipation/ scattering of energy, (after the fact) 35

V.A.R.: Energy Potential as Volts/ Dissipation (0) = Rate of Dissipation (0)

Fields, below radio frequency 20,000 Hertz per second, 40

Waves, radio frequency, above 20,000 Hertz per second, 40

E = M C squared electrically is E = (Volts X Amperes) X Cycles Per Second squared, 61

Electrostatic Volt meter, 65

Hall-Effect devices measures magnetic field, 65

Capacitor formula: $W = .5 X CE^2$ (W = energy in Joules (Watt Seconds), C = capacitance in farads, E = applied potential in volts squared)

Inductor (Coil) formula: $W = .5 X LI^2$ (W = energy in Joules (Watt Seconds), L = inductance in henrys, I = current in amperes squared), 71

Both one henrgy and one farad equal one volt, 71

Comparison of Establishment's Power Generation System with Resonate Induction at 60hz and at 1 million C.P.S., 74-74

Air Core Induction Coil Builder's Guide, 78

Voltage detector having a neon bulb, 79

L. C. R. meter, 79

Negative resistance Trumpet Wave, Transient response of an RLC circuit, 98

Enormously high ambient energy levels are not detected by instruments that use the ambient background as a reference plane, 147

Rate of induction is directly related to the cycles per second, 174

We're talking about Carl Gauss and Oersted, and units of magnetic flux bear their names. We have to relate those to other units of magnetic flux in order to see how this thing all ties together. These maps that you were looking at were made in gammas. They sometimes are in nano Teslas and sometimes other units. But the most common is say gamma. Nano Tesla and a gamma or equal in range. One gamma would be equivalent to ten active volts of electricity, or ten volt-amperes of electricity if you want put it that way. When you get into the other units, Gauss of course is 100,000 volts. it's a unit of measure of flux density and Oersted is 50,000 volts now 1 Tesla is equal to 10,000 Gauss, 177

Plus or minus: one is greater negativity than the other, 182

If any of you are interested in knowing all of the nice formulas that you need to know to understand this sort of thing, this kit comes with a nine page front and back instruction sheet, and it's got all the formulas related to resonance and how that can occur, 198

You have an equal amount of magnetic as well as electrical, so you can use a magnetic detector in order to read the voltage indirectly, 206

LCR meter showing resistors are also inductors and capacitors, 230

Whenever you take your amps and volts apart and you start bringing them back together, you get volts times amps equal watts, 241

Half the energy is in magnetic, and half of it is in electrical, 254

When you actually look into it you find that the earth has an enormous electrical charge on it. It's just that it's everywhere about you. It doesn't vary in a short distance. So all of the instruments that are made to measure electricity are made to measure a reference point and another point. So the electrical flux, when you start measuring electrical flux or magnetic flux, which is the other side of the coin, you find that the earth has an enormous electrical field, 271

That's like my little neon things that I used to check the circuit with. A lot of times I'll put them in different places, and I know up to that point, that everything is okay, 285

If you want to see that, the computer programs for schematics, if you get one that's with nonlinear system, which quite a few of them are, and a open system, which is quite a few of them are, you can actually do the schematics on these things. These programs have an oscilloscope, and all your meters and things are virtual ones built into it. You can build your device with schematics, and you could test it and see how it's working. Then you can play by switching out pieces, until you get the right combination, 286

With Gauss—Hans Christian Oertsed and Gauss were professors over in Europe, and quite some time ago they were trying to measure the electrical value of the Earth's electrical system. Off course one of them did it at night and one of them at day time. One of them got 50,000 volts, and this was an imperfect measurement they made because they did it by flying a kite or you know like a friend that likes to fly kites. The night time value is about half of what it is in the day time. This comes from the fact that in the Earth's rotation, when the earth is spinning, it's spinning into the cosmos so to speak. On that spin inside it has a greater electrical system then on the side that's away from it. That's why the two are different, 287

Benefits of Magnetic Resonance

Louder sound in piano, 2
Light Amplification, 3
Environmentally Friendly Energy, 12
Wars are unnecessary, 14
Provides Safe, Fully Renewable, Unending Source of Energy, 14
Super conductor conditions take over, 37
Non-poluting energy source, 39
Universally available energy source, 39
Makes possible healthy water, 55
Increases food supply, 55

Consequences of Under Unity Systems

Massive taxes, 12
Extinction of all air breathing animals due to lack of oxygen, 12
Disease mechanisms like germs, virus thrive in oxygen deficient situations, 12
Oxygen decline in Earth's Atmosphere, 13
World's Forests disappearance, 13
Nitrous Oxide pollution is produced and Ozone depletion allows infrared penetration causing temperatures to rise and health problems, 14
Enormous amounts of electromagnetic energy overflows (is wasted) and returns into the background unused by conventional methods, 19
A closed circuit world is a necessity for preserving the status quo and keeping the special interest, special, 22

Where Conventional Theory is Mistaken

Converting Earth's electrical field violates laws of nature, 2

Amperage definitions flawed, 5

Coulomb's and Newton's inverse square law is politely ignored, 24

A major obstruction in reference to the correct function of electrical energy is the establishment's incorrect interpretation of Ohm's Law, 35

60 Hertz's iron core transformer system is handcuffed by the inverse relationship of volts to amperes, 47

The dampning process of conventional electrical power generation has all the available electrons randomly bouncing mostly canceling out each other, 71

Enormously high ambient energy levels are not detected by instruments that use the ambient background as a reference plane, 147

Prescription Physics mandates that the Earth's background is of little interest, 148

In convential systems the equal magnetic side of the equation is radiated and wasted, 150

We also talked about the fact that I believe that mechanical advantage is not considered unusual when it comes to pulleys or levers or gears or things of that nature. But if you use a small amount of energy to extract a larger amount of work, regardless of what the books say, haven't you've got something back more than what you put into it? The physics books don't agree with what I just told you. Think about it, the physics books are wrong. The idea was set up back in 1670 or somewhere back like that, 179

Generators don't make electricity. By the coils and the magnets moving in relation to each other they upset the background balance. When that happens, well then you've got energy available, 280

The university people like things that you could do mathematically, and you always end up with the same answer regardless. That way they look good if they're competing with something that's not going to give the answers that they want to be repeated every time. Then from university standpoint they're in danger of needing another job. The University is not interested in people that can't repeat what they're doing, 294

Hindrances to Over Unity

Gray matter, 2

Humans are like the bird sitting unaware on a high voltage line, 2

US Patent office insists that Earth's electrical field is insignificant, and converting this energy violates laws of nature, 2

Bad design by the establishment, allowing hysteresis of the transformer metal plates limits the load, 4

Wind and solar energy depnds on the fancy of nature, 12

Conspiracy benefactors are Industrial, Governmental Agencies, University R&D Programs and Military Complexes, 12

Why Replicators Fail

I think what happens is people change the lengths of the wires, or something is a little bit changed there. That is problematic because depending on that power source there some of them are self-grounded. That's where part of the confusion may come in, because sometimes when you hook that thing up if you're not careful, everything you got, goes there. That's the end of it, so nothing happens out here. As a result of that it didn't work and I was wrong, 290

The problem with that is, that everyone when they buy their wire, and other things, those things are not going to match with what's on the list. All the things I do are generalities, and if you get one of those electronic schematic deals you can fiddle with it all you want to. You might get better results then what I would give you, 290

"Your understanding of electricity is limited to 60 cycle understanding. High frequencies are a different world. See Tesla's Book "My Inventions, Researches and Writings". Any and all of My Devices Work exactly as I state. I can not be responsible the shot comings of others. Regards, Don Smith"

BIBLIOGRAPHY

The following list of books on this first page were referred to by Don in this book. I was able to find pdf's of these. The next seven pages are part of the original book in different places.

Canadian Patent application of the device the MEG copied, CA2172240A
https://patents.google.com/patent/CA2172240A1/en
http://www.tuks.nl/pdf/Reference_Material/Magnetic_Flux_Switching/CA2172240A1.pdf

Canadian Patent application of the Frank Wyatt Prentice device, CA253765A
https://vdocuments.site/frank-wyatt-prentice-patent-ca253765.html

Drake's Radio Cyclopedia, 1928
https://www.americanradiohistory.com/Archive-Early-Radio/Call-Book-1926-Amateur.pdf

Encyclopedia of Electronic Circuits. Vol 3. 9780830673483 1990
https://ia800304.us.archive.org/28/items/E.E.C.Volume3ByRmp/E.E.C.Vol3ByRMP.pdf

Handbook of Electric Tables and Formulas, ISBN 0-672-22469-0

Physics for Scientists and Engineers. By Raymond A. Serway
http://jsnegroup.net/data/GENERALPHYSICS(2)-Book_9thedition-Bookphy1019th-civilianteam.pdf

The Sea of Energy in Which the Earth Floats, John E Moray
http://www.naturalphilosophy.org/pdf/ebooks/Moray-TheSeaofEnergy.pdf

Vacuum Tubes, by Karl R. Spangenberg. 1948
www.tubebooks.org/Books/Spangenberg_vacuum_tubes.pdf

Additional Related Resources

Renaissance Resonance Induction Coupler Kit, Rick Friedrich ISBN 1-932370-84-6
http://www.r-charge.net/product.asp?itemid=305

Rick Friedrich YouTube channel hosting the Don Smith videos transcribed in this book.
https://www.youtube.com/user/rickfriedrich

The YouTube Playlist of Don's videos in chronological order.
https://tinyurl.com/y7wjs44t

Earthing: the most important health discovery ever! Clinton Ober, et al. 2014. ISBN 978-1-59120-374-2. A book that appears to show the important biological side of what we have revealed in this book about the electron cycle loop with Earth Grounding.

Reading List

1. Alston, L.L. (Ed.), 1968, High-Voltage Technology, Oxford Univ. Press, London.
2. Beck, E. 1954, Lightning Protection for Electric Systems, McGraw-Hill, N.Y.
3. Bowdler, G.W. 1973, Measurements in High-voltage Test Circuits Pergamom, Oxford. L.C. 72-86488
4. Brewley, L.V., 1951, Traveling Waves in Transmission Lines Pub: John Wiley, N.Y.
5. Chapman, S. and Bartels, J., 1940, Geomagnetism, two volumes, 1,000 plus pages, Oxford at the Clarendon press, England
6. Craggs, J.D. and Meek, J.M. 1954, High-Voltage Laboratory Technique., Butterworth, London
7. EHV Transmission Line Reference Book, 1968 Edison Electric Institute, N.Y.
8. Farno, R.M., Chu, L.J., and Adler, R.B., 1968, Electromagnetic Fields, Energy and Forces, Pub: M.I.T Press, Cambridge, Mass.
9. Feinberg, R. (Ed.) 1979, Modern Power Transformer Practice, McMillan, London
10. Frungel, F., High Speed Pulse Technology, Vols. 1 and 2, Pub: Academic Press, 1965, N.Y., and London.
11. Gallagher, T.J. and Pearmain, A.J. 1983, ISBN 0-471-90096-6 High Voltage Measurement and Design, John Wiley, N.Y.
12. Hague, B., Alternating-Current Bridge Methods, 5 th. ed. Pub: Sir Iasaac Pitman and Sons, 1959, London.
13. Hawley, W.G. 1959, Impulse-voltage Testing, Chapman and Hill, London.
14. Hayashi, Ch., Nonelinear Oscillations in Physical Systems, Pub: McGraw-Hill, 1964, N.Y.
15. Henny, K. 1933-1959, Editor-in-Chief, Radio Engineering Handbook, five different copyrights, L.C. 58-11174, McGraw Hill, N.Y.
16. Hudlestone, R.H. and Leonard, S.L., Plasma Diagnostics Techniques, Pub: Academic Press, N.Y.
17. Jacobs, J.A., Editor, Geomagnetism, (a massive work) 3 large volumes, Pub: Academic Press, London.
18. Jeans, J.H. 1925, The Mathmatics of Electricity and Magnetism, 5 th. Ed., Chambridge University Press,
19. Jones, B. 1972, New Approaches to the Design and Economics of EHV Transmission Plant, Pergamom, London.
20. Kind, D. 1978, An Introduction to High-voltage Experimental Technique, Viewieg, Braunschweig ISBN 3-528-08383-2
21. Knoepfel, H. 1970, Pulsed High Magnetic Fields, North-Holland, Amsterdam.
22. Kreuger, F.H. 1964, Discharge Detection in High-Voltage Equipment, Temple Press, Heywood, London.
23. Kuffel, E. and Zaengl, W.S. 1984, High-Voltage Engineering, Pergamom, ;London. ISBN 0-08-024213-8
24. Kupfmuller, K. 1957, Introduction to the Scientific Basis of Electrical Engineering, Pub: Spinger, Berlin.

25. Lemon, H.B. and Ference, M. Jr., Analytical Experimental Physics " a major work " from The Ryerson Physical Laboratory at the University of Chicago, numerous copyrights 1933-1944. Printed as a text book by the University of Chicago Press
26. Lewis, I.A.D., and Well, F.H., Millimicrosecond Pulse Techniques Pub: Pergamom Press, N.Y. and London.
27. Malan, D.J. 1963, Physics of Lightning, English University Press, London.
28. Martin, T.L. Jr., Physical Basis for Electrical Engineering, Prentice-Hall, N.J.
29. Matsusta, S. and Campbell, W.H., 1967 Physics of Geomagnetic Phenomena, Massive work in two volumes of 700 pages each, Academic Press, London.
30. National Physical Laboratory, 1956, Notes on Applied Science # 17, High Voltage Impulse Testing, HMSO.
31. Rather, H., 1961, The Electron Avalanche and it's Generation, Pub: Vo. 33, Springer, Berlin.
32. Rokityansky, I.I., 1982, Geoelectromagnetic Investigation of the Earth's Crust and Mantle, Pub: Springer-Verlag, Berlin.
33. Salge, J., Peier, D., Brilka, R., Schneider, D., 1970, Applications of Inductive Energy Storage for the Production of Intense Magnetic Fields, Pub: Proceedure of 6 th. Symp. on Fusion Technology, Aachen.
34. Schulz, E.H., Anderson, L.T., and Leger, R.M., Experiments in Electronics and Communication Engineering, Copyrights, various, from 1943 thru 1954 LC 55-5033 Harper and Brothers, N.Y.
35. Schwab, A.J. 1972, High-Voltage Measurement Techniques, The M.I.T. Press, Cambridge, Mass., ISBN 0262-19096-6
36. Sevin, L., 1965, Field Effect Transistors, McGraw-Hill, London.
37. Silsbee, F.B., 1942, Static Electricity, U.S. Department of Commerce, Washington, D.C., U.S. Government Printing Office, (N.B.S. Circular C 438)
38. Smith, A.A., 1977, Coupling of External Electromagnetic Fields to Transmission Lines, ISBN 0-471-01995-X
39. Terman, F.E., Electronics and Radio Engineering, 4 th. ed., Copyrights 1932, 1937, 1947 and 1955 LC 55-6174 McGraw-Hill Book Co., N.Y.
40. Thomas, R.T., 1970, High-Impulse Current and Voltage Measurement, Trans. I.E.E.E. IM-19, pages 102-107
41. Thompson, Professor W., (titled Lord Kelvin) Reprint of Papers on Electrostatics and Magnetism (original 1872) London.
42. Traister, R., 1983, The Experimenter's Guide to Solid State Diodes, Prentice-Hall, Inc. ISBN 0-13-295444
43. Valley, G.E., and Wallman, H., 1948, Vacuum Tube Amplifiers, McGraw-Hill, N.Y.
44. Weedy, B.M. 1980, Underground Transmission of Electrical Power, Wiley, London.
45. Weeks, W.L. 1981, Transmission and Distrubution of Electrical Energy, Harper and Row, N.Y.
46. Winch, R.P. Electricity and Magnetism (several printings) Composed in Copenhagen, Denmark, Prentice-Hall Physics Series
47. Zijlstra, H., 1957, Experimental Methods in Magnetism, two volumes, North-Holland Publishing Co., Amsterdam.

Additional Reading:

Electricity and Magnetism
By B.I. Bleany and B. Bleany
Oxford University Press 1991
I.S.B.N. 0-19-851172-8

Engineering Electromagnetics
By W. H. Hayt, Jr.
McGraw-Hill 1989
I.S.B.N. 0-07-027406-1

Energy Methods in Electromagnetism
By P. Hammond
Oxford University Press 1986
I.S.B.N. 0-19-859368-6

Energy in Electromagnetism
By H. G. Booker
Institution of Electrical Engineers
by Peter Peregrinus Ltd. 1982
I.S.B.N. 0-906048-59-1

The American Radio Relay League Handbook for 1992 and 1993. 69th and 70th editions.
Published by The American Radio Relay League. (For V.A.R. information)
I.S.B.N. 0-87259-169-7

Electron Paramagnetic Resonance, Techniques & Applications
By R. S. Alger, U. S. Naval Radiological Defense Laboratory, San Francisco, California.
Pub. Interscience Lib. Congress #67-20255

Geomagnetic Diagnosis of the Magnetosphere
By A. Nishida, University of Tokyo 1978
Pub: Springer-Verlag
I.S.B.N. 0-387-08297-2

Energy and The Missing Resource
By I. Dostrovsky
Pub: Cambridge University Press 1988
I.S.B.N. 0-521-26592-4

High Voltage Measurement Techniques
By A. Schwab, M.I.T., 1971
I.S.B.N. 0-262-19096

Continued:

Environmental Magnetism
By R. Thompson & F. Oldfield
Pub: Allen & Unwin, London 1986
I.S.B.N. 0-04-538003-1

Geoelectromagnetic Investigation of the Earth's Crust and Mantle.
Translated from Russian, By I. I. Rokityansky,
Institute of Geophysics, Kiev, U.S.S.R.
Pub: Springer-Verlag 1982.
I.S.B.N. 3-540-10630-8

Electron Paramagnetic Resonace of Transition Ions
By A. Abragam and B. Bleaney
Dover Publications, New York, N.Y. 1986

The Electromagnetic Field
By A. Shadowitz
Dover Publications, New York, N.Y. Date?

Geomagnetism, Several Volumes, Pub. Periodically
By J.A. Jacobs, Institute of Earth Studies, Dyfed, U.K.
Pub: Academic Press 1989-1990's.

Geomagnetism
By S. Chapman and J. Bartels, 3 Volumes
Oxford University Press, 1940

Physics of Geomagnetic Phenomena, Several Volumes
By S. Matsushita and W. H. Campbell
National Center for Atmospheric Research, Boulder, Colorado
Library of Congress #67-23168
Pub: Academic Press, 1967

Continued:

Physics Problems and How to Solve Them
By C. E. Bennett, Professor Emneritus of Physics,
University of Main (Particularly the sections on Electricity and Magnetism, and Units of Measure).
Pub: Harper & Row Date?
I.S.B.N. 0-06-460203-6

Units and Standards for Electromagnetics
By P. Vigoureux, National Physical Laboratory
Pub: Springer-Verlag 1971
I.S.B.N. 0-387-91077-8

Surveyor's Guide to Electromagnetic Distance Measurement, Edited by J. J. Saastamoinen, Canada
Pub: University of Toronto Press Date?

Electromagnetic Distance Measurement
By C. D. Burnside
Pub: Granada, London 1971
I.S.B.N. 0-258-96793-5

Der Magnetische Kreis "The Magnetic Circuit"
By Von Heinz Rieger of Siemens AG. 1970
Berlin and Munchen, Germany
I.S.B.N. 3-8009-4719-6

Electronic Modeling of Power Electronic Converters
By J. A. Ferreira
Pub: Kluwer Academic 1989
33 AH Dordrecht, The Netherlands
I.S.B.N. 0-7923-9034-2

Booker, H.G., " Energy in Electromagnetism ", Institute of Electrical Engineers, Peter Peregrinus, Ltd., 1982, I.S.B.N. 0-906048-59-1

Bleany and Bleany, " Electricity and Magnetism ", Oxford University Press, 1991, I.S.B.N. 0-19-851172-8

Chapman and Bartels, " Geomagnetism ", 3 vol., Oxford University Press, 1940

Hammond, P., " Energy Methods in Electromagnetism ", Oxford University Press, 1986, I.S.B.N. 0-19-859368-6

Matsushita and Campbell, " Physics of Geomagneic Phenomena ", several vols., National Center for Atmospheric Research, Boulder, Colorado, Academic press, 1967

Nashida, A., " Geomagnetic Diagnosis of the Magnetosphere ", University of Tokyo, Springer-Verlag, 1978, I.S.B.N. 0-387-08297-2

Rieger, Von Heinz., " Der Magnetisch Kreis ", Siemens A.G., Berlin and Munchen, Germany, I.S.B.N. 3-8009-4719-6

Rokityansky, I.I., " Geoelectrical investigation of the Earth's Crust and Mantel ", Institute of Geophysics, Kiev, U.S.S.R., Springer-Verlag, 1982, I.S.B.N. 3-540-10630-8

Vigoureux, P., " Units and Standards for Electromagnetics ", National Physical Laboratory, 1971, Springer-Verlag, I.S.B.N. 0-387-91077-8

Finnell, Woosley, " Solar Power Satellite Microwave Transmission and Receiver System. Energy Conversion Conference, Sept. 1981 pp 266-271

Glaser, " Satellite Solar Power Station " The Journal of Solar Energy and Technology, Vol. 12, No. 3., p. 353 .

Denmum et al, "A Microwave Power Transmission System for Space Satellite Power", Energy Conversion Conference Conference, Sept. 1977, pp 162-168

Nalos et al, "Microwave Power Beaming for long range energy transfer" "Proceedings of the 8 th European Microwave Conference" pp 573-578, 4 through 8 th. Sept., 1978

Angrist, S.W., " Direct Energy Conversion ", forth edition, Carnegie-Mellon University, Pub. Allyn and Bacon, Boston, London, Sidney and Toronto, ISBN 0-205-07758-7

Smith, D.L., " An Answer to Americas Energy Defict " , fifth edition, Pub. International Tesla Society, Colorado Springs, Co., 1996

Aspden, H. " The Law of Electrodynamics ", J. Franklin Inst., 287:179, 1969.

Sethian, J.D., " Anomalous Electron-Ion Energy Transfer ", Phys. Rev. Letters, vol. 40, No. 7, pp. 451-454 , 1978.

Westinghouse R. & D., "Electromagnetic Spectrum Chart"., Pub. The Exploratorium, San Francisco, CA 94123, Distributed by Edmond Scientific, Barrington, N.J. 06007 Order # 609-573-6250

Index Glossary of key words, ideas, and people by Don Smith

A

Absorbed wave energy produces heat, 147

Abundant Energy Everywhere Present, User's Guide, 76

AC current, electrons do not travel, 47

Aeromagnetic flux (in gammas) Maps-World Wide, 2, 177, 272

Aerial Magnetometer, entire surface of Earth surveyed by, 29

Aerial Magnetometer Map GP-948

https://crustal.usgs.gov/geophysics/North_America.html 33, 48

Air Core Induction Coil Builder's Guide, 78

Air Coil Resonate Induction System, 36

Air electricity background varies with latitude, 29

Alternating Impact Energy Transfer Analogy, 50

Ambient, 2, 210, 280

Background of Earth is multi billions of volts, 59, 92

... *background*, which where you're sitting is perhaps two hundred thousand volts per square meter, 197

Comes from reflection, deflection and absorption as a result of action at a distance from the above, 147

Disturbed when electron doublets spun or pushed apart, 5

Drops to about one half the daytime value at night, 24

Energy Generator, 76

AEG - DLS (Ambient Energy Generator Donald Lee Smith), 77

Energy greater than is Free Energy, 18

Energy Sources, 59

Is earth ground, 18

Amber, was used to differentiate the non-metals, 27

America's Energy Deficit, An Answer to, 26

Ampere = the rate of dissipation/ scattering of energy, (after the fact) 35, 182, 187

Ohms is load or wasted energy and amperes is the rate of wasting, 71

Antenna, 82

Coils are, 183

Grounding into Earth's electrical system, 289

Kites, 14

Sonar, 80

Antigravity, 351

Atomic batteries used, 11

Atomic Energy an expensive throwback to age of steam engines, 12

Avalanche of electrons, 15

Avalanche devices include dynode bridges, multipactors, masers and lasers, 15

Physics of, 22

Avalanche-regeneration, 19

B

B&W, Barker & Williamson makes many coils Don used.

Back E.M.F not present in Motor-Generator, 16

Barium Titinate Transformer, 163

Barker and Williamson coil manufacturer, 79, 199

Battery that would never run down, 189

Battery self-charging, If the plus connector wire's length is a wave length component of the L-6 or L-7's frequency the battery will recharge itself at the same time it is being used. To see this substitute two opposing LEDs. They show the direction the electrons are moving. DC electrons out and radio frequency electrons into the battery, 69, 188-89

Beam Energy System, 364

Benefits of the Inventor's System, 75

Bloch Wall, 252

Generator, Neutral Generator, 103

Interruptible allows for a smaller motivating force to produce a larger output resulting in a Self-sustainable

Power Source, 16

(*Space*), area of Electron Spin Separation, 63

Bollee, had successful electric automobiles using AC motors, 61

Building an Operational Unit, 6

Bertonee, NPS-12D8 Neon Tube transformer, 10

Brownian movement, Maxwell, Boltzmann and Clausius "the molecules of a gas carry out ceaseless transitional movement in all temperatures above absolute zero." 19

C

Capacitor is an Energy Pump, 77

Carbon Dioxide and Oxygen Cycles, 13

Carbon-related substances give up negative electrical charges, 27

Casimir effect, 330

Center tap, without that, you wouldn't have the amperage available to you, 240

Charged Barrier Technology, 358

Chemistry Vs Physics, 18

Clausius, 19

Closed circuit world is a necessity for preserving the status quo and keeping the special interest, special, 22

Cobalt Nickel Core Toroid Power System for Brazil, Device 6, 157

Cockroft-Walton Device, 164

Cohere of Tesla, precursor of diode and solid state devices helped radio, 15

Coil made of copper, and dipped in stannous chloride, which tinplates them, 199

Coil winding, 6

Coils,

Not matching coil frequencies, Once you get the magnetic flux going, you can have coils that are not tuned to it, that are in that magnetic flux, and they will oscillate at the resonant frequency that they are meant to resonate at. So you can go from one frequency to another and still have your electricity there because actually the electrons are not going through your system, they're just being disturbed by the magnetic flux lines of another electrical system, 234

Paired, One is wound one is wound clockwise and one is wound counterclockwise, 187

Commercial Prototype 45kw, 166

Comparison of Establishment's Power Generation System with Resonate Induction at 60hz and at 1 million C.P.S., 74-74

Computer design program method, 238, 286, 293, 296, 344-346, 353

Conspiracy energy benefactors are Industrial, Governmental Agencies, University R&D Programs and Military Complexes, 12

Containment level, A system shunt in the resonate circuit sets the containment level for energy potential, 36

Conventional under unity method,

In place of the mechanical - coils and magnet system of conventional power generation, visible moving parts are replace by Resonate magnetic induction using radio frequency, 56

Opposite to Resonate system where balls moving in same direction and time, 38

Represented by a room full of ping pong balls randomly bouncing at a high speed, 38

Corona effect, 185

Cosmic magnetic waves, 58, 352, 367

Cosmic, radiation as Ultraviolet magnetic wave, 59

Coulomb, Charles (1736-1806) with Isaac Newton (1642-1727) provide us with the square laws which are based on an open circuit, 23

C.P.S., Cycles Per Second, required formula, 7

Increases in Resonate induction systems and invokes a second round for the law of squares, 24

Crystal Radio converted to Free Energy device, 353

Curies, had high frequency coil systems before Tesla, 61

Current results from the unequal distribution of negativity (electrons), 39

I = Current = Amperes = Right Hand Spin. V = Electrons = Volts = Left Hand Spin, 64

Cycling, The rate of ambient disturbance determines the energy level achieved, 5

Cyclones, 17

D

Δ G or Free Energy Exothermic, additional energy not previously available. These changes occur at elemental and or molecular level. At particle size, as in Electrons there is an overlap into Physics, 17

Dark Ages, 22

D'Arsonval, 242

had high frequency coil systems before Tesla, 61

Professor of Experimental Medicine at the College of France, invented the electrocardiograph, oscilloscope, am and volt meters, thermography, etc., 61

Daylight caused by electron enrichment of upper limit of Earth's Atmosphere in the radiation belt (shear zones), 15 Disease mechanisms like germs, virus thrive in oxygen deficient situations, 12, 34, 210

Denis Lee, 342

Derivation of Magnetic and Electrical Power, 63

Diamagnetic, water is, 92

Dimmer, for Neon Tube Transformers, 237, 239, 240

Dipole Transformer Generator Description, Antenna, 83, 88

Dirigible France, electrically powered airships were in service, 61

Diode, The Electrical Circuit equivalent of the Bloch Wall. Electron Doublets once separated must recouple, resulting in useful energy, 64

Bridge, after L2 coil, 184

The equivalent circuit of a diode is basically *a capacitor with a resistor shunt on it*, 208

Domestic Use Range Module, 51

Don Smith Device-2006 with Ignition Coil and fluxed Metglas transformer, 172

DSE (Don Smith Effect), 77

Dual Field Generator, 103

Dumont, had high frequency coil systems before Tesla, 61

Dynode, 273

Bridge used to adjust the self-powered shutter in your camera receives magnetic flux from light which dislodges electrons from the cathode, reflecting electrons through the dynode bridge to the anode, resulting in billions of more electrons out than in, 3

Avalanche device, 15

Dipole type field, 285

E

$E=MC^2$ electrically: $E=CV^2$ and $E=LA^2$, E = (V X A) X Cycles Per Second squared, 24, 61

Earth's Atmosphere and surface has massive amounts of Non Ionic (Free) Electrons, 9

Earth's electrical field, 2, 3, 59

Difference between night and day, 287

Earth Electrical System II, use of naturally occurring electrical differences in the Earth's Crust and Radio Device in combination for Transmitting of the Energy to Points of Use, 16, 38, 39, 44

Earth grounding is desirable as a voltage limiter and transient spike control, 8

Two necessary, one at capacitor, other at input side of transformer, 8

Earth Rotary movement, 57

Earth Mass, 57

Edison controlled information 20

Edison meeting with J.P. Morgan and Tesla, 60

Efficiency of Induction relates to the square of the cycles per second, 53

EES II, 180

Einstein, Albert (being an insider) took credit for Oliver Heavyside's (1850-1925) work in providing us with Energy equals mass times the square of it's acceleration, 23, 253

Electric fence developing power from distant electrical lines, 288

Electrical Energy diagram, 41

Electrical Energy Generating System, 56

Electrical Energy System, 60

Electrical Energy Systems Methods, 58

Electrical Potential relates to mass and it's accelleration, 2

Electrical Power Generation / Points of Reference, 73

Electrical Power Systems have only the magnetic flux move from point A to point B, 3, 15

Electromagnetic disturbance, 80

Electromagnetic flux, 2

Electromagnetic spectrum, 2

Electromagnetic Spectrum Energy has four physical expressions: radiated, reflected, deflected and absorbed, 146, 149

Electromagnetic stimulation disturbs electrons by avalanche, dislodging adjacent electrons, resulting in more energy out than in, 21

Electromagnetic waves incoming from point "A", the Sun, reacting with point "B", a solar cell which spins the local electrongs sideways, yeilding electricity, 21

Electron (potential) activity, controlling the flux rate controls the, 8

Electron doublets, Since magnetic and amperage are one package this suggests that electrons in a nutural non-ionic state exist as doublets. When pushed apart by agitation one spins and supplies electricity and the other spins and provides magnetic (amperage) energy. When they reunite, we have Volts X Amperage = Watts, 60

Electron donors, 70% of the Earth's crustal surface silicone related non-metals, source of electrical energy, 27

Electron spins

- *Clockwise* producing magnetic impulse, 42
- *Counter-clockwise* back to original position producing electrical impulse, 42

Electrons: 27

- *Avalanche* devices include dynode bridges, multipactors, masers and lasers, 15
- *Any coil system* when fluxed causes electrons to spin and produce useful energy, once converted to the style required, 4
- *Any input system* wherein a lessor amount of electrons disturbs a larger amount of electrons, resulting in the output being greater than the input is desirable, 4
- *Flash of light* occurs when two electrons too close, 34
- *Functions*, become active when placed inside the critical distance allowed by their negativity. Active electrons provide: Electricity, Magnetics, Gravitational thrust, The source of Visiable Light, and it's charge is Negaive, 34
- *Impacted* emit visiable Light and heat, 5
- *Pairing demonstration*, 319
- *The electron as a particle* was postulated by professor J. Thompson in early 1900's, 60
- *Thompson's, JJ, description of the electron originally.* This business of having electrons coupling and being disturbed and pulled apart. And they have to fight to go back together, and when they come back together, you have volts times amps equal watts. That is an extension of JJ Thompson's description of the electron originally. It's expanding onto it, and there's some other things there, 265
- *Move in a closed infinity symbol loop*, half of loop is a magnetic impulse, and return half electrical impulse like sine wave, 34
- *Never deminish*, 5, 14
- *No electrons* pass through from the input side of the isolation transformer to the output side, 4
- *Non-Ionic electron doublets* in vast quantities in universe, 5, 15
- *Origin* is from the emanation of Solar Plasma, 5
- *Radio glass tubes*, electromagnetic stimulation dislodges electrons which avalanche in dislodging many-many more electrons. The excess of electrons at the

Cathode magnify and react with the electrons present at the anode, resulting in more energy out than input, 15

Rate of disturbance (cycling) determines the energy level achieved, 5

Related energy, 5

Remain forever intact for further fluxing, 3

Resonate induction disturbs a large number of adjacent electrons which were not a part of the original source. The Pulsating (pumping effect) then incorporates the newly available extra electrons into the ongoing energy generation source-system. A near unity energy system of reonate air core coils and extra aquired electron-energy source constitutes an over unity system, 66

Travel: In Electrical Systems, Electrons active at point "A" are not the same Electrons active at point "B". The Electons activated at the Electrical Station are not the ones used at your house. When you ground your system by flipping the wall switch, you use your own electrons. In closed energy systems, electrons communicate with and replicate the activity of the overbalanced potential, when provided with Earth and or Air Groundings, 36

Two types of electrons. There's one that's a little bit smaller and weaker then the other. When it spins its direction it becomes the amperage part. The stronger of the two electrons spins and makes the voltage part. In nature they are coupled together and that's the reason why they're inactive and floating about in space without anything happening. It's whenever you disturb them or knock them apart that they try to get back together and that's where you get your useful energy, 252

When disturbed, first produce magnetic flux and then electrical flux when they spin back to their normal position. Therefore, any electron movement produces above ambient energy, being over unity, 24

When spun or pushed apart ambient is disturbed, and yield magnetic and electrical energy. One spins left, yielding (Volts-potential) electricity and other spins right, yielding (amperage-magnetic) energy. One being more negative than the other, 5, 39

Energy Acquired by Magnetic and Electrical Impulsing, Alternating Resonate Electrical Power), 42

Electrostatic voltmeters and magnetic flux methods, 24

EMI measuring instruments, 206

Energy lines of force, 40

Energy shortage, 2

Energy Pump, Capacitor is, 77

Energy Verses Mass, 33

Environmentally Friendly Energy Sources, 12

Equilibrium, when Electrons are disturbed by turbulence-movement of any sort they become active and try to regain equilibrium. This movement towards equilibrium results in magnetic displacement, which causes the motor to rotate. Forward thrust of Electrons results in Magnetic Energy and backwards thrust results in Electrical Energy, 54

Ether, Sea of Continuum, governed by the Law of Squares, 5

Exothermic, additional energy not previously available. This is referred to in all chemistry Texts as Free Energy or (Δ G). These changes occur at elemental and or molecular level. At particle size, as in Electrons there is an overlap into Physics, 17

Explosives, free energy, 19

F

Faraday's Action at a Distance, 58

Farads and henrys of the Resonate system provide the Resonate frequency when pulsed by and external energy system, 36

Farnsworth built upon coher and radio tube, 15, 273

Fibonacci, Leonardo (1170-1230) of Pisa demonstrated very

clearly that, which directly translates from laws of nature, an orderly form of progression, 23

Fibonacci series, 201

Fields, below radio frequency 20,000 Hertz per second, 40

Fitch Impulse Generators, 131

Flux lines, Result both from Induction-henrys-amperage and capacitance-coulombs-volts and define electrical energy, 24

The amount of Energy transferred relates directly to the density of the lines of flux present. Each unit of increase causes a squaring of the flux lines present. One volt, Two Volts = Four, Four Volts = Sixteen. In resonate air core coil energy transfer the increase in flux lines present disturbes more electrons than previous, resulting in over unity energy being present and available, 70

Each turn of L-2 is exposed to the electric flux of the lowered voltage of L-1. Each turn at the bottom end of L-2 acquires the same volts. The flux lines are squared and are additive as the voltage progresses towards the top end of L-2's many turns. A huge amount of flux lines not previously present occur at the top end of L-2. These flux lines excite the electrons nearby in it's earth and air and groundings. This high level of excitement above the abmient causes a large amount of electrons not previously a part of the energy present to become available. At this point overunity is present in large amounts, 73

Flyback transformer, 48

Flying vehicles, 351, 377, 379

Fogal chip/transistor, 358

Forces of nature to operate: Composed essentially of a weak force and a strong force. Each of these forms of energy occur as fields and waves depending on what the resonant frequency is at a particular level, 178

"*Free energy*," has come to mean "something which was never there is being harvested and that this violates the laws of physics," 60

Free Energy or (Δ G) Exothermic, additional energy not previously available. These changes occur at elemental and or molecular level. At particle size, as in Electrons there is an overlap into Physics, 17

Greater than Ambient, 18

Harvesting Requirements, 1. Electron acceleration or deceleration must vary from ambient. 2. Spin rate CPS fluxing provides electrical with magnetic energy. 3. At Resonance (negative resistance), all electrons move in the same direction without interfering with each other. 4. Electrons must Avalanche, pulsing magnetic waves provide three dimensional and nonlinear energy at each cycle, electrons presently available disturb and bring alone many more electrons, resulting in gain, 19

If the value received is greater than the cost, 19

Forests disappearance, 13

Frankilin, Benjamin, demonstrates ectrical flow, 28

French Patent Office, 20

Frequency-timing system, 4

Frequency

Adjust deal, 229

Information, 7

Of Earth with Van Allen radiation belts between 2000 and 8000, 235

Output, determined by the length of L1 wire at the beginning stage of the system. The highest frequency that you have represented in any of the components is going to be the dominant frequency. 182

Not matching coil frequencies, Once you get the magnetic flux going, you can have coils that are not tuned to it, that are in that magnetic flux, and they will oscillate at the resonant frequency that they are meant to resonate at. So you can go from one frequency to another and still have your electricity there because actually the electrons are not going through your system, they're just being disturbed by the magnetic flux lines of another electrical system, 234

Resonate frequency, The farads

and henrys of the Resonate system provide the Resonate frequency when pulsed by and external energy system, 36

G

Galactic Magnetic waves, 58

Galactic, radiation as Infrared magnetic wave, 59

Gamma (one unit = 1/10,000 of a Gauss, 10 volts), 29

Gamma is ordinary every day magnetic flux, 3

One gamma of magnetic flux is equal to that of 100V RMS

Gamma Ray is high impact energy, not flux, 3

Gaseous Tube Electrical Power System, Tesla Coil, Plasma Tube, and Laser Systems, 17

Gasoline gallon combines 200 pounds of Oxygen, 14

Gauss (one unit = 100,000 volts), 29, 358

Gauss, C.F. (1777-1855) define Earth's electrical field, measuring magnetic field flux as an indicator of negative electron energy active, 29, 175, 177

Generator, 4kw, 168

Gold plates in with distilled water to prevent chemical reaction, 227

Gravitational effect,

Caused by Magnetic imbalance. This is evidenced in electric motors by magnet-gravitational displacement of mass which causes the motor to rotate, 39

Spinning mass in the presence of incoming magnetic flux amplifies the gravitational effect, 148

Magnetic lines of force, 40

Gravity, is a component of the movement of energy into the earth. It's a friction factor, related to movement. When some motor is spinning, or the magnets are pushing each other away from each other, that's an electrical reaction there, and that effect between them is the gravitational field, either pushing them apart, or pulling them together, depending. 28, 41, 299

Gravity is the Right Angle component of the Dipole of the Earth's rate of Spin and Mass, which increases with it's depth into the Earth, 384

Accretion masses (Energy Sinks) provides all solid entities with their respective gravitational fields, 147

Flux movement into energy sinks provides the frictional force known as gravity, 148

Frictional drag from inflow current defines gravity, 147

Nullification of, 297

Grounding, is a relative sort of thing. It's that there's more of something here, and less of something here, so it's going to move between the more to the less of something, 290

Air, 175, 321

Amperage comes in, 241

In order to turn this into useful energy it's the way in which you ground it will determine whether it becomes a practical source of useful energy, 312

Into Earth's non-metal crust, 27, 38, 176, 217

Inside the module, and it's coming back through one of these hookup wires, 193

Timing, 222,

which will pulse across the two coils, 223

Shunt, with a negative resistance wave, 175

H

Ham radio operators and people like that were the ones that were successful in replicating 1994 presented system, 231

Harmonic electron activity, 47

Heaveyside, Oliver (1850-1925) provided us with Energy equals mass times the square of it's acceleration. Albert Einstein (being an insider) took credit for *Heavyside's work*, 23

Henrys and farads of the Resonate system provide the Resonate frequency when pulsed by and external energy system, 36

Hertz, had high frequency coil systems before Tesla, 61

High Voltage DC Generators, 130

High Voltage Dividers, 116

High Voltage Devices, 117

Hoover Dam equivalent system fitting on a table top, 203

Hulga Clark, 359

Hysteresis of the transformer metal plates limits the load, 4

I

Ignition coil, 203

Impedance is the sum of the system resistance, which becomes zero at resonance, 24

Impulse Generators, 120

Induced Electrical Energy System, The amount of Energy transferred relates directly to the density of the lines of flux present. Each unit of increase causes a squaring of the flux lines present. One volt, Two Volts = Four, Four Volts = Sixteen. In resonate air core coil energy transfer the increase in flux lines present disturbes more electrons than previous, resulting in over unity energy being present and available, 70

Induction effect, 188

Infinity Symbol, rotation of electrons, 178

Infrared magnetic waves from Galactic radiation, 59

Intermittent Bloch Wall Switching Motor for Brazil and Saudio Arabia, Device 9, 160

Inverse square laws, 23, 24, 29, 62, 209

Inverter, power, how to build a modified wave, 109-111, 193

J

Joe Newman, 208

Joule was a English brewmaster who had all kinds of machinery that ran on steam, and some electrical, and this and that. He had to come up with a system which would make translating one system of energy to another more universal, 202

Joule is 1 volt times 1 amp times 1 second which is a watt second, 202, 214

J.P. Morgan meeting with Edison and Tesla, 60

K

Kapton dialectric, 212

Kinetic energy, two different kinds. We have expanding magnetic energy and it's dominate for that. It gives you your electrons moving apart, pressure decreasing. It gives a cooling effect. That explains why it's two hundred and seventy-six degrees minus out in an open space away from the earth. When the electrons are moving apart or expanding it's cooling, and sometimes very cooling. There's less scattering of energy in an expanding situation. In addition to that you have negative resistance as dominant. If we go into the contraction stage of energy or mass we get electrons moving together. Pressure is increasing, heating effect dominates and you have a scattering of energy, and positive resistance results. You're under unity people work primarily and in that particular set of circumstances, 177

Kickback protection for L-1, 10

Kite radio transmission, 14

Klystrons, Free Energy device, 19

L

L-1 coil, 200

Grounding can be grounded back into the laser module itself, or it can be grounded into the earth, or into the same common ground that L2 is actually grounded to, 184

L-2 coil, 6, 183, 212

Has to be grounded in order to pick up the maximum effect of electrons, 183

Lasers, avalanche, 15

Laser technology, 79

Law of Squares, 5, 23, 24, 202

Laws of Physics, conformity, 14

Laws of conservation, 20, 94, 209

LCR meter, 230

Lee, Denis, 327

Lenz law, 283

Leecher lines, 367

Levassor, had high frequency coil systems before Tesla, 61

Lightning arrestors, used in place of spark gaps, 191, 285

Lightning strikes from the

ground up around 1 trillion gammas, 29, 369

Limiters, Voltage, Amperage, V-1, 8

M

Magnet, potential power is present in a, 63, 385

Magnets are a magnetic antenna system reflecting the Earth's Geomagnetic Structure. A form of Corerer, 378

Magnetic field fluxing (on-off of the electrical source), 47

Magnetic flux,

for each line of electrical flux we have a line of magnetic flux, 206

(only the) moves from point "A" (Sun) to point "B" Earth, 3, 81

Magnetic Flux Amplified, Device 3, 154

Magnetic Energy expancing, as one form of Kinetic energy. It gives you your electrons moving apart, pressure decreasing. It gives a cooling effect. When the electrons are moving apart or expanding it's cooling. There's less scattering of energy. In addition you have negative resistance as dominant, 177

Magnetic imbalance causes the gravitatinonal effect. This is evidenced in electric motors by magnet-gravitational displacement of mass which causes the motor to rotate, 39

Magnetic permeability is the counterpart of negative resistance, 81, 363

Magnetic Resonance Power System, 80

Magnetic spin, recovery, 16

Magnetic turbulence, 47

Magnetic wave induction, 57

Magnetic wave resonance, 146

Magnetic waves from Cosmic, Galactic, Solar and Magnets, 58

Magnetic waves amplified (conjugate), 2

Magnetometer studies, 39, 53

Magnetostrictive material, 80, 81, 342, 344

Permealloy, Sendust, Metglas, Iron with (34%) Cobalt, 81

Piezoelectric in character, 81

The fact that doesn't conduct electricity, it becomes a perfect *conductor for magnetic fields*, 307

Magnetostriction Oscillators work by magnetic resonance, 81, 344

Mandated energy wasting, 19

Marx Generator, Solid State, Power System, 17, 119, 206, 225

Masers, avalanche, 15, 119

Matter is Energy, 18

Maxwell, J.C. (1891), active electron field gives rise to magnetic, 28, 253

Measure, units of, 5

Metglas, 282

Motor-Generator Coils and *Magnet System*, Back E.M.F not present, 16

MEG, (Motionless Electromagnetic Generator), 345, 383

Maxwell, "the molecules of a gas carry out ceaseless transitional movement in all temperatures above absolute zero." 19

Metglas, 350

Metals when subjected to friction, simply conduct the charge, 27

Multi-strand wire causes the magnate flux to multiply in such a way that it is very beneficial. The diameter times the number of strands that are available is a multiplication factor in the coil set, 311

Multipactors, avalanche, 15, 273

N

Natural frequency of coils relates to their being in resonance in relation to their capacitance and inductance, 78

Negative resistance, 307

At resonance Impedance is zero. Adjacent electrons, not a part of the orignal are pulled into the composite, becoming greater energy out than in, 9

Trumpet Wave, Transient response of an RLC circuit, 98, 249

Negantropy from Common Dipole, 363

Neon tube transformer, 8, 85

Newton, Isaac (1642-1727) with the help of Charles Coulomb (1736-1806) provide us with the square laws which are based on an open circuit, 23

Neutral Generator, 103

Newtonian Physics (electrodynamics and laws of conservation), 3

Nishida, A, Earth's AC, 28

Nitrous Oxide pollution is produced, 14

Nonlinear systems, 2

Non-linearity of this system does not obey ohm's law, which is replaced with impedance and reactance for alternating current systems, 24

NORAD, may be in Cheyenne Mountain area because ground has very high potential energy, 177

O

Oersted (one unit = 50,000V), 29

Oerstead, H.C. (1777-1851) define Earth's electrical field, measuring magnetic field flux as an indicator of negative electron energy active, 29, 175, 177

Ohm = Scattering, dissipation of Energy (Load), (after the fact) 35

Ohms is load or wasted energy and amperes is the rate of wasting, 71

Ohm's Law, 9, 181

- *North pole of magnet* is less Negative = Source of Amperes, right hand spin, 63
- *Not apply* to resonate air core radio frequency systems, 37
- *With corrections*, 35

Overunity (see specific index):

- *All useful electricity is over unity* because it has to get back to ambient which is background, 197, 214
- *Electrons*, 5, 36
- *Light Amplification*, 3
- *Negative feedback systems in Op-Amps*, 38
- *Radio tubes*, 38
- *Setups*, where battery charges from the 1/4 wavelength of L1, 188-189

Oudin, had high frequency coil systemsbefore Tesla, 61

Oxygen decline in Earth's Atmosphere, 13, 355

Oxygen Ratio 39% 2000 years ago, 30% in 1900, 18% 1984, 12

Ozone depletion allows infrared penetration causing temperatures to rise and health problems, 14

P

Pancake coil, 283

Panhard-Lavassor, had successful electric automobiles using AC motors, 61

Parallel resonant circuit, Almost all the things that go on here will be essentially in parallel because when you put them in series you get a degradation, 186

Peer Review principal weapon of the establishment in preventing destablilizing to self, 16

Perpetual motion, 19

Peugeot, had successful electric automobiles using AC motors, 61

Piano resonance, 2

Piezoelectric Power System, 17, 291, 312, 318

Ping-pong balls analagy, 180

Plasma Column Dipole, 165

Plasma Globe,

- *Emitting 40,000 volts*, 3
- *Column* (large model) *running at 50,000 C.P.S*, 297
- Device 4, 155, 371

Plasma thru Solids, 9

Politically correct energy generates massive taxes, 12

Potential Energy, 2

Power correcting factor,

- *Adjusting formula*, 7
- *Capacitor* should be as many microfarads as possible to allow lower frequency to match primary side of isolation transformer (correct if needed at transformer), 8
- *Resistor added*, 7
- *System being a capacitor bank* maintains an even flow of flux, 4

Power System, Piezoelectric Solid State Marx Generator, 17

Power System 3, 201

Power Triangle, 66, 200, 234

Program, Computer design, method, 238

Propulsion-Star System, uses Spin System seen in Star Systems, Solar Systems, Tornados, Cyclones and related to Rail Guns, Gyroscopes, and Negative Gravitational Systems, 17

Protons, skeptical of, 178

Q

Q factor, 187

R

Radiated magnetic flux from the Dipole when intercepted by capacitor plates at right angle changes to useful electrical energy, 86

Radiation belt rich with electrons, shear zones 15, 57

Radio broadcast station, mini, 192

Radio Devices, early, could not be explained until after Thompson around 1911, 15

Radio Telescope, 93

Radio tubes, Resonate induction ove runity circuits in conventional tubes (high plate voltage), 38

Radiotelegraphy Pre Civil War, 14

Radios, The number of, running at any one time do not diminish, in any way the electrical output of the source station, 36

Rate of induction is directly related to the cycles per second, 174

Reactance charts makes approximating the required three variables, capacitance, inductance and resistance, much easier. If two of the variables are known, the required missing one is obtained, 7, 96

Reactor (L-1 coil), 36

Remote flux reading requires the squaring of the distance times the remote reading, 24, 62

Renault had successful electric automobiles using AC motors, 61

Resistance, negative, 9

Resistor,

Are *Capacitors and inductors*. If you take your LCR meter and you measure the capacitance and the inductance of almost any resistor you're going to run into, you'll find out that the resistor is a lot better capacitor, and a lot better inductor, than it is a resistor, 199, 211

Choose the right one in the Reactance chart, 7

Resistor R-1 used to set electron pump rate, frequency of the capacitor. Maintains the desired voltage level required to operate the system, 10

Resistors, Voltage - Amperage Limiting required for adjusting output level and desired CPS, 6

Resonance:

Resonant simply means that you're flipping the electrons, 273

Piano, 2

Resonate air core coil system, 6

Resonate Electromagnetic Power System Schematic, 10

Resonate-fluxing of electrons activates potential, 3

Resonate fluxing provides the avalanche effect along with electromagnetic waves when reacting with an electron rich source, 21

Resonate-frequency-flux rate sets the available energy, 3

Resonate induction, 28, 29

fluxing coils, 5

Disturbs a large number of adjacent electrons which were not a part of the original source. The Pulsating (pumping effect) then incorporates the newly available extra electrons into the ongoing energy generation source-system. A near unity energy system of resonate air core coils and extra acquired electron-energy source constitutes an over unity system, 66

Shunt, 183

Transfer from one power system allows other resonate induction systems to duplicate the original source, which in no way diminishes the original source. The number of radio devices or televisions concurrently running in no way diminishes the output source, 66

Resonate Energy Concept, 147

Resonate High Voltage Induction Coil, 84

Resonate magnetic flux is not subject to ohms or Lenz's laws, 383

Resonate Magnetic Waves, Faraday's action at a distance, 147

Resonate Transformer predating the MEG, 167

Roentgen, had high frequency coil systems before Tesla, 61

Ruhmkorff, had high frequency coil systems before Tesla, 61

S

Scooter, motor, 326

Sea of Continuum governed by the Law of Squares, 5

Self-powered:

Battery, due to input wire as an antenna, 188-189, 229

Camera shutter, 3

For Vending Machines in Japan, Device 5, 156

Negative Resistance, 9

Shear zones, electrons, 15

Motor-Generator, 16

If the plus connector wire's length is a wave length component of the L-6 or L-7's frequency the battery will recharge itself at the same time it is being used. To see this substitute two opposing LEDs. They show the direction the electrons are moving. DC electrons out and radio frequency electrons into the battery, 69

Sensitizing of Systems by use of Trace Doping with Radioactive elements, 58

Series resonant circuit, Almost all the things that go on here will be essentially in parallel because when you put them in series you get a degradation, 186

Shunt, as system, in the resonate circuit sets the containment level for energy potential, 36, 175, 183

Skin effect on wires, 188

Small causes large gains, 30, 198, 213

Solar: 93

Depends on the fancy of nature, 12

Electromagnetic waves incoming from point "A", the Sun, reacting with point "B", a solar cell which spins the local electrons sideways, yielding electricity, 21

Only the *magnetic flux* moves from point "A" (Sun) to point "B" Earth, 3

Plasma, 5

Is 65% *None Ionic Electrons*, 9

Quiet day, 1,900 million gamma, 190,000 million volts of electricity, 3

Radiation as Visible Light magnetic waves, 59

Solar Power System relating to Magnetic Waves where domains aligned equates with resonance, 17

Sonar Power System, 80

Sound part electromagnetic spectrum, 2

South pole of magnet is more Negative = Source of Volts, left hand spin, 63

Spark gaps, 8, 14, 206, 225, 357

Controls the energy level that's passing, 204

Not needed because wires and coils have their own inductance and capacitance, and it's going to load and unload based on the frequency that's attributable to that piece of wire. 176, 182

Spike offset, 185

Spin Technology seen in Coils and Magnet Systems, 16

Recovery spin right is magnetic - amperage, 16

Spin the Earth, level of electrical energy is required to, *Earth's mass in Kg is 5.9 X 10 to the 24*

Star Systems, 17

Stainless Steel Transformer, 383

Suitcase Model demonstrated (1996), 11

Sun, only the magnetic flux moves from point "A" (Sun) to point "B" Earth, 3

Surge arresters, 8

T

Televisions, The number of, running at any one time do not diminish, in any way the electrical output of the source station, 36

Terfenol-D rod, 315, 336

Tesla, 211

(one unit = 10,000 Guass), 29

Tesla coil,

14" Prototype, Device 8, 159

Flat Prototype, Device 1, 151

Geometry, 65

Grounding, 184

Is only the *Voltage half* of the coil systems of those before Tesla, 61

Receiver coils multiplied, Device 2, 153

Simple Tesla Coil, 126

The disruptive discharge of the Telsa coil is abscent in the EES II, 53

Tesla meeting with J.P. Morgan and Edison, 60

Tesla system, Advantages of EES II over Tesla System: The disruptive discharge of the Telsa coil is absent. Less complicated; different grounding; smaller; and it's rate of energy production and efficiency is much higher than Tesla. Efficiency of Induction relates to the square of the cycles per second. Tesla coils operate normally below 200,000 C.P.S., and EES II in Mhz range, 53

Thompson, 1911 explained radio devices, 15

Postulated the electron as a particle in early 1900's, 60

Thompson's, JJ, description of the electron originally. This business of having electrons coupling and being disturbed and pulled apart. And they have to fight to go back together, and when they come back together, you have volts times amps equal watts. That is an extension of JJ Thompson's description of the electron originally. It's expanding onto it, and there's some other things there, 265

Time constant (frequency) information in chapter one of *Handbook of Electric Tables and Formulas*, 7

Timing formulas,

T = RC, T= L/R, 10

Timing Circuit, 204, 205, 216, 222

Tornado, 17

Transfer Mechanisms, 58

Transformer,

Conventional, 4

Comes alive when pulsed, 8

Choose the right one in the *Reactance chart*, 7

Feedback transformer, 81

Filter ratings, specify, 113

Isolation type. Only magnetic flux passes from the input side to the output side. 4

Large Transformer for Commercial System, 170

Matrix transformer, 112

Needs to operate at 60 CPS, that's 60 up and 60 down, 120 cycles, 7

Resonant Iron Core Transformer System, 185

Transient response of an RLC circuit, 98

Trumpet Wave negative resistance, 98

Tuning, 367

Turbulence upsets ambient, 2

U

Ultimate Energy Systems, 21

Ultimate Energy Source, 92

Ultraviolet magnetic wave from Cosmic radiation, 59

Useful Energy is Defined as that which is other than Ambient, 2, 53

Useful Electrical Power is Generated when Electrons from Earth and Air Groundings are disturbed by the movement of coils and magnets with reference to each other. The resulting electrical and magnetic energy is then changed to joules (watt seconds, Volts x Amps x Seconds). Each forward electron movement results in a magnetic impulse and each return movement causes and electrical impulse. The composit of the electrical energy impulsed from these electrons yields useful energy (Power). The Energy System here presented consists of a properly adjusted and functional resonate air core coil tank. The electrical Energy is stored in capacitors and magnetic energy in the coil system, 73

USPTO, United States Patent and Trademark Office, 2, 14, 22

Has already issued hundreds of Patents related to Light Amplification, all of which are overunity. 297 direct patents, and over 1000 indirect, 3

V

Van Allen radiation belts are one side of a big capacitor, 235, 364

V.A.R. 174, 213

Volts and Ampere are equal (V.A.R.) until work (load) is introduced, 30

Energy Potential as Volts/ Dissipation (0) = Rate of Dissipation (0), 37

Varactor, in place of diode on battery input wire, 67

Varistors, 8, 10

Blocking device used to connect to ground as either a resitor or capacitor or variation of the two, 176, 185

V-1 limits the voltage as required for the Output *Transformer* L-2, 10

Volts = Energy Available (Potential), 35

Voltage dividers, 206

Volts are basically the water that's behind the dam. Until you poke a hole in the dam, you don't get any flow rate or amperage. Whatever your voltage is it's whatever your potential is, and how you go about letting that thing through the dam, is what's going to give you your useful energy, 209

W

Wavelength,

Wire length, 78, 101

Waves, radio frequency, above 20,000 Hertz per second, 40

Wind energy depnds on the fancy of nature, 12

Wave shapping modules, 192

X

Xerox machines laser modules, 181

W

Wars are unnecessary, 14

Water, distilled, as a dialectric is about 2000, 226

Wave length, if you use a quarter wave system your shortest wave will be the longest length of wire and your shortest length of wire will be your higher frequency, 198, 367

Welding machine, some of the mechanics that are involved in building a welding machine are also involved in building one of these things that will have very high amperage, 203

Z

Zero Point Quantum Physics, 19, 381

www.ingramcontent.com/pod-product-compliance
Lightning Source LLC
LaVergne TN
LVHW070904300926
849162LV00011B/1852
* 9 7 8 1 9 3 2 3 7 0 8 4 3 *